Lecture Notes in Mathematics

Volume 2343

Editors-in-Chief

Jean-Michel Morel, City University of Hong Kong, Kowloon Tong, Hong Kong

Bernard Teissier, IMJ-PRG, Paris, France

Series Editors

Karin Baur, University of Leeds, Leeds, UK

Michel Brion, UGA, Grenoble, France

Rupert Frank, LMU, Munich, Germany

Annette Huber, Albert Ludwig University, Freiburg, Germany

Davar Khoshnevisan, The University of Utah, Salt Lake City, UT, USA

Ioannis Kontoyiannis, University of Cambridge, Cambridge, UK

Angela Kunoth, University of Cologne, Cologne, Germany

Ariane Mézard, IMJ-PRG, Paris, France

Mark Podolskij, University of Luxembourg, Esch-sur-Alzette, Luxembourg

Mark Policott, Mathematics Institute, University of Warwick, Coventry, UK

László Székelyhidi, MPI for Mathematics in the Sciences, Leipzig, Germany

Gabriele Vezzosi, UniFI, Florence, Italy

Anna Wienhard, MPI for Mathematics in the Sciences, Leipzig, Germany

This series reports on new developments in all areas of mathematics and their applications - quickly, informally and at a high level. Mathematical texts analysing new developments in modelling and numerical simulation are welcome. The type of material considered for publication includes:

1. Research monographs
2. Lectures on a new field or presentations of a new angle in a classical field
3. Summer schools and intensive courses on topics of current research.

Texts which are out of print but still in demand may also be considered if they fall within these categories. The timeliness of a manuscript is sometimes more important than its form, which may be preliminary or tentative. Please visit the LNM Editorial Policy ((https://drive.google.com/file/d/1MOg4TbwOSokRnFJ3ZR3ciEeKs9hOnNX_/view?usp=sharing))

Titles from this series are indexed by Scopus, Web of Science, Mathematical Reviews, and zbMATH.

Martin van Beek

Rank 2 Amalgams and Fusion Systems

Martin van Beek
University of Manchester
Manchester, UK

ISSN 0075-8434 ISSN 1617-9692 (electronic)
Lecture Notes in Mathematics
ISBN 978-3-031-54460-6 ISBN 978-3-031-54461-3 (eBook)
https://doi.org/10.1007/978-3-031-54461-3

Mathematics Subject Classification: 20D05, 20D20, 20E06, 20E42, 05C25, 05E18, 20C33, 55R35

© The Editor(s) (if applicable) and The Author(s), under exclusive license to Springer Nature Switzerland AG 2024
This work is subject to copyright. All rights are solely and exclusively licensed by the Publisher, whether the whole or part of the material is concerned, specifically the rights of translation, reprinting, reuse of illustrations, recitation, broadcasting, reproduction on microfilms or in any other physical way, and transmission or information storage and retrieval, electronic adaptation, computer software, or by similar or dissimilar methodology now known or hereafter developed.
The use of general descriptive names, registered names, trademarks, service marks, etc. in this publication does not imply, even in the absence of a specific statement, that such names are exempt from the relevant protective laws and regulations and therefore free for general use.
The publisher, the authors, and the editors are safe to assume that the advice and information in this book are believed to be true and accurate at the date of publication. Neither the publisher nor the authors or the editors give a warranty, expressed or implied, with respect to the material contained herein or for any errors or omissions that may have been made. The publisher remains neutral with regard to jurisdictional claims in published maps and institutional affiliations.

This Springer imprint is published by the registered company Springer Nature Switzerland AG
The registered company address is: Gewerbestrasse 11, 6330 Cham, Switzerland

Paper in this product is recyclable.

Preface

Fusion systems offer a way to isolate the p-local methods used in various facets of finite group theory, modular representation theory, and homotopy theory, and coerce them into a more unified theory. With this, one can extract powerful results from distinct areas of mathematics, and reformulate them into new tools which apply to a broader range of research topics. On the other hand, fusion systems themselves are abstruse, and despite decades of study, there are still a swathe of unanswered questions and conjectures concerning them abstractly. In this text, we task ourselves with understanding certain elementary configurations in fusion systems, showcasing fundamental methods for treating larger classes of fusion systems, and exposing patterns highlighting how, where, and why particularly novel fusion systems appear.

To this end, most of the mathematical analysis in this book concerns certain situations in the setting of *group amalgams*, a topic which already perforates areas such as combinatorics, geometry, topology, and (not necessarily finite) group theory. In this monograph, we demonstrate that group amalgams are an extremely useful viewpoint to adopt for those interested in understanding fusion systems. We explore their untapped potential to answer some of the outstanding problems in fusion systems and the areas where they apply. Using group amalgams, we determine all fusion systems which satisfy a particular minimality condition, distinct from those minimality conditions already examined in the literature. This condition being that our mooted fusion systems have only two *essential subgroups*: the minimal number any "simple" fusion system can have.

In pursuit of this goal, we utilize and develop techniques for dealing with groups with strongly p-embedded subgroups, treating modules for small Lie type groups and some related groups, analyzing essential subgroups in fusion systems, and tackling some intricacies in the *amalgam method*. Hence, throughout this work, we exhibit several of the most important techniques in fusion systems and p-local group theory, while still remaining in configurations small and simple enough as to not obfuscate the utility and subtlety of these methods.

This book serves a broad audience as a reference for structural results for small simple groups and techniques in local group theory, while simultaneously introducing the more contemporary perspective of fusion systems. For those more experienced in the area, this text provides both a wealth of techniques for analyzing fusion systems, and a detailed record of the structure of small fusion systems in a way which should be applicable to understanding larger classes of fusion systems.

This monograph is the culmination of work which began during the author's PhD at the University of Birmingham, and developed in the years which followed. The author gratefully acknowledges the financial support of the Engineering and Physical Sciences Research Council and the Heilbronn Institute for Mathematical Research during this time. The author sincerely thanks Prof. Chris Parker for his direction, advice, and expertise with regard to this topic, and wishes to thank those who read passages from this text in one form or another for their valuable feedback. Finally, the author thanks the referees for their helpful remarks and encouraging words.

Manchester, UK

Martin van Beek

Contents

Chapter 1
Introduction

The second half of the twentieth century was a particularly fruitful time for finite group theory. Permutation group theory and modular representation theory had taken monumental leaps forward, unexpected connections between finite group theory and other areas of mathematics were discovered (and justified), and arguably most importantly and most impressively: the proof of the classification of the finite simple groups was all but complete. A large part of this success may be attributed to the adoption of p-local methods in finite group theory and representation theory (that is, the structure of a finite group's p-subgroups and their normalizers) in this period. Although the first movements toward understanding finite groups by their p-subgroups were far more classical than this, dating back to Sylow's work in the 1800s, this later shift initiated a remarkably unified thrust which group theory, and mathematics as a whole, had never experienced before.

While the classification gave a definitive answer for "understanding" finite groups, the collective effort in proving it provided an abundance of ideas, techniques and machinery which had use beyond this particular application. One key example of this is the *amalgam method*, pioneered by Goldschmidt [31] but heavily influenced by earlier work of Sims. This concept arrived too late to the party to have the kind of influence on the first proof of the classification of the finite simple groups it deserved. However, this methodology has now become pivotal in several of the so called *revision* programs geared towards providing more succinct and uniform approaches the classification.

One such program, spearheaded by Aschbacher, plans to prove some of the more difficult statements in the proof of the classification in the context of *saturated fusion systems*, as opposed to finite groups. For a finite group G and a prime p dividing the order of G, the p-fusion category of G provides a means to concisely express properties of the conjugacy of p-elements within G. Fusion systems may then be viewed as an abstraction of p-fusion categories without the need to specify a finite group G, instead focusing only on the properties of a particular p-group and the potential conjugacy therein. It is already clear that several statements in

© The Author(s), under exclusive license to Springer Nature Switzerland AG 2024

M. van Beek, *Rank 2 Amalgams and Fusion Systems*, Lecture Notes in Mathematics 2343, https://doi.org/10.1007/978-3-031-54461-3_1

finite group theory have more practical and digestible statements in the language of fusion systems than in groups (see, for example, [6, 7] or [56]). In this way, fusion systems seem to be the natural mechanism to distill several of the most applicable techniques in p-local finite group theory, and ready them for use in the greater world of mathematics. It would be remiss to not mention the place held by fusion systems in algebraic topology and representation theory, especially since fusion systems were introduced here long before they were used in the analysis of finite groups, but since this viewpoint is not adopted in the main results of this text, we instead defer to the standard texts on fusion systems for a discussion of this [11, 25].

Aschbacher's program aims to contribute to the revision of the classification in two steps: applying "semisimple" methods from local group theory to fusion systems to classify all *simple* fusion systems, with attention focused at the prime 2; and then using this classification to determine the finite simple groups of "component type." This program is well on its way with regard to the first step [8–10] and some work has already surfaced for the second stage of the program [45].

Alongside the program of Aschbacher, there is another "next generation" scheme to reprove large parts of the classification. This program, headed by Meierfrankenfeld, Stellmacher and Stroth and dubbed the "MSS program," aims to determine the finite simple groups of "local characteristic p" by using mostly "unipotent" methods (see [52] for an overview). Pivotal to this approach is the use of amalgams to identify finite simple groups.

Within the MSS program, there is scope to investigate a larger class of "characteristic p" groups than in the original proof of the classification. Indeed, it may be possible here to determine the finite simple groups which are of *parabolic characteristic p* (but probably only for the prime 2), and this improvement would substantially ease the burden on the treatment of component type groups via Aschbacher's program. Due to the Gorenstein–Walter Dichotomy Theorem, and a suitable analysis of some small cases, the net result of the union of these two programs should be a shortened proof of the classification of finite simple groups.

The results in this text lie somewhere in between these two programs: applying unipotent, or characteristic p, methods from group theory to saturated fusion systems, with a particular focus on the utility of group amalgams in this setting. While some equivalent notion of parabolic characteristic p for fusion systems is not needed here, the results documented would certainly fit more in this framework. Important to note is the *dichotomy theorem* for saturated fusion systems which says that every saturated fusion system is either of "characteristic p-type" or of "component type." Following the proof of this theorem, due to Aschbacher [11, Theorem II.4.3], it is not hard to generalize to a dichotomy theorem partitioning fusion systems into "parabolic characteristic p" and "parabolic component type."

The main remit of this monograph is to classify certain fusion systems which are generated by automorphisms of two subgroups which satisfy certain properties. This is achieved by identifying a rank 2 amalgam within the fusion system, and then utilizing the amalgam method. In this way, the work in this book may be viewed not only as a result about fusion systems, but as a result about (not necessarily finite) groups (see Theorem C). Furthermore, as an application, the work here will aid in classifying certain other families of "small" saturated fusion systems. Indeed, this result has already been used to give a complete description of all saturated fusion systems supported on p-groups isomorphic to a Sylow p-subgroup of a rank 2 simple group of Lie type in characteristic p, see [80].

The methodology for proving this result breaks down as follows. Firstly, the *automizers* of the two distinguished subgroups in our fusion system $\mathcal{F}$ are identified. Here the subgroups in question are *essential* in $\mathcal{F}$ and, using fusion techniques and classification results concerning groups with strongly p-embedded subgroups, we can almost completely describe the outer automorphism groups induced by $\mathcal{F}$. The methods we develop are reasonably generic and applicable outside of this particular area. Then, using the model theorem, we are able to investigate finite groups whose p-fusion categories are isomorphic to normalizer subsystems of the two distinguished essential subgroups. With these two groups in hand, we can then form an amalgam whose faithful completion "realizes" $\mathcal{F}$.

This leads to the second part of the analysis. Here, we employ the amalgam method, building on work began by Goldschmidt [31]. In our interpretation, we closely follow the techniques developed and refined by Delgado and Stellmacher [27] and a large number of the amalgams we investigate are fortunately already classified there. Indeed, several of the amalgams we investigate are unique up to isomorphism and, as it turns out, this is enough to determine the fusion system up to isomorphism. However, in some cases, we do not go so far and instead aim only to bound the order of the p-group on which $\mathcal{F}$ is supported and apply a package in MAGMA [14, 62] which identifies the fusion system. In fact, in two instances there are no finite groups which realize the fusion system appropriately and we uncover two *exotic* fusion systems, one of which was known about previously by work of Parker and Semeraro [63], and another which has been described in [79]. With that said, given the information we gather about the amalgams, it does not seem such a stretch to at least provide a characterization of these amalgams up to some weaker notion of isomorphism.

Within this work, we very often use a $\mathcal{K}$-group hypothesis when investigating automizers of essential subgroups and a $\mathcal{CK}$-system hypothesis on the fusion system we are tasked with identifying, denoted $\mathcal{F}$. Recall that $\mathcal{K}$-group is a finite group in which every simple section is isomorphic to a known finite simple group. A $\mathcal{CK}$-system is then a saturated fusion system in which the induced automorphism groups on all p-subgroups are $\mathcal{K}$-groups. At some stage in the analysis, unfortunately, we make explicit use of the classification of finite simple groups (CFSG), specifically when $\mathcal{F}$ is exotic. However, up to that point, we are still able to determine the size of the p-group on which $\mathcal{F}$ is supported (as well as the local actions) within a $\mathcal{CK}$-system hypothesis and only appeal to the classification to prove that the fusion system is exotic. Thus, this result would still be suitable for use in any investigation of fusion systems in which induction via a minimal counterexample is utilized. The main theorem is as follows:

Main Theorem *Let $\mathcal{F}$ be a local $\mathcal{CK}$-system on a p-group S. Assume that $\mathcal{F}$ has two* $\mathrm{Aut}_{\mathcal{F}}(S)$*-invariant essential subgroups $E_1, E_2 \trianglelefteq S$ such that for $\mathcal{F}_0 := \langle N_{\mathcal{F}}(E_1), N_{\mathcal{F}}(E_2)\rangle_S$ the following conditions hold:*

(a) $O_p(\mathcal{F}_0) = \{1\}$;
(b) *for $G_i := \mathrm{Out}_{\mathcal{F}}(E_i)$ and $T \in \mathrm{Syl}_p(G_i)$, if $G_i/O_{3'}(G_i) \cong \mathrm{Ree}(3)$, G_i is p-solvable or T is generalized quaternion, then $N_{G_i}(T)$ is strongly p-embedded in G_i.*

Then $\mathcal{F}_0$ is saturated and one of the following holds:

(i) *$\mathcal{F}_0 = \mathcal{F}_S(G)$, where $F^*(G)$ is isomorphic to a rank 2 simple group of Lie type in characteristic p;*
(ii) *$\mathcal{F}_0 = \mathcal{F}_S(G)$, where $G \cong$* M_{12}, $\mathrm{Aut}(\mathrm{M}_{12})$, J_2, $\mathrm{Aut}(\mathrm{J}_2)$, $\mathrm{G}_2(3)$ *or* $\mathrm{PSp}_6(3)$ *and $p = 2$;*
(iii) *$\mathcal{F}_0 = \mathcal{F}_S(G)$, where $G \cong$* Co_2, Co_3, McL, Aut(McL), Suz, Aut(Suz) *or* Ly *and $p = 3$;*
(iv) *$\mathcal{F}_0 = \mathcal{F}_S(G)$, where $G \cong$* $\mathrm{PSU}_5(2)$, $\mathrm{Aut}(\mathrm{PSU}_5(2))$, $\Omega_8^+(2)$, $\mathrm{O}_8^+(2)$, $\Omega_{10}^-(2)$, $\mathrm{Sp}_{10}(2)$, $\mathrm{PSU}_6(2)$ *or* $\mathrm{PSU}_6(2).2$ *and $p = 3$;*
(v) *$\mathcal{F}_0$ is a simple fusion system on a Sylow 3-subgroup of* F_3 *and, assuming* CFSG, *$\mathcal{F}_0$ is an exotic fusion system uniquely determined up to isomorphism;*
(vi) *$\mathcal{F}_0 = \mathcal{F}_S(G)$, where $G \cong$* Ly, HN, Aut(HN) *or* B *and $p = 5$; or*
(vii) *$\mathcal{F}_0$ is a simple fusion system on a Sylow 7-subgroup of* $\mathrm{G}_2(7)$ *and, assuming* CFSG, *$\mathcal{F}_0$ is an exotic fusion system uniquely determined up to isomorphism.*

Utilizing the results in [3], [62, 79] and [80], $\mathcal{F}$ is also known up to isomorphism.

We include $\mathrm{G}_2(2)' \cong \mathrm{PSU}_3(3)$, $\mathrm{Sp}_4(2)' \cong \mathrm{Alt}(6)$ and the Tits groups ${}^2\mathrm{F}_4(2)'$ as groups of Lie type in characteristic 2.

Some explanation is required for condition (b) in the above theorem. This assumption ensures that the centralizers of non-central chief factors in a model of the normalizer system $N_{\mathcal{F}}(E_i)$ are p-closed. This is key in our methodology and facilitates the use of various techniques, especially coprime action arguments, which are used to deduce $\mathrm{Out}_{\mathcal{F}}(E_i)$ and its action on E_i. In the case where $\mathrm{Out}_{\mathcal{F}}(E_i)/O_{5'}(\mathrm{Out}_{\mathcal{F}}(E_i)) \cong \mathrm{Sz}(32) : 5$, this could also pose a problem. However, in this situation we are able to apply a transfer argument to dispel this case (see Lemma 5.13). It is easily seen using the Alperin–Goldschmidt theorem that if E_i is not contained in any other essential subgroup of $\mathcal{F}$ (which we term *maximally essential*) then condition (b) holds. We anticipate that in the majority of the applications of the Main Theorem, this hypothesis can be forced.

In the Main Theorem, where $\mathcal{F}_0$ is *realizable* by a finite group, we provide only one example of a group which realizes the fusion system. In several instances, this example is not unique, even among finite simple groups. In particular, if $\mathcal{F}_0$ is realized by a simple group of Lie type in characteristic coprime to p, then there are lots of examples which realize the fusion system, see for instance [17]. Note also that we manage to capture a large number of fusion systems at odd primes associated to sporadic simple groups. Indeed, as can be witnessed in the tables provided in [1], almost all of the p-fusion categories of the sporadic simple groups at odd primes are either constrained, supported on an extraspecial group of exponent p and so are classified in [68], or satisfy the hypothesis of the Main Theorem.

It is surprising that in the conclusion of the Main Theorem there are so few exotic fusion systems. It has seemed that, at least for odd primes, exotic fusion systems were reasonably abundant. Perhaps an explanation for the apparent lack of exotic fusion systems is that the setup from the Main Theorem somehow reflects some of the geometry present in rank 2 groups of Lie type. Additionally, we remark that in the two exotic examples in the classification, the fusion systems are obtained by "pruning" a particular class of essential subgroups, as defined in [62]. Indeed, these essential subgroups, along with their automizers, seem to resemble Aschbacher blocks, the minimal counterexamples to the Local $C(G, T)$-theorem [18]. Most of the exotic fusion systems the author is aware of either have a set of essentials resembling blocks, or are obtained by pruning a class of essentials resembling blocks out of the fusion category of some finite group. For instance, *pearls* in fusion systems, investigated in [35] and [36], are the smallest examples of blocks in fusion systems.

The work we undertake in the proof of the Main Theorem may be regarded as a generalization of some of the results in [2], where only certain configurations at the prime 2 are considered. There, the authors exhibit a situation in which certain subgroups of the automizers of a pair of essential subgroups generate a fusion subsystem, and then describe the possible actions present in the subsystem, utilizing Goldschmidt's pioneering results in the amalgam method. With this in mind, we provide the following corollary along the same lines which, at least with regards to essential subgroups, may also be considered as the minimal situation in which a saturated fusion system is "simple."

Corollary A *Suppose that $\mathcal{F}$ is a saturated fusion system on a p-group S such that $O_p(\mathcal{F}) = \{1\}$. Assume that $\mathcal{F}$ has exactly two essential subgroups E_1 and E_2. Then $N_S(E_1) = N_S(E_2)$ and writing $\mathcal{F}_0 := \langle N_{\mathcal{F}}(E_1), N_{\mathcal{F}}(E_2)\rangle_{N_S(E_1)}$, $\mathcal{F}_0$ is a saturated normal subsystem of $\mathcal{F}$ and either*

(i) *$\mathcal{F} = \mathcal{F}_0$ is determined by the Main Theorem;*
(ii) *p is arbitrary, $\mathcal{F}_0$ is isomorphic to the p-fusion category of H, where $F^*(H) \cong \mathrm{PSL}_3(p^n)$, and $\mathcal{F}$ is isomorphic to the p-fusion category of G where G is the extension of H by a graph or graph-field automorphism;*
(iii) *$p = 2$, $\mathcal{F}_0$ is isomorphic to the 2-fusion category of H, where $F^*(H) \cong \mathrm{PSp}_4(2^n)$, and $\mathcal{F}$ is isomorphic to the 2-fusion category of G where G is the extension of H by a graph or graph-field automorphism; or*
(iv) *$p = 3$, $\mathcal{F}_0$ is isomorphic to the 3-fusion category of H, where $F^*(H) \cong \mathrm{G}_2(3^n)$, and $\mathcal{F}$ is isomorphic to the 3-fusion category of G where G is the extension of H by a graph or graph-field automorphism.*

As intimated earlier in this introduction, we utilize the amalgam method to classify the fusion systems in the statement of the Main Theorem. Here, we work in a purely group theoretic setting and so, as a consequence of this work, we obtain some generic results concerning amalgams of finite groups which apply outside of fusion systems. We operate under the following hypothesis, and note that the relevant definitions are provided in Chap. 4:

Hypothesis B $\mathcal{A} := \mathcal{A}(G_1, G_2, G_{12})$ is a characteristic p amalgam of rank 2 satisfying the following:

1. for $S \in \mathrm{Syl}_p(G_{12})$, $N_{G_1}(S) = N_{G_2}(S) \leq G_{12}$; and
2. writing $\overline{G_i} := G_i/O_p(G_i)$, $\overline{G_i}$ contains a strongly p-embedded subgroup and if $\overline{G_i}/O_{3'}(\overline{G_i}) \cong \mathrm{Ree}(3)$, $\overline{G_i}$ is p-solvable or $\overline{S}$ is generalized quaternion, then $N_{\overline{G_i}}(\overline{S})$ is strongly p-embedded in $\overline{G_i}$ for $S \in \mathrm{Syl}_p(G_{12})$.

It transpires that all the amalgams satisfying Hypothesis B are either weak BN-pairs of rank 2; or $p \leqslant 7$, $|S| \leqslant 2^9$ when $p = 2$, and $|S| \leqslant p^7$ when p is odd. Moreover, in the latter exceptional cases we can generally describe, at least up to isomorphism, the parabolic subgroups of amalgam.

What is remarkable about these results is that amalgams produced have "critical distance" (defined in Notation 5.4) bounded above by 5. In the cases where the amalgam is not a weak BN-pair of rank 2, the critical distance is bounded above by 2, and when this distance is equal to 2, the amalgam is *symplectic* and was already known about by work of Parker and Rowley [61]. We present an undetailed version of the theorem summarizing the amalgam theoretic results below.

Theorem C *Suppose that $\mathcal{A} := \mathcal{A}(G_1, G_2, G_{12})$ satisfies Hypothesis B. Then one of the following occurs:*

(i) *$\mathcal{A}$ is a weak BN-pair of rank 2;*
(ii) *$p = 2$, $\mathcal{A}$ is a symplectic amalgam, $G_1/O_2(G_1) \cong \mathrm{Sym}(3)$, $G_2/O_2(G_2) \cong (3 \times 3) : 2$ and $|S| = 2^6$;*
(iii) *$p = 2$, $\Omega(Z(S)) \trianglelefteq G_2$, $\langle(\Omega(Z(S))^{G_1})^{G_2}\rangle \not\leq O_2(G_1)$, $O^{2'}(G_1)/O_2(G_1) \cong \mathrm{SU}_3(2)'$, $O^{2'}(G_2)/O_2(G_2) \cong \mathrm{Alt}(5)$ and $|S| = 2^9$;*
(iv) *$p = 3$, $\Omega(Z(S)) \trianglelefteq G_2$, $\langle(\Omega(Z(S))^{G_1})\rangle \not\leq O_3(G_2)$, $O_3(G_1) = \langle(\Omega(Z(S))^{G_1})\rangle$ is cubic 2F-module for $O^{3'}(G_1/O_3(G_1))$ and $|S| \leqslant 3^7$; or*
(v) *$p = 5$ or 7, $\mathcal{A}$ is a symplectic amalgam and $|S| = p^6$.*

Much more information about the amalgams is provided where they arise in the proofs. The structure of amalgams documented in Theorem C is exemplified in the appendix (Table A.1) for ease of reading.

The contributions in this work may be viewed as the "rank 2" case of an attempt to classify fusion systems which contain a "parabolic system." For parabolic systems in groups, work of various authors indicates that given sufficient control of the rank 2 residues of a parabolic system, using the theory of chamber systems and buildings, one can identify a BN-pair and, if the group under investigation is finite, a result of Tits implies that the group is a group of Lie type. For a survey of results in this area, see [53]. A result of Onofrei [58] translating parabolic systems in groups to their fusion theoretic counterparts, suggests that it may be possible to use the rank 2 information present in a fusion system $\mathcal{F}$ to identify a "BN-pair" and show that $\mathcal{F}$ is actually isomorphic to the fusion category of a finite group of Lie type.

While this work gives descriptions of what the rank 2 residues *may* look like in such a fusion system setup, it is not quite enough information to apply Onofrei's result immediately and so we will have to gain even more control of the local data before being able to impose such strong geometric conditions in the parabolic system. Again, taking inspiration from the world of finite groups, one can associate a graph in the style of Timmmesfeld [74] using local data, where two points, corresponding to rank 1 parabolic subgroups P_i and P_j, are joined if and only if $O_p(P_i) \cap O_p(P_j)$ is not normal in P_i or P_j. However, this condition can be phrased completely in the amalgam setting and so may be smoothly adapted to fusion systems. Our main result coupled with Proposition 5.27 serves as some indication of what may be possible here. See [71] for how this method is used to gain control in the rank 3 setting when working with groups. Although there are some simple and gratifying corollaries to immediately extract relating to higher rank amalgams from the results in this text, we feel that this work merits its own investigation and so we will not say anymore about it here.

This book is structured as follows. In Chap. 2, we set up the various group theoretic terminologies and results we use throughout the text. Most importantly, we characterize groups with a strongly p-embedded subgroup, groups with associated *FF-modules* and *2F-modules*, groups which contain elements which act quadratically, and exhibit situations in which these phenomena occur. The typical examples of automizers in our investigations are rank 1 groups of Lie type in defining characteristic and, because of this, large parts of this chapter are devoted to the properties of such groups and their "natural" modules. In Chap. 3, we introduce fusion systems and, for the most part, reproduce definitions and properties associated to fusion systems which may be readily found in the literature. We end this chapter by determining the potential automizers of maximally essential subgroups of fusion systems, which may be of independent interest. In Chap. 4 we demonstrate how to identify a rank 2 amalgam given certain hypotheses on a fusion system and begin setting up the group theoretic framework needed for the amalgam method. Here is where we deduce the Main Theorem from Theorem C. If the reader is only interested in the fusion system result, and is willing to take the amalgam result on faith, then they could stop reading at this point.

In Chap. 5 we set up the hypothesis and notations needed for the amalgam method. We demonstrate how to force quadratic or cubic action in this framework, which allows the use of results from Chap. 2. For several arguments, we investigate a minimal counterexample where minimality is imposed on the order of the *models* of the normalizers of essential subgroups. In the amalgam method, the case division separates fairly naturally, following the conventions used in [27]. Then Chaps. 6 and 7 deal with each of the particular cases coming from this division.

Our notation and terminology for groups is reasonably standard and generally follows that used in [4, 32, 49] and [43]. For fusion systems, we follow the notation used in [11] and when working with the amalgam method, our definitions and notation follow [27].

We highlight some particular practices we adopt. With regards to notation concerning simple groups, we will generally follow the Atlas [24], with some caveats regarding the classical groups. We include the prefix "P" to indicate a quotient by the center, and "S" indicates the subgroup of matrices with determinant 1 e.g. we use $\mathrm{PSL}_n(q)$ where the Atlas uses $\mathrm{L}_n(q)$. In addition, we reserve the notations $O_n^+(q)$ and $O_n^-(q)$ for the full orthogonal groups, while $\Omega_n^\varepsilon(q)$ denotes the commutator subgroup of $SO_n^\varepsilon(q)$ for $\varepsilon \in \{+,-\}$. For the sporadic groups, we follow the Atlas with the exception of Thompson's sporadic simple group, which we refer to as F_3 instead of the usual Th. We make this choice to emphasize the connection with "amalgams of type F_3" as defined in [27] and [26].

We denote by $\mathrm{Sym}(n)$ and $\mathrm{Alt}(n)$ the symmetric and alternating groups of degree n, and $\mathrm{Dih}(n)$ represents the dihedral group of order n (so that n is necessarily even). The notation Q_{4n} is used for generalized quaternion groups of order $4n$. When $p = 2$ we let 2_+^{1+2n} denote the extraspecial group obtained by taking the central product of r groups isomorphic to $\mathrm{Dih}(8)$ and $n-r$ groups isomorphic to Q_8 where $n-r$ is even, and 2_-^{1+2n} denote the extraspecial group obtained by a taking the central product of r groups isomorphic to $\mathrm{Dih}(8)$ and $n-r$ groups isomorphic to Q_8 where $n-r$ is odd. For p an odd prime, we reserve the notation p_+^{1+2n} and p_-^{1+2n} for extraspecial p-groups of exponent p and p^2 respectively. We will use Atlas notation for the "shape" of p-groups, often to exhibit the structure of their chief factors in some enveloping group G e.g. q^{1+2} is a group of order q^3 for q some prime power, with some grouped collection of G-chief factors having orders q and q^2. Where unambiguous, we will often present cyclic groups uniquely by their order, and elementary abelian p-groups by their expression as p-powers e.g. $r \times s$ is the direct product of a cyclic group of order r and a cyclic group of order s, and p^n is an elementary abelian group of order p^n.

Throughout, we adopt bar notation for quotients. That is, if G is a group, $H \leq G$ and $U \trianglelefteq G$, then writing $\overline{G} := G/U$, we recognize $\overline{H} := HU/U$. Some clarification is probably also required on the notation used for group extensions. We use $A : B$ to denote the a semidirect product of A and B, where A is normalized by B. We use the notation $A.B$ to denote an arbitrary extension of B by A. That is, A is a normal subgroup of $A.B$ such that the quotient of $A.B$ by A is isomorphic to B. We use the notation $A \cdot B$ to denote a central extension of B by A and the notation $A \circ B$ to denote a central product of A and B, where the intersection of A and B will be clear whenever this arises. Finally, we mention that as the majority of the modules we study occur "internally", we will use multiplicative notation for modules throughout.

Chapter 2
Preliminaries

In this chapter, we collect various results to be used later in the text. Several are well known or elementary, and where possible, we aim to give explicit references or rudimentary proofs. We use [4, 32, 33, 43] and [49] as background texts and also appeal to them for proofs of several results we will use frequently throughout this work. We also make appeals to [34] for known facts about known finite simple groups. The remainder of the citations in this chapter occur as justifications of more isolated results.

2.1 Group Theory Preliminaries

We begin with a selection of definitions and results which should be standard fare for those with a familiarity of the finite simple groups and/or fusion systems.

Definition 2.1 A finite group G is a $\mathcal{K}$-group if, whenever $K \trianglelefteq H \le G$ and H/K is simple, then H/K is isomorphic to a known finite simple group.

Definition 2.2 Let G be a finite group and p a prime dividing $|G|$. Then G is of *characteristic* p if $C_G(O_p(G)) \le O_p(G)$. Equivalently, if $F^*(G) = O_p(G)$.

Proposition 2.3 *Let G be a finite group of characteristic p. If $H \trianglelefteq\trianglelefteq G$ or $O_p(G) \le H \le G$, then H is of characteristic p.*

Proof Assume first that $H \trianglelefteq\trianglelefteq G$. By [49, 6.5.7 (b)], using that $H \trianglelefteq\trianglelefteq G$, we see that $F^*(H) \le F^*(G) = O_p(G)$. Hence, $F^*(H)$ is a normal p-subgroup of H, so that $F^*(H) = O_p(H)$ and H is of characteristic p.

Assume now that $O_p(G) \le H \le G$. Then $O_p(G)$ is a normal p-subgroup of H and so $O_p(G) \le O_p(H)$. Then $C_H(O_p(H)) \le C_G(O_p(H)) \le C_G(O_p(G)) \le O_p(G) \le O_p(H)$ and H is of characteristic p, as desired. □

© The Author(s), under exclusive license to Springer Nature Switzerland AG 2024

M. van Beek, *Rank 2 Amalgams and Fusion Systems*, Lecture Notes in Mathematics 2343, https://doi.org/10.1007/978-3-031-54461-3_2

Definition 2.4 Let G be a finite group and $S \in \mathrm{Syl}_p(G)$. Then G is *p-minimal* if $S \ntrianglelefteq G$ and S is contained in a unique maximal subgroup of G.

Proposition 2.5 (McBride's Lemma) *Let G be a finite group, $S \in \mathrm{Syl}_p(G)$ and $\mathcal{P}_G(S)$ denote the collection of p-minimal subgroups of G over S. Then $G = \langle \mathcal{P}_G(S) \rangle N_G(S)$. Moreover, $O^{p'}(G) = \langle \mathcal{P}_G(S) \rangle$.*

Proof If $G \in \mathcal{P}_G(S)$ then the result holds trivially so assume that G is a counterexample to the first statement with $|G|$ minimal. Since G is not p-minimal over S, there are maximal subgroups M_1, M_2 of G which contain S. But then, since G was a minimal counterexample, $M_i = \langle \mathcal{P}_{M_i}(S) \rangle N_{M_i}(S)$ for $i \in \{1, 2\}$. Since $\mathcal{P}_{M_i}(S) \subseteq \mathcal{P}_G(S)$, $N_{M_i}(S) \leq N_G(S)$ and $G = \langle M_1, M_2 \rangle$, the result holds.

Now, let $P \in \mathcal{P}_G(S)$ and $x \in N_G(S)$. Then for M the unique maximal subgroup of P containing S, we have that M^x is the unique maximal subgroup of P^x containing $S^x = S$ such that $S \ntrianglelefteq P^x$. It follows that $N_G(S)$ normalizes $\langle \mathcal{P}_G(S) \rangle$ and by the definition of $O^{p'}(G)$ and since $G = \langle \mathcal{P}_G(S) \rangle N_G(S)$, we deduce that $O^{p'}(G) \leq \langle \mathcal{P}_G(S) \rangle$. Now, suppose that there is $P \in \mathcal{P}_G(S)$ with $P \not\leq O^{p'}(G)$. Then $O^{p'}(P) \leq P \cap O^{p'}(G) < P$ and so $O^{p'}(P)$ is contained in the unique maximal subgroup of P which contains S. But $N_P(S)$ is also contained in the unique maximal subgroup of P containing S and by the Frattini argument, $P = O^{p'}(P) N_P(S) < P$, a contradiction. Therefore, $\langle \mathcal{P}_G(S) \rangle \leq O^{p'}(G)$ and the lemma holds. □

Proposition 2.6 *Suppose that H is p-minimal over $S \in \mathrm{Syl}_p(H)$ and R is a normal p-subgroup of H. Then H/R is p-minimal.*

Proof Let M be the unique maximal subgroup of H containing S. Then M contains R and so M/R is a maximal subgroup of H/R containing $S/R \in \mathrm{Syl}_p(H/R)$. Let $N \leq H$, be such that NR/R contains S/R and NR/R is maximal in H/R. Then NR is a proper subgroup of H containing S and so, since H is p-minimal, $NR \leq M$ from which we deduce that $NR/R \leq M/R$. By maximality of NR/R, we see that $NR = M$ and M/R is the unique maximal subgroup of H/R containing S/R, as desired. □

In later chapters, we will often (and without explicit reference) make use of a variety of results which may all be grouped together as "coprime action" arguments. These are presented below.

Proposition 2.7 (Coprime Action) *Suppose that a group G acts on a group A coprimely, so that* $\gcd(|A|, |G|) = 1$. *Then, for B a G-invariant subgroup of A, the following hold:*

(i) $C_{A/B}(G) = C_A(G)B/B$;
(ii) *if G acts trivially on A/B and B, then G acts trivially on A;*
(iii) $[A, G] = [A, G, G]$;
(iv) $A = [A, G]C_A(G)$ *and if A is abelian* $A = [A, G] \times C_A(G)$;
(v) *if G acts trivially on $A/\Phi(A)$, then G acts trivially on A;*

(vi) *if p is odd, A is a p-group and G acts trivially on $\Omega(A)$, then G acts trivially on A; and*
(vii) *for $S \in \mathrm{Syl}_p(G)$, if $m_p(S) \geqslant 2$ then $A = \langle C_A(s) : s \in S \setminus \{1\} \rangle$.*

Proof See, for instance [49, Chapter 8]. □

Remark In the definition of coprime action, we are implicitly assuming that one of G or A is solvable. That this holds is a consequence of the Feit–Thompson theorem.

Proposition 2.8 *If a non-cyclic elementary abelian p-group G acts on a p'-group A, then*

$$A = \langle C_A(B) : B \leq G, [G : B] = p \rangle.$$

Proof This is [33, Proposition 11.23]. □

Proposition 2.9 (Burnside) *Let S be a finite p-group. Then $C_{\mathrm{Aut}(S)}(S/\Phi(S))$ is a normal p-subgroup of* $\mathrm{Aut}(S)$.

The final result we describe here which still falls under the umbrella of "coprime action" and is essential to our analysis further on is the A×B-lemma due to Thompson.

Proposition 2.10 (A×B-Lemma) *Let AB be a finite group which acts on a p-group V. Suppose that B is a p-group, $A = O^p(A)$ and $[A, B] = \{1\} = [A, C_V(B)]$. Then $[A, V] = \{1\}$.*

Proof See [4, (24.2)]. □

The following definition is pivotal in understanding saturated fusion systems (as we will see later) and provides the impetus in the amalgam theoretic analysis that makes up the majority of this text.

Definition 2.11 Let G be a finite group and $H < G$. Then H is *strongly p-embedded* in G if and only if $|H|_p > 1$ and $N_G(Q) \leq H$ for each non-trivial p-subgroup Q with $Q \leq H$.

Lemma 2.12 *Suppose that G contains a strongly p-embedded subgroup X. Then the following hold:*

(i) *X contains a Sylow p-subgroup of G;*
(ii) *if $H \leq G$ with $H \not\leq X$ then $H \cap X$ is strongly p-embedded in H, provided $|H \cap X|_p > 1$;*
(iii) *$O^{p'}(G) \cap X$ is strongly p-embedded in $O^{p'}(G)$; and*
(iv) *if $G \neq XO_{p'}(G)$, then $XO_{p'}(G)/O_{p'}(G)$ is strongly p-embedded in $G/O_{p'}(G)$.*

Proof See [64, Lemmas 3.2, 3.3]. □

Proposition 2.13 *If G has a cyclic or generalized quaternion Sylow p-subgroup T and $O_p(G) = 1$, then $N_G(\Omega(T))$ is strongly p-embedded in G.*

Proof *For $X \leq T$ a non-trivial subgroup, X is also cyclic or generalized quaternion and so also has a unique subgroup of order p. Thus, $\Omega(X) = \Omega(T)$ and since $O_p(G) \neq 1$, we have that $N_G(X) \leq N_G(\Omega(X)) = N_G(\Omega(T)) < G$ so that $N_G(\Omega(T))$ is strongly p-embedded in G.* □

Quite remarkably, possessing a strongly p-embedded subgroup is a surprisingly limiting condition. In the following two propositions, we roughly determine the structure of groups with strongly p-embedded subgroups. For $p = 2$, we refer to work of Bender [13], while if p is odd we make use of the classification of finite simple groups.

In the application of these results, groups with strongly p-embedded subgroups will only ever appear in the local analysis of fusion systems. Particularly, these groups appear as automizers of certain p-subgroups and so would fit into the framework of any proofs utilizing a minimal counterexample hypothesis.

Proposition 2.14 *Suppose that $G = O^{p'}(G)$ is a group with a strongly p-embedded subgroup. Let $S \in \mathrm{Syl}_p(G)$ and set $\widetilde{G} := G/O_{p'}(G)$. If $m_p(S) = 1$ then one of the following holds:*

(i) *p is an odd prime, S is cyclic, G is perfect and $\widetilde{G}$ is a non-abelian finite simple group;*
(ii) *$G = SO_{p'}(G)$, S is cyclic or generalized quaternion and G is p-solvable; or*
(iii) *$p = 2$, S is generalized quaternion and $G/\Omega(S)O_{2'}(G) \cong K$, where either $\mathrm{PSL}_2(q) \leq K \leq \mathrm{P\Gamma L}_2(q)$ for q odd, or $K = \mathrm{Alt}(7)$.*

Moreover, in cases (ii) and (iii), $\langle \Omega(S)^G \rangle = \Omega(S)[\Omega(S), O_{p'}(G)]$ is the unique normal subgroup of G which is divisible by p and minimal with respect to this condition.

Proof Since $m_p(S) = 1$, we have that S is either cyclic or generalized quaternion by [32, I.5.4.10 (ii)]. Suppose first that S is cyclic. If $p = 2$, then G has a normal 2-complement (see [32, Theorem 7.4.3]) and (ii) holds. Hence, we may assume that if S is cyclic then p is odd. Notice that $F(\widetilde{G}) = O_p(\widetilde{G})$ since $O_{p'}(\widetilde{G}) = \{1\}$. If $F^*(\widetilde{G}) = F(\widetilde{G}) = O_p(\widetilde{G})$, then $O_p(\widetilde{G})$ is self-centralizing and as $\widetilde{S}$ is abelian, we have that $O_p(\widetilde{G}) = \widetilde{S}$ and $SO_{p'}(G) \trianglelefteq G$. In particular, G is p-solvable, $G = O^{p'}(G) \leq SO_{p'}(G) \leq G$, and (ii) holds.

Suppose now that $\widetilde{G}$ has a component $\widetilde{L}$. If $p \nmid |\widetilde{L}|$, then $\widetilde{L} \leq O_{p'}(E(\widetilde{G})) \leq O_{p'}(\widetilde{G}) = \{1\}$, a clear contradiction. Hence, p divides the order of any component of $\widetilde{G}$. Since $\widetilde{S}$ is cyclic, $\widetilde{L}$ has cyclic Sylow p-subgroups. By [4, Lemma 33.14], $Z(\widetilde{L})$ is a p'-prime group, and so $Z(\widetilde{L}) \leq O_{p'}(E(\widetilde{G})) = \{1\}$ and $\widetilde{L}$ is simple. Notice also that since each component is simple, $E(\widetilde{G})$ is a direct product of components, and since p divides the order of any component, $E(\widetilde{G}) = \widetilde{L}$ is the unique component of $\widetilde{G}$, else $m_p(\widetilde{G}) = m_p(G) > 1$. Since $O_p(\widetilde{G}) \cap E(\widetilde{G}) = \{1\}$, we have that $F^*(\widetilde{G}) = O_p(\widetilde{G}) \times E(\widetilde{G})$ and since $m_p(\widetilde{G}) = 1$, $O_p(\widetilde{G}) = \{1\}$. Therefore, $F^*(\widetilde{G})$ is a non-abelian simple group.

It remains to prove that $\widetilde{S} \leq F^*(\widetilde{G})$ to show that (i) holds. Form the group $\widetilde{H} = F^*(\widetilde{G})\widetilde{S}$ and assume, aiming for a contradiction, that $\widetilde{H} \neq F^*(\widetilde{G})$. Note that by the Frattini argument, $\widetilde{H} = F^*(\widetilde{G})N_{\widetilde{H}}(R)$ for all $R \in \mathrm{Syl}_r(\widetilde{F}^*(\widetilde{G}))$. Moreover, for $r \neq p$ a prime, $\mathrm{Syl}_r(F^*(\widetilde{G})) \subseteq \mathrm{Syl}_r(\widetilde{H})$. Then for $R \in \mathrm{Syl}_r(F^*(\widetilde{G}))$ with $r \neq p$, let $P \in \mathrm{Syl}_p(N_{\widetilde{H}}(R))$ and $T \in \mathrm{Syl}_p(\widetilde{H})$ containing P. Then $F^*(\widetilde{H}) \cap T < T$ and as T is cyclic and $\widetilde{H} = F^*(\widetilde{G})N_{\widetilde{H}}(R)$, we deduce that $P = T$ and $N_{\widetilde{H}}(R)$ contains a Sylow p-subgroup of $\widetilde{H}$. Hence, by conjugacy, $\widetilde{S}$ normalizes a Sylow r-subgroup of $\widetilde{H}$, for all primes r. But then $\widetilde{S}$ normalizes a Sylow r-subgroup of $N_{\widetilde{H}}(\widetilde{S})$ for all r, and so centralizes a Sylow r-subgroup of $N_{\widetilde{H}}(\widetilde{S})$ for all r. Applying [32, Theorem 7.4.3], $\widetilde{H}$ has a normal p-complement, a contradiction since $\widetilde{H}$ contains a component of $\widetilde{G}$. Thus, $\widetilde{S} \leq F^*(\widetilde{G})$ and since $G = O^{p'}(G)$ it follows that $\widetilde{G}$ is a non-abelian simple group. Hence, $\widetilde{G'} = \widetilde{G}$ and so $S \leq G'$. Then $G = O^{p'}(G) \leq G' \leq G$ so that G is perfect and (i) holds.

Assume now that $p = 2$ and S is generalized quaternion. Then by the Brauer-Suzuki theorem [32, Theorem II.12.1.1], we see that $\langle x\rangle O_{2'}(G)$ is a normal subgroup of G, and $\overline{G} := G/\langle x\rangle O_{2'}(G)$ has dihedral Sylow 2-subgroups. If $O_{2'}(\overline{G}) \neq \{1\}$ then for K the preimage in $\widetilde{G}$ of $O_{2'}(\overline{G})$, it is easily seen that $O_{2'}(K) \neq \{1\}$ and so $O_{2'}(\widetilde{G}) \neq \{1\}$, a contradiction. Thus, $\overline{G}$ is described by [32, (III.16.3)] and we are in outcome (iii). Thus, the result holds.

Suppose case (ii) or (iii) occurs and let N be a normal subgroup of G whose order is divisible by p. Then, as $m_p(S) = 1$, $\Omega(S) \leq N$ and so $\Omega(S)[\Omega(S), O_{p'}(G)] = \Omega(S)[\Omega(S), G] = \langle \Omega(S)^G\rangle \leq N$, and the result follows. □

Remark Notice that if H is a non-abelian finite simple group with a cyclic Sylow p-subgroup S then $N_G(\Omega(S))$ is strongly p-embedded in H by Proposition 2.13. Thus, the description in case (i) is best possible up to a better understanding of $O_{p'}(G)$. It is also worth noting that every non-abelian finite simple group has a cyclic Sylow p-subgroup for some odd prime p (although this fact relies on the classification of the finite simple groups).

Proposition 2.15 *Suppose that $G = O^{p'}(G)$ is a $\mathcal{K}$-group with a strongly p-embedded subgroup X. Let $S \in \mathrm{Syl}_p(G)$ and set $\widetilde{G} := G/O_{p'}(G)$. If $m_p(G) \geqslant 2$ then $\widetilde{G}$ is isomorphic to one of:*

(i) $\mathrm{PSL}_2(p^{a+1})$ *or* $\mathrm{PSU}_3(p^b)$ *for p arbitrary, $a \geqslant 1$ and $p^b > 2$;*
(ii) $\mathrm{Sz}(2^{2a+1})$ *for $p = 2$ and $a \geqslant 1$;*
(iii) $\mathrm{Alt}(2p)$ *for $p > 3$;*
(iv) $\mathrm{Ree}(3^{2a+1})$, $\mathrm{PSL}_3(4)$ *or* M_{11} *for $p = 3$ and $a \geqslant 0$;*
(v) $\mathrm{Sz}(32) : 5$, ${}^2\mathrm{F}_4(2)'$, McL *or* Fi_{22} *for $p = 5$; or*
(vi) J_4 *for $p = 11$.*

Proof If $G \neq XO_{p'}(G)$, then this follows from [64, (2.5), (3.3)] which in turn uses [34, Theorem 7.6.1] (with the appropriate erratum). So assume that $G = XO_{p'}(G)$. By coprime action,

$$O_{p'}(G) = \langle C_{O_{p'}(G)}(a) : a \in S^\# \rangle$$

since $m_p(G) \geqslant 2$ and so $O_{p'}(G) \leq X$ and $G = X$, a contradiction. □

2.2 Rank 1 Groups of Lie Type

As witnessed in Proposition 2.15, the generic examples of groups with a strongly p-embedded subgroup are rank 1 simple groups of Lie type in characteristic p. These are the groups which will appear most often in later work, and so we take this opportunity to list some important properties of these groups (and some properties of groups which "resemble" rank 1 groups). While almost all of these results are well known, we aim to provide explicit references and proofs of these results.

Proposition 2.16 *Let $G \cong \mathrm{PSL}_2(p^n)$ or $\mathrm{SL}_2(p^n)$ and $S \in \mathrm{Syl}_p(G)$. Then the following hold:*

(i) *S is elementary abelian of order p^n;*
(ii) $\mathrm{SL}_2(2) \cong \mathrm{Sym}(3)$, $\mathrm{PSL}_2(3) \cong \mathrm{Alt}(4)$ *and* $\mathrm{SL}_2(3)$ *are all solvable;*
(iii) *if $p = 2$, then for $U \leq S$ with $|U| = 4$, there is $x \in G$ such that $G = \langle U, u^x \rangle$ for any $u \in U^\#$;*
(iv) *if $p = 2$, all involutions in S are conjugate and so, for $u \in S^\#$ an involution, there is $x, y \in G$ such that $G = \langle u, u^x, u^y \rangle$;*
(v) *if p is odd, then for $u \in S^\#$, there is $x \in G$ such that $G = \langle u, u^x \rangle$ unless $p^n = 9$ in which case there is $x \in G$ such that $H := \langle u, u^x \rangle < G$ is maximal subgroup of G and $H/Z(H) \cong \mathrm{PSL}_2(5)$;*
(vi) *$N_G(S)$ is a solvable maximal subgroup of G and for K a Hall p'-subgroup of $N_G(S)$, $K/Z(G)$ is cyclic of order $(p^n - 1)/(p^n - 1, 2)$ and K acts fixed point freely on $S \setminus \{1\}$;*
(vii) *if $p^n \geqslant 4$, then G is perfect and if $\widetilde{G}$ is a perfect central extension of G by a group of p'-order, then $\widetilde{G} \cong \mathrm{PSL}_2(p^n)$ or $\mathrm{SL}_2(p^n)$; and*
(viii) *if x is a non-trivial automorphism of G which centralizes S, then $x \in \mathrm{Aut}_S(G)$.*

Proof The proofs of (i)–(vi) are written out fairly explicitly in [43, II.6-II.8]. Detailed information on automorphism groups and Schur multipliers is provided in [34, Theorem 2.5.12] and [34, Theorem 6.1.2]. □

Proposition 2.17 *Let $G \cong \mathrm{PSU}_3(p^n)$ or $\mathrm{SU}_3(p^n)$ and $S \in \mathrm{Syl}_p(G)$. Then the following hold:*

(i) *S is a special p-group of order p^{3n} with $|Z(S)| = p^n$;*
(ii) $\mathrm{SU}_3(2)$ *is solvable, a Sylow 2-subgroup of* $\mathrm{SU}_3(2)$ *is isomorphic to the quaternion group of order 8 and* $\mathrm{SU}_3(2)' \cong 3_+^{1+2} : 2$ *has index 4 in* $\mathrm{SU}_3(2)$*;*
(iii) *for $p^n > 2$, $N_G(S)$ is a solvable maximal subgroup of G and for K a Hall p'-subgroup of $N_G(S)$, $|K/Z(G)| = (p^{2n} - 1)/(p^{2n} - 1, 3)$ and K acts irreducibly on $S/Z(S)$;*
(iv) *for $p^n > 2$, $N_G(Z(S)) = N_G(S)$ and for K a Hall p'-subgroup of $N_G(S)$, $|C_K(Z(S))| = p^n + 1/|Z(K)|$ and $C_K(Z(S))$ acts fixed point freely on $S/Z(S)$;*
(v) *for any $x \in G \setminus N_G(S)$, $\langle Z(S), Z(S)^x \rangle \cong \mathrm{SL}_2(p^n)$ and $G = \langle Z(S), \dot{S}^x \rangle$;*

(vi) *for* $\{1\} \neq U \leq Z(S)$, *unless* $p^n = 9$ *and* $|U| = 3$ *or* $p = 2$ *and* $|U| = 2$, *there is* $x, z \in G$ *such that* $G = \langle U, U^x, U^z \rangle$;
(vii) *for* $\{1\} \neq U \leq Z(S)$, *if* $p^n = 9$ *and* $|U| = 3$ *or* $p = 2 < p^n$ *and* $|U| = 2$, *then there is* $x, y, z \in G$ *such that* $G = \langle U, U^x, U^y, U^z \rangle$;
(viii) *for* $\{1\} \neq U \trianglelefteq S$ *with* $U \not\leq Z(S)$, *if* $p^n \neq 2$ *then there is* $x \in G$ *such that* $G = \langle U, U^x \rangle$;
(ix) *if* $p^n > 2$, *then* G *is perfect and if* $\widetilde{G}$ *is a perfect central extension of* G *by a group of* p'*-order, then* $\widetilde{G} \cong \mathrm{PSU}_3(p^n)$ *or* $\mathrm{SU}_3(p^n)$; *and*
(x) *if* x *is a non-trivial automorphism of* G *which centralizes* S, *then* $x \in \mathrm{Aut}_{Z(S)}(G)$.

Proof The proofs of (i)–(v) may be found in [43, II.10]. Again, information on automorphism groups and Schur multipliers may be found in [34, Theorem 2.5.12, Theorem 6.1.2]. It remains to prove (vi)–(viii).

For (vi) and (vii) suppose that $U \leq Z(S)$, $p^n \neq 2$ and set $H := \langle Z(S), Z(S)^x \rangle \cong \mathrm{SL}_2(p^n)$ for $x \in G \setminus N_G(S)$. By Proposition 2.16 (iv), (v), H is generated by two or three conjugates of U, and by [54], H is contained in a unique maximal subgroup $M \cong \mathrm{GU}_2(p^n) \cong (p^n+1).\mathrm{SL}_2(p^n)$. Since $G = \langle U^G \rangle$, there is z such that $U^z \not\leq M$. It then follows from the maximality of M in G that $G = \langle H, U^z \rangle$ and (vi) and (vii) are proved.

Suppose now that $U \not\leq Z(S)$, $U \trianglelefteq S$ and $p^n \neq 2$. Since $U \not\leq Z(S)$, $\{1\} \neq [U, S] \leq Z(S) \cap U$. Set $C := C_{N_G(S)}(Z(S))$ and observe that C is irreducible on $S/Z(S)$ by (iv). Then, since $[U, S] \leq Z(S)$, $[U, S] = [U, S]^C = [\langle U^C \rangle, \langle S^C \rangle]$. By the irreducibility of C on $S/Z(S)$, $(UZ(S)/Z(S))^C = S/Z(S)$ and so $[\langle U^C \rangle, \langle S^C \rangle] = Z(S) = [U, S] \leq U$. Now, there is $x \in G \setminus N_G(S)$ such that $\langle Z(S), Z(S)^x \rangle \cong \mathrm{SL}_2(p^n)$ is contained in a unique maximal subgroup $M \cong \mathrm{GU}_2(p^n)$. Then, as $U > Z(S)$, $|U| > p^n$, $\langle Z(S), Z(S)^x \rangle < \langle U, U^x \rangle$ and (viii) follows. □

Remark The result in (vii) is certainly not best possible. Indeed, it should be possible to prove that G is generated by just three conjugates of U, using a closer analysis of the maximal subgroups laid out in [54]. However, for the purposes of this work, (vii) in its current form suffices.

Proposition 2.18 *Let* $G \cong \mathrm{Sz}(2^n)$ *and* $S \in \mathrm{Syl}_2(G)$. *Then the following hold:*

(i) n *is odd and* 3 *does not divide the order of* G;
(ii) $\mathrm{Sz}(2) \cong 5 : 4$ *is a Frobenius group*, $\Phi(\mathrm{Sz}(2)) = \{1\}$, $\mathrm{Sz}(2)' \cong C_5$ *and a Sylow* 2*-subgroup of* $\mathrm{Sz}(2)$ *is cyclic of order* 4;
(iii) *if* $n > 1$ *then* $\Phi(S) = Z(S) = \Omega(S)$ *and* $S/\Phi(S) \cong \Phi(S)$ *is elementary abelian of order* 2^n;
(iv) $N_G(S)$ *is a solvable maximal subgroup of* G *and for* K *a Hall* $2'$*-subgroup of* $N_G(S)$, $|K| = 2^n - 1$ *and* K *acts irreducibly on* $S/\Phi(S)$ *and* $\Phi(S)$;
(v) *if* $n > 1$ *then there is* $x \in G$ *such that* $G = \langle Z(S), Z(S)^x \rangle$;
(vi) *all involutions in* S *are conjugate and if* $n > 1$, *for* $u \in Z(S)^\#$, *there is* $x, y \in G$ *such that* $G = \langle u, u^x, u^y \rangle$;

(vii) *for* $U \leq S$ *with* U *elementary abelian of order* 4, *there is* $x \in G$ *such that* $G = \langle U, U^x \rangle$;
(viii) *if* $n > 1$ *then* G *is perfect and has trivial Schur multiplier; and*
(ix) *if* x *is a non-trivial automorphism of* G *which centralizes* S, *then* $x \in \mathrm{Aut}_{Z(S)}(G)$.

Proof Most of the proofs of these facts may be found in [73, Sections 13–16], except the proof of (viii) which may be gleaned from [34, Theorem 6.1.2]. □

Proposition 2.19 *Let* $G \cong \mathrm{Ree}(3^n)$ *and* $S \in \mathrm{Syl}_3(G)$. *Then the following hold:*

(i) n *is odd;*
(ii) *the Sylow* 2*-subgroups of* G *are abelian;*
(iii) *if* $n = 1$ *then* $G \cong \mathrm{PSL}_2(8) : 3$, $G' \cong \mathrm{PSL}_2(8)$, $S \cong 3_-^{1+2}$, $S \cap G'$ *is cyclic of order* 9 *and* $\Omega(S)$ *is elementary abelian of order* 9;
(iv) *if* $n > 1$, *then* S *has order* 3^{3n}, $\Phi(S) = \Omega(S)$ *has order* 3^{2n}, $Z(S) = [S, \Phi(S)]$ *has order* 3^n *and* $S/\Phi(S) \cong \Phi(S)/Z(S) \cong Z(S)$ *is elementary abelian of order* 3^n;
(v) $N_G(S)$ *is a solvable maximal subgroup of* G *and for* K *a Hall* $3'$*-subgroup of* $N_G(S)$, $|K| = 3^n - 1$ *and* K *acts irreducibly on* $S/\Omega(S)$, $\Omega(S)/Z(S)$ *and* $Z(S)$;
(vi) *for* $\{1\} \neq U \trianglelefteq S$, *if* $n > 1$ *then there is* $x, y \in G$ *such that* $G = \langle U, U^x, U^y \rangle$;
(vii) *if* $n > 1$ *then* G *is perfect and has trivial Schur multiplier, and* $\mathrm{Ree}(3)'$ *is perfect and has trivial Schur multiplier; and*
(viii) *if* x *is a non-trivial automorphism of* G *which centralizes* S, *then* $x \in \mathrm{Aut}_{Z(S)}(G)$.

Proof The proofs of (i)–(v) follow from the main theorem of [75] while (vii) and (viii) follow from [34, Theorem 2.5.12, Theorem 6.1.2]. We make use of results in [75] to prove (vi). Since the results when $n = 1$ are easily verified, we assume that $n > 1$ throughout. Note that if $U \not\leq \Omega(S)$, then as $U \trianglelefteq S$ and $\Omega(S) = Z_2(S)$, $Z(S) < Z(S)[U, S] \leq \Omega(S)$ and $\Omega(U) \not\leq Z(S)$. In particular, if $U \not\leq Z(S)$, then there is $u \in \Omega(U) \setminus Z(S)$.

Suppose first that $U \leq Z(S)$. Since K is irreducible on $Z(S)$, there is $y \in N_G(S)$ such that for some $u \in U$, u^y is not represented by elements of a subfield of $\mathrm{GF}(3^n)$. Then $\langle U, U^y \rangle$ is elementary abelian of order at least 9 and contained in a maximal subgroup of G. Considering the maximal subgroup structure of G (as in [46, Theorem C]) and using that the centralizer of an involution in K intersects $Z(S)$ trivially, we deduce that $\langle U, U^y \rangle$ lies in a unique maximal subgroup, namely $N_G(S)$. But now, there is $x \in G \setminus N_G(S)$ such that $U^x \not\leq S$. Thus, $G = \langle U, U^x, U^y \rangle$, as required.

Suppose now that $U \not\leq Z(S)$ so that there is $u \in U$ such that $u \in \Omega(U) \setminus Z(S)$. Then by (v), it follows that $C_{N_G(S)}(u) = \Omega(S)\langle i \rangle$, where $i \in K$ is an involution. Then $u \in C_G(i)$ and by [75], $C_G(i) \cong \langle i \rangle \times L$, where $L \cong \mathrm{PSL}_2(3^n)$, and $C_G(i)$ is a maximal subgroup of G (see also [46, Theorem C]). Since $n > 1$ is odd, there is $x \in L$ such $L = \langle u, u^x \rangle$ by Proposition 2.16 (v). Furthermore, $C_G(i) \cap Z(S) = \{1\}$.

However, $U \trianglelefteq S$ so that $U \cap Z(S) \neq \{1\}$ from which we deduce that $L < \langle U, U^x \rangle$. Since $C_G(i)$ is maximal, we must have that $G = \langle U, U^x \rangle$, as desired. □

Pivotal to the analysis of local actions in the amalgam method and within a fusion system is recognizing $\mathrm{SL}_2(p^n)$ acting on its modules in characteristic p. Below, we list the most important modules (and their properties) for this work.

Definition 2.20 Let $X \cong \mathrm{SL}_2(q)$, $q = p^n$, $k = \mathrm{GF}(q)$ and V a faithful 2-dimensional kX-module.

- $V|_{\mathrm{GF}(p)X}$ is a *natural* $\mathrm{SL}_2(q)$*-module* for X.
- A *natural* $\Omega_3(q)$*-module* for X is the 3-dimensional submodule of $V \otimes_k V$ regarded as a $\mathrm{GF}(p)X$-module by restriction, and is irreducible whenever p is an odd prime.
- If $n = 2a$ for some $a \in \mathbb{N}$, a *natural* $\Omega_4^-(q^{\frac{1}{2}})$*-module* for X is any non-trivial irreducible submodule of $(V \otimes_k V^\tau)|_{\mathrm{GF}(q^{\frac{1}{2}})X}$, where τ is an automorphism of $\mathrm{GF}(q)$ of order 2, regarded as a $\mathrm{GF}(p)X$-module by restriction.
- If $n = 3a$ for some $a \in \mathbb{N}$, a *triality module* for X is any non-trivial irreducible submodule of $(V \otimes V^\tau \otimes V^{\tau^2})|_{\mathrm{GF}(q^{\frac{1}{3}})X}$, where τ is an automorphism of k of order 3, regarded as a $\mathrm{GF}(p)X$-module by restriction.

Proposition 2.21 *Suppose* $G \cong \mathrm{SL}_2(p^n)$, $S \in \mathrm{Syl}_p(G)$ *and* V *is natural* $\mathrm{SL}_2(p^n)$*-module. Then the following hold:*

(i) $[V, S, S] = \{1\}$;
(ii) $|V| = p^{2n}$ *and* $|C_V(S)| = p^n$;
(iii) $C_V(s) = C_V(S) = [V, S] = [V, s] = [v, S]$ *for all* $v \in V \setminus C_V(S)$ *and* $s \in S^\#$;
(iv) $V = C_V(S) \times C_V(S^g)$ *for* $g \in G \setminus N_G(S)$;
(v) *every* p'*-element of* G *acts fixed point freely on* V;
(vi) V *is self dual; and*
(vii) $V/C_V(S)$ *and* $C_V(S)$ *are irreducible* $\mathrm{GF}(p)N_G(S)$*-modules upon restriction.*

Proof See [60, Lemma 4.6]. □

Lemma 2.22 *Suppose that* $G \cong \mathrm{SL}_2(q)$, $q = p^n$, *and* V *is a direct sum of two natural* $\mathrm{SL}_2(q)$*-modules. If* $U \leq C_V(S)$ *is* $N_G(S)$*-invariant and of order* q, *then* $|\langle U^G \rangle| = q^2$.

Proof By [32, (I.3.5.6)], the number of distinct irreducible submodules of V is $q + 1 = (q^2 - 1)/q - 1$. For each W an irreducible submodule, $C_W(S)$ is $N_G(S)$-invariant and of order q. Since $C_V(S)$ may be viewed as a direct sum of two irreducible modules for $N_G(S)$, again applying [32, (I.3.5.6)] we have that there are $q+1$ $N_G(S)$-invariant subgroups of $C_V(S)$ of order q. Hence, each $N_G(S)$-invariant subgroup of $C_V(S)$ of order q is of the form $C_W(S)$ for some irreducible module W and $C_W(S)$ determines W uniquely. Thus, U uniquely determines a submodule $\langle U^G \rangle$ of order q^2. □

Proposition 2.23 *Suppose that $G \cong \mathrm{SL}_2(p^n)$, p an odd prime, $S \in \mathrm{Syl}_p(G)$ and V is a natural $\Omega_3(p^n)$-module for G. Then the following hold:*

(i) $C_G(V) = Z(G)$;
(ii) $[V, S, S, S] = \{1\}$;
(iii) $|V| = p^{3n}$ *and* $|V/[V, S]| = |C_V(S)| = p^n$;
(iv) $[V, S] = [V, s]$ *and* $[V, S, S] = [V, s, s] = C_V(s) = C_V(S)$ *for all* $s \in S^{\#}$;
(v) $[V, S]/C_V(S)$ *is centralized by* $N_G(S)$; *and*
(vi) $V/[V, S]$ *and* $C_V(S)$ *are irreducible* $\mathrm{GF}(p)N_G(S)$*-modules upon restriction.*

Proof See [60, Lemma 4.7]. □

Proposition 2.24 *Let $G \cong \mathrm{SL}_2(p^{2n})$, $S \in \mathrm{Syl}_p(G)$ and V a natural $\Omega_4^-(p^n)$-module for G. Then the following hold:*

(i) $C_G(V) = Z(G)$;
(ii) $[V, S, S, S] = \{1\}$;
(iii) $|V| = p^{4n}$ *and* $|V/[V, S]| = |C_V(S)| = p^n$;
(iv) $|C_V(s)| = |[V, s]| = p^{2n}$ *for all* $s \in S^{\#}$; *and*
(v) $V/[V, S]$ *and* $C_V(S)$ *are irreducible* $\mathrm{GF}(p)N_G(S)$*-modules upon restriction.*

Moreover, for $\{1\} \neq F \leq S$, one of the following occurs:

(a) $[V, F] = [V, S]$ *and* $C_V(F) = C_V(S)$;
(b) $p = 2$, $[V, F] = C_V(F)$ *has order* p^{2n} *and* $|F| \leqslant p^n$; *or*
(c) p *is odd,* $|[V, F]| = |C_V(F)| = p^{2n}$, $[V, S] = [V, F]C_V(F)$, $C_V(S) = C_{[V,F]}(F)$ *and* $|F| \leqslant p^n$.

If $F = \langle x \rangle$ for $x \in S^{\#}$, then either (b) or (c) holds.

Proof See [60, Lemma 4.8] and [61, Lemma 3.15]. □

We require one miscellaneous result concerning the exceptional 1-cohomology of $\mathrm{PSL}_2(9)$ on a natural $\Omega_4^-(3)$-module.

Lemma 2.25 *Suppose that $p \in \{2, 3\}$, $G \cong \mathrm{PSL}_2(p^2)$ and $S \in \mathrm{Syl}_p(G)$. If V is a 5-dimensional $\mathrm{GF}(p)G$-module such that $V/C_V(G)$ is isomorphic to a natural $\Omega_4^-(p)$-module, then either $V = [V, G] \times C_V(G)$; or $p = 3$ and $[V, S, S]$ is 2-dimensional as a $\mathrm{GF}(3)S$-module.*

Proof This follows from direct computation in $\mathrm{GL}_5(p)$. □

Proposition 2.26 *Suppose that $G \cong \mathrm{(P)SL}_2(p^{3n})$, $S \in \mathrm{Syl}_p(G)$ and V is a triality module for G. Then the following hold:*

(i) $[V, S, S, S, S] = \{1\}$;
(ii) $|V| = p^{8n}$, $|V/[V, S]| = |C_V(S)| = |[V, S, S, S]| = p^n$ *and* $|[V, S, S]| = p^{4n}$;
(iii) *if p is odd then* $|V/C_V(s)| = p^{5n}$, *while if* $p = 2$ *then* $|V/C_V(s)| = p^{4n}$, *for all* $s \in S^{\#}$; *and*
(iv) $V/[V, S]$ *and* $C_V(S)$ *are irreducible* $\mathrm{GF}(p)N_G(S)$*-modules upon restriction.*

Proof See [60, Lemma 4.10]. □

We will also need facts concerning the natural modules for $SU_3(p^n)$ and $Sz(2^n)$.

Definition 2.27 The natural module for both $SU_3(p^n)$ and $Sz(2^n)$ is the unique irreducible $GF(p)$-module of smallest dimension (of dimension $6n$ and $4n$ respectively). Equivalently, they may be viewed as the restrictions of a "natural" $SL_3(p^{2n})$-module and $Sp_4(2^n)$-module respectively.

Proposition 2.28 *Suppose $G \cong SU_3(p^n)$, $S \in Syl_p(G)$ and V is a natural module. Then the following hold:*

(i) $C_V(S) = [V, Z(S)] = [V, S, S]$ *is of order* p^{2n}*;*
(ii) $C_V(Z(S)) = [V, S]$ *is of order* p^{4n}*; and*
(iii) $V/[V, S]$, $[V, S]/C_V(S)$ *and* $C_V(S)$ *are irreducible* $GF(p)N_G(S)$*-modules upon restriction.*

Proof See [60, Lemma 4.13]. □

Proposition 2.29 *Suppose $G \cong Sz(2^n)$, $S \in Syl_2(G)$ and V is the natural module. Then the following hold:*

(i) $[V, S]$ *has order* 2^{3n}*;*
(ii) $[V, \Omega(S)] = C_V(\Omega(S)) = [V, S, S]$ *has order* 2^{2n}*;*
(iii) $C_V(S) = [V, S, \Omega(S)] = [V, \Omega(S), S] = [V, S, S, S]$ *has order* 2^n*; and*
(iv) $V/[V, S]$, $[V, S]/C_V(\Omega(S))$, $C_V(\Omega(S))/C_V(S)$ *and* $C_V(S)$ *are all irreducible* $GF(2)N_G(S)$*-modules upon restriction.*

Proof One may calculate in $Sp_4(2^n)$ to obtain these results. In particular, generators for S and $N_G(S)$ as subgroups of $GL_4(2^n)$, as well as many other facts about the 2-modular representation theory of $Sz(2^n)$, are given explicitly in [50]. □

2.3 Recognition Results

Given the descriptions of rank 1 Lie type groups and their modules, we now require ways to identify them. Furthermore, we would like to have ways to completely determine a group G with a strongly p-embedded subgroup, and its actions, given reasonably general hypotheses. We achieve this through characterizations of FF-modules, quadratic action and some Hall–Higman type arguments. We start with an elementary result regarding $GF(p)G$-modules.

Lemma 2.30 *Let G be a group and V be a faithful $GF(p)G$-module. Let $T \in Syl_p(O^p(G))$ and assume that $V = \langle C_V(T)^G \rangle$. Then $V = [V, O^p(G)]C_V(O^p(G))$.*

Proof See [22, Lemma 1.1]. □

We will require some knowledge of the minimal $\mathrm{GF}(p)$-representations of groups with a strongly p-embedded subgroup, especially when their p-rank is at least 2. The following results achieve this.

Proposition 2.31 *Let $G = O^{2'}(G)$ be a group with a strongly 2-embedded subgroup and $m_2(G) > 1$. Assume that V is a faithful $\mathrm{GF}(2)$-module for G, let $S \in \mathrm{Syl}_2(G)$ and $A \leq Z(S)$ of order 4. Then the following hold:*

(i) *if $G/O_{2'}(G) \cong \mathrm{PSL}_2(q)$ then $|V| \geqslant q^2$, $|V/C_V(A)| \geqslant q$ and $|V/C_V(s)| \geqslant q^{\frac{2}{3}}$ for any $s \in Z(S)^\#$;*
(ii) *if $G/O_{2'}(G) \cong \mathrm{Sz}(q)$ then $|V| \geqslant q^4$, $|V/C_V(A)| \geqslant q^2$ and $|V/C_V(s)| \geqslant q^{\frac{4}{3}}$ for any $s \in Z(S)^\#$; and*
(iii) *if $G/O_{2'}(G) \cong \mathrm{PSU}_3(q)$ then $|V| \geqslant q^6$, $|V/C_V(A)| \geqslant q^2$ and $|V/C_V(s)| \geqslant q^{\frac{3}{2}}$ for any $s \in Z(S)^\#$.*

Proof The bounds on V follow from the proof of [57, Lemma 1.7]. Set d_A be the number of conjugates of A required to generate $G/O_{2'}(G)$ and d_s the number of conjugates of s required to generate $G/O_{2'}(G)$. Then we form $H \leq G$ from d_A conjugates of A or d_s conjugates of s with the property $G = HO_{2'}(G)$. Moreover, H acts faithfully on V and $|V/C_V(H)| \leqslant \min(|V/C_V(A)|^{d_A}, |V/C_V(s)|^{d_s})$. Since H is also a group with a strongly 2-embedded subgroup, $|V/C_V(H)| \geqslant q^2, q^4$ or q^6 respectively. Applying Proposition 2.16 (iii), (iv) when $G/O_{2'}(G) \cong \mathrm{PSL}_2(q)$, Proposition 2.18 (vi), (vii) when $G/O_{2'}(G) \cong \mathrm{Sz}(q)$ and Proposition 2.17 (vi), (vii) when $G/O_{2'}(G) \cong \mathrm{PSU}_3(q)$, we deduce the bounds on $|V/C_V(A)|$ and $|V/C_V(s)|$. □

Proposition 2.32 *Let $G = O^{p'}(G)$ be a $\mathcal{K}$-group with a strongly p-embedded subgroup, $m_p(G) > 1$ and p an odd prime. Assume that V is a faithful $\mathrm{GF}(p)$-module for G, let $S \in \mathrm{Syl}_p(G)$ and $A \leq Z(S)$ of order p^2. Then the following hold:*

(i) *if $G/O_{p'}(G) \cong \mathrm{PSL}_2(q)$ then $|V| \geqslant q^2$, $|V/C_V(A)| \geqslant q$ and $|V/C_V(s)| \geqslant q^{\frac{2}{3}}$ for any $s \in S^\#$;*
(ii) *if $G/O_{3'}(G) \cong \mathrm{Ree}(q)'$ then $|V| \geqslant \max(q^6, 3^7)$ and $|V/C_V(s)| \geqslant q^2$ for any $s \in Z(S)^\#$;*
(iii) *if $G/O_{p'}(G) \cong \mathrm{PSU}_3(q)$ then $|V| \geqslant q^6$, $|V/C_V(A)| \geqslant q^2$ and $|V/C_V(s)| \geqslant q^{\frac{3}{2}}$ for any $s \in Z(S)^\#$; and*
(iv) *$|V| \geqslant p^5$ otherwise.*

Proof Suppose first that $G/O_{p'}(G)$ is not isomorphic to the derived subgroup of a rank 1 finite group of Lie type, including Ree(3). Then one can check, comparing with Proposition 2.15, that $|G/O_{p'}(G)|$ does not divide the order of $\mathrm{GL}_4(p)$. Hence, $G/O_{p'}(G)$ is isomorphic to a rank 1 finite group of Lie type and we appeal to the proof of [62, Lemma 4.6] to deduce the bounds on V. For $G/O_{3'}(G) \cong \mathrm{Ree}(3)$, we simply calculate that such a group does not embed in $\mathrm{SL}_6(3)$ using MAGMA [14]. For the determination of the bounds on $|V/C_V(A)|$ and $|V/C_V(s)|$, we proceed as in Proposition 2.31, instead using Proposition 2.16 (v), Proposition 2.17 (vi), (vii) and Proposition 2.19 (vi) for the values of d_A and d_s. □

We require, at least when p is an odd prime, a way to distinguish between $\mathrm{SL}_2(p^n)$ and $\mathrm{PSL}_2(p^n)$ from a strongly p-embedded hypothesis. Additionally, as can be seen from the Main Theorem, none of the configurations we are interested in have Ree groups as their automizers, so we will also have to dispel of this case later on. Generally, we achieve this using quadratic action.

Definition 2.33 Let G be a finite group and V an $\mathrm{GF}(p)G$-module. If $A \leq G$ satisfies $[V, A, A] = \{1\} \neq [V, A]$, then A acts *quadratically* on V and if $[V, A, A, A] = \{1\}$ and A is not quadratic or trivial on V, then A acts *cubically*.

Proposition 2.34 *Suppose that V is an irreducible* $\mathrm{GF}(p)$*-module for* $G \cong \mathrm{Ree}(3^n)$ *or* $G \cong \mathrm{PSL}_2(p^n) \not\cong \mathrm{SL}_2(p^n)$. *If there is a non-trivial subgroup A of G with $[V, A, A] = \{1\}$, then $[V, A] = [V, G] = \{1\}$.*

Proof Since the Sylow 2-subgroups of $\mathrm{PSL}_2(p^n)$ are either abelian or dihedral and the Sylow 2-subgroups of $\mathrm{Ree}(3^n)$ are abelian, this follows from [32, (I.3.8.4)]. □

For $p \geqslant 5$, the pairs (G, V) where G is a group acting faithfully on a module V such that G is generated by elements which act quadratically on V were classified by Thompson. Thompson's results were extended to the prime 3 by work of Ho. It seems imperative to emphasize that these works predate the classification of the finite simple groups. For convenience, the version we use here is by Chermak and utilizes the classification of finite simple groups, although as we stressed earlier, these groups will only ever appear as local subgroups in any arguments.

Lemma 2.35 *Suppose $G = O^{p'}(G)$ is a $\mathcal{K}$-group which has a strongly p-embedded subgroup for p an odd prime and let V be a faithful* $\mathrm{GF}(p)$*-module for G. Suppose there is a p-subgroup $A \leq G$ such that $[V, A, A] = \{1\}$. Then one of the following occurs:*

(i) $G \cong \mathrm{SL}_2(p^n)$ *where p is any odd prime;*
(ii) $G \cong \mathrm{(P)SU}_3(p^n)$ *where p is any odd prime;*
(iii) G, $G/C_G(U) \cong 4 \circ 2^{1+4}_{-}.\mathrm{Alt}(6)$ *for some non-trivial irreducible composition factor U of V, $|S| = 9$ and $p = |A| = 3$;*
(iv) G, $G/C_G(U) \cong 2 \cdot \mathrm{Alt}(5)$ *or* $2^{1+4}_{-}.\mathrm{Alt}(5)$ *for every non-trivial irreducible composition factor U of V and $p = |S| = |A| = 3$; or*
(v) $G/C_G(U) \cong \mathrm{SL}_2(3)$ *for every non-trivial irreducible composition factor U of V and $p = |S| = |A| = 3$.*

Moreover, in case (v), G is 3-solvable.

Proof The list of groups follows from [20, 21], Proposition 2.34 and a comparison with the groups listed in Propositions 2.14 and 2.15. In particular, if G is not p-solvable then $G/O_{p'}(G)$ is non-abelian simple and so we ascertain that all the outcomes are exclusive and in case (v) G is 3-solvable.

Suppose that case (v) holds. By Proposition 2.14, we have that S is cyclic and $G = SO_{p'}(G)$. Since A acts quadratically on V, and V is faithful, we deduce that A is elementary abelian and so $A = \Omega(S)$. Assume that S has order at least 9. Then for every non-trivial composition factor U of V we have that $|S/C_S(U)| = 3$ from which it follows that $A \leq C_S(U)$. Indeed, $L \leq C_G(U)$ and so for $r \in L$ of order prime to 3, by coprime action, we have that r centralizes V. But V is a faithful module, and so we must have that L is a normal 3-subgroup of G, impossible since G has a strongly 3-embedded subgroup. Hence, $|S| = 3$ and $S = A$, as desired. □

Remark There are two conjugacy classes of maximal subgroups of $\mathrm{SU}_4(3)$ isomorphic to the group $4 \circ 2_-^{1+4}.\mathrm{Alt}(6)$. Importantly for future use, we note that when $G/C_G(U) \cong 4 \circ 2_-^{1+4}.\mathrm{Alt}(6)$ in the above classification, a Sylow 3-subgroup of G does *not* act quadratically on U.

As the typical example in our case will be central extensions of a rank 1 simple groups of Lie type, we also provide a generic result regarding these particular actions.

Proposition 2.36 *Suppose that G is a rank 1 group of Lie type, $S \in \mathrm{Syl}_p(G)$ and let $\{1\} \neq A \leq S$ with $[V, A, A] = \{1\}$ for some faithful $\mathrm{GF}(p)G$-module V. Then $A \leq \Omega(Z(S))$ and if $|A| \geqslant 3$, for any non-trivial element $a \in A$, the following hold:*

(i) *if $G \cong \mathrm{SL}_2(p^n)$, then $|V/C_V(a)| \geqslant p^n$; and*
(ii) *if $G \cong \mathrm{Sz}(2^n)$ or $\mathrm{(P)SU}_3(p^n)$ then $|V/C_V(a)| \geqslant p^{2n}$.*

Proof See [27, (5.9)] and [27, (5.10)]. □

More than just a quadratic module, the natural module for $\mathrm{SL}_2(p^n)$ provides the minimal example of an *FF-module*. FF-modules are named due to how they arise as counterexamples to *Thompson factorization* (see [4, 32.11]), which aims to factorize a group into two p-local subgroups. One of these p-local subgroups is the normalizer of the *Thompson subgroup* of a fixed Sylow p-subgroup.

Definition 2.37 Let S be a finite p-group. Set $\mathcal{A}(S)$ to be the set of all elementary abelian subgroups of S of maximal rank. Then the *Thompson subgroup* of S is defined as $J(S) := \langle A : A \in \mathcal{A}(S)\rangle$.

Proposition 2.38 *Let S be a finite p-group. Then the following hold:*

(i) *$J(S)$ is a non-trivial characteristic subgroup of S;*
(ii) *for $A \in \mathcal{A}(S)$, $A = \Omega(C_S(A))$;*
(iii) *$\Omega(C_S(J(S))) = \Omega(Z(J(S))) = \bigcap_{A\in\mathcal{A}(S)} A$; and*
(iv) *if $J(S) \leq T \leq S$, then $J(S) = J(T)$.*

Proof See [49, 9.2.8]. □

Definition 2.39 Let G be a finite group and V a GF$(p)G$-module. If there exists $A \leq G$ such that

(i) $A/C_A(V)$ is an elementary abelian p-group;
(ii) $[V, A] \neq \{1\}$; and
(iii) $|V/C_V(A)| \leqslant |A/C_A(V)|$

then V is a *failure to factorize module* (abbrev. FF-module) for G and A is an *offender* on V.

The following proposition describes a fairly natural situation in which one can identify an FF-module from a group failing to satisfy Thompson factorization. This result is well known and the proof is standard (see [49, 9.2]).

Proposition 2.40 *Let G be a finite group with $S \in \mathrm{Syl}_p(G)$ and $F^*(G) = O_p(G)$. Set $V := \langle \Omega(Z(S))^G \rangle$. Then $O_p(G) = O_p(C_G(V))$ and $O_p(G/C_G(V)) = \{1\}$. Furthermore, if $\Omega(Z(S)) < V$ and $J(S) \not\leq C_S(V)$ then V is an FF-module for $G/C_G(V)$.*

As a counterpoint to the determination of groups with a strongly p-embedded subgroup, whenever a group with a strongly p-embedded subgroup has an associated FF-module, we can almost completely determine the group and its action without the need for a $\mathcal{K}$-group hypothesis. Indeed, the following lemma relies only on a specific case in the Local $C(G, T)$-theorem [18].

Lemma 2.41 *Suppose $G = O^{p'}(G)$ has a strongly p-embedded subgroup and $V = \langle C_V(S)^G \rangle$ is (dual to) a faithful FF-module for G. Then $G \cong \mathrm{SL}_2(p^n)$ and $V/C_V(G)$ is the natural module.*

Proof See [42, Theorem 5.6]. □

Given a way to characterize the natural $\mathrm{SL}_2(p^n)$-module, it is a natural to ask whether we can characterize some other modules, particularly those irreducible modules associated to Lie type groups of rank 1.

Lemma 2.42 *Let $G \cong \mathrm{SL}_2(p^n)$ and $S \in \mathrm{Syl}_p(G)$. Suppose that V is a GF$(p)G$-module such that $[V, S, S] = \{1\}$ and such that $[V, O^p(G)] \neq \{1\}$. Then $[V/C_V(O^p(G)), O^p(G)]$ is a direct sum of natural modules for G.*

Proof See [21, Lemma 2.2]. □

Lemma 2.43 *Let $G \cong \mathrm{SL}_2(p^n)$, $S \in \mathrm{Syl}_p(G)$ and V an irreducible GF$(p)G$-module. If $|V| \leqslant p^{3n}$ then both $C_V(S)$ and $V/[V, S]$ are irreducible as $N_G(S)$-modules, $|C_V(S)| = |V/[V, S]|$ and either*

(i) *V is natural $\mathrm{SL}_2(p^n)$-module, $|V| = p^{2n}$ and $|C_V(S)| = p^n$;*
(ii) *V is natural $\Omega_4^-(p^{\frac{n}{2}})$, n is even, $|V| = p^{2n}$ and $|C_V(S)| = p^{\frac{n}{2}}$;*
(iii) *V is natural $\Omega_3(p^n)$, p is odd, $|V| = p^{3n}$ and $|C_V(S)| = p^n$; or*
(iv) *V is a triality module, $n = 3r$ for some $r \in \mathbb{N}$, $|V| = p^{\frac{8n}{3}}$ and $|C_V(S)| = p^{\frac{n}{3}}$.*

Proof This is [23, Lemma 2.6]. □

Lemma 2.44 *Assume that* $G = O^{p'}(G)$, $G/O_{p'}(G) \cong \mathrm{PSL}_2(q)$ *and* V *is a faithful, irreducible* $\mathrm{GF}(p)$*-module for* $p \in \{2, 3\}$, *where* $q = p^n$ *and* $n > 1$. *If* $|V| \leqslant q^3$, *then either*

(i) $O_{p'}(G) \leq Z(G)$ *and both* G *and* V *are determined in Lemma 2.43; or*
(ii) $p = 3$, $q = 9$, $G \cong 2^5 : \mathrm{Alt}(6)$ *and* V *is unique up to isomorphism.*

In particular, in (ii) no element of $S \in \mathrm{Syl}_3(G)$ *acts quadratically on* V.

Proof Assume that G is a minimal counterexample to the lemma and let $S \in \mathrm{Syl}_p(G)$. If $[x, R] = \{1\}$ for all $R \in \mathrm{Syl}_r(O_{p'}(G))$ then $[x, O_{p'}(G)] = \{1\}$ and taking normal closures, we would have that $[G, O_{p'}(G)] = \{1\}$. In this case, since $G = O^{p'}(G)$, we deduce that G is quasisimple and by Lemma 2.43, (i) holds and G is not a minimal counterexample.

Hence, for $x \in S^\#$, there is some $R \in \mathrm{Syl}_r(O_{p'}(G))$ with $[x, R] \neq \{1\}$. By the Frattini argument, $G = O_{p'}(G)N_G(R)$ and so we may assume that x normalizes R. If $N_G(R) < G$, then by minimality, $O^{p'}(N_G(R))$ is quasisimple. But then $[O^{p'}(N_G(R)), R] = \{1\}$ a contradiction since $[x, R] \neq \{1\}$. Hence, $R \trianglelefteq G$ and so $R = O_r(G)$. Moreover, as $[x, O_r(G)] \neq \{1\}$ we certainly have that $[G, O_r(G)] \neq \{1\}$. Since $G/O_{p'}(G)$ is simple and $G = O^{p'}(G)$, we see that $[O_r(G), s] \neq \{1\}$ for all $s \in S^\#$. Choose $K \trianglelefteq G$ minimally by inclusion such that $K \leq O_r(G)$, $[K, O_r(G)] \leq Z(K)$ and $[K, x] \neq \{1\}$ for all $s \in S^\#$. Such a subgroup exists by the critical subgroup theorem (see Theorem 2.59). Furthermore, if r is an odd prime, or if K is abelian and $r = 2$, then by minimality we deduce that $\Omega(K) = K$.

Let T be a proper, non-trivial characteristic subgroup of K. By minimality, we have that there is $x \in S^\#$ such that $[x, T] = \{1\}$ from which it follows that $T \leq Z(G)$. In particular, T is abelian. By Clifford theory (see [33, Theorem 9.7]), $V|_T = \bigoplus_{i \in I} W_i$ where G acts transitively on the set $\{W_i\}$ and each W_i is irreducible for T. Since T is abelian, by Schur's lemma, we have that $T/C_T(W_i)$ is a cyclic r-group and since $C_T(W_i) \leq Z(G)$, we conclude that $C_T(W_i) \leq C_T(V) = \{1\}$. Hence, T is cyclic.

Assume first that K is non-abelian. Then K contains no non-cyclic characteristic abelian subgroups and is of class two. By [32, Theorem 5.4.9], K is a central product of an extraspecial group and a cyclic group. Note also that $\Phi(K)$ is cyclic and that $K = \Omega(K)$ if p is odd. Hence, we deduce that K is extraspecial if p is odd; while K is either extraspecial, or the central product of an extraspecial group and a cyclic group of order 4 when $p = 2$. Let $\widehat{K}$ be a subgroup of K which is extraspecial of maximal possible order, and let m be such that $|\widehat{K}| = r^{2m+1}$. Since $\widehat{K}$ acts faithfully on V, by [33, Lemma 9.5], we deduce that $|V| \geqslant p^{r^m}$ so that $3n \geqslant r^m$. Since S acts faithfully on K and $G/O_{p'}(G)$ is simple, by [76], S embeds in $\mathrm{Sp}_{2m}(r)$. Applying [61, Corollary 13.23], if $p = 3$ then $r = 2$ and $m \leqslant 3$. We calculate that $\mathrm{PSL}_2(27)$ does not arise as a section of $\mathrm{Sp}_6(2)$ and so we have that $r = 2$, $n = 2$ and $m = 2$. In particular, $|V| \leqslant 3^6$ and G embeds as a subgroup of $\mathrm{SL}_6(3)$. We verify the result computationally in this case using MAGMA [14]. Likewise, if $p = 2$, we apply [61,

Corollary 13.23] so that $r = 3$ and $m = 2$, or $m = 1$. In the latter case, we observe that $\mathrm{Sp}_2(r) \cong \mathrm{SL}_2(r)$ has a unique involution and so as $n > 1$, S cannot embed in $\mathrm{Sp}_2(r)$ and we have a contradiction. In the former case, we see that $3 \leqslant n \leqslant 4$ and comparing the orders of $G/O_{2'}(G)$ and $\mathrm{Sp}(4, 3)$ yields another contradiction. Hence, we must have that K is abelian.

Assume now that K is elementary abelian. By Clifford theory, $V|_K = \bigoplus_{i \in I} W_i$ where G acts transitively on the set $\{W_i\}$ and each W_i is irreducible for K. Write $I(W_i)$ for the stabilizer in G of W_i. Note that $C_K(W_i) \trianglelefteq I(W_i)$ for all i. If $C_K(W_i) \trianglelefteq G$, then $C_K(W_i) \leq C_K(V) = \{1\}$. But then, K acts faithfully on each W_i and by Schur's lemma, we deduce that $|K| = r$, which is impossible since $|S| > p^2$ and S acts faithfully on K. Hence, $I(W_i) \leq N_G(C_K(W_i)) \leq M < G$ for M some maximal subgroup. If $G = I(W_i)O_{p'}(G)$ then for V_i an irreducible, non-trivial $O^{p'}(I(W_i))$ composition factor of W_i, we have by minimality that $|V_i| \geqslant q^2$ from which it follows that $V = W_i$, a contradiction. Thus, we can select M such that $O_{p'}(G) < M$. Thus, either $q = 9$ and $[G : M] = 6$, or $[G : M] \geqslant q + 1$.

Aiming for a contradiction, suppose first that K is elementary abelian and $[G : M] \geqslant q + 1$. Then $|W_i|^{q+1} \leqslant |V| \leqslant q^3$. Since K acts non-trivially on each W_i, we have that $|W_i| \geqslant 3$ and as $n > 1$, we arrive at the desired contradiction.

Hence, if K is elementary abelian then $q = 9$ and $[G : M] = 6$ so that $M/O_{3'}(G) \cong \mathrm{Alt}(5)$. Then we have that $|W_i| = 3$, $I(W_i) = M$ and $|M/C_M(W_i)| = 2$. In particular, $|O_{3'}(G)/C_{O_{3'}(G)}(W_i)| = 2$ and as $O_{3'}(G)$ is faithful on V, $O_{3'}(G)$ is an elementary abelian 2-group of order at most 2^6. Since $|V| = 3^6$ and $m_2(\mathrm{SL}_6(3)) = 5$, we have that $|O_{3'}(G)| \leqslant 2^5$, and since $G/O_{3'}(G) \cong \mathrm{Alt}(6)$, we see that $|K| \geqslant 2^4$. Calculating in MAGMA [14], we find that $K = O_{3'}(G)$ is elementary abelian of order 2^5, $|Z(G)| = 2$ and $N_{\mathrm{SL}_6(3)}(K) \cong 2^5 : \mathrm{Sym}(6)$. One can check that the index 2 subgroup isomorphic to $2^5 : \mathrm{Alt}(6)$ has a unique faithful, irreducible module of dimension 6 over GF(3) and that no element acts quadratically on V. Another way to see this is to note that the restriction of V to an Alt(6) complement has an $\Omega_4^-(3)$-module as a factor (the restriction of V is the GF(3)-permutation module for an Alt(6) complement). Hence, we are in case (ii) and so the proof is now complete. □

We may relax the restrictions in the definition of an FF-module to allow for a greater class of module setups. An an example, the natural modules for $\mathrm{SU}_3(p^n)$ and $\mathrm{Sz}(2^n)$ are not FF-modules but satisfy the ratio $|V/C_V(A)| \leqslant |A/C_A(V)|^2$ for A some elementary abelian p-subgroup. Such modules are referred to as *2F-modules*.

Definition 2.45 Let G be a finite group and V a GF$(p)G$-module. If there exists $A \leq G$ such that

(i) $A/C_A(V)$ is an elementary abelian p-group;
(ii) $[V, A] \neq \{1\}$; and
(iii) $|V/C_V(A)| \leqslant |A/C_A(V)|^2$

then V is *2F-module* for G.

If G is an almost quasisimple group with a 2F module V, then both G and V are known by work of Guralnick and Malle [38, 39] and Guralnick et al. [37]. Importantly for applications in this work, even when G is not almost quasisimple, we have a good idea of the structure of the groups which have a strongly p-embedded subgroup and an associated 2F-module which admits a quadratically acting element.

First we introduce two groups that have associated GF(p)-modules which exhibit 2F-action and arise heavily in the local actions in later chapters. In addition, we provide some "characterizations" of these groups, and some structural properties of the groups and the associated 2F-module we are interested in.

Proposition 2.46 *There is a unique group G of shape $(3 \times 3) : 2$ which has a faithful quadratic 2F-module V, namely the generalized dihedral group of order* 18. *Moreover, for $S \in \mathrm{Syl}_2(G)$ and V an associated faithful quadratic 2F-module, the following hold:*

(i) $|V| = 2^4$ *and G is unique up to conjugacy in* $\mathrm{GL}_4(2)$;
(ii) $\{G, \mathrm{Dih}(18)\} = \{H : |H| = 18, O_2(H) = \{1\}$ *and* $H = O^{2'}(H)\}$;
(iii) *there are exactly four overgroups of S in G which are isomorphic to* $\mathrm{Sym}(3)$, *any two of which generate G; and*
(iv) $C_{\mathrm{GL}_4(2)}(G) = \{1\}$ *and* $|\mathrm{Out}_{\mathrm{GL}_4(2)}(G)| = 4$.

Proof This follows directly from calculations in MAGMA [14], working explicitly with matrices in $\mathrm{GL}_4(2)$ and comparing with the Small Groups Library. □

Indeed, in the above lemma G is also isomorphic to $\mathrm{PSU}_3(2)'$ and is listed in the Small Groups Library as *SmallGroup*(18, 4).

Proposition 2.47 *There is a unique group G of shape $(Q_8 \times Q_8) : 3$ which has a faithful quadratic 2F-module V. Moreover, for $S \in \mathrm{Syl}_3(G)$ and V an associated faithful quadratic 2F-module, the following hold:*

(i) $|V| = 3^4$ *and G is determined uniquely up to conjugacy in* $\mathrm{GL}_4(3)$;
(ii) *G is the unique group of order $2^4.3$ or $2^6.3$ such that $O_3(G) = \{1\}$, $Z(G) \neq \{1\}$, $G = O^{3'}(G)$ and, if the order is $2^6.3$, there exists at least two distinct normal subgroups of G of order* 8;
(iii) *there are exactly five overgroups of S in G which are isomorphic to* $\mathrm{SL}_2(3)$, *any two of which generate G;*
(iv) $N_{O_2(G)}(S) = Z(G) \cong 2 \times 2$;
(v) $\mathrm{Aut}(G) = \mathrm{Aut}_{\mathrm{GL}_4(3)}(G)$, $C_{\mathrm{GL}_4(3)}(G) = Z(G)$ *and* $|\mathrm{Out}(G)| = 2^2.3$; *and*
(vi) *if $U < V$ is $N_G(S)$-invariant and $|U| = 3$, then $|\langle U^G \rangle| = 9$.*

Proof This follows directly from calculations in MAGMA [14], working explicitly with matrices in $\mathrm{GL}_4(3)$ and comparing with the Small Groups Library. □

The above group is listed in the Small Groups Library as *SmallGroup*(192, 1022).

We now give an important characterization of certain "small" groups which have an associated non-trivial quadratic 2F-module. The proof of this result will be broken up over a series of lemmas.

Lemma 2.48 *Assume that $G = O^{p'}(G)$ is a $\mathcal{K}$-group that has a strongly p-embedded subgroup, $S \in \mathrm{Syl}_p(G)$, V is a faithful* GF(p)*-module with $C_V(O^p(G)) = \{1\}$ and $V = \langle C_V(S)^G \rangle$. Furthermore, assume that $m_p(S) \geqslant 2$. If there is a p-element $x \in S^\#$ such that $[V, x, x] = \{1\}$ and $|V/C_V(x)| = p^2$ then either:*

(i) *p is odd, $G \cong (\mathrm{P})\mathrm{SU}_3(p)$ and V is the natural module;*
(ii) *p is arbitrary, $G \cong \mathrm{SL}_2(p^2)$ and V is the natural module; or*
(iii) *$p = 2$, $G \cong \mathrm{PSL}_2(4)$ and V is a natural $\Omega_4^-(2)$-module.*

Proof Applying the characterization in Proposition 2.15 and using Lemma 2.35 when p is odd, we deduce that $G/O_{p'}(G) \cong \mathrm{PSL}_2(p^{n+1})$, $\mathrm{PSU}_3(p^n)$ or $\mathrm{Sz}(2^{2n+1})$ for $n \geqslant 1$. Assume that G is a minimal counterexample to the lemma and let P be a p-minimal subgroup G containing S with the property $G = PO_{p'}(G)$. If $G = P$ then by [22, Theorem 1] and Lemma 2.30, the lemma holds, a contradiction since G was an assumed counterexample. Hence, we may assume that $P < G$. Set V^P to be some non-trivial P-composition factor of V. By Lemma 2.41, using that $m_p(S) \geqslant 2$, we have that $|V^P/C_{V^P}(x)| = p^2$ and $[V, x, x] = \{1\}$. By minimality, P is described by (i), (ii) or (iii). Since $C_V(P) \leq C_V(x)$, we can arrange that $V/C_V(P) = V^P$.

If $P \cong \mathrm{SL}_2(p^2)$ and $V/C_V(P)$ is a natural module, then for each $y \in S^\#$, we have that $C_V(y) = [V, y]C_V(P) = [V, S]C_V(P)$ from which it follows that $C_V(S) = [V, S]C_V(P)$ has index p^2 in V. Then Lemma 2.41 yields that $G = P$, a contradiction since we assumed $P < G$. If $G \cong (\mathrm{P})\mathrm{SU}_3(p)$ and $V/C_V(P)$ is the natural module, then $\langle x \rangle = Z(S)$. Moreover, for $A \leq S$ with $|A| = p^2$ and A elementary abelian, for $y, z \in A$ with $y, z \notin Z(S)$ and $\langle y \rangle \neq \langle z \rangle$, we have that $C_V(y) = C_V(z) = C_V(A)$. Furthermore, $[V, y]C_V(y) = [V, z]C_V(z) = [V, A]C_V(A) = C_V(x)$ and by coprime action, $C_V(x)$ is normalized by $O_{p'}(G)$. Then $H := \langle x^{O_{p'}(G)} \rangle$ centralizes $C_V(x)$ and by coprime action, $V = [V, O_{p'}(H)] \times C_V(O_{p'}(H))$ is a $\langle x \rangle$ invariant decomposition. Since $C_V(x) \leq C_V(O_{p'}(H))$, we deduce that $O_{p'}(H)$ centralizes V so that $O_{p'}(H) = \{1\}$ and $O_{p'}(G)$ normalizes $\langle x \rangle = Z(S)$. Hence, $[G, O_{p'}(G)] = [\langle Z(S)^G \rangle, O_{p'}(G)] = \{1\}$, G is quasisimple and $G = P$, again a contradiction. Finally, if $P \cong \mathrm{PSL}_2(4)$ and $V/C_V(P)$ is a natural $\Omega_4^-(2)$-module, then we select $x \in G$ and $s \in S^\#$ with the property $s^x \not\leq P$. Then $H := \langle P, s^x \rangle \ncong \mathrm{PSL}_2(4)$ and by minimality, we have that $G = H$. Since $|V/C_V(s^x)| = 4$, we deduce that $|V| = |V/C_V(G)| \leqslant 2^6$ and so G embeds in $\mathrm{SL}_6(2)$. Comparing with [15] yields that $G = P$, and we arrive at a final contradiction. □

Lemma 2.49 *Assume that $G = O^{p'}(G)$ is a $\mathcal{K}$-group, $S \in \mathrm{Syl}_p(G)$, V is a faithful* GF(p)*-module with $C_V(O^p(G)) = \{1\}$ and $V = \langle C_V(S)^G \rangle$. Furthermore, assume that $m_p(S) = 1$, G is not p-solvable and if $p = 2$ and S is generalized quaternion* *then $N_G(S)$ is strongly 2-embedded in G. If there is a p-element $x \in S^\#$ such that $[V, x, x] = \{1\}$ and $|V/C_V(x)| = p^2$ then either:*

(i) *$p = 3$, $G \cong 2\cdot\mathrm{Alt}(5)$ or $2_-^{1+4}.\mathrm{Alt}(5)$ and V is the unique irreducible quadratic $2F$-module of dimension 4; or*
(ii) *$p \geqslant 5$, $G \cong \mathrm{SL}_2(p)$ and V is the direct sum of two natural $\mathrm{SL}_2(p)$-modules.*

Proof Suppose first that $p = 2$. Applying Proposition 2.14, we deduce that S is generalized quaternion and $G = O_{2'}(G)C_G(\Omega(S))$. But now, $C_G(\Omega(S)) = N_G(\Omega(Z(S))) = N_G(S)$ is solvable so that G itself is solvable, a contradiction to the initial hypothesis.

Therefore, p is odd. Applying Lemma 2.35 and using that G is not p-solvable, we deduce that for $L := \langle x^G\rangle$, $L/C_L(U) \cong \mathrm{SL}_2(p)$ for $p \geqslant 5$, or $p = 3$ and $L/C_L(U) \cong 2 \cdot \mathrm{Alt}(5)$ or $2_-^{1+4}.\mathrm{Alt}(5)$ for U some non-trivial irreducible composition factor of $V|_L$. Indeed, applying Proposition 2.14, $G = L$ and $C_G(U)$ is a p'-group. Now, by coprime action, $V = C_V(C_G(U)) \times [V, C_G(U)]$ and $U \leq C_V(C_G(U))$. Throughout, we set $V_1 := C_V(C_G(U))$ and $V_2 := [V, C_G(U)]$.

Applying Lemma 2.41, if $L/C_L(U) \cong 2 \cdot \mathrm{Alt}(5)$ or $2_-^{1+4}.\mathrm{Alt}(5)$ when $p = 3$, we have that $|U/C_U(s)| = 3^2$ so that $V_2 \leq C_V(s)$. Then $V_2 \leq C_V(G) = \{1\}$ and as V is a faithful module, $C_G(U) = \{1\}$. Indeed, by Lemma 2.30 and using that $C_V(G) = \{1\}$, $V = U$ is an irreducible module and outcome (i) holds.

Hence, we may assume that $G/C_G(U) \cong \mathrm{SL}_2(p)$ and $p \geqslant 5$. Then V_1 is a quadratic module for $G/C_G(U)$ and by Lemma 2.42, and using that $C_V(G) = \{1\}$, V_1 is a direct sum of at most two natural $\mathrm{SL}_2(p)$-modules.

Suppose first that V_1 is a natural $\mathrm{SL}_2(p)$-module so that $U = V_1$ and $|U/C_U(s)| = p$. Then $|V_2/C_{V_2}(s)| = p$ and applying Lemma 2.41, we deduce that $G/C_G(V_2) \cong \mathrm{SL}_2(p)$ and V_2 is a natural $\mathrm{SL}_2(p)$-module. Since V_2 is acted upon non-trivially by $C_G(U)$ and $C_G(U)$ is a p'-group, we conclude that $C_G(V_2)C_G(U)/C_G(U) = Z(G/C_G(U))$ and $C_G(V_2)C_G(U)/C_G(V_2) = Z(G/C_G(V_2))$ so that $G/(C_G(V_2) \cap C_G(U))$ is a central extension of $\mathrm{PSL}_2(p)$ by a fours group. Since the 2-part of the Schur multiplier of $\mathrm{PSL}_2(p)$ has order 2, G is perfect and $G = O^{p'}(G)$, this is a contradiction. Hence, V_1 is not a natural module.

Suppose now that V_1 is a direct sum of two natural $\mathrm{SL}_2(p)$-modules. Then $|V_1/C_{V_1}(s)| = p^2$ and we deduce that $V_2 \leq C_V(s)$ so that $V_2 \leq C_V(G) = \{1\}$. As V is a faithful module, $C_G(U) = \{1\}$ and outcome (ii) holds. □

Lemma 2.50 *Assume that $G = O^{p'}(G)$, $S \in \mathrm{Syl}_p(G)$, V is a faithful* GF(p)-*module with $C_V(O^p(G)) = \{1\}$ and $V = \langle C_V(S)^G\rangle$. Furthermore, assume that $m_p(S) = 1$, $N_G(S)$ is strongly p-embedded in G, and G is p-solvable. If there is a p-element $x \in S^\#$ such that $[V, x, x] = \{1\}$ and $|V/C_V(x)| = p^2$ then, setting $L := \langle x^G\rangle$, one of the following holds:*

(i) *$p \leqslant 3$, $G = L \cong \mathrm{SL}_2(p)$ and V is the direct sum of two natural $\mathrm{SL}_2(p)$-modules;*
(ii) *$p = 2$, $L \cong \mathrm{SU}_3(2)'$, G is isomorphic to a subgroup of $\mathrm{SU}_3(2)$ which contains $\mathrm{SU}_3(2)'$ and V is a natural $\mathrm{SU}_3(2)$-module viewed as an irreducible $\mathrm{GF}(2)G$-module by restriction;*

(iii) $p = 2$, $L \cong \mathrm{Dih}(10)$, $G \cong \mathrm{Dih}(10)$ *or* $\mathrm{Sz}(2)$ *and* V *is a natural* $\mathrm{Sz}(2)$*-module viewed as an irreducible* $\mathrm{GF}(2)G$*-module by restriction;*

(vi) $p = 3$, $G = L \cong (Q_8 \times Q_8) : 3$ *and* $V = V_1 \times V_2$ *where* V_i *is a natural* $\mathrm{SL}_2(3)$*-module for* $G/C_G(V_i) \cong \mathrm{SL}_2(3)$*;*

(v) $p = 2$, $G = L \cong (3 \times 3) : 2$ *and* $V = V_1 \times V_2$ *where* V_i *is a natural* $\mathrm{SL}_2(2)$*-module for* $G/C_G(V_i) \cong \mathrm{Sym}(3)$*; or*

(vi) $p = 2$, $L \cong (3 \times 3) : 2$, $G \cong (3 \times 3) : 4$, V *is irreducible as a* $\mathrm{GF}(2)G$*-module and* $V|_L = V_1 \times V_2$ *where* V_i *is a natural* $\mathrm{SL}_2(2)$*-module for* $L/C_L(V_i) \cong \mathrm{Sym}(3)$.

Proof Let $L := \langle x^G \rangle$ so that $L = [\Omega(S), O_{p'}(G)]\Omega(S)$ by Proposition 2.14. Since $N_G(S) = N_G(\Omega(S))$, we deduce that $G = LS$ so that $O^p(L) = O^p(G) = [\Omega(S), O_{p'}(G)]$ and $C_V(O^p(L)) = \{1\}$. Moreover, any element of S centralizes $\Omega(Z(S)) \in \mathrm{Syl}_p(L)$ but does not centralize L, for otherwise, since S contains a unique subgroup of order p, $[\Omega(Z(S)), L] = \{1\}$ and $\Omega(Z(S)) \trianglelefteq G$. Thus, $S/\Omega(S)$ embeds into $\mathrm{Out}(L)$. Finally, using Lemma 2.30, $V = [V, O^p(L)]$ and so both L and V are determined in [22, Lemma 4.3]. We examine each of the cases individually, using MAGMA [14] for the explicit calculation in $\mathrm{Out}(L)$.

First, if $L \cong \mathrm{SL}_2(p)$ then it follows from Proposition 2.16 (viii) that $\mathrm{Out}_S(L) = \{1\}$, $L = G$ and V is a direct sum of two natural modules. This is (i). If $L \cong \mathrm{Dih}(10)$ then $\mathrm{Aut}(L) \cong \mathrm{Sz}(2)$ and it follows that $G = \mathrm{Dih}(10)$ or $\mathrm{Sz}(2)$, V is the restriction of a natural $\mathrm{Sz}(2)$-module to G, and (iii) holds.

Suppose that $L \cong \mathrm{SU}_3(2)'$. Then a Sylow 2-subgroup of $\mathrm{Aut}(L)$ is isomorphic to a semidihedral group of order 16 and since $m_p(S) = 1$, $|S| \leqslant 8$ and S is either cyclic or quaternion. Moreover, $54 \leqslant |G| \leqslant 216$ and $|G| = 54$ if and only if $G = L \cong \mathrm{SU}_3(2)'$. Suppose that $|G| = 216$ and S is cyclic. Utilizing the small group library in MAGMA [14], we identify a unique group H such that $\langle \Omega(S)^H \rangle \cong \mathrm{SU}_3(2)'$. But in such a group, $N_H(T) < N_H(\Omega(T))$ for $T \in \mathrm{Syl}_2(H)$, a contradiction to our hypothesis. Employing similar methods when $|G| = 108$, or when $|G| = 216$ and S is quaternion, gives that G is isomorphic to any index 2 subgroup of $\mathrm{SU}_3(2)$ resp. $G \cong \mathrm{SU}_3(2)$. In all cases, V is the restriction of a natural $\mathrm{SU}_3(2)$-module to G. This is outcome (ii).

Suppose that $L \cong (Q_8 \times Q_8) : 3$. Since G acts faithfully on V, of order 3^4, G embeds into $\mathrm{GL}_4(3)$ and since the embedding of L is uniquely determined up to conjugacy in $\mathrm{GL}_4(3)$, it follows that G embeds into its normalizer in $\mathrm{GL}_4(3)$. For H the image of L in $\mathrm{GL}_4(3)$, we have that a Sylow 3-subgroup of $N_{\mathrm{GL}_4(3)}(H)$ is elementary abelian of order 9. Since $m_p(S) = 1$, we have that $G = L$ in this case and V is as described in [22, Lemma 4.3], and (iv) holds.

Finally, suppose that $L \cong (3 \times 3) : 2$. Since G acts faithfully on V, of order 2^4, G embeds into $\mathrm{GL}_4(2)$ and since the embedding of L is uniquely determined up to conjugacy in $\mathrm{GL}_4(2)$, it follows that G embeds into the normalizer of its image. For H the image of L in $\mathrm{GL}_4(2)$, we have that a Sylow 2-subgroup of $N_{\mathrm{GL}_4(2)}(H)$ is a dihedral group of order 8 and there is a unique proper overgroup of H in $N_{\mathrm{GL}_4(2)}(H)$ with a cyclic Sylow 2-subgroup. Moreover, this group is irreducible in $\mathrm{GL}_4(2)$, is defined uniquely up to conjugacy in $\mathrm{GL}_4(2)$ and is isomorphic to any index 2 subgroup of $\mathrm{PSU}_3(2)$. We denote this group $(3 \times 3) : 4$. It follows that either $G = L \cong (3 \times 3) : 2$ or $G \cong (3 \times 3) : 4$, and V is as given in [22, Lemma 4.3]. This is outcomes (v) and (vi) respectively. □

The following proposition is the summation of the previous three lemmas. This situation occurs frequently throughout the later chapters of this work.

Proposition 2.51 *Assume that $G = O^{p'}(G)$ is a $\mathcal{K}$-group that has a strongly p-embedded subgroup, $S \in \mathrm{Syl}_p(G)$, V is a faithful* GF(p)*-module with $C_V(O^p(G)) = \{1\}$ and $V = \langle C_V(S)^G \rangle$. Furthermore, assume that if $m_p(G) = 1$ and G is p-solvable or S is generalized quaternion, then $N_G(S)$ is strongly p-embedded in G. Suppose that there is a p-element $x \in S^\#$ such that $[V, x, x] = \{1\}$ and $|V/C_V(x)| = p^2$. Setting $L := \langle x^G \rangle$ one of the following holds:*

(i) *p is odd, $G = L \cong (\mathrm{P})\mathrm{SU}_3(p)$ and V is the natural module;*
(ii) *p is arbitrary, $G = L \cong \mathrm{SL}_2(p^2)$ and V is the natural module;*
(iii) *$p = 2$, $G = L \cong \mathrm{PSL}_2(4)$ and V is a natural $\Omega_4^-(2)$-module;*
(iv) *$p = 3$, $G = L \cong 2 \cdot \mathrm{Alt}(5)$ or $2_-^{1+4}.\mathrm{Alt}(5)$ and V is the unique irreducible quadratic $2F$-module of dimension 4;*
(v) *p is arbitrary, $G = L \cong \mathrm{SL}_2(p)$ and V is the direct sum of two natural $\mathrm{SL}_2(p)$-modules;*
(vi) *$p = 2$, $L \cong \mathrm{SU}_3(2)'$, G is isomorphic to a subgroup of $\mathrm{SU}_3(2)$ which contains $\mathrm{SU}_3(2)'$ and V is a natural $\mathrm{SU}_3(2)$-module viewed as an irreducible* GF(2)*G-module by restriction;*
(vii) *$p = 2$, $L \cong \mathrm{Dih}(10)$, $G \cong \mathrm{Dih}(10)$ or $\mathrm{Sz}(2)$ and V is a natural $\mathrm{Sz}(2)$-module viewed as an irreducible* GF(2)*G-module by restriction;*
(viii) *$p = 3$, $G = L \cong (Q_8 \times Q_8) : 3$ and $V = V_1 \times V_2$ where V_i is a natural $\mathrm{SL}_2(3)$-module for $G/C_G(V_i) \cong \mathrm{SL}_2(3)$;*
(ix) *$p = 2$, $G = L \cong (3 \times 3) : 2$ and $V = V_1 \times V_2$ where V_i is a natural $\mathrm{SL}_2(2)$-module for $G/C_G(V_i) \cong \mathrm{Sym}(3)$; or*
(x) *$p = 2$, $L \cong (3 \times 3) : 2$, $G \cong (3 \times 3) : 4$, V is irreducible as a* GF(2)*G-module and $V|_L = V_1 \times V_2$ where V_i is a natural $\mathrm{SL}_2(2)$-module for $L/C_L(V_i) \cong \mathrm{Sym}(3)$.*

While most of the groups and modules above have been described earlier in this chapter, we list some properties of the groups and modules occurring in (iv) and (x) above.

Lemma 2.52 *Suppose that $G \cong 2 \cdot \mathrm{Alt}(5)$ or $2_-^{1+4}.\mathrm{Alt}(5)$, $S \in \mathrm{Syl}_3(G)$ and V is an associated faithful quadratic $2F$-module. Then $C_V(S) = [V, S]$ has order 3^2 and $V/[V, S]$ and $[V, S]$ are irreducible as* GF(3)*$N_G(S)$-modules.*

Proof This follows directly from calculations in MAGMA [14], working explicitly with the matrices in $\mathrm{Sp}_4(3)$. □

Lemma 2.53 *Suppose that $G \cong (3 \times 3) : 4$, $S \in \mathrm{Syl}_2(G)$ and V is an associated faithful quadratic $2F$-module. Then the following hold:*

(i) $[V, S]$ *has order* 2^3;
(ii) $[V, \Omega(S)] = C_V(\Omega(S)) = [V, S, S]$ *has order* 2^2; *and*
(iii) $C_V(S) = [V, S, \Omega(S)] = [V, \Omega(S), S] = [V, S, S, S]$ *has order* 2.

Proof This follows directly from calculations in MAGMA [14], working explicitly with the matrices in $\mathrm{GL}_4(2)$. □

Lemma 2.54 *Suppose that* (G, V) *satisfies the hypothesis of Proposition 2.51. In addition, assume that* V *is generated as a* $\mathrm{GF}(p)G$*-module by an* $N_G(S)$*-invariant subspace of order* p. *Then* G *is isomorphic to one of* $\mathrm{PSL}_2(4)$, $\mathrm{Dih}(10)$, $\mathrm{Sz}(2)$, $(3 \times 3) : 2$ *or* $(3 \times 3) : 4$ *and* V *is as described in Proposition 2.51.*

Proof We apply Proposition 2.51 to get the list of candidates for G and V. By Proposition 2.28 (iii), Proposition 2.21 (vi) and Lemma 2.52, if (G, V) satisfy (i), (ii), (iv) or (vi), then there are no $N_G(S)$-invariant subspaces of order p. By Lemma 2.22 and Proposition 2.47 (vi), if (G, V) satisfy (v) or (viii) then V is not generated by a subspace of order p. This leaves outcomes (iii), (vii), (ix) and (x), as required. □

Following on from Lemma 2.35 and Proposition 2.51, we now recognize some groups with a strongly p-embedded subgroup and p-rank greater than 2 which have a quadratic 2F-module.

Lemma 2.55 *Suppose that* G *is a finite group such that* $G = O^{p'}(G)$ *and* $G/O_{p'}(G) \cong \mathrm{PSL}_2(p^n)$, $\mathrm{Sz}(2^n)$ *or* $\mathrm{PSU}_3(p^n)$ *for* $n > 1$. *Assume that* V *is a faithful* $\mathrm{GF}(p)$*-module for* G *such that there is* $A \leq S \in \mathrm{Syl}_p(G)$ *with* $|A| > p^{\frac{n}{2}}$, $[V, A, A] = \{1\}$ *and* $|V/C_V(A)| \leqslant |A|^2$. *Then* $G \cong \mathrm{SL}_2(p^n)$, $\mathrm{Sz}(2^n)$ *or* $\mathrm{SU}_3(p^n)$ *and* $V/C_V(G)$ *is a direct sum of at most* r *natural modules for* G, *where* $r = 2$ *if* $G \cong \mathrm{SL}_2(p^n)$ *and* $r = 1$ *otherwise.*

Proof Assume that G is a finite group satisfying the hypothesis of the lemma with $|G|$ minimal. Since A acts quadratically on V we have that A is elementary abelian and by Lemma 2.35 and Proposition 2.36 when p is odd, we deduce that $A \leq Z(S)$.

Let P be a p-minimal subgroup of G such that $G = PO_{p'}(G)$. If $P < G$, then by minimality, $P \cong \mathrm{SL}_2(p^n)$, $\mathrm{Sz}(2^n)$ or $\mathrm{SU}_3(2^n)$ and $V/C_V(P)$ is as described in the lemma. Since $[V, A]C_V(P) \leq C_V(A)$ we deduce that $C_V(Z(S)) = C_V(A) = C_V(z)$ for all $z \in Z(S)^\#$. By coprime action, $O_{p'}(G)$ normalizes $C_V(Z(S))$ and forming $H := \langle Z(S)^{O_{p'}(G)} \rangle$, we have that H acts trivially on $C_V(Z(S))$. But then $V = [VO_{p'}(H)] \times C_V(O_{p'}(H))$ is a $Z(S)$-invariant decomposition and since $C_V(Z(S)) \leq C_V(O_{p'}(H))$, we conclude that $V = C_V(O_{p'}(H))$ and $Z(S)$ is normalized by $O_{p'}(G)$. But then $\{1\} = [Z(S), O_{p'}(G)]^G = [G, O_{p'}(G)]$ so that $O_{p'}(G) \leq Z(G)$ and G is quasisimple. A consideration of Schur multipliers contradicts that G is a minimal counterexample.

Thus, we have that G is p-minimal. We apply [22, Theorem 1] to $V/C_V(G)$. Since $|A| > p^{\frac{n}{2}}$ and $[V, A, A] = \{1\}$, $V/C_V(G)$ is not a natural $\Omega_4^-(2^{\frac{n}{2}})$-module for G. Thus, $G \cong \mathrm{SL}_2(p^n)$, $\mathrm{Sz}(2^n)$ or $\mathrm{SU}_3(p^n)$ and $V/C_V(G)$ is a direct sum of natural modules, a contradiction since G was an assumed counterexample. □

We now generalize even further than quadratic or cubic action by investigating the *minimal polynomial* of p-elements in a representation, noticing that in quadratic and cubically acting elements, the minimal polynomial is of degree 2 and 3 respectively. We cannot hope to make such strong statements as in the earlier cases, but for larger primes and solvable groups, we have decent control due to the Hall–Higman theorem.

Theorem 2.56 (Hall–Higman Theorem) *Suppose that G is p-solvable group with $O_p(G) = \{1\}$ and V a faithful* GF(p)*-module for G. If $x \in G$ has order p^n and $[V, x; r] = \{1\}$ then one of the following holds:*

(i) $r = p^n$;
(ii) *p is a Fermat prime, the Sylow 2-subgroups of G are non-abelian and $r \geqslant p^n - p^{n-1}$; or*
(iii) *$p = 2$, the Sylow q-subgroups of G are non-abelian for some Mersenne prime $q = 2^m - 1 < 2^n$ and $r \geqslant 2^n - 2^{n-m}$.*

Proof See [40, Theorem B]. □

As witnessed in the Hall-Higman theorem, cubic action is far less powerful tool in our application whenever $p \in \{2, 3\}$. We remedy this with the following strengthening in [51].

Definition 2.57 Let G be a finite group and V an GF(p)G-module. If $A \leq G$ acts cubically on V and $[V, A]C_V(A) = [v\mathrm{GF}(p), A]C_V(A)$ for all $v \in V \setminus [V, A]C_V(A)$ then A is said to act *nearly quadratically* on V.

Theorem 2.58 *Suppose $G = O^{p'}(G)$ is a group which has a strongly p-embedded subgroup and let V be a faithful, irreducible* GF(p)*-module for G. If G is generated by elementary abelian p-subgroups which act nearly quadratically, but not quadratically, on V then either $F^*(G) = Z(G)K$ where K is a component of G, or one of the following holds:*

(i) *$p = 3$ and $G \cong \mathrm{Frob}(39)$ and $|V| = 3^3$; or*
(ii) *$p = 3$, $G \cong (2 \wr \mathrm{Sym}(n))' = O^{3'}(2 \wr \mathrm{Sym}(n))$ for $3 \leqslant n \leqslant 5$ and $|V| = 3^n$.*

Proof See [51, Theorem 2]. □

Remark In the above, we have that $(2 \wr \mathrm{Sym}(3))' \cong \mathrm{PSL}_2(3)$ and V is a natural $\Omega_3(3)$-module.

Whenever $p \geqslant 5$, by applying the Hall–Higman theorem to the situation where the group G has a strongly p-embedded subgroup and some associated cubic module, we can generally characterize G completely. Thus, we pursue ways which in which to force cubic action. The final concept in this chapter is that of *critical subgroups*, which first arose in the proof of the Feit–Thompson theorem. Originally in this work, critical subgroups provided a means to control the automizer of some p-group Q whenever $p \geqslant 5$. In the context of the amalgam method, they force cubic action on some faithful section of Q and from there, one can apply Hall–Higman type results to deduce information about Q and its automizer. Where this methodology was previous employed, we now have methods to treat these cases uniformly across all primes and so critical subgroups now play a far lesser role in this work. However, we believe they still provide some interesting consequences in the amalgam method and we still include some of these consequences (see Proposition 5.12). We present the *critical subgroup theorem*, due to Thompson, below.

Theorem 2.59 *Let Q be a p-group. Then there exists $C \leq Q$ such that the following hold:*

(i) *C is characteristic in Q;*
(ii) *$\Phi(C) \leq Z(C)$ so that C has class at most 2;*
(iii) *$[C, Q] \leq Z(C)$;*
(iv) *$C_Q(C) \leq C$; and*
(v) *C is coprime automorphism faithful.*

Proof This is [32, (I.5.3.11)]. □

We call such a subgroup $C \leq Q$ a *critical subgroup* of Q.

Corollary 2.60 *Suppose that $G = O^{p'}(G)$ is a $\mathcal{K}$-group which has a strongly p-embedded subgroup, $S \in \mathrm{Syl}_p(G)$ and V is a faithful $\mathrm{GF}(p)$-module. Suppose that $p \geqslant 5$ and there is $s \in S$ of order p^n such that $[V, s, s, s] = \{1\}$. Then either:*

(i) *$G \cong (\mathrm{P})\mathrm{SL}_2(p^n)$ or $(\mathrm{P})\mathrm{SU}_3(p^n)$ for any prime $p \geqslant 5$; or*
(ii) *$p = 5$, G is isomorphic to one of $3 \cdot \mathrm{Alt}(6)$ or $3 \cdot \mathrm{Alt}(7)$ and for W some irreducible composition factor of V, $|W| \geqslant 5^6$.*

Proof Suppose first that $m_p(S) = 1$. Then, by [32, I.5.4.10 (ii)], S is cyclic and so we may as well assume that $[V, \Omega(S), \Omega(S), \Omega(S)] = \{1\}$. Suppose first that G is p-solvable. Since $p^n - p^{n-1} = p^{n-1}(p-1) \geqslant 4$, the Hall–Higman theorem implies that $O_p(G) \neq \{1\}$, a contradiction since G has a strongly p-embedded subgroup.

Suppose now that $m_p(S) = 1$ and G is not p-solvable. Since $G = O^{p'}(G)$, by Proposition 2.14 we have that $G/O_{p'}(G)$ is a simple group with a cyclic Sylow p-subgroup. Form $X := \Omega(S)O_{p'}(G)$. Then X is a p-solvable group and V is a faithful module for X by restriction. Then $p^n - p^{n-1} = p^{n-1}(p-1) \geqslant 4$ since $p \geqslant 5$ and by the Hall–Higman theorem we deduce that $O_p(X) \neq \{1\}$. In particular, $\Omega(S) \trianglelefteq X$ and $[O_{p'}(G), \Omega(S)] \leq O_{p'}(G) \cap \Omega(S) = \{1\}$. But then, since $G/O_{p'}(G)$ is simple, $[O_{p'}(G), G] = [O_{p'}(G), \langle \Omega(S)^G \rangle] = \{1\}$ and $O_{p'}(G) \leq Z(G)$. Hence, G is a quasisimple group with a cyclic Sylow p-subgroup such that the degree of the minimal polynomial of some p-element is 3. Such groups and their associated modules are determined in [77]. This gives (ii).

Suppose that $m_p(S) \geqslant 2$ so that $G/O_{p'}(G)$ is determined by Proposition 2.15, and let $X = O_{p'}(G)\Omega(Z(S))$. Unless $G/O_{5'}(G) \cong \mathrm{Sz}(32) : 5$, we have that for any $s \in \Omega(S)^{\#}$, $G = \langle s^G \rangle$. In this case, forming $X := \langle s \rangle O_{p'}(G)$, we have that X acts faithfully on V with s acting cubically, and by the Hall–Higman theorem, $\langle s \rangle \trianglelefteq X$. But then $[s, O_{p'}(G)] \leq \langle s \rangle \cap O_{p'}(G) = \{1\}$. Thus, $[G, O_{p'}(G)] = [\langle s^G \rangle, O_{p'}(G)] = [s, O_{p'}(G)]^G = \{1\}$ and $O_{p'}(G) \leq Z(G)$. Since $G = O^{p'}(G)$ is perfect, G is a perfect central extension of $G/O_{p'}(G)$. If $G/O_{p'}(G)$ is isomorphic to a rank 1 simple group of Lie type in characteristic p, then outcome (i) follows from Proposition 2.16 (vii) and Proposition 2.17 (ix). If $G/O_{p'}(G) \cong \mathrm{Alt}(2p)$ then, as $p \geqslant 5$, G has no faithful modules which witness cubic action by [48]. Hence, by Proposition 2.15, we are left with a finite number of perfect p'-central extensions of simple groups. We again utilize the classification in [77]. That is, in each case, we identify a quasisimple subgroup H of G having $\langle x \rangle$ as a Sylow p-subgroup and, with the aim of forcing a contradiction, observe that if G has a non-trivial cubic module then so too does H. We will make liberal use of the conjugacy of p-elements and the maximal subgroups of $G/O_{p'}(G)$ as can be found in the ATLAS [24].

Suppose that $G/O_{5'}(G) \cong \mathrm{Sz}(32):5$. Then, for $s \in \Omega(S)$, we have that for $H := \langle s^G \rangle$, $H/O_{5'}(H) \cong \mathrm{Sz}(32)$ and following the reasoning applied in the Lie type case, we have that $O_{5'}(H) \leq Z(H)$. Since the Schur multiplier of Sz(32) is trivial we have that $H \cong \mathrm{Sz}(32)$. But H acts faithfully on V, with $s \in S \cap H$ acting cubically. Since Sz(32) has a cyclic Sylow 5-subgroup and no cubic modules by [77], we have a contradiction. Hence, $G/O_{5'}(G) \not\cong \mathrm{Sz}(32):5$.

Suppose that G is isomorphic to ${}^2\mathrm{F}_4(2)'$ or a central extension of Fi_{22}. Then a Sylow 5-subgroup is elementary abelian subgroup of order 25 and G has one conjugacy class of 5-elements. Thus, we may verify the claim for any 5-element. Taking $H = O^{5'}(H)$ such that $H \cong \mathrm{Alt}(6)$ or $HZ(G)/Z(G) \cong \mathrm{M}_{12}$ respectively, we see that H is a quasisimple group with a cyclic Sylow 5-subgroup and has no cubic module by [48] and [77] respectively. Hence, G is not isomorphic to ${}^2\mathrm{F}_4(2)'$ or an extension of Fi_{22}.

Suppose that $G/Z(G) \cong \mathrm{McL}$ so that a Sylow 5-subgroup of G is extraspecial of order 5^3 and exponent 5. Then G has two classes of 5-elements: those which are central in some Sylow 5-subgroup of G, and those which are not. In the former case, we verify in MAGMA [14] that such an element is contained in $H = O^{5'}(H)$ with the property $HZ(G)/Z(G) \cong 2 \cdot \mathrm{Alt}(8)$. Then H is a quasisimple group with a cyclic Sylow 5-subgroup and has no cubic module by [48]. In the latter case, we verify that x is contained in $H = O^{5'}(H)$ with the property $HZ(G)/Z(G) \cong \mathrm{M}_{11}$. Then H is a quasisimple group with a cyclic Sylow 5-subgroup and has no cubic module by [77]. Hence, $G/Z(G) \not\cong \mathrm{McL}$.

Finally, suppose that $G \cong \mathrm{J}_4$ so that a Sylow 11-subgroup of G is extraspecial of order 11^3 and exponent 11. Then G has two classes of 11-elements: those which are central in some Sylow 11-subgroup of G, and those which are not. We apply [47, Section 6.2] to see that the 11-elements which are central in some Sylow 11-subgroup are contained in maximal subgroups of G of shape $\mathrm{PSL}_2(32):5$. Since $\mathrm{PSL}_2(32)$ is a a simple group with a cyclic Sylow 11-subgroup with no cubic modules by [77], we rule out this case. On the other hand, by [47, Section 6.3], a non-central 11-element is contained in a subgroup of G isomorphic to M_{24}. Since M_{24} is a a simple group with a cyclic Sylow 11-subgroup with no cubic modules by [77], we also rule out this case. Hence, $G \not\cong \mathrm{J}_4$ and the proof is complete. □

Chapter 3
Fusion Systems

In this chapter, we set up notations and terminology, and list some properties of fusion systems. The standard references for the study of fusion systems are [11] and [25] and most of what follows may be gleaned from these texts.

3.1 Fusion Systems

Definition 3.1 Let G be a finite group with $S \in \mathrm{Syl}_p(G)$. The *p-fusion category* of G over S, written $\mathcal{F}_S(G)$, is the category with object set $\mathrm{Ob}(\mathcal{F}_S(G)) := \{Q : Q \leq S\}$ and for $P, Q \leq S$, $\mathrm{Mor}_{\mathcal{F}_S(G)}(P, Q) := \mathrm{Hom}_G(P, Q)$, where $\mathrm{Hom}_G(P, Q)$ denotes maps induced by conjugation by elements of G.

Definition 3.2 Let S be a p-group. A fusion system $\mathcal{F}$ over S is a category with object set $\mathrm{Ob}(\mathcal{F}) := \{Q : Q \leq S\}$ and whose morphism set satisfies the following properties for $P, Q \leq S$:

- $\mathrm{Hom}_S(P, Q) \subseteq \mathrm{Mor}_{\mathcal{F}}(P, Q) \subseteq \mathrm{Inj}(P, Q)$; and
- each $\phi \in \mathrm{Mor}_{\mathcal{F}}(P, Q)$ is the composite of an $\mathcal{F}$-isomorphism followed by an inclusion,

where $\mathrm{Inj}(P, Q)$ denotes injective homomorphisms between P and Q.

To motivate the group analogy, we write $\mathrm{Hom}_{\mathcal{F}}(P, Q) := \mathrm{Mor}_{\mathcal{F}}(P, Q)$ and $\mathrm{Aut}_{\mathcal{F}}(P) := \mathrm{Hom}_{\mathcal{F}}(P, P)$. This latter group is referred to as the *automizer* of P.

Two subgroups of S are said to be *$\mathcal{F}$-conjugate* if they are isomorphic as objects in $\mathcal{F}$. We write $Q^{\mathcal{F}}$ for the set of all $\mathcal{F}$-conjugates of Q.

Definition 3.3 Let $\mathcal{F}$ be a fusion system on a p-group S. Then $\mathcal{H}$ is a *subsystem* of $\mathcal{F}$, written $\mathcal{H} \leq \mathcal{F}$, on a p-group T if $T \leq S$, $\mathcal{H} \subseteq \mathcal{F}$ as sets and $\mathcal{H}$ is itself a fusion system. Then, for $\mathcal{F}_1, \mathcal{F}_2$ subsystems of $\mathcal{F}$ supported on S, write $\langle \mathcal{F}_1, \mathcal{F}_2 \rangle$ for the smallest subsystem of $\mathcal{F}$ containing $\mathcal{F}_1$ and $\mathcal{F}_2$.

© The Author(s), under exclusive license to Springer Nature Switzerland AG 2024

M. van Beek, *Rank 2 Amalgams and Fusion Systems*, Lecture Notes in Mathematics 2343, https://doi.org/10.1007/978-3-031-54461-3_3

Definition 3.4 Let $\mathcal{F}$ be a fusion system on a p-group S. Then for $\alpha : S \to T$ a group isomorphism, we define $\mathcal{F}^\alpha$ to be the fusion system over T with morphism sets $\mathrm{Hom}_{\mathcal{F}^\alpha}(P\alpha, Q\alpha) = \{\alpha^{-1}\theta\alpha : \theta \in \mathrm{Hom}_{\mathcal{F}}(P, Q)\}$ where $P, Q \leq S$.

Definition 3.5 Let $\mathcal{F}$ be a fusion system over a p-group S and let $Q \leq S$. Say that Q is

- *fully $\mathcal{F}$-normalized* if $|N_S(Q)| \geq |N_S(P)|$ for all $P \in Q^{\mathcal{F}}$;
- *fully $\mathcal{F}$-centralized* if $|C_S(Q)| \geq |C_S(P)|$ for all $P \in Q^{\mathcal{F}}$;
- *fully $\mathcal{F}$-automized* if $\mathrm{Aut}_S(Q) \in \mathrm{Syl}_p(\mathrm{Aut}_{\mathcal{F}}(Q))$;
- *receptive* in $\mathcal{F}$ if for each $P \leq S$ and each $\phi \in \mathrm{Iso}_{\mathcal{F}}(P, Q)$, setting

$$N_\phi = \{g \in N_S(P) : c_g^\phi \in \mathrm{Aut}_S(Q)\},$$

  there is $\overline{\phi} \in \mathrm{Hom}_{\mathcal{F}}(N_\phi, S)$ such that $\overline{\phi}|_P = \phi$;
- *$\mathcal{F}$-centric* if $C_S(P) = Z(P)$ for all $P \in Q^{\mathcal{F}}$;
- *$\mathcal{F}$-radical* if $O_p(\mathrm{Out}_{\mathcal{F}}(Q)) = \{1\}$; and
- *strongly closed* in $\mathcal{F}$ if $x\alpha \leq Q$ for all $\alpha \in \mathrm{Hom}_{\mathcal{F}}(\langle x \rangle, S)$ whenever $x \in Q$.

If it is clear which fusion system we are working in, we will refer to subgroups as being fully normalized (centralized, centric etc.) without the $\mathcal{F}$ prefix.

Definition 3.6 Let $\mathcal{F}$ be a fusion system on a p-group S. Say that Q is *$\mathcal{F}$-essential* if Q is $\mathcal{F}$-centric, fully $\mathcal{F}$-normalized and $\mathrm{Out}_{\mathcal{F}}(Q)$ contains a strongly p-embedded subgroup.

Note that essential subgroups of S are also $\mathcal{F}$-radical subgroups by definition.

Definition 3.7 Let $\mathcal{F}$ be a fusion system over a p-group S. Then $\mathcal{F}$ is *saturated* if the following conditions hold:

(i) Every fully $\mathcal{F}$-normalized subgroup is also fully $\mathcal{F}$-centralized and fully $\mathcal{F}$-automized.
(ii) Every fully $\mathcal{F}$-centralized subgroup is receptive in $\mathcal{F}$.

By a theorem of Puig [66], the p-fusion category of a finite group $\mathcal{F}_S(G)$ is a saturated fusion system.

Definition 3.8 We say a saturated fusion system is *realizable* if there exists a finite group G with $S \in \mathrm{Syl}_p(G)$ and $\mathcal{F} = \mathcal{F}_S(G)$. Otherwise, the fusion system is said to be *exotic*.

Exotic fusion systems are interesting phenomena which, at the time of writing, are still not particularly well understood. While we now have a large number of examples of exotic fusion systems (particular when p is an odd prime) to analyze, there is still no justification of precisely why, where and how exotic fusion systems arise, nor are there any "good" techniques for showing that an arbitrary saturated fusion system is exotic, save for using the classification of the finite simple groups.

From this point on we focus more on saturated fusion systems, as they most closely parallel the group phenomenon. Throughout we will often adopt a *local CK-system* hypothesis.

Definition 3.9 A *local CK-system* is a saturated fusion $\mathcal{F}$ such that $\mathrm{Aut}_{\mathcal{F}}(P)$ is a $\mathcal{K}$-group for all $P \leq S$.

Local $\mathcal{CK}$-systems provide a means to apply the results from Chap. 2 which relied on a $\mathcal{K}$-group hypothesis. This allows for minimal counterexample arguments in fusion systems and provides a link between fusion systems and the classification of finite simple groups. That is, if G is a finite group which is a counterexample to the classification with $|G|$ minimal subject to these constraints, then $\mathcal{F}_S(G)$ is a local $\mathcal{CK}$-system for $S \in \mathrm{Syl}_p(G)$.

We now present arguably the most important tool in classifying saturated fusion systems. Because of this, we need only investigate the local actions for a relatively small number of p-subgroups to obtain a global characterization of a saturated fusion system.

Theorem 3.10 (Alperin–Goldschmidt Fusion Theorem) *Let $\mathcal{F}$ be a saturated fusion system over a p-group S. Then*

$$\mathcal{F} = \langle \mathrm{Aut}_{\mathcal{F}}(Q) : Q \text{ is essential or } Q = S \rangle.$$

That is, every morphism in $\mathcal{F}$ as a composite of restrictions of maps in $\mathrm{Aut}_{\mathcal{F}}(Q)$ *for Q described above.*

Proof See [11, Theorem I.3.5]. □

Along these lines, another important notion is for a p-subgroup to be *normal* in a saturated fusion system.

Definition 3.11 Let $\mathcal{F}$ be a fusion systems over a p-group S and $Q \leq S$. Say that Q is *normal* in $\mathcal{F}$ if $Q \trianglelefteq S$ and for all $P, R \leq S$ and $\phi \in \mathrm{Hom}_{\mathcal{F}}(P, R)$, ϕ extends to a morphism $\overline{\phi} \in \mathrm{Hom}_{\mathcal{F}}(PQ, RQ)$ such that $\overline{\phi}(Q) = Q$.

It may be checked that the product of normal subgroups is itself normal. Thus, we may talk about the largest normal subgroup of $\mathcal{F}$ which we denote $O_p(\mathcal{F})$ (and occasionally refer to as the p-core of $\mathcal{F}$). Further, it follows immediately from the saturation axioms that any subgroup normal in S is fully normalized and fully centralized.

Definition 3.12 Let $\mathcal{F}$ be a fusion system over a p-group S and let $Q \leq S$. The *normalizer fusion subsystem* of Q, denoted $N_{\mathcal{F}}(Q)$, is the largest subsystem of $\mathcal{F}$, supported over $N_S(Q)$, in which Q is normal.

It is clear from the definition that if $\mathcal{F}$ is the fusion category of a group G i.e. $\mathcal{F} = \mathcal{F}_S(G)$, then $N_{\mathcal{F}}(Q) = \mathcal{F}_{N_S(Q)}(N_G(Q))$. The following result is originally attributed to Puig [65].

Theorem 3.13 *Let $\mathcal{F}$ be a saturated fusion system over a p-group S. If $Q \leq S$ is fully $\mathcal{F}$-normalized then $N_{\mathcal{F}}(Q)$ is saturated.*

Proof See [11, Theorem I.5.5]. □

Definition 3.14 Let $\mathcal{F}$ be a fusion systems over a p-group S and $P \leq Q \leq S$. Say that P is *$\mathcal{F}$-characteristic* in Q if $\mathrm{Aut}_{\mathcal{F}}(Q) \leq N_{\mathrm{Aut}(Q)}(P)$.

Plainly, if $Q \trianglelefteq \mathcal{F}$ and P is $\mathcal{F}$-characteristic in Q, then $P \trianglelefteq \mathcal{F}$.

We now present a link between normal subgroups of a saturated fusion system $\mathcal{F}$ and its essential subgroups.

Proposition 3.15 *Let $\mathcal{F}$ be a saturated fusion system over a p-group S. Then Q is normal in $\mathcal{F}$ if and only if Q is contained in each essential subgroup, Q is $\mathrm{Aut}_{\mathcal{F}}(E)$-invariant for any essential subgroup E of $\mathcal{F}$ and Q is $\mathrm{Aut}_{\mathcal{F}}(S)$-invariant.*

Proof See [11, Proposition I.4.5]. □

As for finite groups, we desire a more global sense of normality in fusion systems, not just restricted to p-subgroups. That is, we are interested in subsystems of a fusion system $\mathcal{F}$ which are *normal.*

Definition 3.16 Let $\mathcal{F}$ be a saturated fusion system over a p-group S. A fusion system $\mathcal{E}$ is *weakly normal* in $\mathcal{F}$ if the following conditions hold:

(i) $\mathcal{E}$ is a saturated subsystem of $\mathcal{F}$ over $T \leq S$;
(ii) T is strongly closed in S;
(iii) $\mathcal{E}^{\alpha} = \mathcal{E}$ for all $\alpha \in \mathrm{Aut}_{\mathcal{F}}(T)$; and
(iv) for each $P \leq T$ and each $\phi \in \mathrm{Hom}_{\mathcal{F}}(P, T)$ there are $\alpha \in \mathrm{Aut}_{\mathcal{F}}(T)$ and $\phi_0 \in \mathrm{Hom}_{\mathcal{E}}(P, T)$ such that $\phi = \alpha \circ \phi_0$.

A fusion system $\mathcal{E}$ is *normal* in $\mathcal{F}$, denoted $\mathcal{E} \trianglelefteq \mathcal{F}$, if $\mathcal{E}$ is weakly normal in $\mathcal{F}$ and each $\alpha \in \mathrm{Aut}_{\mathcal{E}}(T)$ extends to some $\overline{\alpha} \in \mathrm{Aut}_{\mathcal{F}}(TC_S(T))$ which fixes every coset of $Z(T)$ in $C_S(T)$.

Conditions (iii) and (iv) are referred to as the invariance condition and Frattini condition respectively. As one would hope, for a p-subgroup Q, if $Q \trianglelefteq \mathcal{F}$, then $\mathcal{F}_Q(Q) \trianglelefteq \mathcal{F}$. As is the case with groups, we refer to a saturated fusion system as *simple* if it contains no proper non-trivial normal subsystems.

We shall describe some important subsystems associated to a saturated fusion which have a natural analogues in finite group theory. More details on the construction of such subsystems may be found in Section I.7 of [11].

Definition 3.17 Let $\mathcal{F}$ be a saturated fusion system on a p-group S. Say a subsystem $\mathcal{E}$ has *index prime to p* in $\mathcal{F}$ if $\mathcal{E}$ is a fusion system on S and $\mathrm{Aut}_{\mathcal{E}}(P) \geq O^{p'}(\mathrm{Aut}_{\mathcal{F}}(P))$ for all $P \leq S$.

By [11, Theorem I.7.7], there is a unique minimal saturated fusion system of index prime to p in $\mathcal{F}$ denoted by $O^{p'}(\mathcal{F})$. Moreover, $O^{p'}(\mathcal{F})$ is a normal subsystem of $\mathcal{F}$.

Definition 3.18 Let $\mathcal{F}$ be a saturated fusion system on a p-group S. Then the *hyperfocal subgroup* $\mathfrak{hyp}(\mathcal{F})$ of $\mathcal{F}$ is defined as

$$\mathfrak{hyp}(\mathcal{F}) := \langle g^{-1}(g\alpha) : g \in P \le S, \alpha \in O^p(\mathrm{Aut}_{\mathcal{F}}(P))\rangle.$$

A subsystem $\mathcal{E}$ has *p-power index* in $\mathcal{F}$ if $\mathcal{E}$ is a fusion system on $T \ge \mathfrak{hyp}(\mathcal{F})$ and $\mathrm{Aut}_{\mathcal{E}}(P) \ge O^p(\mathrm{Aut}_{\mathcal{F}}(P))$ for all $P \le S$.

Moreover, by [11, Theorem I.7.4], there is a unique minimal saturated fusion subsystem of p-power index in $\mathcal{F}$ denoted by $O^p(\mathcal{F})$. Moreover, $O^p(\mathcal{F})$ is supported on $\mathfrak{hyp}(\mathcal{F})$ and $O^p(\mathcal{F})$ is a normal subsystem of $\mathcal{F}$.

Definition 3.19 A saturated fusion system is *reduced* if $O_p(\mathcal{F}) = \{1\}$ and $\mathcal{F} = O^p(\mathcal{F}) = O^{p'}(\mathcal{F})$.

Naturally, an important consideration in fusion systems is the notion of isomorphism. After defining what isomorphism means in the context fusion systems, it follows readily that the "sensible" properties hold, which we state below.

Definition 3.20 Let $\mathcal{F}$ be a fusion system on a p-group S and $\mathcal{E}$ a fusion system on a p-group T. A *morphism* $\phi : \mathcal{F} \to \mathcal{E}$ is a tuple $(\phi_S, \phi_{P,Q} : P, Q \le S)$ such that $\phi_S : S \to T$ is a group homomorphism and $\phi_{P,Q} : \mathrm{Hom}_{\mathcal{F}}(P, Q) \to \mathrm{Hom}_{\mathcal{E}}(P\phi, Q\phi)$ is such that $\alpha\phi_S = \phi_S(\alpha\phi_{P,Q})$ for all $\alpha \in \mathrm{Hom}_{\mathcal{F}}(P, Q)$.

Say that ϕ is *injective* if $\phi_S \colon S \to T$ is injective, and ϕ is *surjective* if ϕ_S is surjective and, for all $P, Q \le S$, $\phi_{P_0,Q_0} \colon \mathrm{Hom}_{\mathcal{F}}(P_0, Q_0) \to \mathrm{Hom}_{\mathcal{E}}(P\phi, Q\phi)$ is surjective, where P_0, Q_0 denote the preimages in S of $P\phi, Q\phi$. Then, ϕ is an *isomorphism* of fusion systems if $\phi : \mathcal{F} \to \mathcal{E}$ is an injective, surjective morphism.

Proposition 3.21 *Let $G \cong H$ be finite groups with $S \in \mathrm{Syl}_p(G)$ and $T \in \mathrm{Syl}_p(H)$. Then $\mathcal{F}_S(G) \cong \mathcal{F}_T(H)$.*

The following lemma is one reason why fusion systems have found purpose in p-local arguments in finite group theory and representation theory.

Proposition 3.22 *Let $\mathcal{F} = \mathcal{F}_S(G)$ be a saturated fusion system and set $\overline{G} = G/O_{p'}(G)$. Then $\mathcal{F}_S(G) \cong \mathcal{F}_{\overline{S}}(\overline{G})$.*

In order to investigate the local actions in a saturated fusion systems, and in particular in its normalizer subsystems, it will often be convenient to work in a purely group theoretic context. The *model theorem* guarantees that we may do this for a certain class of p-subgroups of a saturated fusion system $\mathcal{F}$.

Definition 3.23 Let $\mathcal{F}$ be a saturated fusion system on a p-group S. Then a *model* for $\mathcal{F}$ is a finite group G with $S \in \mathrm{Syl}_p(G)$, $F^*(G) = O_p(G)$ and $\mathcal{F} = \mathcal{F}_S(G)$.

Theorem 3.24 (Model Theorem) *Let $\mathcal{F}$ be a saturated fusion system over a p-group S. Suppose there is $Q \leq S$ which is $\mathcal{F}$-centric and normal in $\mathcal{F}$. Then the following hold:*

(i) *There are models for $\mathcal{F}$.*
(ii) *If G_1 and G_2 are two models for $\mathcal{F}$, then there is an isomorphism $\phi : G_1 \to G_2$ such that $\phi|_S = \mathrm{Id}_S$.*
(iii) *For any finite group G containing S as a Sylow p-subgroup such that $Q \leq G$, $C_G(Q) \leq Q$ and $Aut_G(Q) = Aut_{\mathcal{F}}(Q)$, there is $\beta \in \mathrm{Aut}(S)$ such that $\beta|_Q = \mathrm{Id}_Q$ and $\mathcal{F}_S(G) = \mathcal{F}^\beta$. Thus, there is a model for $\mathcal{F}$ which is isomorphic to G.*

Proof See [11, Theorem I.4.9]. □

3.2 Essential Subgroups

In order to make use of the Alperin–Goldschmidt fusion theorem, and given some restrictions on the number of essential subgroups in our fusion systems, we must also determine the induced automorphism groups on the essential subgroups by $\mathcal{F}$. The first result along these lines determines the potential automizer $\mathrm{Aut}_{\mathcal{F}}(E)$ of an essential subgroup E whenever some non-central chief factor of E is an FF-module. It is important to note that this theorem does not rely on a $\mathcal{K}$-group hypothesis, and it is essentially the fusion theoretic equivalent of Lemma 2.41.

Theorem 3.25 *Suppose that E is an essential subgroup of a saturated fusion system $\mathcal{F}$ over a p-group S, and assume that there is an $\mathrm{Aut}_{\mathcal{F}}(E)$-invariant subgroup $V \leq \Omega(Z(E))$ such that V is an FF-module for $G := \mathrm{Out}_{\mathcal{F}}(E)$. Then, writing $L := O^{p'}(G)$, we have that $L/C_L(V) \cong \mathrm{SL}_2(p^n)$, $C_L(V)$ is a p'-group and $V/C_V(L)$ is a natural $\mathrm{SL}_2(q)$-module.*

Proof This is [42, Theorem 1.2]. □

Armed with the analysis of groups with strongly p-embedded subgroups from Chap. 2, we have reasonable limitations on the structure of $\mathrm{Out}_{\mathcal{F}}(E)$ for E an essential subgroup of $\mathcal{F}$. We now substantiate the claim in the introduction that the restriction we impose whenever an essential automizer has quotient Ree(3), is p-solvable or has a generalized quaternion Sylow 2-subgroup is implied by the maximality of the essential subgroup. We also record some generic results for *maximally essential* subgroups for application elsewhere.

Definition 3.26 Suppose that $\mathcal{F}$ is a saturated fusion system on a p-group S. Then $E \leq S$ is *maximally essential* in $\mathcal{F}$ if E is essential and, if $F \leq S$ is essential in $\mathcal{F}$ and $E \leq F$, then $E = F$.

Coupled with saturation arguments and the Alperin–Goldschmidt theorem, this definition drastically limits the possibilities for $\mathrm{Out}_{\mathcal{F}}(E)$.

Lemma 3.27 *Let $\mathcal{F}$ be a saturated fusion systems on a p-group S with E a maximally essential subgroup of $\mathcal{F}$. Then $N_{\mathrm{Out}_{\mathcal{F}}(E)}(\mathrm{Out}_S(E))$ is strongly p-embedded in $\mathrm{Out}_{\mathcal{F}}(E)$.*

Proof Let $T \leq N_S(E)$ with $E < T$. Now, since E is receptive, for all $\alpha \in N_{\mathrm{Aut}_{\mathcal{F}}(E)}(\mathrm{Aut}_T(E))$, α lifts to a morphism $\widehat{\alpha} \in \mathrm{Hom}_{\mathcal{F}}(N_\alpha, S)$ with $N_\alpha > E$. Since E is maximally essential, applying the Alperin–Goldschmidt theorem, $\widehat{\alpha}$ is the restriction of a morphism $\overline{\alpha} \in \mathrm{Aut}_{\mathcal{F}}(S)$. But then, α normalizes $\mathrm{Aut}_S(E)$ and so $N_{\mathrm{Aut}_{\mathcal{F}}(E)}(\mathrm{Aut}_T(E)) \leq N_{\mathrm{Aut}_{\mathcal{F}}(E)}(\mathrm{Aut}_S(E))$. This induces the inclusion $N_{\mathrm{Out}_{\mathcal{F}}(E)}(\mathrm{Out}_T(E)) \leq N_{\mathrm{Out}_{\mathcal{F}}(E)}(\mathrm{Out}_S(E)$. Since this holds for all $T \leq N_S(E)$ with $E < T$, we infer that $N_{\mathrm{Out}_{\mathcal{F}}(E)}(\mathrm{Out}_S(E))$ is strongly p-embedded in $\mathrm{Out}_{\mathcal{F}}(E)$, as required. □

Hence, Lemma 3.27 implies that maximally essential subgroups of $\mathcal{F}$ which are $\mathrm{Aut}_{\mathcal{F}}(S)$-invariant satisfy the extra conditions in the hypothesis of the Main Theorem. As in the earlier analysis of groups with strongly p-subgroups, we divide into two cases, where $m_p(\mathrm{Out}_S(E)) = 1$ or $m_p(\mathrm{Out}_S(E)) \geqslant 2$.

Proposition 3.28 *Let $\mathcal{F}$ be a saturated fusion systems on a p-group S with E a maximally essential subgroup of $\mathcal{F}$, and set $G = \mathrm{Out}_{\mathcal{F}}(E)$. If $m_p(G) = 1$ then either*

(i) $\mathrm{Out}_S(E)$ *is cyclic or generalized quaternion and*

$$\begin{aligned} O^{p'}(G) &= \mathrm{Out}_S(E)[O_{p'}(O^{p'}(G)), \Omega(\mathrm{Out}_S(E))] \\ &= \mathrm{Out}_S(E)\langle \Omega(\mathrm{Out}_S(E))^{O^{p'}(G)}\rangle \end{aligned}$$

is p-solvable; or

(ii) $O^{p'}(G)/O_{p'}(O^{p'}(G))$ *is a non-abelian simple group, p is odd and* $\mathrm{Out}_S(E)$ *is cyclic.*

Proof Since G has a strongly p-embedded subgroup, so does $O^{p'}(G)$ and we apply Proposition 2.14 and (ii) follows immediately. In the other cases of Proposition 2.14, since $\Omega(\mathrm{Out}_S(E))[O_{p'}(O^{p'}(G)), \Omega(\mathrm{Out}_S(E))] \trianglelefteq O^{p'}(G)$, by the Frattini argument,

$$\begin{aligned} O^{p'}(G) &= N_{O^{p'}(G)}(\Omega(\mathrm{Out}_S(E)))[O_{p'}(O^{p'}(G)), \Omega(\mathrm{Out}_S(E))] \\ &= N_{O^{p'}(G)}(\Omega(\mathrm{Out}_S(E)))\langle \Omega(\mathrm{Out}_S(E))^{O^{p'}(G)}\rangle. \end{aligned}$$

Since E is maximally essential, applying Lemma 3.27,

$$N_{O^{p'}(G)}(\Omega(\mathrm{Out}_S(E))) \leq N_G(\Omega(\mathrm{Out}_S(E))) = N_G(\mathrm{Out}_S(E)).$$

But then $\mathrm{Out}_S(E)[O_{p'}(O^{p'}(G)), \Omega(\mathrm{Out}_S(E))] \trianglelefteq O^{p'}(G)$ and by the definition of $O^{p'}(G)$, we have that $O^{p'}(G) = \mathrm{Out}_S(E)[O_{p'}(O^{p'}(G)), \Omega(\mathrm{Out}_S(E))]$. □

Proposition 3.29 *Let $\mathcal{F}$ be a local $\mathcal{CK}$-system on a p-group S with E a maximally essential subgroup of $\mathcal{F}$ and set $G = \mathrm{Out}_{\mathcal{F}}(E)$. If $m_p(G) \geqslant 2$ then $O^{p'}(G)$ is isomorphic to a central extension by a group of p'-order of one of the following groups:*

(i) $\mathrm{PSL}_2(p^{a+1})$ *or* $\mathrm{PSU}_3(p^b)$ *for p arbitrary, $a \geqslant 1$ and $p^b > 2$;*
(ii) $\mathrm{Sz}(2^{2a+1})$ *for $p = 2$ and $a \geqslant 1$;*
(iii) $\mathrm{Ree}(3^{2a+1})$, $\mathrm{PSL}_3(4)$ *or* M_{11} *for $p = 3$ and $a \geqslant 0$;*
(iv) $\mathrm{Sz}(32) : 5$, ${}^2\mathrm{F}_4(2)'$ *or* McL *for $p = 5$; or*
(v) J_4 *for $p = 11$.*

Furthermore, either $O^{p'}(G)$ is a perfect central extension, or $O^{p'}(G) \cong \mathrm{Ree}(3)$ resp. $\mathrm{Sz}(32) : 5$ and $p = 3$ resp. $p = 5$.

Proof Set $\widetilde{G} = G/O_{p'}(G)$ and $K = O^{p'}(G)$. By Lemma 3.27, $N_G(\mathrm{Out}_S(E))$ is strongly p-embedded in G. In particular, we deduce that $N_K(\mathrm{Out}_S(E))$ is strongly p-embedded in K. Let $A \leq \mathrm{Out}_S(E)$ be elementary abelian of order p^2. By coprime action, $O_{p'}(K) = \langle C_{O_{p'}(K)}(a) : a \in A^\# \rangle$. Since $N_K(\mathrm{Out}_S(E))$ is strongly p-embedded in K, we have that $O_{p'}(K) \leq N_K(\mathrm{Out}_S(E))$ so that $[O_{p'}(K), \mathrm{Out}_S(E)] = \{1\}$. Then

$$[O_{p'}(K), K] = [O_{p'}(K), \langle \mathrm{Out}_S(E)^K \rangle] = [O_{p'}(K), \mathrm{Out}_S(E)]^K = \{1\}$$

and $O_{p'}(K) \leq Z(K)$.

Now, $\widetilde{K} \cong K/O_{p'}(K)$ is determined as in Proposition 2.15. Moreover, $\widetilde{N_K(\mathrm{Out}_S(E))} = N_{\widetilde{K}}(\widetilde{\mathrm{Out}_S(E)})$ is strongly p-embedded in $\widetilde{K}$ and applying [34, Theorem 7.6.2], $\widetilde{K} \not\cong \mathrm{Alt}(2p)$ or Fi_{22}. Unless $\widetilde{K} \cong \mathrm{Ree}(3)$ or $\mathrm{Sz}(32) : 5$, using that $\widetilde{K}$ is simple and $K = O^{p'}(K)$, K is perfect central extension of $\widetilde{K}$ by a group of p'-order. If $\widetilde{K} \cong \mathrm{Ree}(3)$ or $\mathrm{Sz}(32) : 5$ then $O^{p'}(O^p(K))$ is a perfect central extension of $\mathrm{Ree}(3)' \cong \mathrm{PSL}_2(8)$ resp. $\mathrm{Sz}(32)$ by a p'-group so that $O^{p'}(O^p(K)) \cong \mathrm{PSL}_2(8)$ resp. $\mathrm{Sz}(32)$. Since $O_{p'}(K) \leq N_K(\mathrm{Out}_S(E))$ and $K = O^{p'}(O^p(K))O_{p'}(K)\mathrm{Out}_S(E)$, we conclude that $O_{p'}(K) = \{1\}$ and $K = O^{p'}(K) = O^{p'}(O^p(K))\mathrm{Out}_S(E) \cong \mathrm{Ree}(3)$ resp. $\mathrm{Sz}(32) : 5$. □

The following elementary example gives a flavor of what can happen without the restriction on maximality of essentials.

Example 3.30 *Let V be a 4-dimensional vector space over* GF(2) *and let* Dih(10) *act irreducibly on it. In its embedding in* $\mathrm{GL}_4(2)$, Dih(10) *is centralized by a 3-element and so we may form a subgroup of* $\mathrm{GL}_4(2)$ *of shape* $\mathrm{Dih}(10) \times 3$. *This group is normalized by an element t of order 4 such that* $\langle \mathrm{Dih}(10), t \rangle \cong \mathrm{Sz}(2)$, $t^2 \in \mathrm{Dih}(10)$ *and t inverts the 3-element which centralizes* Dih(10). *Thus, we may construct a group H of shape* Dih(10).Sym(3) *in* $\mathrm{GL}_4(2)$.

Form the semidirect product $G := V : H$ and consider the 2-fusion category of G over some Sylow 2-subgroup S. Since H has cyclic Sylow 2-subgroups and $O_2(H) = \{1\}$, we have that V is essential in the 2-fusion category of G. Moreover, *for s the unique involution in $H \cap S$, we have that $E := V\langle s\rangle$ has order 2^5 and $N_G(E)/E \cong \mathrm{Sym}(3)$. Therefore, E is also an essential subgroup which properly contains another essential subgroup V.*

It is easy to construct other examples in which smaller essentials are contained in some larger essential, even when imposing the condition that the essential subgroups are $\mathrm{Aut}_{\mathcal{F}}(S)$-invariant. But it is reasonable to ask whether such examples actually occur in an amalgam setting motivated by the hypothesis of the Main Theorem.

To this end, let E be an $\mathrm{Aut}_{\mathcal{F}}(S)$-invariant essential subgroup of a saturated fusion system $\mathcal{F}$ on a p-group S, let G be a model for $N_{\mathcal{F}}(E)$ and suppose that $\Omega(Z(S)) \ntrianglelefteq G$. In the midst of the amalgam method, to determine $\mathrm{Out}_{\mathcal{F}}(E)$ and its actions, we work "from the bottom up" by determining $\mathrm{Out}_{\mathcal{F}}(E)$-chief factors of E, starting with those in $\langle \Omega(Z(S))^G\rangle$ and taking progressively larger subgroups of E, so working "up."

Taking Example 3.30 as inspiration, one might imagine a situation in which $\mathrm{Out}_{\mathcal{F}}(E)$ induces a Sym(3)-action on almost all $\mathrm{Out}_{\mathcal{F}}(E)$-chief factors in E. Without examining an ever increasing sequence of subgroups and chief factors, it may be hard to eventually uncover a chief factor which witnesses non-trivial action by a 5-element (although this would probably only happen for amalgams with large "critical distance", see Notation 5.4, and even then it seems unlikely). It seems some additional tricks and techniques (or perhaps an even more granular case division) are required to treat these types of examples uniformly.

Chapter 4
Amalgams in Fusion Systems

In this chapter, we introduce amalgams and demonstrate their connections with and applications to saturated fusion systems. We will only make use of elementary definitions and facts regarding amalgams as can be found in [27, Chapter 2].

4.1 Group Amalgams

Definition 4.1 An *amalgam* of rank n is a tuple $\mathcal{A} = \mathcal{A}(G_1, \ldots, G_n, B, \phi_1, \cdots, \phi_n)$ where B is a group, each G_i is a group and $\phi_i : B \to G_i$ is an injective group homomorphism. A group G is a *faithful completion* of $\mathcal{A}$ if there exists injective group homomorphisms $\psi_i : G_i \to G$ such that for all $i, j \in \{1, \ldots, n\}$, we have that $\phi_i\psi_i = \phi_j\psi_j$, $G = \langle \mathrm{Im}(\psi_1), \ldots, \mathrm{Im}(\psi_n)\rangle$ and no non-trivial subgroup of $B\phi_i\psi_i$ is normal in G. Under these circumstances, we identify $G_1, \ldots, G_n, B$ with their images in G and opt for the notation $\mathcal{A} = \mathcal{A}(G_1, \ldots, G_n, B)$.

For almost all of the work towards the Main Theorem, we reduce to the case where the amalgam is of rank 2 and the groups G_1 and G_2 are finite groups. In this setting, we may always realize $\mathcal{A}$ in a faithful completion, namely the free amalgamated product of G_1 and G_2 over B, denoted $G_1 *_B G_2$. This completion is universal in that every faithful completion occurs as some quotient of this free amalgamated product. Generally, whenever we work in the setting of rank 2 amalgams we will opt to work in this free amalgamated product which we will often denote G and, in an abuse of terminology, refer to G as an amalgam. In particular, we may as well assume the following:

(i) $G = \langle G_1, G_2\rangle$, G_i is a finite group and $G_i < G$ for $i \in \{1, 2\}$;
(ii) no non-trivial subgroup of B is normal in G; and
(iii) $B = G_1 \cap G_2$.

© The Author(s), under exclusive license to Springer Nature Switzerland AG 2024

M. van Beek, *Rank 2 Amalgams and Fusion Systems*, Lecture Notes in Mathematics 2343, https://doi.org/10.1007/978-3-031-54461-3_4

Definition 4.2 Let $\mathcal{A} = \mathcal{A}(G_1, G_2, B, \phi_1, \phi_2)$ and $\mathcal{B} = \mathcal{B}(H_1, H_2, C, \psi_1, \psi_2)$ be two rank 2 amalgams. Then $\mathcal{A}$ and $\mathcal{B}$ are *isomorphic* if, up to permuting indices, there are isomorphisms $\theta_i : G_i \to H_i$ and $\xi : B \to C$ such that the following diagram commutes for $i \in \{1, 2\}$:

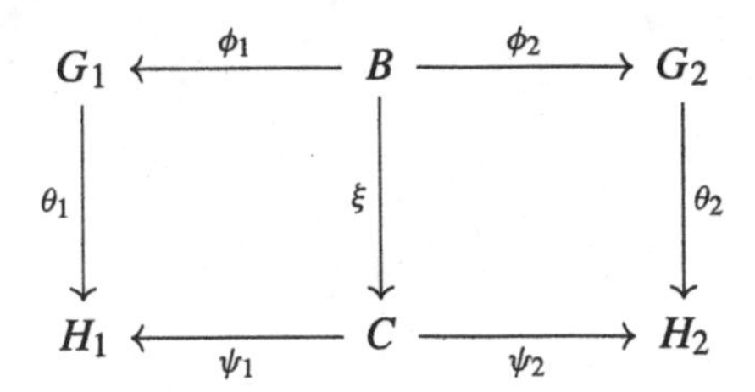

Often, for some finite group H arising as a faithful completion of some rank 2 amalgam $\mathcal{B}$, we will often say a completion G of $\mathcal{A}$ is *locally isomorphic* to H, by which we mean $\mathcal{A}$ is isomorphic to $\mathcal{B}$.

An important observation in this definition is that the faithful completions of two isomorphic amalgams coincide. In fact, two amalgams being isomorphic is equivalent to demanding that $G_1 *_B G_2 \cong H_1 *_C H_2$.

Say that $\mathcal{A} = \mathcal{A}(G_1, G_2, B)$ and $\mathcal{B} = \mathcal{B}(H_1, H_2, C)$ are *parabolic isomorphic* if, up to permuting indices, $G_i \cong H_i$ and $B \cong C$ as abstract groups.

We provide the following elementary example with regard to isomorphisms of amalgams.

Example 4.3 For $G = \mathrm{J}_2$, there are two maximal subgroups M_1, M_2 containing $N_G(S)$ for $S \in \mathrm{Syl}_2(G)$. Furthermore, $M_1/O_2(M_1) \cong \mathrm{SL}_2(4)$, $M_2/O_2(M_2) \cong \mathrm{Sym}(3) \times 3$ and $|N_G(S)/S| = 3$. Thus, G gives rise to the amalgam $\mathcal{A} := \mathcal{A}(M_1, M_2, N_G(S))$.

For $H = \mathrm{J}_3$ and $T \in \mathrm{Syl}_2(H)$, $S \cong T$ and H contains two maximal subgroups N_1, N_2 containing $N_G(T)$ such that $N_i \cong M_i$ for $i \in \{1, 2\}$. Thus, H gives rise to the amalgam $\mathcal{B} := \mathcal{B}(N_1, N_2, N_H(T))$.

We have that $\mathcal{A}$ is isomorphic to $\mathcal{B}$.

Definition 4.4 Let $\mathcal{A} = \mathcal{A}(G_1, G_2, B)$ be an amalgam of rank 2. Then $\mathcal{A}$ is a *characteristic p amalgam of rank* 2 if the following hold for $i \in \{1, 2\}$:

(i) G_i is a finite group;
(ii) $\mathrm{Syl}_p(B) \subseteq \mathrm{Syl}_p(G_1) \cap \mathrm{Syl}_p(G_2)$; and
(iii) G_i is of characteristic p.

For any faithful completion G of an amalgam $\mathcal{A}$, G is not necessarily a finite group and so we must define generally what a Sylow p-subgroup is. We say that P is a Sylow p-subgroup of a group G if every finite p-subgroup of G is conjugate in G to some subgroup of P.

We now collect some results using the *amalgam method* which are relevant to this work. With the application to fusion systems in mind, we are particular interested in the case where the local action involves strongly p-embedded subgroups.

Definition 4.5 Let $\mathcal{A} := \mathcal{A}(G_1, G_2, G_{12})$ be a characteristic p amalgam of rank 2 such that there is $G_i^* \trianglelefteq G_i$ satisfying the following for $i \in \{1, 2\}$:

(i) $O_p(G_i) \leq G_i^*$ and $G_i = G_i^* G_{12}$;
(ii) $G_i^* \cap G_{12}$ is the normalizer of a Sylow p-subgroup of G_i^*; and
(iii) $G_i^*/O_p(G_i) \cong \mathrm{PSL}_2(p^n), \mathrm{SL}_2(p^n), \mathrm{PSU}_3(p^n), \mathrm{SU}_3(p^n), \mathrm{Sz}(2^n), \mathrm{Dih}(10), \mathrm{Ree}(3^n)$ or $\mathrm{Ree}(3)'$.

Then $\mathcal{A}$ is a *weak BN-pair of rank* 2. For G a faithful completion of $\mathcal{A}$, we say that G is a group with a weak BN-pair of rank 2.

We define the set of groups

$$\bigwedge = \{\mathrm{PSL}_3(q), \mathrm{PSp}_4(q), \mathrm{PSU}_4(q), \mathrm{PSU}_5(q), \mathrm{G}_2(q), {}^3\mathrm{D}_4(q), {}^2\mathrm{F}_4(2^n), \mathrm{G}_2(2)', {}^2\mathrm{F}_4(2)', \mathrm{M}_{12}, \mathrm{J}_2, \mathrm{F}_3 : q = p^n, p \text{ a prime}\}$$

and associate a distinguished prime in each case. For ${}^2\mathrm{F}_4(2^n), \mathrm{G}_2(2)', {}^2\mathrm{F}_4(2)', \mathrm{M}_{12}, \mathrm{J}_2$ the prime is 2, for F_3 the prime is 3 and for the other cases, the prime is p where $q = p^n$.

For $X \in \bigwedge$, let $\mathrm{Aut}^0(X) = \mathrm{Aut}(X)$ unless $X = \mathrm{PSL}_3(q), \mathrm{PSp}_4(2^n), \mathrm{G}_2(3^n)$ in which case $\mathrm{Aut}^0(X)$ is group generated by all inner, diagonal and field automorphisms of X so that $\mathrm{Aut}^0(X)$ is of index 2 in $\mathrm{Aut}(X)$ and $\mathrm{Aut}(X) = \langle \mathrm{Aut}^0(X), \phi \rangle$ where ϕ is a graph automorphism. Finally, define

$$\bigwedge^0 = \{Y : \mathrm{Inn}(X) \leq Y \leq \mathrm{Aut}^0(X), X \in \bigwedge\}.$$

For the remainder of this work, whenever we describe a group as being locally isomorphic to $Y \in \bigwedge^0$, we will always mean that Y is a faithful completion of the rank 2 amalgam given by amalgamating two non-conjugate maximal parabolic subgroups of Y which share a common Borel subgroup, sometimes referred to as the *Lie amalgam* of Y. It is straightforward to check that this amalgam is a weak BN-pair of rank 2.

Theorem 4.6 *Suppose that G is a group with a weak BN-pair of rank* 2. *Then one of the following holds:*

(i) *G is locally isomorphic to Y for some $Y \in \bigwedge^0$;*
(ii) *G is parabolic isomorphic to* $\mathrm{G}_2(2)'$, J_2, $\mathrm{Aut}(\mathrm{J}_2)$, M_{12}, $\mathrm{Aut}(\mathrm{M}_{12})$ *or* F_3.

Proof This follows from [27, Theorem A], [26] and [28]. □

In the above theorem, using a lemma of Goldschmidt [31, (2.7)], it may be checked (e.g. using MAGMA [14]) that the groups with a weak BN-pair of rank

2 are actually determined up to local isomorphism in all cases. However, we will not require this fact here.

Notice that all the candidates for $G_i^*/O_p(G_i)$ in the definition of a weak BN-pair of rank 2 have strongly p-embedded subgroups. Indeed, the fusion categories of groups which possess a weak BN-pair of rank 2 form the majority of the examples stemming from the hypothesis in the Main Theorem. Another important class of amalgams which provide examples in the Main Theorem and Theorem C are *symplectic amalgams*.

Definition 4.7 Let $\mathcal{A} := \mathcal{A}(G_1, G_2, G_{12})$ be a characteristic p amalgam of rank 2. Then $\mathcal{A}$ is a *symplectic amalgam* if, up to interchanging G_1 and G_2, the following hold:

(i) $O^{p'}(G_1)/O_p(G_1) \cong \mathrm{SL}_2(p^n)$;
(ii) for $W := \langle ((O_p(G_1) \cap O_p(G_2))^{G_1})^{G_2} \rangle$, $G_2 = G_{12}W$ and $O^p(O^{p'}(G_2)) \le W$;
(iii) for $S \in \mathrm{Syl}_p(G_{12})$, $G_{12} = N_{G_1}(S)$;
(iv) $\Omega(Z(S)) = \Omega(Z(O^{p'}(G_2)))$ for $S \in \mathrm{Syl}_p(G_{12})$; and
(v) for $Z_1 := \langle \Omega(Z(S))^{G_1} \rangle$, $Z_1 \le O_p(G_2)$ and there is $x \in G_2$ such that $Z_1^x \not\le O_p(G_1)$.

Theorem 4.8 *Suppose that $\mathcal{A} := \mathcal{A}(G_1, G_2, G_{12})$ is a symplectic amalgam such that $G_2/O_p(G_2)$ has a strongly p-embedded subgroup and for $S \in \mathrm{Syl}_p(G_{12})$, $G_{12} = N_{G_1}(S) = N_{G_2}(S)$. Assume further than G_i is a $\mathcal{K}$-group for $i \in \{1, 2\}$. Then one of the following holds, where $\mathcal{A}_k$ corresponds to the listing given in [61, Table 1.8]:*

(i) *$\mathcal{A}$ has a weak BN-pair of rank 2 of type* ${}^3\mathrm{D}_4(p^n)$ ($\mathcal{A}_{27}$), $\mathrm{G}_2(p^n)$ ($\mathcal{A}_2$, $\mathcal{A}_6$ *and* $\mathcal{A}_{26}$ *when* $p \ne 3$), $\mathrm{G}_2(2)'$ ($\mathcal{A}_1$), J_2 ($\mathcal{A}_{41}$) *or* $\mathrm{Aut}(\mathrm{J}_2)$ ($\mathcal{A}_{41}^1$)*;*
(ii) $p = 2$, $\mathcal{A} = \mathcal{A}_4$, $|S| = 2^6$, $O_2(L_2) \cong 2_+^{1+4}$ *and* $L_2/O_2(L_2) \cong (3 \times 3) : 2$*;*
(iii) $p = 5$, $\mathcal{A} = \mathcal{A}_{20}$, $|S| = 5^6$, $O_5(L_2) \cong 5_+^{1+4}$ *and* $L_2/O_5(L_2) \cong 2_-^{1+4}.5$*;*
(iv) $p = 5$, $\mathcal{A} = \mathcal{A}_{21}$, $|S| = 5^6$, $O_5(L_2) \cong 5_+^{1+4}$ *and* $L_2/O_5(L_2) \cong 2_-^{1+4}.\mathrm{Alt}(5)$*;*
(v) $p = 5$, $\mathcal{A} = \mathcal{A}_{46}$, $|S| = 5^6$, $O_5(L_2) \cong 5_+^{1+4}$ *and* $L_2/O_5(L_2) \cong 2 \cdot \mathrm{Alt}(6)$*; or*
(vi) $p = 7$, $\mathcal{A} = \mathcal{A}_{48}$, $|S| = 7^6$, $O_7(L_2) \cong 7_+^{1+4}$ *and* $L_2/O_7(L_2) \cong 2 \cdot \mathrm{Alt}(7)$.

Proof We apply the classification in [61] and upon inspection of the tables there, we need only rule out $\mathcal{A}_3$, $\mathcal{A}_5$ and $\mathcal{A}_{45}$ when $p = 3$; and $\mathcal{A}_{42}$ when $p = 2$. Set $Q_i := O_3(G_i)$ and $L_i := O^{3'}(L_i)$. With regards to $\mathcal{A}_{45}$, it is proved in [61, Theorem 11.4] that $N_{G_2}(S) \not\le G_1$. In $\mathcal{A}_3$, we have that $L_2/O_3(L_2) \cong \mathrm{SL}_2(3)$ and $|S| = 3^6$. In particular, if $G_{12} = N_{G_1}(S) = N_{G_2}(S)$ then G has a weak BN-pair but comparing with the configurations in [27], we have a contradiction.

Suppose that we are in the situation of $\mathcal{A}_5$ so that L_2 is of shape $3.((3^2 : Q_8) \times (3^2 : Q_8)) : 3$. Furthermore, by [61, Lemma 6.21], we have that $Q_2 = \langle \Omega(Z(Q_1))^{G_2} \rangle$. Let K_2 be a Hall $2'$-subgroup of $L_2 \cap N_G(S)$. Then K_2 is elementary abelian of order 4. By hypothesis, K_2 normalizes Q_1 and so K_2 normalizes $\Omega(Z(Q_1))$. Moreover, K_2 centralizes $\Omega(Z(S)) = \Omega(Z(L_2)) = \Phi(Q_2)$

and since $|\Omega(Z(Q_1))/\Omega(Z(L_2))| = 3$ by [61, Lemma 6.21], it follows that there is $k \in K$ an involution which centralizes $\Omega(Z(Q_1))$. Since $\langle kQ_2\rangle \trianglelefteq G_2$, we infer that $\Omega(Z(S)) = [\langle kQ_2\rangle, \Omega(Z(Q_1))]^{G_2} = [\langle kQ_2\rangle, \langle \Omega(Z(Q_1))^{G_2}\rangle]$. But $Q_2 = \langle \Omega(Z(Q_1))^{G_2}\rangle$ by [61, Lemma 6.21] so that k centralizes $Q_2/\Phi(Q_2)$, a contradiction as G_2 is of characteristic 3.

In the situation of $\mathcal{A}_{42}$ when $p = 2$, we have that $L_2/Q_2 \cong \mathrm{Alt}(5) \cong \mathrm{SL}_2(4)$ so that G has a weak BN-pair of rank 2. Since $|S| = 2^9$ in this case, comparing with [27], we have a contradiction.

Hence, the amalgams $\mathcal{A}_3$, $\mathcal{A}_5$, $\mathcal{A}_{42}$ and $\mathcal{A}_{45}$ do not satisfy the hypothesis, and the theorem holds. □

Remark The symplectic amalgams $\mathcal{A}_3$, $\mathcal{A}_5$ and $\mathcal{A}_{45}$ where $G_2/O_3(G_2)$ has a strongly p-embedded subgroup have as example completions $\Omega_8^+(2) : \mathrm{Sym}(3)$, $\mathrm{F}_4(2)$ and HN. Indeed, in these configurations $|S|$ is bounded and one can employ [62] to get a list of candidate fusion systems supported on S. It transpires that the only appropriate fusion systems supported on S are exactly the fusion categories of the above examples, but in each case there are three essentials, all normal in S, one of which is $\mathrm{Aut}_{\mathcal{F}}(S)$-invariant while the other two are fused under the action of $\mathrm{Aut}_{\mathcal{F}}(S)$.

Remark In a later section, we come across an amalgam which satisfies almost all of the properties of $\mathcal{A}_{42}$. Indeed, this amalgam contains $\mathcal{A}_{42}$ as a subamalgam and we show that the fusion system supported from this configuration is the 2-fusion system of $\mathrm{PSp}_6(3)$. Indeed, $\mathrm{PSp}_6(3)$ is listed as an example completion of $\mathcal{A}_{42}$ in [61] and in $\mathrm{PSp}_6(3)$ itself, there is a choice of generating subgroups G_1, G_2 such that $(G_1, G_2, G_1 \cap G_2)$ is a symplectic amalgam. However, the fusion subsystem generated by the fusion systems of the groups G_1 and G_2 fails to generate the fusion system of $\mathrm{PSp}_6(3)$. In fact, such a subsystem fails to be saturated.

4.2 Amalgams in Fusion Systems

The following theorem provides the connection between amalgams and fusion systems. Indeed, the original application of this theorem demonstrates that any saturated fusion system may be realized by a (possibly infinite) group.

Theorem 4.9 *Let p be a prime, G_1, G_2 and G_{12} be groups with $G_{12} \leq G_1 \cap G_2$. Assume that $S_1 \in \mathrm{Syl}_p(G_1)$ and $S_2 \in \mathrm{Syl}_p(G_{12}) \cap \mathrm{Syl}_p(G_2)$ with $S_2 \leq S_1$. Set*

$$G = G_1 *_{G_{12}} G_2$$

to be the free amalgamated product of G_1 and G_2 over G_{12}. Then $S_1 \in \mathrm{Syl}_p(G)$ and

$$\mathcal{F}_{S_1}(G) = \langle \mathcal{F}_{S_1}(G_1), \mathcal{F}_{S_2}(G_2)\rangle.$$

Proof This is [67, Theorem 1]. □

In other words, the above theorem implies that given two fusion systems which give rise to two rank 2 amalgams, and the data from these amalgams "generate" the fusion system, then provided that the amalgams are isomorphic, the fusion systems are isomorphic.

However, there are some key differences in the group theoretic applications of amalgams, and the fusion theoretic applications. Consider the configurations from Example 4.3. The two amalgams there, $\mathcal{A}$ and $\mathcal{B}$, are isomorphic. In this way, we can actually embed a copy of the 2-fusion system of J_2 inside the 2-fusion system of J_3, but J_2 is certainly not a subgroup of J_3. Indeed, the 2-fusion system of J_3 contains an additional class of essential subgroups arising from different maximal subgroups of J_3 of shape $2^4 : (3 \times SL_2(4))$ not involved in the amalgams.

Thus, there are some important considerations demonstrated in Example 4.3 that one should be aware of. One is that for a group G with two maximal subgroups M_1 and M_2 containing a Sylow p-subgroup of G, even though $G = \langle M_1, M_2\rangle$ there are situations in which $\mathcal{F}_S(G) \neq \langle \mathcal{F}_S(M_1), \mathcal{F}_S(M_2)\rangle$. The second is that one must be very careful in choosing the "correct" completion when working with amalgams in the context of fusion systems. Indeed, most of the time, this often requires knowledge of the fusion systems (and in particular the essential subgroups) of the completions of the amalgam.

We now state the main hypothesis of this work with regard to fusion systems.

Hypothesis 4.10 $\mathcal{F}$ is a local $\mathcal{CK}$-system on a p-group S with two $\mathrm{Aut}_{\mathcal{F}}(S)$-invariant essential subgroups $E_1, E_2 \trianglelefteq S$ such that for $\mathcal{F}_0 := \langle N_{\mathcal{F}}(E_1), N_{\mathcal{F}}(E_2)\rangle$, we have that $O_p(\mathcal{F}_0) = \{1\}$. Furthermore, for $G_i := \mathrm{Out}_{\mathcal{F}}(E_i)$ and $T \in \mathrm{Syl}_p(G_i)$, if $G_i/O_{3'}(G_i) \cong \mathrm{Ree}(3)$, G_i is p-solvable or T is generalized quaternion, then $N_{G_i}(T)$ is a strongly p-embedded in G_i.

We now recognize a characteristic p amalgam of rank 2 present in $\mathcal{F}_0$. Namely, we take the models G_1 and G_2 of $N_{\mathcal{F}}(E_1)$ and $N_{\mathcal{F}}(E_2)$. By the uniqueness of models in Theorem 3.24, it follows quickly that we may assume that $N_{G_1}(E_2) = N_{G_2}(E_1) =: G_{12}$. By Theorem 4.9, we have that $\mathcal{F}_0 = \mathcal{F}_S(G)$ where $G = G_1 *_{G_{12}} G_2$. We take the liberty of recognizing G_1, G_2 and G_{12} as subgroups of G. Since E_i is $\mathrm{Aut}_{\mathcal{F}}(S)$-invariant, we deduce that $N_{G_i}(S) \le N_{G_i}(E_{3-i}) = G_{12}$ for $i \in \{1, 2\}$.

We now have a hypothesis in purely amalgam theoretic terms. Indeed, G is a characteristic p amalgam of rank 2 such that the following hold:

Hypothesis 4.11 $\mathcal{A} := \mathcal{A}(G_1, G_2, G_{12})$ is a characteristic p amalgam of rank 2 with faithful completion G satisfying the following:

(i) for $S \in \mathrm{Syl}_p(G_{12})$, $N_{G_1}(S) = N_{G_2}(S) \le G_{12}$; and
(ii) writing $\overline{G_i} := G_i/O_p(G_i)$, $\overline{G_i}$ contains a strongly p-embedded subgroup and if $\overline{G_i}/O_{3'}(\overline{G_i}) \cong \mathrm{Ree}(3)$, $\overline{G_i}$ is p-solvable or $\overline{S}$ is generalized quaternion, then $N_{\overline{G_i}}(\overline{S})$ is strongly p-embedded in $\overline{G_i}$.

Due to the strongly p-embedded hypothesis, whenever $\mathcal{F}_0$ is the p-fusion category of a finite group G with $F^*(G)$ one of the groups in $\bigwedge$, we are almost always able to deduce that $F^*(G) = O^{p'}(G)$.

Lemma 4.12 *Suppose that $\mathcal{F}$ satisfies Hypothesis 4.10. Let G_i be a model for $N_{\mathcal{F}}(E_i)$ such that $S \in \mathrm{Syl}_p(G_i)$ with $i \in \{1, 2\}$. If the amalgam $\mathcal{A} := \mathcal{A}(G_1, G_2, G_{12})$ extracted from $\mathcal{F}_0$ is a weak BN-pair of rank 2 and is locally isomorphic to $Y \in \bigwedge^0$ with $F^*(Y) \not\cong \mathrm{G}_2(2)', \mathrm{PSp}_4(2)', {}^2\mathrm{F}_4(2)', \mathrm{M}_{12}$ or J_2, then $F^*(Y) = O^{p'}(Y)$.*

Proof Suppose that $\mathcal{A}$ is locally isomorphic to $Y \in \bigwedge^0$. If $m_p(S/O_p(G_i)) > 1$, then $O^{p'}(G_i/O_p(G_i))$ is determined by Proposition 2.15 and comparing with the structure of the parabolic subgroups of Y, we deduce that $F^*(Y) = O^{p'}(G)$. If $m_p(S/O_p(G_i)) = 1$ then Y is generated by only inner and diagonal automorphisms so that $F^*(Y) = O^{p'}(Y)$ as required. □

Proposition 4.13 *Suppose that $\mathcal{F}$ satisfies Hypothesis 4.10. Let G_i be a model for $N_{\mathcal{F}}(E_i)$ such that $S \in \mathrm{Syl}_p(G_i)$ with $i \in \{1, 2\}$. If the amalgam $\mathcal{A} := \mathcal{A}(G_1, G_2, G_{12})$ extracted from $\mathcal{F}_0$ is a weak BN-pair of rank 2 then either:*

(i) *$\mathcal{F}_0$ is the p-fusion category of Y for some $Y \in \bigwedge^0$; or*
(ii) *$\mathcal{A}$ is parabolic isomorphic* F_3.

Proof By Theorem 4.6 we have that $\mathcal{A}$ is locally isomorphic to Y for some $Y \in \bigwedge^0$; or parabolic isomorphic to $\mathrm{G}_2(2)'$, ${}^2\mathrm{F}_4(2)'$, J_2, $\mathrm{Aut}(\mathrm{J}_2)$, M_{12}, $\mathrm{Aut}(\mathrm{M}_{12})$ or F_3. Assume first that $\mathcal{A}$ is parabolic isomorphic to $\mathrm{G}_2(2)'$, ${}^2\mathrm{F}_4(2)'$, J_2, $\mathrm{Aut}(\mathrm{J}_2)$, M_{12}, $\mathrm{Aut}(\mathrm{M}_{12})$. Then the possible fusion systems on S with trivial 2-core may be enumerated using the fusion systems package in MAGMA [14, 62]. Indeed, $\mathcal{F} = \mathcal{F}_0$ or $\mathcal{F}$ is the 2-fusion category of J_3 or $\mathrm{Aut}(\mathrm{J}_3)$ in which case $\mathcal{F}_0$ is isomorphic to the 2-fusion category of J_2 resp. $\mathrm{Aut}(\mathrm{J}_2)$. Alternatively, except in the case of ${}^2\mathrm{F}_4(2)'$, one could apply [3] to the reduction of $\mathcal{F}$ and using the *tameness* of the candidate systems, both $\mathcal{F}$ and $\mathcal{F}_0$ can be calculated from the automorphisms groups of certain small simple groups.

Assume now that $\mathcal{A}$ is locally isomorphic to Y where $F^*(Y) = O^{p'}(Y)$ is a simple group of Lie type. By [34, Corollary 3.1.6], the p-fusion category of Y has only two essential subgroups, namely the unipotent radical subgroups of the maximal parabolic subgroups in $F^*(Y)$, which we label P_1 and P_2. We identify S with a Sylow p-subgroup of Y contained in P_1 and P_2. Hence, by the Alperin–Goldschmidt theorem we have that $\mathcal{F}_S(Y) = \langle N_{\mathcal{F}_S(Y)}(O_p(P_1)), N_{\mathcal{F}_S(Y)}(O_p(P_2))\rangle$. By Robinson's result, $\mathcal{F}_S(Y)$ is isomorphic to the fusion system of the free amalgamated product of $N_Y(O_p(P_1))$ and $N_Y(O_p(P_2))$ over $N_Y(S)$ and, by the uniqueness of the amalgam, $\mathcal{F}_S(Y) = \mathcal{F}_0$, as desired. □

The case where S is isomorphic to a Sylow 3-subgroup of F_3 has been treated in [79]. Indeed, this is another instance where $\mathcal{F}_0 \subset \mathcal{F}$. We also mention that for $\mathcal{F}$ the 5-fusion category of Co_1, we have that $\mathcal{F}_0$ is isomorphic to the 5-fusion category of $\mathrm{PSp}_4(5)$, yet another example of this phenomena.

4.3 The Main Results

The main work in this text is in proving Theorem C, which we list below for convenience.

Theorem C *Suppose that* $\mathcal{A} = \mathcal{A}(G_1, G_2, G_{12})$ *satisfies Hypothesis 4.11. Then one of the following occurs:*

(i) $\mathcal{A}$ *is a weak BN-pair of rank* 2;
(ii) $p = 2$, $\mathcal{A}$ *is a symplectic amalgam,* $|S| = 2^6$, $G_1/O_2(G_1) \cong \mathrm{Sym}(3)$ *and* $G_2/O_2(G_2) \cong (3 \times 3) : 2$;
(iii) $p = 2$, $\Omega(Z(S)) \trianglelefteq G_2$, $\langle(\Omega(Z(S))^{G_1})^{G_2}\rangle \not\leq O_2(G_1)$, $|S| = 2^9$, $O^{2'}(G_1)/O_2(G_1) \cong \mathrm{SU}_3(2)'$ *and* $O^{2'}(G_2)/O_2(G_2) \cong \mathrm{Alt}(5)$;
(iv) $p = 3$, $\Omega(Z(S)) \trianglelefteq G_2$, $\langle(\Omega(Z(S))^{G_1})\rangle \not\leq O_2(G_2)$, $|S| \leqslant 3^7$ *and* $O_3(G_1) = \langle(\Omega(Z(S))^{G_1})\rangle$ *is cubic* 2*F-module for* $G_1/O_3(G_1)$; *or*
(v) $p = 5$ *or* 7, $\mathcal{A}$ *is a symplectic amalgam and* $|S| = p^6$.

More information may be extracted than what we have provided here, but with the application of fusion systems in mind and the available results classifying fusion systems supported on p-groups of small order, we stop short of completely describing G_1 and G_2 up to isomorphism, although this seems possible in most cases.

We now prove the Main Theorem assuming the validity of Theorem C. The necessary structural details for some of the cases we treat are documented later in Propositions 7.57, 7.61, and 7.62.

To verify that two of the fusion systems uncovered are exotic, the classification of the finite simple groups is invoked (see [79] and [63]). This is the only occasion in this work where we apply the classification in its full strength and not in an inductive context. Without the classification, outcome (v) below would instead read "$\mathcal{F}$ is a simple fusion system on a Sylow 3-subgroup of F_3 which is not isomorphic to the 3-fusion category of F_3" and outcome (vii) would read "$\mathcal{F}$ is a simple fusion system on a Sylow 7-subgroup of $\mathrm{G}_2(7)$ which is not isomorphic to 7-fusion category of $\mathrm{G}_2(7)$ or M."

Main Theorem *Let* $\mathcal{F}$ *be a local* $\mathcal{CK}$*-system on a* p*-group* S*. Assume that* $\mathcal{F}$ *has two* $\mathrm{Aut}_{\mathcal{F}}(S)$*-invariant essential subgroups* $E_1, E_2 \trianglelefteq S$ *such that for* $\mathcal{F}_0 := \langle N_{\mathcal{F}}(E_1), N_{\mathcal{F}}(E_2)\rangle_S$ *the following conditions hold:*

(a) $O_p(\mathcal{F}_0) = \{1\}$;
(b) *for* $G_i := \mathrm{Out}_{\mathcal{F}}(E_i)$ *and* $T \in \mathrm{Syl}_p(G_i)$, *if* $G_i/O_{3'}(G_i) \cong \mathrm{Ree}(3)$, G_i *is* p*-solvable or* T *is generalized quaternion, then* $N_{G_i}(T)$ *is strongly* p*-embedded in* G_i.

Then $\mathcal{F}_0$ *is saturated and one of the following holds:*

(i) $\mathcal{F}_0 = \mathcal{F}_S(G)$, *where* $F^*(G)$ *is isomorphic to a rank* 2 *simple group of Lie type in characteristic* p;

(ii) $\mathcal{F}_0 = \mathcal{F}_S(G)$, *where* $G \cong \mathrm{M}_{12}, \mathrm{Aut}(\mathrm{M}_{12}), \mathrm{J}_2, \mathrm{Aut}(\mathrm{J}_2), \mathrm{G}_2(3)$ *or* $\mathrm{PSp}_6(3)$ *and* $p = 2$;
(iii) $\mathcal{F}_0 = \mathcal{F}_S(G)$, *where* $G \cong \mathrm{Co}_2, \mathrm{Co}_3, \mathrm{McL}, \mathrm{Aut}(\mathrm{McL}), \mathrm{Suz}, \mathrm{Aut}(\mathrm{Suz})$ *or* Ly *and* $p = 3$;
(iv) $\mathcal{F}_0 = \mathcal{F}_S(G)$, *where* $G \cong \mathrm{PSU}_5(2), \mathrm{Aut}(\mathrm{PSU}_5(2)), \Omega_8^+(2), \mathrm{O}_8^+(2), \Omega_{10}^-(2), \mathrm{Sp}_{10}(2), \mathrm{PSU}_6(2)$ *or* $\mathrm{PSU}_6(2).2$ *and* $p = 3$;
(v) $\mathcal{F}_0$ *is a simple fusion system on a Sylow* 3*-subgroup of* F_3 *and, assuming* CFSG, $\mathcal{F}_0$ *is an exotic fusion system uniquely determined up to isomorphism;*
(vi) $\mathcal{F}_0 = \mathcal{F}_S(G)$, *where* $G \cong \mathrm{Ly}, \mathrm{HN}, \mathrm{Aut}(\mathrm{HN})$ *or* B *and* $p = 5$; *or*
(vii) $\mathcal{F}_0$ *is a simple fusion system on a Sylow* 7*-subgroup of* $\mathrm{G}_2(7)$ *and, assuming* CFSG, $\mathcal{F}_0$ *is an exotic fusion system uniquely determined up to isomorphism.*

Proof Let $\mathcal{A}(G_1, G_2, G_{12})$ be the amalgam determined by $\mathcal{F}_0$. If $\mathcal{A}$ is a weak BN-pair of rank 2 then by Proposition 4.13, we either have that $\mathcal{F}_0$ satisfies one of the conclusions of the theorem; or $\mathcal{A}$ is parabolic isomorphic to F_3 and S is isomorphic to a Sylow 3-subgroup of F_3. In the latter case, we refer to [79] where both $\mathcal{F}$ and $\mathcal{F}_0$ are determined.

Suppose now that $\mathcal{A}$ is a symplectic amalgam but not a weak BN-pair of rank 2. If $p = 2$ and $|S| = 2^6$ as in case (ii) of Theorem 4.8, then it follows from [61, Lemma 6.21] that $S = (O^2(G_1) \cap S)(O^2(G_2) \cap S)$ so that $O^2(\mathcal{F}) = \mathcal{F}$ by [11, Theorem I.7.4]. Checking against the lists provided in [3, Theorem 4.1], $\mathcal{F}$ is isomorphic to the 2-fusion category of $\mathrm{G}_2(3)$, which has two essential subgroups both of which are $\mathrm{Aut}_{\mathcal{F}}(S)$-invariant. By the Alperin–Goldschmidt theorem, we have that $\mathcal{F}_0 = \mathcal{F}$ so that $\mathcal{F}_0$ is saturated. If $p \in \{5, 7\}$ and $\mathcal{A}$ satisfies (iii)–(vi) of Theorem 4.8, then $|S| \leqslant p^6$ and again we deduce that $O^p(\mathcal{F}) = \mathcal{F}$. Then using the results follows from the tables provided in [62], we deduce that if $p = 5$ then $\mathcal{F}$ has only two essential subgroups, both of which are $\mathrm{Aut}_{\mathcal{F}}(S)$-invariant, and $\mathcal{F} = \mathcal{F}_0$ is as described. If $p = 7$, again applying [62], then S is isomorphic to a Sylow 7-subgroup of $\mathrm{G}_2(7)$ and $\mathcal{F}$ is determined in [63]. Indeed, it can be gleaned from [63, Table 1] that $\mathcal{F}_0$ is either isomorphic to the 7-fusion category of $\mathrm{G}_2(7)$ or $\mathcal{F}_0$ is a simple exotic fusion system on S. In both cases, there exist fusion systems which satisfy the hypothesis of $\mathcal{F}$ with $\mathcal{F}_0 \subset \mathcal{F}$.

Suppose that outcome (iii) of Theorem C holds. Comparing with the structure provided in Proposition 6.23, it follows that $S = (O^2(G_1) \cap S)(O^2(G_2) \cap S)$ from which it follows that $O^2(O^{2'}(\mathcal{F})) = O^{2'}(\mathcal{F})$ so $O^{2'}(\mathcal{F})$ is reduced. Comparing with the lists in [3], $O^{2'}(\mathcal{F})$ is isomorphic to the 2-fusion system of $\mathrm{PSp}_6(3)$. Furthermore, by [3, Proposition 6.4], the only saturated fusion system supported on a Sylow 2-subgroup of $\mathrm{PSp}_6(3)$ with $O_2(\mathcal{F}) = \{1\}$ is the fusion category of $\mathrm{PSp}_6(3)$, which has exactly two essential subgroups, both of which are $\mathrm{Aut}_{\mathcal{F}}(S)$-invariant. By the Alperin–Goldschmidt theorem, we have that $\mathcal{F}_0 = \mathcal{F} = O^{2'}(\mathcal{F})$ and the result holds.

Finally, assume that outcome (iv) of Theorem C holds and $\mathcal{A}$ is not a weak BN-pair of rank 2. Applying the structural results provided in Propositions 7.57, 7.61, and 7.62, we deduce that $S = (S \cap O^3(G_1))(S \cap O^3(G_2))$ and so $\mathcal{F} = O^3(\mathcal{F})$ and $O_3(\mathcal{F}) = \{1\}$. Since $|S| \leqslant 3^7$, $\mathcal{F}$ is determined by the work in [62] and in all cases, there are only two essential subgroups of $\mathcal{F}$, both of which are $\mathrm{Aut}_{\mathcal{F}}(S)$-invariant so that $\mathcal{F}_0 = \mathcal{F}$ is saturated. □

In the above theorem, whenever $\mathcal{F} \neq \mathcal{F}_0$, $\mathcal{F}_0$ is obtained from $\mathcal{F}$ by pruning [62, 23] a class of *pearls* (see [35]). Indeed, $\mathcal{F}_0$ occurs as the fusion theoretic equivalent of the $C(G, S)$ subgroup [18], and $\mathcal{F}$ is an odd prime equivalent to an *obstruction to pushing up* in the sense of Aschbacher [5].

Given the conclusions of the main theorem, it is not hard to prove Corollary A.

Corollary A *Suppose that $\mathcal{F}$ is a saturated fusion system on a p-group S such that $O_p(\mathcal{F}) = \{1\}$. Assume that $\mathcal{F}$ has exactly two essential subgroups E_1 and E_2. Then $N_S(E_1) = N_S(E_2)$ and writing $\mathcal{F}_0 := \langle N_{\mathcal{F}}(E_1), N_{\mathcal{F}}(E_2)\rangle_{N_S(E_1)}$, $\mathcal{F}_0$ is a saturated normal subsystem of $\mathcal{F}$ and either*

(i) *$\mathcal{F} = \mathcal{F}_0$ is determined by the Main Theorem;*
(ii) *p is arbitrary, $\mathcal{F}_0$ is isomorphic to the p-fusion category of H, where $F^*(H) \cong \mathrm{PSL}_3(p^n)$, and $\mathcal{F}$ is isomorphic to the p-fusion category of G where G is the extension of H by a graph or graph-field automorphism;*
(iii) *$p = 2$, $\mathcal{F}_0$ is isomorphic to the 2-fusion category of H, where $F^*(H) \cong \mathrm{PSp}_4(2^n)$, and $\mathcal{F}$ is isomorphic to the 2-fusion category of G where G is the extension of H by a graph or graph-field automorphism; or*
(iv) *$p = 3$, $\mathcal{F}_0$ is isomorphic to the 3-fusion category of H, where $F^*(H) \cong \mathrm{G}_2(3^n)$, and $\mathcal{F}$ is isomorphic to the 3-fusion category of G where G is the extension of H by a graph or graph-field automorphism.*

Proof Note that if both E_1 and E_2 are $\mathrm{Aut}_{\mathcal{F}}(S)$-invariant then (appealing to Proposition 5.15 to verify that E_1 and E_2 are maximally essential) $\mathcal{F} = \mathcal{F}_0$ is determined by the Main Theorem. Assume throughout that at least one of E_1 and E_2 is not $\mathrm{Aut}_{\mathcal{F}}(S)$-invariant, and without loss of generality, E_1 is not $\mathrm{Aut}_{\mathcal{F}}(S)$-invariant. Then $N_S(E_1)\alpha \leq N_S(E_1\alpha)$ and since E_1 is fully $\mathcal{F}$-normalized, it follows that $N_S(E_1)\alpha = N_S(E_1\alpha)$. Moreover, $E_1\alpha$ is also essential in $\mathcal{F}$ and so $E_1\alpha = E_2$. By a similar reasoning, $E_2\alpha = E_1$, $\alpha^2 \in N_{\mathcal{F}}(E_1) \cap N_{\mathcal{F}}(E_2)$ and both E_1 and E_2 are maximally essential.

Suppose first that p is odd. Then $S = N_S(E_1) = N_S(E_2)$ and by [11, Lemma I.7.6(b)] and the Alperin–Goldschmidt theorem, $\mathcal{F}_0$ is a saturated subsystem of $\mathcal{F}$ of index 2 and by [11, Theorem I.7.7], $\mathcal{F}_0$ is normal in $\mathcal{F}$. Hence, $O_p(\mathcal{F}_0)$ is normalized by $\mathcal{F}$ and as $O_p(\mathcal{F}) = \{1\}$, $O_p(\mathcal{F}_0) = \{1\}$ and $\mathcal{F}_0$ is determined by the Main Theorem.

Since there is $\alpha \in \mathrm{Aut}_{\mathcal{F}}(S)$ such that $E_1\alpha = E_2$, we must have that $E_1 \cong E_2$ as abstract p-groups. Thus, comparing with the Main Theorem, $\mathcal{F}_0$ is isomorphic to the p-fusion category of H where $F^*(H)$ is one of $\mathrm{PSL}_3(p^n)$ or $\mathrm{G}_2(3^n)$ (where $p > 2$ is arbitrary or $p = 3$ respectively). Indeed, since $\mathcal{F}_0 \trianglelefteq \mathcal{F}$, there is $\mathcal{F}^0 \trianglelefteq \mathcal{F}$ with $\mathcal{F}^0$ isomorphic to the p-fusion category of $F^*(H)$ and supported on S. At this point, we can either apply [16, Theorem A]; or recognize that the possible fusion systems correspond exactly to the p-fusion categories of certain subgroups of $\mathrm{Aut}(F^*(H))$ containing $\mathrm{Inn}(F^*(H))$ by applying [11, Theorem I.7.7].

Suppose now that $p = 2$. Then $N_S(E_1) = N_S(E_2)$ has index 2 in S and $E_1E_2 \trianglelefteq S$. Let G_i be a model for $N_{\mathcal{F}}(E_i)$ for $i \in \{1,2\}$. Note that if there is $Q \leq N_S(E_1)$ with Q normal in $N_{\mathcal{F}}(E_i)$ then $Q\alpha$ is normal in $N_{\mathcal{F}}(E_i)\alpha = N_{\mathcal{F}}(E_i\alpha) = N_{\mathcal{F}}(E_{3-i})$ for $i \in \{1,2\}$. Hence, if $Q \leq N_S(E_1)$ is normal in both $N_{\mathcal{F}}(E_1)$ and $N_{\mathcal{F}}(E_2)$, then $Q \trianglelefteq \mathcal{F}$. Since $O_2(\mathcal{F}) = \{1\}$, we deduce that Q is trivial. Moreover, applying [5, (2.2.4)] and using the Alperin–Goldschmidt and the receptiveness of E_1E_2, $N_{G_1}(E_2) = N_{G_1}(E_1E_2) = N_{G_1}(N_S(E_2))$ is isomorphic to $N_{G_2}(E_1) = N_{G_2}(N_S(E_2)$ by an isomorphism fixing $N_S(E_1)$.

Hence, suppressing the necessary inclusion maps, we form the rank 2 amalgam $\mathcal{A} := \mathcal{A}(G_1, G_2, G_{12}^*)$ writing G_{12}^* for the group gained by identifying $N_{G_1}(N_S(E_1))$ with $N_{G_2}(N_S(E_2))$ in the previously described isomorphism. Then $\mathcal{F}_0 = \langle \mathcal{F}_{N_S(E_1)}(G_1), \mathcal{F}_{N_S(E_2)}(G_2)\rangle = \mathcal{F}_{N_S(E_1)}(G_1 *_{G_{12}^*} G_2)$ by Theorem 4.9, $N_S(E_1)$ is a Sylow of $\mathcal{F}_0$ and $O_2(\mathcal{F}_0) = \{1\}$. Moreover, $\mathcal{A}$ satisfies Hypothesis 5.1 and since $E_2 = E_1\alpha$, E_1 and E_2 are isomorphic as abstract 2-groups. Then $G_1 *_{G_{12}^*} G_2$ is locally isomorphic to H where $H \in \bigwedge^0$ is as described after Definition 4.5, and $F^*(H) \cong \mathrm{PSL}_3(2^n)$ or $\mathrm{PSp}_4(2^n)$. Then by Proposition 4.13, $\mathcal{F}_0$ is isomorphic to the 2-fusion category of H and so $\mathcal{F}_0$ is saturated. Moreover, applying [11, Theorem I.7.4] and the Alperin–Goldschmidt theorem, $\mathcal{F}_0$ is a normal subsystem of index 2 in $\mathcal{F}$. Again, there is $\mathcal{F}^0 \trianglelefteq \mathcal{F}$ with $\mathcal{F}^0$ isomorphic to the p-fusion category of $F^*(H)$ and supported on $N_S(E_1)$ and we can either apply [16, Theorem A]; or recognize that the possible fusion systems correspond exactly to the 2-fusion categories of certain subgroups of $\mathrm{Aut}(F^*(H))$ containing $\mathrm{Inn}(F^*(H))$ by applying [11, Theorem I.7.7]. □

Chapter 5
The Amalgam Method

By the previous chapter, in order to prove the Main Theorem, it suffices to prove Theorem C. We classify configurations under the following hypothesis, which we fix for the remainder of this work.

Hypothesis 5.1 $\mathcal{A} := \mathcal{A}(G_1, G_2, G_{12})$ is a characteristic p amalgam of rank 2 with faithful completion G satisfying the following:

(i) for $S \in \mathrm{Syl}_p(G_{12})$, $N_{G_1}(S) = N_{G_2}(S) \leq G_{12}$; and
(ii) writing $\overline{G_i} := G_i/O_p(G_i)$, $\overline{G_i}$ contains a strongly p-embedded subgroup and if $\overline{G_i}/O_{3'}(\overline{G_i}) \cong \mathrm{Ree}(3)$, $\overline{G_i}$ is p-solvable or $\overline{S}$ is generalized quaternion, then $N_{\overline{G_i}}(\overline{S})$ is strongly p-embedded in $\overline{G_i}$.

We remark that a consequence of this hypothesis is that when $m_p(S/O_p(G_i)) = 1$ then, for $i \in \{1, 2\}$, writing $L_i := O^{p'}(G_i)$ we either have that p is odd, $\overline{S}$ is cyclic and $\overline{L_i}/O_{p'}(\overline{L_i})$ is non-abelian finite simple group; or G_i is p-solvable and $\overline{L_i} = \overline{S}O_{p'}(\overline{L_i})$. This follows from a similar argument to that in Proposition 3.28.

From this point, our methodology is completely based in group theory. At various stages of the analysis, we refer to $\mathcal{A}$ or G as being a minimal counterexample to Theorem C. By this, we mean a counterexample in each case chosen such that $|G_1|+|G_2|$ is as small as possible.

5.1 The Amalgam Method

We assume Hypothesis 5.1 and fix the following notation for the remainder of this work.

In this section, we provide a slew of results which are generic in the amalgam method several of which do not depend on the strongly p-embedded condition in

© The Author(s), under exclusive license to Springer Nature Switzerland AG 2024

M. van Beek, *Rank 2 Amalgams and Fusion Systems*, Lecture Notes in Mathematics 2343, https://doi.org/10.1007/978-3-031-54461-3_5

Hypothesis 5.1. We will, however, assume Hypothesis 5.1 and leave the exercise of determining what hypothesis are needed for which results to the reader.

We let $G = G_1 *_{G_{12}} G_2$ and Γ be the (right) coset graph of G with respect to G_1 and G_2, with vertex set $V(\Gamma) = \{G_i g : g \in G, i \in \{1, 2\}\}$ and $(G_i g, G_j h)$ an edge if $G_i g \neq G_j h$ and $G_i g \cap G_j h \neq \emptyset$ for $\{i, j\} = \{1, 2\}$. It is clear that G operates on Γ by right multiplication. Throughout, we identify Γ with its set of vertices, let $d(\cdot, \cdot)$ be the usual distance on Γ and observe the following notations.

Notation 5.2 Set $\delta \in \Gamma$ an arbitrary vertex.

- $\Delta^{(n)}(\delta) = \{\lambda \in \Gamma : 0 < d(\delta, \lambda) \leqslant n\}$. In particular, we write $\Delta(\delta) := \Delta^{(1)}(\delta)$.
- For $\lambda \in \Delta(\delta)$, we let G_δ be the stabilizer in G of δ and $G_{\delta,\lambda}$ be the stabilizer in G of the edge $\{\delta, \lambda\}$.
- $G_\delta^{(n)}$ is the largest normal subgroup of G_δ which fixes $\Delta^{(n)}(\delta)$ element-wise. By convention, $G_\delta = G_\delta^{(0)}$.

The following proposition is elementary and its proof may be found in [27, Chapter 3].

Proposition 5.3 *The following facts hold:*

(i) $G_{G_i g} = G_i^g$ *so that every vertex stabilizer is conjugate in G to either G_1 or G_2. In particular, G has finite vertex stabilizers.*
(ii) *Each edge stabilizer of Γ is conjugate in G to G_{12} in its action on Γ.*
(iii) *Γ is a tree.*
(iv) *G acts faithfully and edge transitively on Γ, but does not act vertex transitively.*
(v) *For each edge $\{\lambda_1, \lambda_2\}$, $G = \langle G_{\lambda_1}, G_{\lambda_2}\rangle$.*
(vi) *For $\delta \in \Gamma$ such that $G_\delta = G_i^g$, we have that $\Delta(\delta)$ and G_δ / G_{12}^g are equivalent as G_δ-sets. In particular, G_δ is transitive on $\Delta(\delta) \setminus \{\delta\}$.*
(vii) *G_δ is of characteristic p for all $\delta \in \Gamma$.*
(viii) *If δ and λ are adjacent vertices, then* $\mathrm{Syl}_p(G_{\delta,\lambda}) \subseteq \mathrm{Syl}_p(G_\delta) \cap \mathrm{Syl}_p(G_\lambda)$.
(ix) *If δ and λ are adjacent vertices, then for $S \in \mathrm{Syl}_p(G_{\delta,\lambda})$, $N_{G_\delta}(S) = N_{G_\lambda}(S) \leq G_{\delta,\lambda}$.*

The above proposition, along with Proposition 5.5 and Lemma 5.6, holds in far greater generality than the situation enforced by Hypothesis 5.1.

The following notations will be used extensively throughout the rest of this work.

Notation 5.4 Set $\delta \in \Gamma$ to be an arbitrary vertex and $S \in \mathrm{Syl}_p(G_\delta)$.

- $L_\delta := O^{p'}(G_\delta)$.
- $Q_\delta := O_p(G_\delta) = O_p(L_\delta)$.
- $\overline{G_\delta} := L_\delta / Q_\delta$.
- $Z_\delta := \langle \Omega(Z(S))^{G_\delta} \rangle$.
- For $n \in \mathbb{N}$, $V_\delta^{(n)} := \langle Z_\lambda : d(\lambda, \delta) \leqslant n \rangle \trianglelefteq G_\delta$, with the additional conventions $V_\delta^{(0)} = Z_\delta$ and $V_\delta := V_\delta^{(1)}$.

- $b_\delta := \min_{\lambda \in \Gamma}\{d(\delta, \lambda) : Z_\delta \not\leq G_\lambda^{(1)}\}$.
- $b := \min_{\delta \in \Gamma}\{b_\delta\}$.

We refer to b as the *critical distance* of the amalgam. Indeed, as G acts edge transitively on Γ it follows that $b = \min\{b_\delta, b_\lambda\}$ where δ and λ are any adjacent vertices in Γ. A *critical pair* is any pair (δ, λ) such that $Z_\delta \not\leq G_\lambda^{(1)}$ and $d(\delta, \lambda) = b$. This definition is not symmetric and so (λ, δ) is not necessarily a critical pair in this case.

It is clear from the definition that symplectic amalgams have critical distance 2. It is remarkable that in all the examples we uncover, $b \leqslant 5$ and if G does not have a weak BN-pair, then $b \leqslant 2$.

Proposition 5.5 *The following facts hold:*

(i) *$b \geqslant 1$ is finite.*
(ii) *We may choose $\{\alpha, \beta\}$ such that $\{G_\alpha, G_\beta\} = \{G_1, G_2\}$ and $G_{\alpha,\beta} = G_{12} \geq N_{G_i}(S)$ for $i \in \{1, 2\}$.*
(iii) *If $N \leq G_{\alpha,\beta}$, $N_{G_\alpha}(N)$ operates transitively on $\Delta(\alpha)$ and $N_{G_\beta}(N)$ operates transitively on $\Delta(\beta)$ then $N = \{1\}$.*
(iv) *For $\delta \in \Gamma$, $\lambda \in \Delta(\delta)$ and $T \in \mathrm{Syl}_p(G_{\delta,\lambda})$, no subgroup of T is normal in $\langle L_\delta, L_\lambda\rangle$.*
(v) *For $\delta \in \Gamma$ and $\lambda \in \Delta(\delta)$, there does not exist a non-trivial element $g \in G_{\delta,\lambda}$ with $gQ_\delta/Q_\delta \in Z(L_\delta/Q_\delta)$ and $gQ_\lambda/Q_\lambda \in Z(L_\lambda/Q_\lambda)$.*
(vi) *For $\delta \in \Gamma$ and $\lambda \in \Delta(\delta)$, $V_\lambda^{(i)} = \langle (V_\delta^{(i-1)})^{G_\lambda}\rangle$.*

For the remainder of this work, we will often fix a critical pair (α, α'). As Γ is a tree, we may set β to be the unique neighbor of α with $d(\beta, \alpha') = b - 1$. Then we label each vertex along the path from α to α' additively e.g. $\beta = \alpha + 1$, $\alpha' = \alpha + b$. In this way we also see that β may be written as $\alpha' - b + 1$ and so we will often write vertices on the path from α' to α subtractively with respect to α'. The following diagram better explains the situation (the reader may want to reverse the diagram depending on their handedness).

$$\cdots \ \alpha'+1 \ — \ \alpha' \ — \ \alpha'-1 \ — \ \alpha'-2 \ — \cdots — \ \alpha+2 \ — \ \beta \ — \ \alpha \ — \ \alpha-1$$

Lemma 5.6 *Let $\delta \in \Gamma$, (α, α') be a critical pair, $T \in \mathrm{Syl}_p(G_\alpha)$ and $S \in \mathrm{Syl}_p(G_{\alpha,\beta})$. Then*

(i) *$Q_\delta \leq G_\delta^{(1)}$;*
(ii) *$Z_{\alpha'} \leq G_\alpha$, $Z_\alpha \leq G_{\alpha'}$ and $[Z_\alpha, Z_{\alpha'}] \leq Z_\alpha \cap Z_{\alpha'}$;*
(iii) *$Z_\alpha \neq \Omega(Z(T))$; and*
(iv) *if $\Omega(Z(S))$ is centralized by $L \leq G_\beta$ such that L acts transitively on $\Delta(\beta)$, then $Z(L_\alpha) = \{1\}$.*

Proof For all $\lambda \in \Delta(\delta)$, we have that $Q_\delta \leq T_\lambda \in \mathrm{Syl}_p(G_\lambda \cap G_\delta)$ and $Q_\delta \leq G_\lambda$. Since $Q_\delta \trianglelefteq G_\delta$, it follows immediately that $Q_\delta \leq G_\delta^{(1)}$ and (i) holds. By the minimality of b, we have that $Z_{\alpha'} \leq G_\beta^{(1)} \leq G_\alpha$ and similarly $Z_\alpha \leq G_{\alpha'-1}^{(1)} \leq G_{\alpha'}$. In particular, Z_α normalizes $Z_{\alpha'}$ and vice versa, so that $[Z_\alpha, Z_{\alpha'}] \leq Z_\alpha \cap Z_{\alpha'}$ and the proof of (ii) is complete.

Aiming for a contradiction, suppose that $Z_\alpha = \Omega(Z(T))$. Then $Z_\alpha = \Omega(Z(S))$ by the transitivity of G_α. By definition and minimality of b, $Z_\alpha \leq Z_\beta \leq G_{\alpha'}^{(1)}$, a contradiction since (α, α') is a critical pair. Thus, (iii) holds.

Finally, suppose that $\Omega(Z(S))$ is centralized by $L \leq G_\beta$ such that L acts transitively on $\Delta(\beta)$. Since Q_α is self-centralizing, it follows that $Z(L_\alpha)$ is a p-group and so $\Omega(Z(L_\alpha)) \leq \Omega(Z(S))$ and L centralizes $\Omega(Z(L_\alpha))$. Then Proposition 5.5 (iii) implies that $\Omega(Z(L_\alpha)) = \{1\}$, and so $Z(L_\alpha) = \{1\}$. This is (iv), and the proof is complete. □

Proposition 5.7 *Suppose that $b > 2n$. Then $V_\delta^{(n)}$ is abelian for all $\delta \in \Gamma$.*

Proof Since $b > 2n$, for all $\lambda, \mu \in \Delta^{(n)}(\delta)$ we have that $Z_\lambda \leq G_\mu^{(1)}$ by the minimality of b. Thus, $Z_\lambda \leq Q_\mu$, Z_λ centralizes Z_μ and since $V_\delta^{(n)} = \langle Z_\mu : \mu \in \Delta^{(n)}(\delta) \rangle$, it follows that $V_\delta^{(n)}$ is abelian. □

Lemma 5.8 *$V_\lambda^{(n)}/[V_\lambda^{(n)}, Q_\lambda]$ contains a non-central chief factor for L_λ for all $n \geqslant 1$ such that $V_\lambda^{(n)} \leq Q_\lambda$.*

Proof Set $V_\mu^{(0)} = Z_\mu$ for all $\mu \in \Gamma$ and suppose that $O^p(L_\lambda)$ centralizes $V_\lambda^{(n)}/[V_\lambda^{(n)}, Q_\lambda]$. Observe that $V_\lambda^{(n)} = \langle (V_\mu^{(n-1)})^{L_\lambda} \rangle$ for $\mu \in \Delta(\lambda)$ so that $V_\mu^{(n-1)} \not\leq [V_\lambda^{(n)}, Q_\lambda] < V_\lambda^{(n)}$. Moreover, $V_\mu^{(n-1)}[V_\lambda^{(n)}, Q_\lambda] \trianglelefteq L_\lambda$ so that $V_\lambda^{(n)} = V_\mu^{(n-1)}[V_\lambda^{(n)}, Q_\lambda]$. Set $V_i := [V_\lambda^{(n)}, Q_\lambda; i]$. In particular, $V_0 = V_\lambda^{(n)}$ and $V_1 = [V_0, Q_\lambda] = [V_\mu^{(n-1)}, Q_\lambda]V_2$. Notice that $V_\lambda^{(n)} \neq V_\mu^{(n-1)}$ and let k be maximal such that $V_\lambda^{(n)} = V_\mu^{(n-1)}V_k$. Then $V_1 = [V_\mu^{(n-1)}, Q_\mu]V_{k+1} \leq V_\mu^{(n-1)}V_{k+1}$. But $V_\lambda^{(n)} = V_\mu^{(n-1)}V_1 = V_\mu^{(n-1)}V_{k+1}$, contradicting the maximal choice of k. Thus, $O^p(L_\lambda)$ does not centralize $V_\lambda^{(n)}/[V_\lambda^{(n)}, Q_\lambda]$, as required. □

We will use the following lemma often in the amalgam method and without reference. Recall also that if $U, V \trianglelefteq G$ with $V < U$ then, in our setup and using coprime action, U/V does not contain a non-central chief factor for G if and only if $O^p(G)$ centralizes U/V.

Lemma 5.9 *For any $\lambda \in \Gamma$, $V_\lambda^{(n)}/V_\lambda^{(n-2)}$ contains a non-central chief factor for L_λ for all $n \geqslant 2$ such that $V_\lambda^{(n)} \leq Q_\lambda$.*

Proof With the intention of forcing a contradiction, assume that $V_\lambda^{(n)}/V_\lambda^{(n-2)}$ contains only central chief factors for L_λ so that $O^p(L_\lambda)$ centralizes $V_\lambda^{(n)}/V_\lambda^{(n-2)}$. Since $V_\lambda^{(n-2)} < V_\mu^{(n-1)} < V_\lambda^{(n)}$ for all $\mu \in \Delta(\lambda)$, we have that $V_\mu^{(n-1)} \trianglelefteq O^p(L_\lambda)G_{\lambda,\mu} = G_\lambda$ by a Frattini argument. But then $V_\mu^{(n-1)} \trianglelefteq \langle G_\mu, G_\lambda \rangle$, a contradiction. Thus, $V_\lambda^{(n)}/V_\lambda^{(n-2)}$ contains a non-central chief factor, as required. □

As described at the end of Chap. 2, we can guarantee cubic action on a faithful module for $\overline{L_\delta}$ for δ at least one of α, β. We use critical subgroups to achieve this and refer to Theorem 2.59 for their properties. We describe this in the following proposition.

Proposition 5.10 *There is $\lambda \in \Gamma$ such that there is a $\overline{G_\lambda}$-module V on which p'-elements of $\overline{G_\lambda}$ act faithfully and a non-trivial p-subgroup C of $\overline{G_\lambda}$ such that $[V, C, C, C] = \{1\}$.*

Proof Let $(\alpha, \dots, \alpha')$ be a path in Γ with (α, α') a critical pair. For each $\lambda \in (\alpha, \dots, \alpha')$, set K_λ to be a critical subgroup of Q_λ. Since $Z_\alpha \leq K_\alpha$, we must have that $K_\alpha \not\leq Q_{\alpha'+1}$. Set $c := \{\min(d(\mu, \lambda)) : K_\mu \not\leq Q_\lambda \text{ for } \mu, \lambda \in (\alpha, \dots, \alpha' + 1)\}$. Choose a pair (μ, λ) such that $K_\mu \not\leq Q_\lambda$ and $d(\mu, \lambda) = c$. Then, by minimality of c, $K_\mu \leq G_\lambda$ but $K_\mu \not\leq Q_\lambda$ and from the definition of a critical subgroup, p'-elements of $\overline{G_\lambda}$ act faithfully on the $\overline{G_\lambda}$-module $K_\lambda/\Phi(K_\lambda)$. Moreover, again by minimality, K_λ normalizes K_μ so that $[K_\lambda, K_\mu, K_\mu, K_\mu] \leq [K_\mu, K_\mu, K_\mu] = \{1\}$, as required. □

Remark It seems likely that the above observation could also be proved using the results on Glauberman's K-infinity groups [30, Theorem A]. Furthermore, following the above proof, we see that either K_μ acts quadratically on K_λ; or $K_\lambda \not\leq Q_\mu$ and K_λ acts cubically on K_μ.

We will also makes use of the qrc-lemma, although where it is applied there are certainly more elementary arguments which would suffice. In this way, we do not use the lemma in its full capacity and instead, it serves as a way to reduce the length of some of our arguments. This lemma first appeared in [69] but only for the prime 2. We use the extension to all primes presented in [72, Theorem 3].

Theorem 5.11 (qrc Lemma) *Let (H, M) be an amalgam such that both H, M are of characteristic p and contain a common Sylow p-subgroup. Set $Q_X := O_p(X)$ for $X \in \{H, M\}$, $Z = \langle \Omega(Z(S))^H \rangle$ and $V := \langle Z^M \rangle$. Suppose that M is p-minimal and $Q_H = C_S(Z)$. Then one of the following occurs:*

(i) *$Z \not\leq Q_M$;*
(ii) *Z is an FF-module for $H/C_H(Z)$;*
(iii) *the dual of Z is an FF-module for $H/C_H(Z)$;*
(iv) *Z is a 2F-module with quadratic offender and V contains more than one non-central chief factor for M; or*
(v) *M has exactly one non-central chief factor in V, $Q_H \cap Q_M \trianglelefteq M$, $[V, O^p(M)] \leq Z(Q_M)$ and contains some non-trivial p-reduced module.*

We note now that case (v) of the qrc-lemma will be ruled out in our later analysis (Proposition 5.27) and in cases (ii) and (iii), Lemma 2.41 implies that $H/C_H(Z) \cong \mathrm{SL}_2(q)$, for q some power of p.

5.2 Strongly p-Embedded Subgroups in the Amalgam Method

We now reintroduce the strongly p-embedded conditions present in Hypothesis 5.1. This will allows us the necessary control to properly utilize the amalgam method.

By Proposition 5.10, we are in a good position to apply Hall–Higman style arguments whenever $p \geqslant 5$. We get the following fact almost immediately from Corollary 2.60.

Proposition 5.12 *Suppose that $p \geqslant 5$, and $\overline{L_\alpha}$ and $\overline{L_\beta}$ have strongly p-embedded subgroups. Then, for some $\lambda \in \{\alpha, \beta\}$, one of the following holds:*

(i) *$p \geqslant 5$ is arbitrary and $\overline{L_\lambda} \cong \mathrm{PSL}_2(p^n), \mathrm{SL}_2(p^n), \mathrm{PSU}_3(p^n)$ or $\mathrm{SU}_3(p^n)$ for $n \in \mathbb{N}$; or*

(ii) *$p = 5$ and $\overline{L_\lambda} \cong 3 \cdot \mathrm{Alt}(6)$ or $3 \cdot \mathrm{Alt}(7)$.*

Proof By Proposition 5.10, there is a p-element $x \in \overline{L_\lambda}$ which acts cubically on $K_\lambda/\Phi(K_\lambda)$. Suppose there is $y \in L_\lambda$ such that $[y, K_\lambda] \leq \Phi(K_\lambda)$. Since K_λ is a critical subgroup, by coprime action, y is a p-element so that $C_{L_\lambda}(K_\lambda/\Phi(K_\lambda))$ is a normal p-subgroup. In particular, $\overline{L_\lambda}$ acts faithfully on $K_\lambda/\Phi(K_\lambda)$ and so we may apply Corollary 2.60 and the result holds. □

In the following proposition, we essentially rid ourselves of the possibility that one of G_α or G_β involves $\mathrm{Sz}(32) : 5$. This group is problematic in much the same way as the Ree(3) and p-solvable cases we exclude in the hypothesis of the Main Theorem in that there may arise situations in which the centralizer of a non-central chief factor is not p-closed. The following proposition asserts that this is not the case. We also believe that this result could also be extended to get rid of the p-solvable case whenever $p \geqslant 5$. The Ree(3) case when $p = 3$ would rely on some control of transfer results at the prime 3. The author is unaware of any usable results in the literature for this.

Lemma 5.13 *For any $\lambda \in \Gamma$ and $S \in \mathrm{Syl}_5(G_\lambda)$, if $\overline{L_\lambda}/O_{5'}(\overline{L_\lambda}) \cong \mathrm{Sz}(32) : 5$, then $O^5(L_\lambda) \leq \langle x^{G_\lambda} \rangle$ for any $x \in S \setminus Q_\lambda$.*

Proof Assume that G is a minimal counterexample to Theorem C with L_λ as described in the setup. Set $G_1 = G_\lambda$, $G_2 = G_\mu$ for some $\mu \in \Delta(\lambda)$, $Q_i = O_5(G_i)$ for $i \in \{1, 2\}$ and $S \in \mathrm{Syl}_5(G_{\lambda,\mu})$. We may fix $j \in \{1, 2\}$ such that $K^\infty(S) \ntrianglelefteq G_j$, where $K^\infty(S)$ is defined in [30]. Writing $L_j = O^{5'}(G_j)$, we have that $K^\infty(S) \ntrianglelefteq L_j$. Then, by [30, Theorem 4], there is $x \in S \setminus Q_i$ and a chief factor X/Y of L_j contained in Q_i such that $[X, x, x] \leq Y$ but $[X, x] \nleq Y$. We may as well take x with xQ_j/Q_j of order 5.

Suppose first that L_j is 5-solvable. By the Hall–Higman theorem, $O_p(L_j/C_{L_j}(X/Y)) \neq \{1\}$ and so by the Frattini argument, we have that $L_j = C_{L_j}(X/Y)N_{L_j}(\langle x \rangle Q_j)$. But then, since $N_{L_j}(S/Q_j)$ is strongly 5-embedded in L_j/Q_j, we see that $L_j = C_{L_j}(X/Y)N_{L_j}(S)$ and as $L_j = O^{5'}(L_j)$, we conclude that $L_j = C_{L_j}(X/Y)S$. But then, $[X, S]Y \trianglelefteq L_j$ and since X/Y is a chief factor,

$[X, S] \le Y$. But then $[X, L_j] \le Y$, a contradiction since $[X, x] \not\le Y$. Hence, L_j is not 5-solvable.

Set $H_j := \langle x^{G_j} \rangle Q_j$ so that by Propositions 2.14 and 2.15, either $H_j = L_j$ or for $\overline{G_j} := G_j/Q_j$, $\overline{L_j}/O_{p'}(\overline{L_j}) \cong \mathrm{Sz}(32) : 5$ in which case, perhaps $\overline{H_j}/O_{5'}(\overline{H_j}) \cong \mathrm{Sz}(32)$. Note that in all cases, $Q_j = O_5(H_j)$. Let $Y \le V \le U \le X$ with U/V a non-central chief factor for H_j in Q_j. We observe by Propositions 2.14 and 2.15 that $\langle x^{H_j} \rangle Q_j = O^{5'}(H_j) \trianglelefteq G_j$. Since U/V is a non-central $H-j$-chief factor, we have that $[U, x] \not\le V$ but $[U, x, x] \le V$. Now, applying Lemma 2.35, we deduce that $H_j/C_{H_j}(U/V) \cong \mathrm{SL}_2(5^n)$ or $\mathrm{SU}_3(5^n)$. Then Propositions 2.14 and 2.15 reveal that $H_j = L_j$.

Applying Proposition 5.12 we have that $\overline{L_j} \cong \mathrm{SL}_2(5^n)$ or $\mathrm{SU}_3(5^n)$ so that $G_j = G_\mu$. Then $G_{\lambda,\mu} = N_{G_\mu}(S) = N_{G_\mu}(K^\infty(S))$. Applying [30, Theorem B], $G_\mu/O^5(G_\mu) \cong G_{\lambda,\mu}/O^5(G_{\lambda,\mu})$ from which it follows that $Q_\mu/Q_\mu \cap O^5(G_\mu) \cong Q_\mu/Q_\mu \cap O^5(G_{\lambda,\mu})$ so that $Q_\mu \cap O^5(G_\mu) = Q_\mu \cap O^5(G_{\lambda,\mu})$. But now, since $\mathrm{Sz}(32) : 5$ has no outer automorphisms, we deduce that $\overline{G_\lambda}/O_{5'}(\overline{G_\lambda}) \cong \mathrm{Sz}(32) : 5$ so that $(O^5(G_{\lambda,\mu}) \cap S)Q_\lambda < S$. In particular, writing $S^* := (Q_\mu \cap O^5(G_\mu))Q_\lambda$, $L_\delta^* := \langle (S^*)^{G_\delta} \rangle\rangle$ and $G_\delta^* := L_\delta^* K$ for $\delta \in \{\lambda, \mu\}$ and K a Hall $5'$-subgroup of $G_{\lambda,\mu}$, we have that $L_\delta^* = O^{5'}(G_\delta^*)$, $G_\delta^* \trianglelefteq G_\delta$ and $|S/S^*| = |L_\delta/L_\delta^*| = |G_\delta/G_\delta^*| = 5$ for $\delta \in \{\lambda . \mu\}$. Hence, any subgroup of $G_\lambda^* \cap G_\mu^*$ which is normal in $\langle G_\lambda^*, G_\mu^* \rangle$ is also normalized by G, so is trivial. Thus, the triple $(G_\lambda^*, G_\mu^*, G_\lambda^* \cap G_\mu^*)$ satisfies Hypothesis 5.1, and by minimality, and since $\overline{L_\lambda^*}/O_{5'}(\overline{L_\lambda^*}) \cong \mathrm{Sz}(32)$, we have a contradiction. Hence, such a G does not exist and the lemma holds. □

Lemma 5.14 *Suppose that $N \trianglelefteq G_\delta$ with N not p-closed and set $S \in \mathrm{Syl}_p(G_\delta)$. Then the following hold:*

(i) *If L_δ is not p-solvable, then $O^p(L_\delta) \le N$.*
(ii) *If L_δ is p-solvable, then $K \le NQ_\delta$, where $\overline{K}$ is the unique normal subgroup of $\overline{L_\delta}$ which is divisible by p and minimal with respect to this constraint.*
(iii) *$G_\delta = NN_{G_\delta}(S)$ and N is transitive on $\Delta(\delta)$.*
(iv) *For U/V any non-central chief factor for L_δ inside of Q_δ, we have that $Q_\delta \in \mathrm{Syl}_p(C_{L_\delta}(U/V))$.*

Proof Suppose L_δ is not p-solvable and let $A \in \mathrm{Syl}_p(N)$. Notice that as N is not p-closed, $A \not\le Q_\delta$ and since $\overline{L_\delta}$ has a strongly p-embedded subgroup, by Hypothesis 5.1 we have that $\widetilde{L_\delta} := \overline{L_\delta}/O_{p'}(\overline{L_\delta})$ is isomorphic to a non-abelian simple group; $\mathrm{Sz}(32) : 5$ or $\mathrm{Ree}(3)$. If $\widetilde{L_\delta}$ is a non-abelian simple group then $\widetilde{L_\delta} = \langle \widetilde{A}^{\widetilde{L_\delta}} \rangle$. In particular, $\overline{S} \le \overline{\langle A^{L_\delta} \rangle}$ and so $S \le \langle A^{L_\delta} \rangle Q_\delta \le L_\delta$ and since $L_\delta = O^{p'}(L_\delta)$, $L_\delta = \langle A^{L_\delta} \rangle Q_\delta$. It then follows that $O^p(L_\delta) \le \langle A^{L_\delta} \rangle \le N$. If $\widetilde{L_\delta} \cong \mathrm{Ree}(3)$, then as $N_{\overline{L_\delta}}(\overline{S})$ is strongly 3-embedded in $\overline{L_\delta}$ and $m_3(\overline{S}) = 2$, by coprime action we have that $O_{3'}(\overline{L_\delta})$ normalizes $\overline{S}$. Hence, $\{1\} = [O_{3'}(\overline{L_\delta}), \overline{S}] = [O_{3'}(\overline{L_\delta}), \overline{L_\delta}]$. Then $O^3(\overline{L_\delta})$ is isomorphic to a central extension of $\mathrm{PSL}_2(8)$ by a $3'$-group so that $O^{3'}(O^3(\overline{L_\delta})) \cong \mathrm{PSL}_2(8)$. But now, since $O_{3'}(\overline{L_\delta})$ normalizes $\overline{S}$, $O^{3'}(O^3(\overline{L_\delta}))\overline{S} \trianglelefteq \overline{L_\delta}$ and $\overline{L_\delta} \cong \mathrm{Ree}(3)$ and so (i) holds in this case. By Lemma 5.13, (i) holds when $\widetilde{L_\delta} \cong \mathrm{Sz}(32) : 5$. Thus, (i) holds in all cases.

By the Frattini argument $G_\delta = L_\delta N_{G_\delta}(S) = O^p(L_\delta)N_{G_\delta}(S) = \langle A^{G_\delta}\rangle N_{G_\delta}(S)$. Since $\langle A^{G_\delta}\rangle \le N$, (iii) follows whenever L_δ is not p-solvable.

Suppose now that L_δ is p-solvable and let $\overline{K}$ be the unique minimal normal subgroup of $\overline{L_\delta}$ divisible by p. Again, we let $A \in \mathrm{Syl}_p(N)$ and remark that since N is not p-closed $A \not\le Q_\delta$. Hence, $p \mid |\overline{N}|$ so that $\overline{K} \le \overline{N}$ and $K \le NQ_\delta$, completing the proof of (ii). By Proposition 2.14, $\overline{L_\delta} = N_{\overline{L_\delta}}(\overline{S})\overline{K} \le \overline{N_{G_\delta}(S)N}$ so that $G_\delta = L_\delta N_{G_\delta}(S) \le NN_{G_\delta}(S) \le G_\delta$, completing the proof of (iii).

For (iv), choose any non-central chief factor U/V for L_δ inside Q_δ. Then U/V is a faithful, irreducible module for $L_\delta/C_{L_\delta}(U/V)$. Since $[Q_\delta, U] \trianglelefteq L_\delta$ and $[Q_\delta, U] < U$, $Q_\delta \le C_{L_\delta}(U/V)$. Moreover, as $C_{L_\delta}(U/V)$ is normal in L_δ, we deduce that $O_p(C_{L_\delta}(U/V)) = Q_\delta$. If $C_{L_\delta}(U/V)$ is not p-closed, then $L_\delta = C_{L_\delta}(U/V)N_{L_\delta}(S)$ and it follows that U/V is irreducible for $N_{L_\delta}(S)$. But then $[U/V, S] = \{1\}$ from which it follows that $\{1\} = [U/V, \langle S^{L_\delta}\rangle] = [U/V, L_\delta]$, a contradiction. Hence, (iv). □

Of course, the following lemma is superfluous if we initially assumed that the essentials E_1, E_2 of $\mathcal{F}$ were maximally essential.

Proposition 5.15 *For all $\delta \in \Gamma$ and $\lambda \in \Delta(\delta)$, $Q_\delta \not\le Q_\lambda$.*

Proof Aiming for a contradiction, suppose that there is $\delta \in \Gamma$ and $\lambda \in \Delta(\delta)$ with $Q_\delta \le Q_\lambda$ and let $S \in \mathrm{Syl}_p(G_{\delta,\lambda})$. Then $J(Q_\lambda) \not\le Q_\delta$ for otherwise, by Proposition 2.38 (v), $J(Q_\lambda) = J(Q_\delta) \trianglelefteq \langle G_\lambda, G_\delta\rangle$. Furthermore, since $C_S(Q_\delta) \le Q_\delta$, $\Omega(Z(Q_\lambda)) < \Omega(Z(Q_\delta))$. Let $V := \langle \Omega(Z(Q_\lambda))^{G_\delta}\rangle \le \Omega(Z(Q_\delta))$ and choose $A \in \mathcal{A}(Q_\lambda) \setminus \mathcal{A}(Q_\delta)$. If $Q_\delta < C_S(V)$, then by Lemma 5.14 (iii), $G_\delta = \langle C_S(V)^{G_\delta}\rangle N_{G_\delta}(S) = C_{G_\delta}(V)N_{G_\delta}(S)$ normalizes $\Omega(Z(Q_\lambda))$, a contradiction. Hence, $Q_\delta = C_S(V)$.

By the choice of A, $|A| \geqslant |C_A(V)V| = |C_A(V)||V|/|V \cap C_A(V)| = |C_A(V)||V|/|V \cap A|$. Since $A = \Omega(C_S(A))$, we have that $A \cap V = C_V(A)$ and rearranging we conclude that $|A|/|C_A(V)| \geqslant |V|/|C_V(A)|$ and $A/C_A(V) \cong AQ_\delta/Q_\delta$ is an offender on the FF-module V. By Lemma 2.41, $L_\delta/C_{L_\delta}(V) \cong \mathrm{SL}_2(p^n)$ and $V/C_V(L_\delta)$ is a natural $\mathrm{SL}_2(q)$-module. But $Q_\lambda/Q_\delta < S/Q_\delta$ is a $G_{\lambda,\delta}$-invariant subgroup of S/Q_δ, a contradiction by Proposition 2.16 (vi). □

Lemma 5.16 *Let $\delta \in \Gamma$, (α, α') be a critical pair and $S \in \mathrm{Syl}_p(G_{\alpha,\beta})$. Then*

(i) *$Q_\delta \in \mathrm{Syl}_p(G_\delta^{(1)})$ and $G_\delta^{(1)}/Q_\delta$ is centralized by L_δ/Q_δ;*
(ii) *either $Q_\delta \in \mathrm{Syl}_p(C_{L_\delta}(Z_\delta))$ or $Z_\delta = \Omega(Z(L_\delta))$;*
(iii) *$Z_\alpha \not\le Q_{\alpha'}$; and*
(iv) *$C_S(Z_\alpha) = Q_\alpha$, and $C_{G_\alpha}(Z_\alpha)$ is p-closed and p-solvable.*

Proof By Lemma 5.6 (i), aiming for a contradiction, we may assume that $Q_\delta < T$ for $T \in \mathrm{Syl}_p(G_\delta^{(1)})$. Since $G_\delta^{(1)} \trianglelefteq G_\delta$ it follows that $O_p(G_\delta^{(1)}) = Q_\delta$ and so $G_\delta^{(1)}$ is not p-closed. But by Lemma 5.14 (iii), then $G_\delta^{(1)}$ would be transitive on $\Delta(\delta)$, a clear contradiction. Thus, $Q_\delta \in \mathrm{Syl}_p(G_\delta^{(1)})$. Letting $P \in \mathrm{Syl}_p(G_\delta)$, $[P, G_\delta^{(1)}] \le P \cap G_\delta^{(1)} = Q_\delta$ so that $[L_\delta, G_\delta^{(1)}] \le Q_\delta$, and so (i) holds.

If $Q_\delta \notin \mathrm{Syl}_p(C_{L_\delta}(Z_\delta))$ then by Lemma 5.14 (iii), $G_\delta = C_{L_\delta}(Z_\delta)N_{G_\delta}(S)$ and so $Z_\delta = \langle \Omega(Z(S))^{G_\delta}\rangle = \Omega(Z(S))$. But then $\{1\} = [Z_\delta, S]^{G_\delta} = [Z_\delta, L_\delta]$ and so $Z_\delta \le Z(L_\delta)$. Since Q_δ is self-centralizing, $Z(L_\delta)$ is a p-group and $Z_\delta = \Omega(Z(S)) = \Omega(Z(L_\delta))$, so that (ii) holds.

If $Z_\alpha \le Q_{\alpha'}$ then $Z_\alpha \le G_{\alpha'}^{(1)}$, against the definition of a critical pair, and so (iii) holds. Since $Z_\alpha \ne \Omega(Z(S))$ by Lemma 5.6 (iii), $C_S(Z_\alpha) = Q_\alpha \trianglelefteq C_{G_\alpha}(Z_\alpha)$ so that $C_{G_\alpha}(Z_\alpha)$ is p-closed and p-solvable. □

By the above lemma, we can reinterpret the minimal distance b as $b = \min_{\delta\in\Gamma}\{b_\delta\}$ where $b_\delta := \min_{\lambda\in\Gamma}\{d(\delta, \lambda) : Z_\delta \not\le Q_\lambda\}$.

Lemma 5.17 *Let (α, α') be a critical pair. Then*

(i) *if $Z_{\alpha'} \le Z(L_{\alpha'})$ then α is not conjugate to α'; and*
(ii) *$C_{Z_\alpha}(Z_{\alpha'}) \ne Z_\alpha \cap Q_{\alpha'}$ if and only if $Z_{\alpha'} = \Omega(Z(L_{\alpha'}))$ and (α', α) is not a critical pair.*

Proof Suppose $Z_{\alpha'} \le Z(L_{\alpha'})$. By Lemma 5.16 (ii), $Z_{\alpha'} = \Omega(Z(L_{\alpha'}))$. If α and α' were conjugate, then $Z_\alpha = \Omega(Z(L_\alpha))$, a contradiction to Lemma 5.6 (iii). Thus (i) holds.

Suppose that $Z_{\alpha'} = \Omega(Z(L_{\alpha'}))$. Since $Z_\alpha \not\le Q_{\alpha'}$ but $Z_\alpha \le L_{\alpha'}$, we infer that $Z_\alpha = C_{Z_\alpha}(Z_{\alpha'}) \ne Z_\alpha \cap Q_{\alpha'}$. Suppose conversely that $C_{Z_\alpha}(Z_{\alpha'}) \ne Z_\alpha \cap Q_{\alpha'}$. Then $C_{L_{\alpha'}}(Z_{\alpha'})$ is not p-closed and by Lemma 5.16 (ii), we have that $Z_{\alpha'} = \Omega(Z(L_{\alpha'}))$. □

We now introduce some notation which is non-standard in the amalgam method and is tailored for our purposes.

Notation 5.18

- If $Z_\delta \ne \Omega(Z(L_\delta))$, then $R_\delta = C_{L_\delta}(Z_\delta)$.
- If $Z_\delta = \Omega(Z(L_\delta))$ and $b > 1$, then $R_\delta = C_{L_\delta}(V_\delta / C_{V_\delta}(O^p(L_\delta)))$.
- If $Z_\delta = \Omega(Z(L_\delta))$ and $b > 1$, then $C_\delta = C_{Q_\delta}(V_\delta)$.

Lemma 5.19 *Suppose that $Z_\delta = \Omega(Z(L_\delta))$, $b > 1$ and let $T \in \mathrm{Syl}_p(G_\delta)$. Then $R_\delta \cap T \le Q_\delta$ and $C_T(V_\delta) = C_\delta$.*

Proof Suppose for a contradiction that $R_\delta \cap T \not\le Q_\delta$. Then R_δ is not p-closed and by Lemma 5.14 (iii), $G_\delta = R_\delta N_{G_\delta}(T)$. Let $\mu \in \Delta(\delta)$ with $T \in \mathrm{Syl}_p(G_{\delta,\mu})$. Then $Z_\mu \le V_\delta$ so that $Z_\mu \trianglelefteq \langle G_\delta, G_\mu\rangle$, a contradiction.

Suppose now that $C_T(V_\delta) > Q_\delta$ so that $C_{G_\delta}(V_\delta)$ is not p-closed and is normal in G_δ. As above, by Lemma 5.14 (iii), we quickly get that $G_\delta = C_{G_\delta}(V_\delta)G_{\delta,\mu}$ normalizes Z_μ for $\mu \in \Delta(\delta)$ with $T \in \mathrm{Syl}_p(G_{\delta,\mu})$, a contradiction. Hence, the result. □

Lemma 5.20 *Suppose that there is $X \le L_\delta$ such that $\overline{X}$ is strongly p-embedded in $\overline{L_\delta}$ for $\delta \in \Gamma$. Then XR_δ/R_δ is strongly p-embedded in L_δ/R_δ.*

Proof By Lemmas 5.16 and 5.19, we have that $\overline{R_\delta}$ is a p'-group. Then, if XR_δ/R_δ is not strongly p-embedded in L_δ/R_δ, Lemma 2.12 (iv) implies that $L_\delta = XR_\delta$.

By Proposition 2.15, we clearly have that $m_p(\overline{S}) = 1$ where $S \in \mathrm{Syl}_p(L_\delta)$. Hence, $X = N_{L_\delta}(\Omega(\overline{S}))$ and so by Proposition 2.14 and our additional assumptions, we see that $L_\delta = SR_\delta$ is p-solvable. Thus, $Z_\delta = \Omega(Z(L_\delta))$ and it quickly follows that for any $\lambda \in \Delta(\delta)$, $Z_\lambda \trianglelefteq L_\delta$, a contradiction. □

Lemma 5.21 *Suppose that $L_\delta/R_\delta \cong \mathrm{SL}_2(p^n)$, $Q_\delta \in \mathrm{Syl}_p(R_\delta)$ and $R_\delta \leq G_{\delta,\lambda}$ for some $\lambda \in \Delta(\delta)$. Then $\overline{L_\delta} \cong \mathrm{SL}_2(p^n)$.*

Proof Since $R_\delta \leq G_{\delta,\lambda}$ and $R_\delta \trianglelefteq G_\delta$, $\overline{R_\delta} \leq Z(\overline{L_\delta})$ is a p'-group by Lemma 5.16. If $p^n > 3$, then as $L_\delta = O^{p'}(L_\delta)$, it follows from Proposition 2.16 (vii) that $\overline{L_\delta} \cong \mathrm{SL}_2(p^n)$.

If $L_\delta/R_\delta \cong \mathrm{SL}_2(2) \cong \mathrm{Sym}(3)$ and $R_\delta \neq Q_\delta$, then $\overline{R_\delta}$ is a non-trivial 3-group since $L_\delta = O^{2'}(L_\delta)$ and $O_r(\overline{R_\delta})$ is complemented in $\overline{L_\delta}$ for any prime $r \notin \{2, 3\}$ by Schur–Zassenhaus. But now, since $\overline{R_\delta}$ is maximal and central in $O_3(\overline{L_\delta})$, $O_3(\overline{L_\delta})$ is abelian. By coprime action, $O_3(\overline{L_\delta}) = [O_3(\overline{L_\delta}), \overline{S}] \times C_{O_3(\overline{L_\delta})}(\overline{S})$ and $\overline{R_\delta}$ is complemented in $\overline{L_\delta}$ by $[O_3(\overline{L_\delta}), \overline{S}]\overline{S} \cong \mathrm{Sym}(3)$. Since $L_\delta = O^{2'}(L_\delta)$, we have that $\overline{R_\delta} = \{1\}$, a contradiction. Hence, $\overline{L_\delta} \cong \mathrm{SL}_2(2)$.

If $L_\delta/R_\delta \cong \mathrm{SL}_2(3)$ with $R_\delta \neq Q_\delta$ then $\overline{R_\delta}$ is a non-trivial 2-group since $L_\delta = O^{3'}(L_\delta)$ and $O_r(\overline{R_\delta})$ is complemented in $\overline{L_\delta}$ for any prime $r \notin \{2, 3\}$. Let A be a maximal subgroup of $\overline{R_\delta}$. Then $|O_2(\overline{L_\delta})/A| = 16$. By Gaschütz' theorem [49, (3.3.2)], we may assume that $\overline{R_\delta}/A$ is not complemented in $O_2(\overline{L_\delta})/A$. We see that $O_2(\overline{L_\delta})/A$ is a non-abelian group of order 16 with center of order at most 4. Checking the Small Groups Library in MAGMA [14] for groups of order 48 with a quotient by a central involution isomorphic to $\mathrm{SL}_2(3)$ and a Sylow 2-subgroup satisfying the required properties, we have a contradiction. Thus, $R_\delta = Q_\delta$ and $\overline{L_\delta} \cong \mathrm{SL}_2(3)$. □

Lemma 5.22 *Suppose that $\delta \in \Gamma$, $Z_{\delta-1} = Z_{\delta+1}$, $Q_\delta \in \mathrm{Syl}_p(R_\delta)$ and $i \in \mathbb{N}$. If $Q_{\delta-1}Q_\delta \in \mathrm{Syl}_p(L_\delta)$, L_δ/R_δ is generated by any two distinct Sylow p-subgroups and $O^p(R_\delta)$ normalizes $V_{\delta-1}^{(i-1)}$, then $V_{\delta-1}^{(i-1)} = V_{\delta+1}^{(i-1)}$.*

Proof Since $Q_{\delta-1}Q_\delta \in \mathrm{Syl}_p(L_\delta)$, if $Q_{\delta-1}R_\delta \neq Q_{\delta+1}R_\delta$, then $Z_{\delta+1} = Z_{\delta-1} \trianglelefteq L_\delta = \langle R_\delta, Q_{\delta-1}, Q_{\delta+1}\rangle$, a contradiction. Thus, $Q_{\delta-1}R_\delta = Q_{\delta+1}R_\delta$. As $Q_{\delta-1}Q_\delta \in \mathrm{Syl}_p(Q_{\delta-1}R_\delta)$, there is $r \in R_\delta$ such that $Q_{\delta-1}^rQ_\delta = (Q_{\delta-1}Q_\delta)^r = (Q_{\delta+1}Q_\delta) = Q_{\delta+1}Q_\delta$. Since $Q_{\delta-1}Q_\delta$ is the unique Sylow p-subgroup of $G_{\delta-1,\delta}$, it follows that $G_{\delta,\delta-1}^r = G_{\delta,\delta+1} = N_{G_\delta}(Q_\delta Q_{\delta+1})$. Set $\theta = (\delta-1)\cdot r \in \Delta(\delta)$. Then by properties of Γ, $G_{\delta,\delta+1} = G_{\delta,\delta-1}^r = G_{\delta,\delta-1\cdot r} = G_{\delta,\theta}$ and so $(\delta-1)\cdot r = \delta+1$. Since r acts as a graph automorphism on Γ, r preserves i neighborhoods of vertices in the graph and it follows immediately that $V_{\delta-1\cdot r}^{(i-1)} = (V_{\delta-1}^{(i-1)})^r$ so that, as $V_{\delta-1}^{(i-1)}$ is normalized by $R_\delta = O^p(R_\delta)Q_\delta$, $V_{\delta+1}^{(i-1)} = V_{\delta-1}^{(i-1)}$, completing the proof. □

We record one further generic lemma concerning the action of R_γ for $\gamma \in \Gamma$.

Lemma 5.23 *Let $\gamma \in \Gamma$ and fix $\delta \in \Delta(\gamma)$. Assume that one of Z_γ or Z_δ is equal to $\Omega(Z(T))$ for $T \in \mathrm{Syl}_p(G_{\gamma,\delta})$. Then for $n < b$, $\langle V_\mu^{(n)} : Z_\mu = Z_\delta, \mu \in \Delta(\gamma)\rangle \trianglelefteq R_\gamma Q_\delta$.*

Proof Set $U^\gamma := \langle V_\mu^{(n)} : Z_\mu = Z_\delta, \mu \in \Delta(\gamma)\rangle$ and let $r \in R_\gamma Q_\delta$. Since r is a graph automorphism, for $\mu \in \Delta(\gamma)$ such that $Z_\mu = Z_\delta$, $(V_\mu^{(n)})^r = V_{\mu\cdot r}^{(n)}$. But now, $Z_{\mu\cdot r} = Z_\mu^r = Z_\delta^r = Z_\delta$ and so $(V_\mu^{(n)})^r \leq U^\gamma$. Thus, $U^\gamma \trianglelefteq R_\gamma Q_\delta$, as required. □

5.3 The "Pushing Up" Case

We now deal with the so called "pushing up" case of the amalgam method. The proof breaks up over a series of lemmas, culminating in Proposition 5.27. This is a generalization of [27, (6.5-6.6)] and as the result there underpinned the generic study of higher rank amalgams in characteristic p (see for instance [71]), we believe Proposition 5.27 will be applicable in future works on fusion systems and amalgams. The proofs in this section are technical and so the reader may be better served by skipping some of the details of the proofs on the first pass.

Throughout, let $\lambda \in \Gamma$, $\mu \in \Delta(\lambda)$ and $S \in \mathrm{Syl}_p(G_{\lambda,\mu})$.

Lemma 5.24 *Suppose that $Q_\lambda \cap Q_\mu \trianglelefteq G_\lambda$. Then, writing $L := \langle Q_\mu^{G_\lambda}\rangle$, we have that $Q_\mu \in \mathrm{Syl}_p(L)$, $O_p(L) = Q_\mu \cap Q_\lambda$, $Z_\lambda/Z(L_\lambda)$ is a natural $\mathrm{SL}_2(q)$-module for $L_\lambda/R_\lambda \cong \mathrm{SL}_2(q)$ and no non-trivial characteristic subgroup of Q_μ is normal in L.*

Proof Set $L := \langle Q_\mu^{G_\lambda}\rangle \trianglelefteq L_\lambda$ and let $V := Z_\lambda$ if $Z_\lambda \neq \Omega(Z(S))$, and $V := V_\lambda/C_{V_\lambda}(O^p(L_\lambda))$ if $Z_\lambda = \Omega(Z(S))$ and $b > 1$. Since $L \trianglelefteq L_\lambda$, we have that $C_L(O_p(L)) \leq O_p(L)$ and since $Q_\mu \not\leq Q_\lambda$, it follows by Lemma 5.14 that $L/O_p(L)$ has a strongly p-embedded subgroup and $L_\lambda = LS$. Note that if $J(Q_\mu) \leq O_p(L)$, then $J(Q_\mu) \leq Q_\mu \cap Q_\lambda \leq Q_\mu$ and so, by Proposition 2.38 (v), $J(Q_\mu) = J(Q_\mu \cap Q_\lambda) \trianglelefteq L_\lambda$, a contradiction. Hence, $J(Q_\mu) \not\leq O_p(L)$.

Suppose first that $b = 1$ and $Z_\lambda = \Omega(Z(S))$. Then $Z_\lambda \leq Q_\mu$ and we may as well assume that $Z_\mu \not\leq Q_\lambda$. But then Z_μ centralizes $Q_\lambda/Q_\lambda \cap Q_\mu$ and $Q_\lambda \cap Q_\mu$. Since $\langle Z_\mu^{G_\lambda}\rangle$ contains elements of p'-order, using coprime action and observing that G_λ is of characteristic p, we have a contradiction. Hence, V is defined in all cases.

Now, if $V := Z_\lambda$ then $O_p(L) = C_{S\cap L}(V)$ and by Proposition 2.40 and Lemma 2.41, $L/C_L(V) \cong \mathrm{SL}_2(q)$. On the other hand, if $Z_\lambda = \Omega(Z(S)) < V$, then $Q_\lambda \cap Q_\mu = C_\lambda$ and we may assume that μ belongs to a critical pair (μ, μ') with $d(\lambda, \mu) = b-1$. Then b is odd, otherwise $\mu' - 1 \in \lambda^G$ and $Z_\mu \leq Q_{\mu'-1} \cap Q_{\mu'-2} = Q_{\mu'-1} \cap Q_{\mu'} \leq Q_{\mu'}$. Thus, $V_{\mu'} \cap Q_\lambda \leq C_\lambda$ and $V_\lambda \cap Q_{\mu'} \leq C_{\mu'}$. Without loss of generality, assume that $|V_{\mu'}/(V_{\mu'} \cap Q_\lambda)| \leqslant |V_\lambda/(V_\lambda \cap Q_{\mu'})|$. A straightforward calculation ensures that $V_\lambda Q_{\mu'}/Q_{\mu'}$ is an offender on $V_{\mu'}/[V_{\mu'}, Q_{\mu'}]$, $[V_{\mu'}, Q_{\mu'}] \leq C_{V_{\mu'}}(O^p(L_{\mu'}))$ and by Lemma 2.41, $L_{\mu'}/C_{L_{\mu'}}(V_{\mu'}/C_{V_{\mu'}}(O^p(L_{\mu'}))) \cong \mathrm{SL}_2(q)$.

Either way, it follows from Proposition 2.16 (vi) that $L_\lambda/C_{L_\lambda}(V) \cong L/C_L(V) \cong \mathrm{SL}_2(q)$, $S = Q_\lambda Q_\mu$ and

$$Q_\mu O_p(L) = Q_\mu(Q_\lambda \cap L) = Q_\lambda Q_\mu \cap L = S \cap L \in \mathrm{Syl}_p(L).$$

Since $[O_p(L), Q_\mu] \leq [Q_\lambda, Q_\mu] \leq Q_\lambda \cap Q_\mu \leq O_p(L)$ it follows that

$$[O_p(L), L] = [O_p(L), \langle Q_\mu^{L_\lambda} \rangle] = [O_p(L), Q_\mu]^{L_\lambda} \leq Q_\lambda \cap Q_\mu$$

and so $\widehat{L} := L/(Q_\lambda \cap Q_\mu)$ is a central extension of $L/O_p(L)$ by $\widehat{O_p(L)}$. But $Q_\mu \cap O_p(L) = Q_\mu \cap Q_\lambda$ and so $\widehat{Q_\mu}$ is complement to $\widehat{O_p(L)}$ in $\widehat{S \cap L}$. It follows by Gaschütz' theorem [49, (3.3.2)] that there is a complement in $\widehat{L}$ to $\widehat{O_p(L)}$. Now, letting K_λ be a Hall p'-subgroup of $N_L(S \cap L)$, unless $q \in \{2, 3\}$, we deduce that $\widehat{Q_\mu} \leq [\widehat{S \cap L}, K_\lambda]$ is contained in a complement to $\widehat{O_p(L)}$ and since $L = \langle Q_\mu^{G_\lambda} \rangle$, it follows that $\widehat{O_p(L)} = \{1\}$ and $Q_\mu \in \mathrm{Syl}_p(L)$. If $q \in \{2, 3\}$, then $\widehat{L} \cong p \times \mathrm{SL}_2(p)$, $|\widehat{Q_\mu}| = p$ and one can check that $\langle \widehat{Q_\mu}^{\widehat{L}} \rangle \cong \mathrm{SL}_2(p)$, contradicting the initial definition of L. Thus $Q_\mu \in \mathrm{Syl}_p(L)$ and $O_p(L) = Q_\mu \cap Q_\lambda$. Since $L_\lambda = LQ_\lambda$, there is no non-trivial characteristic subgroup of Q_μ which is normal in L, for such a subgroup would then be normal in $\langle G_\lambda, G_\mu \rangle$.

It remains to show that $V := Z_\lambda$. Aiming for a contradiction, suppose that $Z_\lambda = \Omega(Z(S))$ and $V = V_\lambda/C_{V_\lambda}(O^p(L_\lambda))$. Then, $Z(L_\mu) = \{1\}$ by Lemma 5.6 (iv), $O_p(L) = C_\lambda$, $b > 1$ is odd and V_λ is abelian. Let R_L be the preimage in L of $O_{p'}(L/O_p(L))$ and suppose that R_L is not a p-group. Then $V_\lambda = [V_\lambda, R_L] \times C_{V_\lambda}(R_L)$ is an S-invariant decomposition, and since $Z_\lambda = \Omega(Z(S)) \leq C_{V_\lambda}(R_L)$, V_λ is centralized by R_L. Since V_λ is an FF-module for $L/O_p(L)$, unless $q = 2^n > 2$, using coprime action and Proposition 2.21 (v) we infer that $C_{V_\lambda}(R_L) = C_{V_\lambda}(O^p(L))$ so that Z_μ is centralized by L and normalized by $\langle L, G_\mu \rangle$, a contradiction.

Suppose now that $q = 2^n > 2$ and let K' be a Hall $2'$-subgroup of $(G_{\lambda,\mu} \cap L)/O_2(L)$ complementing $Q_\mu/O_2(L)$. Then for K the preimage if K' in L, we have that $[K, Q_\lambda] \leq O_2(L)$. In particular, $[V_\lambda, K]$ is normalized by Q_λ and as $V_\lambda = [V_\lambda, K] \times C_{V_\lambda}(K)$ by coprime action, $[V_\lambda, K, Q_\lambda] = \{1\}$ since $C_{V_\lambda}(K) = C_{V_\lambda}(L)$ and $V_\lambda/C_{V_\lambda}(L)$ is a natural $\mathrm{SL}_2(2^n)$-module for $L/R_L \cong \mathrm{PSL}_2(2^n)$. But then $[Z_\mu, K, Q_\lambda] = \{1\}$ so that $\{1\} \neq [Z_\mu, K]$ is centralized by $S = Q_\lambda Q_\mu$ and $[Z_\mu, K] \leq Z_\lambda$. However, by coprime action $[Z_\mu, K] = [Z_\mu, K, K] = [Z_\lambda, K] = \{1\}$, a contradiction. Therefore, $V = Z_\lambda$ and the result holds. □

Lemma 5.25 *Suppose that $Q_\lambda \cap Q_\mu \trianglelefteq L_\lambda$. Then writing $L := \langle Q_\mu^{L_\lambda} \rangle$, $L/O_p(L) \cong L_\lambda/Q_\lambda \cong \mathrm{SL}_2(q)$, $b = 2$ and $O_p(L)$ contains a unique non-central chief factor for L. Moreover, there is $\lambda' \in \Delta(\mu)$ such that both (λ, λ') and (λ', λ) are critical pairs.*

Proof Suppose that $b = 1$. Then $\Omega(Z(S)) \leq Q_\lambda \cap Q_\mu = O_p(L) \trianglelefteq G_\lambda$ and it follows from the definition of Z_λ that $Z_\lambda \leq O_p(L) \leq Q_\mu$. Thus, we may as well assume that $Z_\mu \not\leq Q_\lambda$. But then Z_μ centralizes $O_p(L)$ and so $O^p(L)$ centralizes $O_p(L)$, a contradiction since L is of characteristic p. Thus, we conclude that $b > 1$.

Suppose that (λ, δ) is not a critical pair for any $\delta \in \Gamma$. Then there is some μ' such that (μ, μ') is a critical pair and $d(\lambda, \mu') = b - 1$. Then $Z_\mu \neq \Omega(Z(S)) \neq Z_\lambda$, $C_{G_{\mu'}}(Z_{\mu'})$ is p-closed and $Z_{\mu'} \leq Q_{\mu+2} \cap Q_\lambda = Q_\lambda \cap Q_\mu$. But then, $[Z_\mu, Z_{\mu'}] = \{1\}$, a contradiction for then $Z_\mu \leq Q_{\mu'}$. Thus, we may assume λ belongs to a critical pair (λ, λ') with $d(\mu, \lambda') = d(\lambda, \lambda') - 1$. Suppose that b is odd. Then $Z_\lambda \leq Q_{\lambda'-1}$ and $\lambda' - 1 \in \lambda^G$. But then $Z_\lambda \leq Q_{\lambda'-1} \cap Q_{\lambda'-2} = Q_{\lambda'-1} \cap Q_{\lambda'} \leq Q_{\lambda'}$, a contradiction. Thus, b is even. Moreover, since $C_S(Z_\lambda) = Q_\lambda \in \mathrm{Syl}_p(G_\lambda^{(1)})$ and $[Z_\lambda, Z_{\lambda'}] \neq \{1\}$, (λ', λ) is also a critical pair.

We observe if $b \geqslant 4$, then $V_\lambda^{(2)} \leq O_p(L)$ and $V_\lambda^{(2)}/Z_\lambda$ contains a non-central chief factor. Hence, if $O_p(L)$ contains a unique non-central chief factor for L then $b = 2$. So in order to prove the lemma, aiming for a contradiction, in the sequel we will assume that $O_p(L)$ contains more the one non-central chief factor for L.

Assume first that $O_p(L)$ contains more than one non-central chief factor for L and that p is odd. If $b = 2$, then $O_p(L) = Q_\lambda \cap Q_\mu = Z_\lambda(Q_\lambda \cap Q_\mu \cap Q_{\lambda'})$, a contradiction since $O_p(L)$ contains more than one non-central chief factor. Thus, we may assume that $b \geqslant 4$ and b is even. Set T_λ to be a Hall p'-subgroup of the preimage in L_λ of $Z(L_\lambda/R_\lambda)$. Note also that since p is odd, we may apply coprime action along with Proposition 2.21 (v) so that $Z_\lambda = [Z_\lambda, T_\lambda] \times C_{Z_\lambda}(T_\lambda) = [Z_\lambda, L_\lambda] \times Z(L_\lambda)$.

Choose $\lambda - 1 \in \Delta(\lambda)$ such that $\Omega(Z(L_{\lambda-1})) \times Z(L_\lambda) \neq \Omega(Z(L_\mu)) \times Z(L_\lambda)$ and set $U = \langle V_\gamma : \Omega(Z(L_{\lambda-1})) = \Omega(Z(L_\gamma)), \gamma \in \Delta(\lambda)\rangle$. Let $r \in R_\lambda Q_{\lambda-1} \leq C_{L_\lambda}(\Omega(Z(L_{\lambda-1})))$. Since r is an automorphism of the graph, it follows that for $V_\gamma \leq U$, $V_\gamma^r = V_{\gamma \cdot r}$. But $\Omega(Z(L_{\gamma \cdot r})) = \Omega(Z(L_\gamma))^r = \Omega(Z(L_{\lambda-1}))^r = \Omega(Z(L_{\lambda-1}))$ and so $V_\gamma^r \leq U$ and $U \trianglelefteq R_\lambda Q_{\lambda-1}$. Note that if $U \leq Q_{\lambda'-2}$ then $U \leq Q_{\lambda'-2} \cap Q_{\lambda'-3} = Q_{\lambda'-2} \cap Q_{\lambda'-1} \leq Q_{\lambda'-1}$ and so, $U = Z_\lambda(U \cap Q_{\lambda'})$. Thus, $Z_{\lambda'}$ centralizes U/Z_λ and since $L_\lambda = \langle R_\lambda, Z_{\lambda'}, Q_{\lambda-1}\rangle$, it follows that $O^p(L_\lambda)$ centralizes U/Z_λ and so normalizes $V_{\lambda-1}$, a contradiction.

Therefore, $U \not\leq Q_{\lambda'-2}$ so that there is some $\lambda - 2 \in \Delta^{(2)}(\lambda)$ such that $(\lambda-2, \lambda'-2)$ is also a critical pair. Since $Z_\lambda = [Z_\lambda, L_\lambda] \times \Omega(Z(L_\lambda))$, to deliver a contradiction it suffices to prove that $[Z_\lambda, Z_{\lambda'}] = \Omega(Z(L_\mu)) = \Omega(Z(L_{\lambda'-1}))$ and that this holds for any critical pair, since then, as there is $\lambda - 2 \in \Delta(\lambda - 1)$ with $(\lambda - 2, \lambda' - 2)$ a critical pair, $Z_\lambda = \Omega(Z(L_{\lambda'-1})) \times \Omega(Z(L_{\lambda'-3})) \times \Omega(Z(L_\lambda))$ which is contained in $Q_{\lambda'}$ since $b > 2$.

Suppose that $Z_\mu = \Omega(Z(S)) = \Omega(Z(L_\mu))$. In particular, $Z(L_\lambda) = \{1\}$ and Z_λ is irreducible. Since Z_λ is a natural $\mathrm{SL}_2(q)$-module, $Z_{\lambda'-1} = [Z_\lambda, Z_{\lambda'}] = Z_\mu$, as required.

Assume now that $Z_\mu \neq \Omega(Z(S))$. Then $[Z_\lambda, Z_{\lambda'}] = C_{[Z_\lambda, L_\lambda]}(S) = \Omega(Z(S)) \cap [Z_\lambda, L_\lambda]$. Since $\Omega(Z(S)) = \Omega(Z(L_\lambda)) \times \Omega(Z(L_\mu))$ and T_λ normalizes $\Omega(Z(L_\mu))$, we have that $\Omega(Z(L_\mu)) \geqslant [\Omega(Z(L_\mu)), T_\lambda] = [\Omega(Z(S)), T_\lambda] = \Omega(Z(S)) \cap [Z_\lambda, L_\lambda]$. Comparing orders, we conclude that $\Omega(Z(L_\mu)) = [\Omega(Z(S)), T_\lambda] = [Z_\lambda, Z_{\lambda'}]$. By symmetry, we have that $\Omega(Z(L_{\lambda'-1})) = [Z_\lambda, Z_{\lambda'}]$, as desired. Hence, we have forced a contradiction and we conclude that if p is odd, then $b = 2$ and $O_p(L)$ contains a unique non-central chief factor for L.

Assume now that $p = 2$ and $O_2(L)$ contains more than one non-central chief factor for L. Choose $1 < m < b/2$ minimal such that $V_\lambda^{(2m)} \leq Q_{\lambda'-2m}$. Notice by the minimal choice of m that $V_\lambda^{(2(m-k))} Q_{\lambda'-2(m-k)} \in \mathrm{Syl}_p(L_{\lambda'-2(m-k)})$ for all $1 \leqslant k \leqslant m$. Then $V_\lambda^{(2m)} \leq Q_{\lambda'-2m} \cap Q_{\lambda'-2m-1} \leq Q_{\lambda'-2m+1}$ and, extending further, $V_\lambda^{(2m)} = V_\lambda^{(2m-2)}(V_\lambda^{(2m)} \cap Q_{\lambda'})$. But then, $O^p(L_\lambda)$ centralizes $V_\lambda^{(2m)}/V_\lambda^{(2m-2)}$, a contradiction. Thus, no such m exists. Even still an index q subgroup of $V_\lambda^{(2k)}/V_\lambda^{(2k-2)}$ is centralized by $Z_{\lambda'}$ for all $k < b/2$ and it follows that for all $1 < l < b/2$, $V_\lambda^{(2l)}/V_\lambda^{(2l-2)}$ contains a unique non-central chief factor and this factor is an FF-module for L_λ/Q_λ. Note that for R_1, R_2 the centralizers in $L/O_2(L)$ of distinct non-central chief factors in $V_\lambda^{(2l)}$ for $1 < l < b/2$, we deduce that R_1R_2/R_i is an odd order normal subgroup of $L_i/R_i \cong \mathrm{SL}_2(q)$ for $i \in \{1,2\}$. Thus, unless $q = 2$, we have that $L/O_2(L)C_L(V_\lambda^{(2l)}) \cong \mathrm{SL}_2(q)$. In a similar manner to the calculation above, we deduce that $V_\lambda^{(2l)}C_{Q_\lambda}(V_\lambda^{(2l)}) = V_\lambda^{(2l)}(V_\lambda^{(2l)}C_{Q_\lambda}(V_\lambda^{(2l)}) \cap Q_{\lambda'})$ so that $O^2(L)$ centralizes $V_\lambda^{(2l)}C_{Q_\lambda}(V_\lambda^{(2l)})/V_\lambda^{(2l)}$. An application of the three subgroup lemma, using that $O_2(L)$ is self-centralizing in L, ensures that $L/O_2(L) \cong \mathrm{SL}_2(q)$.

Since no non-trivial characteristic subgroup of Q_β is normal in L, we may apply pushing up arguments from [55, Theorem B] when $L/O_2(L) \cong \mathrm{SL}_2(q)$. Thus, Q_μ has class 2 and there is a unique non-central chief factor for L within $O_2(L)$. It is clear that $Z_\lambda/Z(L_\lambda)$ is the unique non-central chief factor for L inside $O_2(L)$ and is isomorphic to the natural module for $L/O_2(L) \cong \mathrm{SL}_2(q)$. Therefore, the lemma holds in this case.

Thus, we are reduced to examining the case where $q = p = 2$. Since no non-trivial characteristic subgroup of Q_β is normal in L, we may apply [29, Theorem 4.3] to see that Q_μ has nilpotency class 2 and exponent 4. Notice that if $b \geqslant 4$, then $V_\lambda^{(2)}$ is contained in Q_μ and $[V_\lambda^{(2)}, Q_\mu] \leq \Omega(Z(Q_\mu))$. But $\langle(\Omega(Z(Q_\mu))^L)\rangle$ is an FF-module for $L/O_2(L)$ by Proposition 2.40, and contains $[Z_\lambda, L_\lambda]$ as its unique non-central chief factor. Thus, it follows that $[V_\lambda^{(2)}, L] \leq Z_\lambda$ and $V_\mu \trianglelefteq \langle L, G_\mu\rangle$, a contradiction. Hence, when $q = 2$, we conclude that $b = 2$ so that $O_2(L)$ contains a unique non-central chief factor, and the proof is complete. □

Lemma 5.26 *Suppose that $Q_\lambda \cap Q_\mu \trianglelefteq L_\lambda$. Then $Z_\mu \neq \Omega(Z(S))$.*

Proof By Lemma 5.25, there is a unique non-central chief factor for L_λ contained in $Q_\mu \cap Q_\lambda$ and, as a consequence, $L/O_p(L) \cong L_\lambda/Q_\lambda \cong \mathrm{SL}_2(q)$. Aiming for a contradiction, assume throughout the proof that $Z_\mu = \Omega(Z(S)) = \Omega(Z(L_\mu))$. Then $Z(L_\lambda) = \{1\}$ by Lemma 5.6 (iv). Hence, Z_λ is the unique non-central chief factor within $Q_\lambda \cap Q_\mu$. In particular, Z_λ is isomorphic to a natural $\mathrm{SL}_2(q)$-module and $[O^p(L_\lambda), Q_\lambda] = Z_\lambda$.

If $\Phi(Q_\lambda) \neq \{1\}$, then the irreducibility of Z_λ implies that $Z_\lambda \leq \langle(\Phi(Q_\lambda) \cap \Omega(Z(S)))^{L_\lambda}\rangle \leq \Phi(Q_\lambda)$. But then $O^p(L)$ acts trivially on $Q_\lambda/\Phi(Q_\lambda)$, a contradiction by coprime action. Thus, $\Phi(Q_\lambda) = \{1\}$ and Q_λ is elementary abelian. If p is odd or $q = 2$, then for T_λ the preimage in L_λ of $O_{p'}(\overline{L_\lambda})$, we have that $Q_\lambda = [Q_\lambda, T_\lambda] \times C_{Q_\lambda}(T_\lambda) = Z_\lambda \times C_{Q_\lambda}(T_\lambda)$ is an S-invariant decomposition and since $\Omega(Z(S)) \leq Z_\lambda$, we have that $C_{Q_\lambda}(T_\lambda) = \{1\}$ and $Q_\lambda = Z_\lambda$. But then $Z_\mu = Z_\lambda \cap Q_\mu = Q_\lambda \cap Q_\mu \trianglelefteq L_\lambda$, a contradiction.

If $q > 2$ is even, then since $S \leq N_{G_\mu}(O_2(L))$, we have that $[G : N_{G_\mu}(O_2(L))]$ is odd and applying [70, Theorem 3], $V_\mu \trianglelefteq G = \langle L, G_\mu\rangle$, a contradiction. □

Finally, we complete the first main "pushing up" result in this work, relying on all the lemmas proved so far in this section.

Proposition 5.27 *Let $S \in \mathrm{Syl}_p(G_\lambda \cap G_\mu)$ for $\lambda \in \Gamma$ and $\mu \in \Delta(\lambda)$. Then $Q_\lambda \cap Q_\mu$ is not normal in L_λ. Moreover, if $Z_\lambda Z_\mu \trianglelefteq L_\lambda$ then $Z_\mu = \Omega(Z(S)) \le Z_\lambda$.*

Proof Suppose that $Z_\lambda Z_\mu \trianglelefteq L_\lambda$ but $Z_\mu \neq \Omega(Z(S))$. By Lemma 5.16 (ii), we have that $C_S(Z_\mu) = Q_\mu$ and so $C_{Q_\lambda}(Z_\lambda Z_\mu) = Q_\lambda \cap C_S(Z_\mu) = Q_\lambda \cap Q_\mu$ and it follows that $Q_\lambda \cap Q_\mu \trianglelefteq L_\lambda$. Thus, we may suppose that $Q_\lambda \cap Q_\mu \trianglelefteq L_\lambda$, and derive a contradiction to complete the proof.

Under this assumption, Z_λ contains the unique non-central chief factor for L inside $Q_\mu \cap Q_\lambda$ and $Z_\mu \neq \Omega(Z(S))$. Moreover, $b = 2$ and there is $\lambda' \in \Delta(\mu)$ such that $Z_\lambda \not\le Q_{\lambda'}$ and $Z_{\lambda'} \not\le Q_\lambda$. Since $L_\lambda/Q_\lambda \cong \mathrm{SL}_2(q_\lambda)$ and $Z_\lambda/Z(L_\lambda)$ is a natural module, we get that $Q_\mu = (Q_{\lambda'} \cap Q_\mu \cap Q_\lambda)Z_\lambda Z_{\lambda'}$ and $Q_\lambda \cap Q_\mu = (Q_{\lambda'} \cap Q_\mu \cap Q_\lambda)Z_\lambda$. Then $(Q_\lambda \cap Q_\mu)/\Phi(Q_{\lambda'} \cap Q_\mu \cap Q_\lambda)$ is elementary abelian and it follows that $\Phi(Q_\lambda \cap Q_\mu) = \Phi(Q_{\lambda'} \cap Q_\mu) = \Phi(Q_{\lambda'} \cap Q_\mu \cap Q_\lambda)$. Set $F := \Phi(Q_\lambda \cap Q_\mu)$. Since Q_λ contains a unique non-central chief factor for L_λ, we infer that F is centralized by $O^p(L)$ and as Q_μ has class 2, $F \le Z(L)$. Let Z_μ^* be the preimage in Q_μ of $Z(Q_\mu/F)$. Since F is normal in both G_λ and $G_{\lambda'}$, we deduce that $Z_\mu^* \trianglelefteq \langle G_{\lambda,\mu}, G_{\mu,\lambda'}\rangle$. Moreover, since $Q_\mu = (Q_{\lambda'} \cap Q_\mu \cap Q_\lambda)Z_\lambda Z_{\lambda'}$, we have that $Q_\mu \cap Q_\lambda \cap Q_{\lambda'} \le Z_\mu^*$. Since $[Z_\mu^*, Z_\lambda] \le F \le Z(L)$, we have that $Z_\mu^* \le Q_\lambda$ and by symmetry, $Z_\mu^* = Q_\mu \cap Q_\lambda \cap Q_{\lambda'}$.

Suppose that p is odd and let $H_{\lambda,\mu}$ be a Hall p'-subgroup of $G_{\lambda,\mu} \cap L_\lambda$. By Proposition 2.16 (vi), $H_{\lambda,\mu}$ is cyclic of order $q_\lambda - 1$. Furthermore, $H_{\lambda,\mu}$ normalizes Q_μ, F and Z_μ^*, and acts non-trivially on Q_μ/Z_μ^*. Now, for t_λ the unique involution in $H_{\lambda,\mu}$, t_λ centralizes $Q_\mu/Q_\lambda \cap Q_\mu$ and inverts $Q_\lambda \cap Q_\mu/Z_\mu^* = Z_\lambda Z_\mu^*/Z_\mu^*$. By coprime action, $Q_\mu/Z_\mu^* = Z_\lambda Z_\mu^*/Z_\mu^* \times C_{Q_\mu/Z_\mu^*}(t_\lambda)$ is a Q_μ-invariant decomposition. Since $[S, t_\lambda] \le Q_\lambda \cap Q_\mu$ the previous decomposition is S-invariant. But then $[Q_\lambda, C_{Q_\mu/Z_\mu^*}(t_\lambda)] \le (Q_\mu \cap Q_\lambda)/Z_\mu^* = Z_\lambda Z_\mu^*/Z_\mu^*$ and we deduce that Q_λ centralizes Q_μ/Z_μ^*. Hence, Q_λ normalizes $Q_{\lambda'} \cap Q_\mu$. Let $M = \langle Q_\lambda, Q_{\lambda'}, Q_\mu\rangle \le G_\mu$. Then there is an $m \in M$ such that $(Q_\lambda Q_\mu)^m = Q_{\lambda'} Q_\mu$ and since $Q_{\lambda'} Q_\mu$ is the unique Sylow p-subgroup of $G_{\mu,\lambda'}$, it follows that $\lambda \cdot m = \lambda'$. But then $(Q_\lambda \cap Q_\mu)^m = Q_{\lambda'} \cap Q_\mu$ and as M normalizes $Q_{\lambda'} \cap Q_\mu$, we have that $Q_\mu \cap Q_\lambda = Q_\lambda \cap Q_\mu$, absurd since $Z_\lambda \le Q_\lambda \cap Q_\mu$.

Suppose that $p = 2$. Since $(Q_\lambda \cap Q_\mu)/F$ and $(Q_\mu \cap Q_{\lambda'})/F$ are elementary abelian, by [61, Lemma 2.29], every involution in Q_μ/F is contained in $(Q_\lambda \cap Q_\mu)/F$ or $(Q_\mu \cap Q_{\lambda'})/F$. Indeed, for A any other elementary abelian subgroup of Q_μ/F and B the preimage of A in Q_μ, we must have that $B = (B \cap Q_\lambda) \cup (B \cap Q_{\lambda'})$. If $B \not\le Q_\lambda$, then $F \cap Z_\lambda = C_{Z_\lambda}(B) = Z_\lambda \cap B$ and it follows that $B \cap Q_\lambda = F$. By symmetry, we have shown that $\mathcal{A}(Q_\mu/F) = \{(Q_\lambda \cap Q_\mu)/F, (Q_\mu \cap Q_{\lambda'})/F\}$.

Set $M = \langle Q_\lambda, Q_{\lambda'}, Q_\mu\rangle \le G_\mu$ so that M normalizes Q_μ, Z_μ^* and F. Thus, all elements of M which do not normalize $Q_\mu \cap Q_\lambda$, conjugate $Q_\mu \cap Q_{\lambda'}$ to $Q_\mu \cap Q_\lambda$, and vice versa. Thus all odd order elements normalize $Q_\mu \cap Q_\lambda$. There is an $m \in M$ such that $(Q_\lambda Q_\mu)^m = Q_{\lambda'} Q_\mu$ and since $Q_{\lambda'} Q_\mu$ is the unique Sylow 2-subgroup of $G_{\mu,\lambda'}$, it follows that $\lambda \cdot m = \lambda'$. Since $M = O^p(M)Q_\lambda Q_\mu$, we may as well choose m of order coprime to p. But then $(Q_\lambda \cap Q_\mu)^m = Q_{\lambda'} \cap Q_\mu$ and as m normalizes

$Q_{\lambda'} \cap Q_\mu$, we conclude that $Q_\mu \cap Q_{\lambda'} = Q_\lambda \cap Q_\mu$, a final contradiction since $Z_\lambda \leq Q_\lambda \cap Q_\mu$. This completes the proof. □

We can now prove a result analogous to Lemma 5.9, instead working "down" through chief factors. Again, we will apply this lemma often and without reference throughout this work.

Lemma 5.28 *Let $\lambda \in \Gamma$ and $\mu \in \Delta(\lambda)$, $b > 1$ and $n \geqslant 2$. If $V_\lambda^{(n)} \leq Q_\lambda$, then $C_{Q_\lambda}(V_\lambda^{(n-2)})/C_{Q_\lambda}(V_\lambda^{(n)})$ contains a non-central chief factor for L_λ.*

Proof Observe that as $V_\lambda^{(n)} \leq Q_\lambda$, we have that $Z(Q_\lambda) \leq C_{Q_\lambda}(V_\lambda^{(n)}) \leq C_{Q_\lambda}(V_\mu^{(n-1)}) \leq C_{Q_\lambda}(V_\lambda^{(n-2)})$. In particular, $C_{Q_\lambda}(V_\mu^{(n-1)})$ is non-trivial. If $C_{Q_\lambda}(V_\lambda^{(n-2)})/C_{Q_\lambda}(V_\lambda^{(n)})$ contains only central chief factors for L_λ, $O^p(L_\lambda)$ centralizes $C_{Q_\lambda}(V_\lambda^{(n-2)})/C_{Q_\lambda}(V_\lambda^{(n)})$ and normalizes $C_{Q_\lambda}(V_\mu^{(n-1)})$. Thus, $C_{Q_\lambda}(V_\mu^{(n-1)}) \trianglelefteq O^p(L_\lambda)G_{\lambda,\mu} = G_\lambda$. In order to force a contradiction, we need only show that $C_{Q_\lambda}(V_\mu^{(n-1)}) = C_{Q_\mu}(V_\mu^{(n-1)})$.

Let $S \in \mathrm{Syl}_p(G_{\lambda,\mu})$. Since $n \geqslant 2$, $Z_\lambda \leq V_\lambda^{(n-2)}$ is centralized by $C_S(V_\mu^{(n-1)})$ and unless $n = 2$ and $V_\lambda^{(n-2)} = Z_\lambda = \Omega(Z(S))$, applying Lemmas 5.16 (ii) and 5.19, we have that $C_S(V_\mu^{(n-1)}) \leq Q_\lambda \cap Q_\mu$ and $C_{Q_\lambda}(V_\mu^{(n-1)}) = C_{Q_\mu}(V_\mu^{(n-1)})$, as desired. If $V_\lambda^{(n-2)} = \Omega(Z(S))$, then $V_\mu^{(n-1)} = Z_\mu$ and $C_S(Z_\mu) = Q_\mu$. But then, $C_{Q_\lambda}(Z_\mu) = Q_\lambda \cap Q_\mu \trianglelefteq G_\lambda$, a contradiction by Proposition 5.27. □

5.4 Subamalgams and More Pushing Up Arguments

We use the section to document some more pushing arguments we rely on in the proof of Theorem C. The results proved in this section mostly focus on identifying *subamalgams* in a minimal counterexample to Theorem C. Since we are operating in a minimal counterexample, by induction, this subamalgam is described in Theorem C and its structure is completely transparent. This forces severe restrictions on the minimal counterexample in question. In the later analysis, this is one of our primary tools in forcing contradictions in the pursuit of proving Theorem C. We will first require a result on FF-modules for weak BN-pairs.

Theorem 5.29 *Suppose that G is an outcome of Theorem C where L_α and L_β are p-solvable and let $S \in \mathrm{Syl}_p(L_\alpha) \cap \mathrm{Syl}_p(L_\beta)$. Assume that $G = \langle S^G \rangle$ and V is an FF-module for G such that $C_S(V) = \{1\}$. Then G has a weak BN-pair of rank 2 and is locally isomorphic to one of $\mathrm{SL}_3(p)$, $\mathrm{Sp}_4(p)$, or $\mathrm{G}_2(2)$. Moreover, if G is locally isomorphic to $\mathrm{G}_2(2)$, then $G/C_G(V) \cong \mathrm{G}_2(2)$.*

Proof If G does not have a weak BN-pair of rank 2, comparing with Theorem C, we see that $p = b = 2$, $L_\alpha/Q_\alpha \cong \mathrm{Sym}(3)$ and $L_\beta/Q_\beta \cong (3 \times 3) : 2$. Moreover, there is $P_\beta \leq L_\beta$ such that P_β contains S, $P_\beta/Q_\beta \cong \mathrm{Sym}(3)$ and Q_β contains two non-central chief factors for P_β. Indeed, no non-trivial subgroup of S is normalized by both L_α and P_β and by [28], (L_α, P_β, S) is locally isomorphic to M_{12}. Setting $X := \langle L_\alpha, P_\beta\rangle$ and applying [23], V is an FF-module for X upon restriction and applying [23, Lemma 3.12], we have a contradiction.

Hence, G has a weak BN-pair of rank 2 and the result follows from [23, Theorem A, Theorem B, Corollary 1]. □

While the results proved in the remainder of this section will be vital later in our analysis, the proofs are long, technical and not particularly illuminating. However, this is the first instance where one may appreciate the utility in analysing a minimal counterexample to Theorem C.

As in the previous section, the upcoming results are technical and mired in notation, and the proofs long and abstruse. The reader is encouraged to work through them slowly, and may consider skipping some of the details in a first pass.

Proposition 5.30 *Suppose that G is a minimal counterexample to Theorem C, $\{\lambda, \delta\} = \{\alpha, \beta\}$ and the following conditions hold:*

(i) *$L_\alpha/R_\alpha \cong \mathrm{SL}_2(q) \cong L_\beta/R_\beta$, and Z_α and $V_\beta/C_{V_\beta}(O^p(L_\beta))$ are natural $\mathrm{SL}_2(q)$-modules;*
(ii) *there is a non-central chief factor U/W for G_λ such that, as an $\overline{L_\lambda}$-module, U/W is an FF-module, $C_{L_\lambda}(U/W) \neq R_\lambda$, and $C_{L_\lambda}(U/W) \cap R_\lambda$ normalizes $Q_\alpha \cap Q_\beta$; and*
(iii) *if $q = p$ then $Z(Q_\alpha) = Z_\alpha$ is of order p^2 and $Z(Q_\beta) = Z_\beta = \Omega(Z(S))$ is of order p.*

Then $q \in \{2, 3\}$ and one of the following holds:

(a) *there is $H_\lambda \leq G_\lambda$ containing $G_{\alpha,\beta}$ such that $(H_\lambda, G_\delta, G_{\alpha,\beta})$ is a weak BN-pair of rank 2, $b \leqslant 5$ and if $b > 3$, then $(H_\lambda, G_\delta, G_{\alpha,\beta})$ is parabolic isomorphic to F_3 and $V_\alpha^{(2)}/Z_\alpha$ is not acted on quadratically by S;*
(b) *$p = 3$, $\lambda = \alpha$, neither $C_{L_\alpha}(U/W)$ nor R_α normalizes $Q_\alpha \cap Q_\beta$ and there does not exist $P_\alpha \leq L_\alpha$ such that $S(C_{L_\alpha}(U/W) \cap R_\alpha) \leq P_\alpha$, P_α is $G_{\alpha,\beta}$-invariant, $P_\alpha/C_{L_\alpha}(U/W) \cap R_\alpha \cong \mathrm{SL}_2(p)$, $L_\alpha = P_\alpha R_\alpha = P_\alpha C_{L_\alpha}(U/W)$ and $Q_\alpha \cap Q_\beta \ntrianglelefteq P_\alpha$;*
(c) *$\lambda = \beta$ and neither R_β nor $C_{L_\beta}(U/W)$ normalizes $V_\alpha^{(2)}$; or*
(d) *there is $H_\lambda \leq G_\lambda$ containing $G_{\alpha,\beta}$ such that for $X := \langle H_\lambda, G_\delta\rangle$ and $V := \langle Z_\beta^X\rangle$, we have that $V_\beta \leq V \leq S$, $C_S(V) \trianglelefteq X$ and for $\widetilde{X} := X/C_X(V)$, either $\widetilde{X}$ is locally isomorphic to $\mathrm{SL}_3(p)$, $\mathrm{Sp}_4(p)$ or $\mathrm{G}_2(2)$; or $p = 3$ and there is an involution x in $G_{\alpha,\beta}$ such that $\widetilde{X}/\widetilde{\langle x\rangle}$ is locally isomorphic to $\mathrm{PSp}_4(3)$.*

Moreover, in outcome (d), if $\widetilde{Q}_\mu$ contains more than one non-central chief factor for $\widetilde{L}_\mu$ then $\widetilde{Q}_\mu$ contains two non-central chief factors, $\widetilde{Q}_\nu$ contains a unique non-central chief factor for $\widetilde{L}_\nu$, and $\widetilde{X} \cong \mathrm{G}_2(2)$ where $\{\mu, \nu\} = \{\alpha, \beta\}$.

Proof It follows from (i), (ii) and Lemma 2.41 that $L_\lambda/C_{L_\lambda}(U/W) \cong L_\lambda/R_\lambda \cong \mathrm{SL}_2(q)$ and $\mathrm{Syl}_p(C_{L_\lambda}(U/W)) = \mathrm{Syl}_p(R_\lambda) = \{Q_\lambda\}$. Thus, $C_{L_\lambda}(U/W)R_\lambda/Q_\lambda$ is a non-trivial normal p'-subgroup of L_λ/Q_λ.

Assume that $q \geqslant 4$ and $C_{L_\lambda}(U/W) \neq R_\lambda$. Then $C_{L_\lambda}(U/W)R_\lambda/C_{L_\lambda}(U/W) = Z(L_\lambda/C_{L_\lambda}(U/W))$ and $C_{L_\lambda}(U/W)R_\lambda/R_\lambda = Z(L_\lambda/R_\lambda)$. In particular, p is odd and $L_\lambda/(C_{L_\lambda}(U/W) \cap R_\lambda)$ is isomorphic to a central extension of $\mathrm{PSL}_2(q)$ by an elementary abelian group of order 4. Since $O^{p'}(L_\lambda) = L_\lambda$ and the p'-part of the Schur multiplier of $\mathrm{PSL}_2(q)$ is of order 2 by Proposition 2.16 (vii), we have a contradiction. Thus, we have that

(1) $q \in \{2, 3\}$.

Indeed, G_α and G_β are p-solvable and by condition (iii), $Z(Q_\alpha) = Z_\alpha$ is of order p^2 and $Z(Q_\beta) = Z_\beta = \Omega(Z(S))$ is of order p. By Propositions 2.46 (ii) and 2.47 (ii), $L_\lambda/(C_{L_\lambda}(U/W) \cap R_\lambda) \cong (3 \times 3) : 2$ if $p = 2$, and $L_\lambda/(C_{L_\lambda}(U/W) \cap R_\lambda) \cong (Q_8 \times Q_8) : 3$ if $p = 3$.

Suppose that $p = 2$. By Proposition 2.46 (iii) there are $P_1, \ldots, P_4 \leq L_\lambda$ such that $S(C_{L_\lambda}(U/W) \cap R_\lambda) \leq P_i$ and $P_i/(C_{L_\lambda}(U/W) \cap R_\lambda) \cong \mathrm{Sym}(3)$. Indeed, $C_{L_\lambda}(U/W)S$ and $R_\lambda S$ are non-equal and satisfy this condition. Moreover, P_i is $G_{\alpha,\beta}$-invariant for all i. Since any two P_i generate L_λ, we may choose $P_\lambda = P_j \neq R_\lambda S$ such that $Q_\alpha \cap Q_\beta \ntrianglelefteq P_\lambda$ and $O^2(P_\lambda)$ does not centralize U/W. Set $H_\lambda := P_\lambda G_{\alpha,\beta}$, $X := \langle H_\lambda, G_\delta\rangle$ and $V := \langle Z_\beta^X\rangle$. By (ii) and (iii), we have that $V_\beta \leq V$.

Suppose that $p = 3$. By Proposition 2.47 (iii), there is $P_1, \ldots, P_5 \leq L_\lambda$ such that $S(C_{L_\lambda}(U/W) \cap R_\lambda) \leq P_i$ and $P_i/(C_{L_\lambda}(U/W) \cap R_\lambda) \cong \mathrm{SL}_2(3)$. Again, $C_{L_\lambda}(U/W)S$ and $R_\lambda S$ are non-equal and satisfy this condition, and for any $i \neq j$, $L_\lambda = \langle P_i, P_j\rangle$. Since $C_{L_\lambda}(U/W)S$ and $R_\lambda S$ are $G_{\alpha,\beta}$-invariant there is at least one other P_i which is $G_{\alpha,\beta}$-invariant. Notice that $R_\beta S$ normalizes $Q_\alpha \cap Q_\beta$ and as any two P_i generate, by Proposition 5.27, if $\lambda = \beta$ there is a choice of $P_\lambda = P_i$ such that $Q_\alpha \cap Q_\beta \ntrianglelefteq P_\lambda$, P_λ is $G_{\alpha,\beta}$-invariant and $O^3(P_\lambda)$ does not centralize U/W or V_β. If $\lambda = \alpha$, then unless outcome (b) holds, we may choose $P_\lambda = P_j \neq R_\lambda S$ such that $Q_\alpha \cap Q_\beta \ntrianglelefteq P_\lambda$ and $O^3(P_\lambda)$ does not centralize U/W. Again, we set $H_\lambda := P_\lambda G_{\alpha,\beta}$, $X := \langle H_\lambda, G_\delta\rangle$ and $V := \langle Z_\beta^X\rangle$, remarking that $V_\beta \leq V$. Thus, for $p \in \{2, 3\}$ we have that

(2) there is $S \leq P_\lambda \leq L_\lambda$ such that $L_\lambda = \langle P_\lambda, R_\lambda\rangle$, $G_{\alpha,\beta} \leq N_{G_\lambda}(P_\lambda)$, $P_\lambda/(C_{L_\lambda}(U/W) \cap R_\lambda) \cong \mathrm{SL}_2(p)$, $Q_\alpha \cap Q_\beta \ntrianglelefteq P_\lambda$, $O^p(P_\lambda)$ does not centralize U/W and $V_\beta \leq V := \langle Z_\beta^{\langle P_\lambda, G_\delta\rangle}\rangle$.

For $p = 2$ or 3, $O_p(P_\lambda) = Q_\lambda$ and P_λ/Q_λ has a strongly p-embedded subgroup. Moreover, P_λ is of characteristic p, $C_S(V) \leq C_\beta \leq Q_\alpha \cap Q_\beta$ so that $C_S(V) = C_{Q_\alpha}(V) = C_{Q_\beta}(V) \trianglelefteq X$. We observe that V is not necessarily contained in S and that $C_S(V)$ may be trivial. If no non-trivial subgroup of $G_{\alpha,\beta}$ is normal in X, then X satisfies Hypothesis 5.1 and since both H_λ and G_δ are p-solvable, by minimality, $(H_\lambda, G_\delta, G_{\alpha,\beta})$ is a weak BN-pair of rank 2; or $p = 2$, X is a symplectic amalgam, $|S| = 2^6$ and exactly one of $\overline{H_\lambda}$ or $\overline{G_\delta}$ is isomorphic to $(3 \times 3) : 2$.

In the latter case, we get that Q_λ and Q_δ are non-abelian subgroups of order 2^5 and $\overline{G_\delta}$ and $\overline{G_\lambda}$ are isomorphic to subgroups of $\mathrm{GL}_4(2)$. But $\overline{S} \leq \overline{H_\lambda}$ and every subgroup of $\mathrm{GL}_4(2)$ properly containing a subgroup isomorphic to $(3 \times 3) : 2$ has a Sylow 2-subgroup of order at least 4. It follows that $\overline{G_\delta} \cong (3 \times 3) : 2$ and $\overline{H_\lambda} \cong$

$\mathrm{Sym}(3)$. Now, for some $\gamma \in \{\lambda, \delta\}$, $|Q_\gamma/\Phi(Q_\gamma)| = 2^3$ so that $\overline{G_\gamma}$ is isomorphic to a subgroup of $\mathrm{GL}_3(2)$. Since $\mathrm{GL}_3(2)$ has no subgroup isomorphic to $(3 \times 3) : 2$, we deduce that $\overline{G_\lambda}$ is isomorphic to a subgroup of $\mathrm{GL}_3(2)$ and has a strongly 2-embedded subgroup. But then $\overline{G_\lambda} = \overline{H_\lambda}$, $G_\lambda = H_\lambda$ and $G = X$, a contradiction.

Hence, if no non-trivial subgroup of $G_{\alpha,\beta}$ is normal in X then $(H_\lambda, G_\delta, G_{\alpha,\beta})$ is a weak BN-pair and we may associate a critical distance to it. Since $\langle (V_\delta^{(n)})^{H_\lambda}\rangle \le \langle (V_\delta^{(n)})^{G_\lambda}\rangle$, it follows that the critical distance associated to $(H_\lambda, G_\delta, G_{\alpha,\beta})$ is greater than or equal to b. Comparing with the results in [27], we have that $b \leqslant 5$ and $b \leqslant 3$ unless $(H_\lambda, G_\delta, G_{\alpha,\beta})$ is parabolic isomorphic to F_3. That $V_\alpha^{(2)}/Z_\alpha$ is not acted on quadratically by S is a consequence of the structure of an F_3-type amalgam. This case is outcome (a). Hence, for the remainder of the proof we may assume that

(3) some non-trivial subgroup of $G_{\alpha,\beta}$ is normal in $X = \langle P_\lambda, G_\delta\rangle$.

Let K be the largest subgroup by inclusion satisfying this condition. Since S is the unique Sylow p-subgroup of $G_{\alpha,\beta}$, K normalizes S so that $O_p(K) = S \cap K \trianglelefteq X$. If $O_p(K) = \{1\}$, then K is a p'-group which is normal in G_δ, impossible since $F^*(G_\delta) = Q_\delta$ is self-centralizing in G_δ. Thus, there is a finite p-group which is normal in X. Since $O_p(K) \trianglelefteq S$, $Z_\beta \le O_p(K)$. Then, by definition, $V \le O_p(K)$. Indeed, as $[O_p(K), V] = [O_p(K), \langle Z_\beta^X\rangle] = \{1\}$, we conclude that $V \le \Omega(Z(O_p(K)))$ and $O_p(K) \le C_S(V)$. By an earlier observation, $C_S(V) \trianglelefteq X$ so that

(4) $C_S(V) = O_p(K)$.

Set $\widetilde{X} := X/C_X(V)$ so that $\widetilde{X} = \langle \widetilde{H_\lambda}, \widetilde{G_\delta}\rangle$ and $\widetilde{H_\lambda} \cong H_\lambda/C_{H_\lambda}(V)$ is a finite group. Additionally, $\widetilde{G_\delta} \cong G_\delta/C_{G_\delta}(V)$ is a finite group. Since $C_S(V) \in \mathrm{Syl}_p(C_{H_\lambda}(V) \cap C_{G_\delta}(V))$, $C_S(V) \le C_\beta$ and H_λ does not normalize $Q_\alpha \cap Q_\beta$ we deduce that $\widetilde{Q_\lambda} = O_p(\widetilde{H_\lambda})$ and $\widetilde{H_\lambda}/\widetilde{Q_\lambda}$ has a strongly p-embedded subgroup. Similarly, $\widetilde{Q_\delta} = O_p(\widetilde{G_\delta})$ and $\widetilde{G_\delta}/\widetilde{Q_\delta}$ has a strongly p-embedded subgroup.

In order to show that the triple $(\widetilde{H_\lambda}, \widetilde{G_\delta}, \widetilde{G_{\alpha,\beta}})$ satisfies Hypothesis 5.1, we need to show that $\widetilde{H_\lambda}$ and $\widetilde{G_\delta}$ are of characteristic p, $\widetilde{G_{\alpha,\beta}} = \widetilde{H_\lambda} \cap \widetilde{G_\delta} = N_{\widetilde{H_\lambda}}(\widetilde{S}) = N_{\widetilde{G_\delta}}(\widetilde{S})$ and no non-trivial subgroup of $\widetilde{G_{\alpha,\beta}}$ is normal in both $\widetilde{H_\lambda}$ and $\widetilde{G_\delta}$. In the following, we often examine the "preimage in H_λ" of some subgroup of $\widetilde{H_\lambda}$, by which we mean the preimage in H_λ of the isomorphic image in $H_\lambda/C_{H_\lambda}(V)$.

Notice that if $\widetilde{H_\lambda}$ is not of characteristic p then $F^*(\widetilde{H_\lambda}) \neq \widetilde{Q_\lambda}$. Then, as $\widetilde{H_\lambda}$ is p-solvable, $O_{p'}(\widetilde{H_\lambda}) \neq \{1\}$ so that for $\mathcal{D}_\lambda$ the preimage in H_λ of $O_{p'}(\widetilde{H_\lambda})$, $[\mathcal{D}_\lambda, Q_\lambda, V] = \{1\}$. For $r \in \mathcal{D}_\lambda$ of order coprime to p, it follows from the A×B-lemma that if r centralizes $C_V(Q_\lambda)$, then $\widetilde{r} = 1$. Since Q_λ is self-centralizing in S, we have that $C_V(Q_\lambda) \le Z(Q_\lambda)$. Similarly, if $\widetilde{G_\delta}$ is not of characteristic p, defining $\mathcal{D}_\delta$ analogously, by the A×B-lemma we have that $\mathcal{D}_\delta$ acts faithfully on $C_V(Q_\delta) \le Z(Q_\delta)$.

Suppose that $\lambda = \beta$. Then $|Z(Q_\beta)| = p$ and so, either $\widetilde{H_\beta}$ is of characteristic p; or $p = 3$, $|\widetilde{\mathcal{D}_\beta}| = 2$ and $\mathcal{D}_\beta$ acts non-trivially on Z_β. Assume that the latter case holds. Then $\widetilde{\mathcal{D}_\beta} \le Z(\widetilde{H_\beta})$ so that $[\mathcal{D}_\beta, S] \le C_{H_\beta}(V)$. Moreover, by coprime action, we have that $V = [V, \mathcal{D}_\beta] \times C_V(\mathcal{D}_\beta)$ is an S-invariant decomposition and as $\widetilde{\mathcal{D}_\beta}$ acts non-trivially on Z_β, it follows that $V = [V, \mathcal{D}_\beta]$ is inverted by $\widetilde{\mathcal{D}_\beta}$. By the Frattini argument, $\mathcal{D}_\beta S = C_{H_\beta}(V) S(G_{\alpha,\beta} \cap \mathcal{D}_\beta)$ and we may as well assume that there is $x \in G_{\alpha,\beta} \cap \mathcal{D}_\beta$ such that $\widetilde{\langle x \rangle} = \widetilde{\mathcal{D}_\beta}$. But then $[x, Q_\alpha] \leq [x, S] \leq C_S(V)$ and as $x \in G_{\alpha,\beta} \leq G_\alpha$

(5) if $\lambda = \beta$ and $\widetilde{H_\beta}$ is not of characteristic p, then $p = 3$, $|\widetilde{\mathcal{D}_\beta}| = 2$ and $\widetilde{G_\alpha}$ is not of characteristic p.

Consider $\mathcal{D}_\alpha$, the preimage in G_α of $O_{p'}(\widetilde{G_\alpha})$. If $\widetilde{G_\alpha}$ is not of characteristic p, then applying the A×B-lemma, $\mathcal{D}_\alpha \cap C_{G_\alpha}(Z_\alpha) \leq C_{G_\alpha}(V)$ and $\widetilde{\mathcal{D}_\alpha}$ is isomorphic to a normal p'-subgroup of $\mathrm{GL}_2(p)$. Our first aim will be to reduce to the case where $p = 3$ and $|\widetilde{\mathcal{D}_\alpha}| \leqslant 2$.

To this end, we may suppose that $|\widetilde{\mathcal{D}_\alpha}| = 3$ if $p = 2$, or $\widetilde{\mathcal{D}_\alpha} \cong Q_8$ if $p = 3$. Noticing that $[S, C_G(Z_\alpha)] \leq [L_\alpha, C_{G_\alpha}(Z_\alpha)] \leq R_\alpha$, by the Frattini argument, $C_{G_\alpha}(Z_\alpha) G_{\alpha,\beta} = R_\alpha G_{\alpha,\beta}$ and $G_\alpha = R_\alpha G_{\alpha,\beta} \mathcal{D}_\alpha$. By Proposition 5.27, since $\mathcal{D}_\alpha G_{\alpha,\beta}$ normalizes $Q_\alpha \cap Q_\beta$, we will have a contradiction whenever R_α is shown to normalize $Q_\alpha \cap Q_\beta$.

Hence, if $\widetilde{G_\alpha}$ is not of characteristic p and $|\widetilde{\mathcal{D}_\alpha}| \not\leqslant 2$ then R_α does not normalize $Q_\alpha \cap Q_\beta$. Under these assumptions, let $M_\alpha := C_{G_\alpha}(Z_\alpha) G_{\alpha,\beta}$. Then, $C_{G_\alpha}(Z_\alpha) \not\leq G_{\alpha,\beta}$ so that $Q_\alpha = O_p(M_\alpha)$. Reapplying the A×B-lemma yields $\widetilde{M_\alpha \cap \mathcal{D}_\alpha} = \{1\}$ if $p = 2$ and $|\widetilde{M_\alpha \cap \mathcal{D}_\alpha}| \leqslant 2$ if $p = 3$. In the latter case, suppose that $\widetilde{M_\alpha \cap \mathcal{D}_\alpha}$ is non-trivial and choose $y \in M_\alpha \cap \mathcal{D}_\alpha$ with $[y, V] \neq \{1\}$. Indeed, $\widetilde{\langle y \rangle} = \widetilde{M_\alpha \cap \mathcal{D}_\alpha}$ is central in $\widetilde{M_\alpha}$. It follows that $[y, S] \leq C_{M_\alpha}(V)$. Now, by the Frattini argument, $(\mathcal{D}_\alpha \cap M_\alpha) S = C_{M_\alpha}(V) S(G_{\alpha,\beta} \cap \mathcal{D}_\alpha)$ and we may as well assume that $y \in G_{\alpha,\beta}$ so that $[y, S] \leq C_S(V)$. But then $[y, Q_\beta] \leq C_X(V)$ and so $\widetilde{H_\beta}$ is not of characteristic 3. Indeed, we can arrange that $\langle y \rangle C_{H_\beta}(V) = \mathcal{D}_\beta$.

Now, we may form $M_\alpha^* := C_{G_\alpha}(Z_\alpha)(L_\beta \cap G_{\alpha,\beta})$ and $H_\beta^* := (H_\beta \cap L_\beta)(M_\alpha^* \cap G_{\alpha,\beta})$ and arguing as above, we infer that $\widetilde{M_\alpha^*}$ and $\widetilde{H_\beta^*}$ are both of characteristic p. Moreover, by construction and since R_α does not normalize $Q_\alpha \cap Q_\beta$, we deduce that $\widetilde{Q_\alpha} = O_p(\widetilde{M_\alpha^*})$ and $\widetilde{M_\alpha^*}/\widetilde{Q_\alpha}$ has a strongly p-embedded subgroup. Similarly, $\widetilde{Q_\beta} = O_p(\widetilde{H_\beta^*})$ and $\widetilde{H_\beta^*}/\widetilde{Q_\beta}$ also has a strongly p-embedded subgroup. Set $Y := \langle M_\alpha^*, H_\beta^* \rangle$ and write $G_{\alpha,\beta}^* := M_\alpha^* \cap G_{\alpha,\beta}$. Since $\widetilde{S} = \widetilde{Q_\alpha} \widetilde{Q_\beta}$, it is easily checked that $\widetilde{G_{\alpha,\beta}^*} = N_{\widetilde{M_\alpha^*}}(\widetilde{S}) = N_{\widetilde{H_\beta^*}}(\widetilde{S}) = \widetilde{M_\alpha^*} \cap \widetilde{H_\beta^*}$.

Suppose there exists $K^* \leq \widetilde{G_{\alpha,\beta}^*}$ such that $K^* \trianglelefteq \langle \widetilde{M_\alpha^*}, \widetilde{H_\beta^*} \rangle = \widetilde{Y}$. Since $\widetilde{M_\alpha^*}$ and $\widetilde{H_\beta^*}$ are both of characteristic p, we may assume that K^* is not a p'-group, and since $K^* \leq \widetilde{G_{\alpha,\beta}^*}$, $O_p(K^*) = K^* \cap \widetilde{S} \neq \{1\}$. Let K_α denote the preimage of $O_p(K^*)$ in M_α^* and K_β denote the preimage of $O_p(K^*)$ in H_β^*. Then, $T_\alpha := Q_\alpha \cap K_\alpha$ is a normal p-subgroup of M_α^* and, likewise, $T_\beta := Q_\beta \cap K_\beta$ is a normal p-subgroup of H_β^*. Since $\widetilde{T_\alpha T_\beta} = \widetilde{T_\alpha} = \widetilde{T_\beta}$, a comparison of orders yields $T_\alpha T_\beta = T_\alpha = T_\beta \trianglelefteq Y$. Moreover, $T_\alpha > C_S(V)$ and as Y is normalized by $G_{\alpha,\beta}$, T_α is normalized by $G_{\alpha,\beta}$. But now, $G_\alpha = G_{\alpha,\beta} \mathcal{D}_\alpha M_\alpha^*$ and as $\mathcal{D}_\alpha$ centralizes $Q_\alpha / C_S(V)$, $T_\alpha \trianglelefteq$

$\langle G_\alpha, H_\beta \rangle = X$, a contradiction since $C_S(V)$ is the largest p-subgroup of $G_{\alpha,\beta}$ which is normalized by X. Hence,

(6) if $\lambda = \beta$, $\widetilde{G_\alpha}$ is not of characteristic p and $|\widetilde{\mathcal{D}_\alpha}| \neq 2$, then the triple $(\widetilde{M_\alpha^*}, \widetilde{H_\beta^*}, \widetilde{G_{\alpha\beta}^*})$ with completion $\widetilde{Y}$ satisfies Hypothesis 5.1.

Since $C_S(V) \leq Q_\alpha \cap Q_\beta$ and $C_S(V)$ is the largest subgroup of S which is normal in Y, we have that $J(S) \not\leq C_S(V)$ and a elementary calculation yields that $\Omega(Z(C_S(V)))$ is an FF-module for $\widetilde{Y}$. Moreover, by construction, $Y = \langle S^Y \rangle$ and, by minimality and since $\widetilde{M_\alpha^*}$ and $\widetilde{H_\beta^*}$ are p-solvable, using Theorem 5.29 we have that

(7) if $\lambda = \beta$, $\widetilde{G_\alpha}$ is not of characteristic p and $|\widetilde{\mathcal{D}_\alpha}| \neq 2$ then $\widetilde{Y}$ is locally isomorphic to one of $\mathrm{SL}_3(p)$, $\mathrm{Sp}_4(p)$ or $\mathrm{G}_2(2)$.

Moreover, since $\lambda = \beta$ and $V_\beta \leq V \trianglelefteq X$, we have that $V_\alpha^{(2)} \leq V$ so that $C_S(V) \leq C_{Q_\alpha}(V_\alpha^{(2)})$. If $\widetilde{Y}$ is locally isomorphic to $\mathrm{SL}_3(p)$, then C_β is the largest normal subgroup of H_β contained in $Q_\alpha \cap Q_\beta$, it follows that $C_\beta \leq C_S(V) \leq C_{Q_\alpha}(V_\alpha^{(2)})$, a contradiction for then $C_\beta \trianglelefteq \langle G_\alpha, G_\beta \rangle$.

If $\widetilde{Y}$ is locally isomorphic to $\mathrm{Sp}_4(p)$, then it follows that $|\widetilde{C_\beta}| \leqslant p$. We may as well assume that $C_S(V) = C_{Q_\alpha}(V_\alpha^{(2)})$ has index p in C_β, else we obtain a contradiction as before. Since $C_S(V) \trianglelefteq X$ and $G_\beta = \langle H_\beta, R_\beta \rangle = \langle H_\beta, C_{L_\beta}(U/W) \rangle$, it follows that neither R_β nor $C_{L_\beta}(U/W)$ normalizes $V_\alpha^{(2)}$ and conclusion (c) holds.

If $\widetilde{Y} \cong \mathrm{G}_2(2)$, then one can calculate in a similar manner that $C_S(V) = C_{Q_\alpha}(V_\alpha^{(2)})$ and again we retrieve outcome (c). Therefore, we may assume that

(8) if $\lambda = \beta$ and $\widetilde{G_\alpha}$ is not of characteristic p, then $p = 3$ and $|\widetilde{\mathcal{D}_\alpha}| = 2$.

Then $[\widetilde{S}, \widetilde{\mathcal{D}_\alpha}] = \{1\}$ and, again applying the Frattini argument, we have that $\mathcal{D}_\alpha S = C_{G_\alpha}(V)S(G_{\alpha,\beta} \cap \mathcal{D}_\alpha)$. Choose $y \in G_{\alpha,\beta} \cap \mathcal{D}_\alpha$ with $[y, V] \neq \{1\}$ so that $\widetilde{\langle y \rangle} = \widetilde{\mathcal{D}_\alpha}$. Indeed, $[y, S] \leq C_S(V)$ and it follows that $\widetilde{H_\beta}$ is not of characteristic 3. Hence, we have reduced to a situation such that

(9) if $\lambda = \beta$ then $\widetilde{H_\beta}$ is not of characteristic p if and only if $\widetilde{G_\alpha}$ is not of characteristic p.

Moreover, there is $x \in G_{\alpha,\beta}$ such that $\widetilde{\langle x \rangle} = \widetilde{\mathcal{D}_\alpha} = \widetilde{\mathcal{D}_\beta}$.

If $\widetilde{G_\alpha}$ is not of characteristic p, then set $\widehat{X} := \widetilde{X}/\widetilde{\langle x \rangle}$ so that both $\widehat{H_\beta}$ and $\widehat{G_\alpha}$ are of characteristic 3. Moreover, $\widehat{L_\alpha}/\widehat{R_\alpha} \cong \mathrm{PSL}_2(3)$ and $\widehat{O^{p'}(H_\beta)}/(R_\beta \cap \widehat{O^{p'}(H_\beta)}) \cong \mathrm{SL}_2(3)$. As in the construction of $\widetilde{Y}$ above, it is easily checked that $\widehat{G_{\alpha,\beta}} = N_{\widehat{G_\alpha}}(\widehat{S}) = N_{\widehat{H_\beta}}(\widehat{S}) = \widehat{G_\alpha} \cap \widehat{H_\beta}$ and no non-trivial subgroup of $\widehat{G_{\alpha,\beta}}$ is normal in $\widehat{X}$. Thus, by minimality, the triple $(\widehat{G_\alpha}, \widehat{H_\beta}, \widehat{G_{\alpha,\beta}})$ is a weak BN-pair. Indeed, $\widehat{L_\alpha} = O^{3'}(\widehat{G_\alpha})$ and $\widehat{L_\alpha} \cong \mathrm{PSL}_2(3)$ or $\mathrm{SL}_2(3)$. If $\widehat{L_\alpha} \cong \mathrm{SL}_2(3)$, then a Sylow 2-subgroup of $\widehat{L_\alpha}$ is of order 16, and arguing as in Lemma 5.21, we force a contradiction. Thus, $\widehat{L_\alpha} \cong \mathrm{PSL}_2(3)$ and $\widehat{X}$ is locally isomorphic to $\mathrm{PSp}_4(3)$.

Then, using that C_β is the largest normal subgroup of H_β which is contained in $Q_\alpha \cap Q_\beta$ and $C_{Q_\alpha}(V_\alpha^{(2)})$ is the largest subgroup of C_β normal in G_α, it follows that $C_S(V) = C_{Q_\alpha}(V_\alpha^{(2)}) \trianglelefteq X$. Since $G_\beta = \langle H_\beta, R_\beta \rangle = \langle H_\beta, C_{L_\beta}(U/W) \rangle$, it follows that neither R_β nor $C_{L_\beta}(U/W)$ normalizes $V_\alpha^{(2)}$ and conclusion (c) holds. Thus, we may as well assume that

(10) whenever $\lambda = \beta$, $\widetilde{X}$ satisfies Hypothesis 5.1 and acts faithfully on V.

Suppose now that $\lambda = \alpha$ so that $H_\alpha / C_{H_\alpha}(Z_\alpha)$ is isomorphic to a subgroup of $\mathrm{GL}_2(p)$. If $\widetilde{H_\alpha}$ is not of characteristic p then, by the A×B-lemma, $\mathcal{D}_\alpha \not\leq C_{H_\alpha}(Z_\alpha)$ and so $\mathcal{D}_\alpha C_{H_\alpha}(Z_\alpha)/C_{H_\alpha}(Z_\alpha)$ is isomorphic to a normal p'-subgroup of $\mathrm{GL}_2(p)$. If $p = 2$ or $|\mathcal{D}_\alpha C_{H_\alpha}(Z_\alpha)/C_{H_\alpha}(Z_\alpha)| > 2$ and $p = 3$, using the Frattini argument it follows that $H_\alpha = C_{P_\alpha}(Z_\alpha)\mathcal{D}_\alpha G_{\alpha,\beta} = (R_\alpha \cap C_{L_\alpha}(U/W))\mathcal{D}_\alpha G_{\alpha,\beta}$ which normalizes $Q_\alpha \cap Q_\beta$, a contradiction. Thus, $p = 3$ and $|\widetilde{\mathcal{D}_\alpha}| = 2$ so that $[\mathcal{D}_\alpha, S] \leq C_X(V)$. Additionally, by coprime action, $V = [V, \mathcal{D}_\alpha] \times C_V(\mathcal{D}_\alpha)$ and as $\widetilde{\mathcal{D}_\alpha}$ does not centralize Z_β we deduce that $V = [V, \mathcal{D}_\alpha]$ is inverted by $\widetilde{\mathcal{D}_\alpha}$. Then, by the Frattini argument, $S\mathcal{D}_\alpha = SC_{H_\alpha}(V)(G_{\alpha,\beta} \cap \mathcal{D}_\alpha)$ and we may choose $x \in G_{\alpha,\beta} \cap \mathcal{D}_\alpha$ with $[x, V] \neq \{1\}$ so that $\widetilde{\langle x \rangle} = \widetilde{\mathcal{D}_\alpha}$ and $[x, S] \leq C_S(V)$. It follows that $\widetilde{G_\beta}$ is not of characteristic 3.

If $\widetilde{G_\beta}$ is not of characteristic p then, by the A×B-lemma, $\mathcal{D}_\beta$ does not centralize Z_β. In particular, $p = 3$ and $|\widetilde{\mathcal{D}_\beta}| = 2$. Then applying coprime action, $\widetilde{\mathcal{D}_\beta}$ inverts V and we see that there is $x \in G_{\alpha,\beta}$ with $\widetilde{\langle x \rangle} = \widetilde{\mathcal{D}_\alpha} = \widetilde{\mathcal{D}_\beta}$. Hence,

(11) if $\lambda = \alpha$ then $\widetilde{H_\alpha}$ is of characteristic p if and only if $\widetilde{G_\beta}$ is of characteristic p.

If $\widetilde{G_\beta}$ is not of characteristic p, then set $\widehat{X} := \widetilde{X}/\widetilde{\langle x \rangle}$ so that $\widehat{H_\alpha}$ and $\widehat{G_\beta}$ are both of characteristic 3, $\widehat{O^{p'}(H_\alpha)}/\widehat{O^{p'}(H_\alpha) \cap R_\alpha} \cong \mathrm{PSL}_2(3)$ and $\widehat{L_\beta}/\widehat{R_\beta} \cong \mathrm{SL}_2(3)$. As in the above, it quickly follows that $\widehat{X}$ satisfies Hypothesis 5.1 and by minimality, the triple $(\widehat{H_\alpha}, \widehat{G_\beta}, \widehat{G_{\alpha,\beta}})$ is a weak BN-pair of rank 2. Indeed, $\widehat{O^{p'}(H_\alpha)} \cong \mathrm{PSL}_2(3)$ and $\widehat{X}$ is locally isomorphic to $\mathrm{PSp}_4(3)$, and the outstanding case in (d) is satisfied. We may as well assume that

(12) whenever $\lambda = \alpha$, $\widetilde{X}$ has satisfies Hypothesis 5.1 and acts faithfully on V.

Finally, we treat the case where: for either $\lambda = \alpha$ or $\lambda = \beta$, $\widetilde{X}$ satisfies Hypothesis 5.1 and acts faithfully on V. Moreover, since $J(S) \not\leq C_S(V)$ an elementary argument (as in the proof of Proposition 2.40) implies that V is an FF-module for $\widetilde{X}$. By minimality, $\widetilde{X}$ satisfies Hypothesis 5.1 and since both $\widetilde{H_\lambda}$ and $\widetilde{G_\delta}$ are p-solvable, $\widetilde{X}$ is determined by Theorem 5.29. Counting the number of non-central chief factors in amalgams locally isomorphic to $\mathrm{SL}_3(p)$, $\mathrm{Sp}_4(p)$ or $\mathrm{G}_2(2)$ (as can be gleaned from [27]), outcome (d) is satisfied. □

The hypothesis of Proposition 5.30 exhibits a common situation we encounter in the work ahead: where $Z_\beta = Z(Q_\beta)$ is of order p, and both $Z(Q_\alpha) = Z_\alpha$ and $V_\beta/C_{V_\beta}(O^p(L_\beta))$ are natural $\mathrm{SL}_2(p)$-modules for $L_\alpha/R_\alpha \cong \mathrm{SL}_2(p) \cong L_\beta/R_\beta$. Upon first glance, it seems that we have very little control over the action of R_λ for $\lambda \in \{\alpha, \beta\}$. Throughout this section we strive to force situations in which the full hypotheses of Proposition 5.30 are satisfied. In applying Proposition 5.30, the outcomes there will often force contradictions and the conclusion we draw is that $O^p(R_\lambda)$ centralizes U/W. In this situation, Lemma 5.22 becomes a powerful tool in dispelling a large number of cases. Motivated by this, we make the following hypothesis and record a number of lemmas controlling the actions of R_λ for $\lambda \in \Gamma$.

Hypothesis 5.31 The following conditions hold:

(i) $L_\alpha/R_\alpha \cong \mathrm{SL}_2(q) \cong L_\beta/R_\beta$, and Z_α and $V_\beta/C_{V_\beta}(O^p(L_\beta))$ are natural $\mathrm{SL}_2(q)$-modules; and
(ii) if $q = p$ then $Z(Q_\alpha) = Z_\alpha$ is of order p^2 and $Z(Q_\beta) = Z_\beta = \Omega(Z(S))$ is of order p.

As a first consequence of this hypothesis, we make the following observation, gaining some control over the order of V_β and the number of non-central chief factors in $V_\alpha^{(2)}$.

Lemma 5.32 *Suppose that $b > 2$ and Hypothesis 5.31 is satisfied. Then, for $\lambda \in \alpha^G$ and $\delta \in \Delta(\lambda)$, exactly one of the following occurs:*

(i) $|V_\delta| = q^3$ and $[V_\lambda^{(2)}, Q_\lambda] = Z_\lambda$; or
(ii) $C_{V_\delta}(O^p(L_\delta)) \neq Z_\delta$ and for $V^\lambda := \langle C_{V_\delta}(O^p(L_\delta))^{G_\lambda}\rangle$, $[V^\lambda, Q_\lambda] = Z_\lambda$, both $V_\lambda^{(2)}/V^\lambda$ and V^λ/Z_λ contain a non-central chief factor for L_λ, and $V^\lambda V_\delta \ntrianglelefteq L_\delta$.

Moreover, whenever $Z_{\delta+1}C_{V_\delta}(O^p(L_\delta)) = Z_{\delta-1}C_{V_\delta}(O^p(L_\delta))$ for $\delta \in \Gamma^G$, we have that $Z_{\delta+1} = Z_{\delta-1}$.

Proof Suppose first that $|V_\delta| = q^3$. Then $[Q_\lambda, V_\lambda^{(2)}] = [Q_\lambda, V_\delta]^{G_\lambda} = Z_\lambda$ and (i) holds. Hence, we assume for the remainder of the proof that $C_{V_\delta}(O^p(L_\delta)) \neq Z_\delta$.

Suppose that $q = p \in \{2, 3\}$. Then by coprime action, $V_\delta/Z_\delta = [V_\delta/Z_\delta, O^p(L_\delta)] \times C_{V_\delta/Z_\delta}(O^p(L_\delta))$ and $C_{V_\delta/Z_\delta}(O^p(L_\delta)) = C_{V_\delta}(O^p(L_\delta))/Z_\delta$. We infer from the definition of V_δ that $V_\delta = [V_\delta, O^p(L_\delta)]Z_\lambda$. In particular, $[Q_\lambda, C_{V_\delta}(O^p(L_\delta))] \leq [V_\delta, O^p(L_\delta)] \cap C_{V_\delta}(O^p(L_\delta)) = Z_\delta$. In all other cases of q and p, we have that $L_\delta = O^p(L_\delta)Q_\delta$, so that $[Q_\lambda, C_{V_\delta}(O^p(L_\delta))] \leq [L_\delta, C_{V_\delta}(O^p(L_\delta))] \leq Z_\delta$.

Note that if $[Q_\lambda, C_{V_\delta}(O^p(L_\delta))] = \{1\}$, then as $Q_\delta/(Q_\delta \cap Q_\lambda)$ is $G_{\delta,\lambda}$-irreducible, we would either have that $Q_\lambda \cap Q_\delta = C_{Q_\delta}(C_{V_\delta}(O^p(L_\delta))) \trianglelefteq L_\delta$, a contradiction by Proposition 5.27; or that $C_{V_\delta}(O^p(L_\delta)) \leq Z(Q_\delta)$. In the latter case, since $Z(Q_\delta) = Z_\delta$ when $q = p$ and since $L_\delta = O^p(L_\delta)Q_\delta$ otherwise, we deduce that $C_{V_\delta}(O^p(L_\delta)) = Z_\delta$, another contradiction. Hence, $[Q_\lambda, C_{V_\delta}(O^p(L_\delta))] \neq \{1\}$.

Let $V^\lambda := \langle C_{V_\delta}(O^p(L_\delta))^{G_\lambda}\rangle$. Since $\{1\} \neq [Q_\lambda, C_{V_\delta}(O^p(L_\delta)] \leq Z_\lambda$, we deduce that $[Q_\lambda, V^\lambda] = Z_\lambda < V^\lambda \leq V_\lambda^{(2)}$. Suppose that V^λ/Z_λ does not contain a non-central chief factor for L_λ. Then L_λ normalizes $Z_\lambda C_{V_\delta}(O^p(L_\delta))$ so that $V^\lambda = Z_\lambda C_{V_\delta}(O^p(L_\delta))$ and $[Q_\lambda, Z_\lambda C_{V_\delta}(O^p(L_\delta))] \trianglelefteq L_\lambda$. But $[Q_\lambda, Z_\lambda C_{V_\delta}(O^p(L_\delta))] \leq Z_\delta$ and we deduce that Q_λ centralizes $C_{V_\delta}(O^p(L_\delta))$, a contradiction. Thus, V^λ/Z_λ contains a non-central chief factor for L_λ.

Assume that $V_\lambda^{(2)}/V^\lambda$ does not contain a non-central chief factor for L_λ. Similarly to above, we deduce in this case that $V_\lambda^{(2)} = V^\lambda V_\delta$ so that

$$Z_\lambda \le [V_\lambda^{(2)}, Q_\lambda] \le [V^\lambda, Q_\lambda][V_\delta, Q_\lambda] \le Z_\lambda C_{V_\delta}(O^p(L_\delta)).$$

Indeed, $[V_\lambda^{(2)}, Q_\lambda] = Z_\lambda([V_\lambda^{(2)}, Q_\lambda] \cap C_{V_\delta}(O^p(L_\delta)))$. But now, $[Q_\lambda, [V_\lambda^{(2)}, Q_\lambda] \cap C_{V_\delta}(O^p(L_\delta))]$ is a normal subgroup of L_δ contained in $[Q_\lambda, C_{V_\delta}(O^p(L_\delta))] \le Z_\delta$. Hence, $[V_\lambda^{(2)}, Q_\lambda] \cap C_{V_\delta}(O^p(L_\delta))$ is centralized by Q_λ. But $[V_\lambda^{(2)}, Q_\lambda] \cap C_{V_\delta}(O^p(L_\delta)) \trianglelefteq L_\delta = SO^p(L_\delta)$ so that $[L_\delta, [V_\lambda^{(2)}, Q_\lambda] \cap C_{V_\delta}(O^p(L_\delta))] = [\langle Q_\lambda{}^{L_\delta}\rangle, [V_\lambda^{(2)}, Q_\lambda] \cap C_{V_\delta}(O^p(L_\delta))] = \{1\}$ and $[V_\lambda^{(2)}, Q_\lambda] \cap C_{V_\delta}(O^p(L_\delta)) = Z_\delta$. Hence, $[V_\lambda^{(2)}, Q_\lambda] = Z_\lambda$ so that $[V_\delta, Q_\lambda] \le Z_\lambda$. Hence $Z_\lambda Z_\mu \trianglelefteq L_\delta = \langle Q_\lambda, Q_\mu, R_\delta\rangle$ for any $\mu \in \Delta(\delta)$ with $Q_\lambda R_\delta \ne Q_\mu R_\delta$. By the definition of V_δ, $V_\delta = Z_\lambda Z_\mu$ is of order q^3 and $C_{V_\delta}(O^p(L_\delta)) = Z_\delta$, a contradiction. Thus, $V_\lambda^{(2)}/V^\lambda$ contains a non-central chief factor for L_λ.

Suppose that $Z_\lambda C_{V_\delta}(O^p(L_\delta)) = Z_\mu C_{V_\delta}(O^p(L_\delta))$ for some $\mu \in \Delta(\delta)$. In particular, $Z_\lambda C_{V_\delta}(O^p(L_\delta))$ is normalized by $Q_\lambda R_\delta$ and $Q_\mu R_\delta$. If $Q_\lambda R_\delta \ne Q_\mu R_\delta$ then $Z_\lambda C_{V_\delta}(O^p(L_\delta)) \trianglelefteq L_\delta = \langle Q_\lambda, Q_\mu, R_\delta\rangle$, and from the definition of V_δ, $V_\delta = Z_\lambda C_{V_\delta}(O^p(L_\delta))$ is centralized, modulo Z_δ, by Q_λ, a contradiction by Lemma 5.19. Thus, $Q_\lambda R_\delta = Q_\mu R_\delta$. Then, there is $r \in R_\delta$ such that $Q_\lambda^r Q_\delta = (Q_\lambda Q_\delta)^r = Q_\mu Q_\delta = Q_\mu Q_\delta$ and we may as well pick r of order coprime to p. Moreover, since $O^p(R_\delta)$ centralizes Q_δ/C_δ, it follows that $Q_\lambda \in \mathrm{Syl}_p(Q_\lambda O^p(R_\delta))$. But then $Q_\mu \in \mathrm{Syl}_p(Q_\lambda O^p(R_\delta))$. Since r centralizes Q_δ/C_δ we conclude that $Q_\lambda \cap Q_\delta = Q_\mu \cap Q_\delta$. Therefore, $Z_\lambda = Z(Q_\lambda \cap Q_\delta) \cap V_\delta = Z(Q_\mu \cap Q_\delta) \cap V_\delta = Z_\mu$.

Suppose for a contradiction that $V^\lambda V_\delta \trianglelefteq L_\delta$ and choose $\mu \in \Delta(\mu)$ such that $Z_\lambda \ne Z_\mu$ and $Q_\delta = (Q_\lambda \cap Q_\delta)(Q_\mu \cap Q_\delta)$. In particular, $Z_\lambda C_{V_\delta}(O^p(L_\delta)) \ne Z_\mu C_{V_\delta}(O^p(L_\delta))$. Moreover, as $V^\lambda V_\delta \trianglelefteq L_\delta$, $V^\lambda V_\delta = V^\mu V^\delta$. Now,

$$Z_\delta \le [Q_\delta, V^\lambda V_\delta] = [Q_\lambda \cap Q_\delta, V^\lambda V^\delta][Q_\mu \cap Q_\delta, V^\mu V^\delta] \le Z_\lambda Z_\mu$$

and $[Q_\delta, V^\lambda V_\delta] \trianglelefteq L_\delta$. Set $W := [Q_\delta, V^\lambda V_\delta]$. If $W \not\le C_{V_\delta}(O^p(L_\delta))$, then $V_\delta = W C_{V_\delta}(O^p(L_\delta))$ and we deduce that $|W| \geqslant q^3$. Since $W \le Z_\lambda Z_\mu$ we conclude that $W = Z_\lambda Z_\mu \trianglelefteq L_\delta$ and has order q^3. By definition, $V_\delta = W$, a contradiction since $C_{V_\delta}(O^p(L_\delta)) \ne Z_\delta$. Hence, $W \le C_{V_\delta}(O^p(L_\delta))$. Then $W \le (Z_\lambda Z_\mu) \cap C_{V_\delta}(O^p(L_\delta))$ and since $Z_\lambda C_{V_\delta}(O^p(L_\delta)) \ne Z_\mu C_{V_\delta}(O^p(L_\delta))$ we deduce that $Z_\delta \le W \le (Z_\lambda Z_\mu) \cap C_{V_\delta}(O^p(L_\delta)) = Z_\delta$ so that $W = Z_\delta$. But then Q_δ centralizes V^λ/Z_λ, a contradiction since V^λ/Z_λ contains a non-central chief factor for L_λ. Hence, $V^\lambda V_\delta \ntrianglelefteq L_\delta$. This completes the proof. □

Lemma 5.33 *Suppose that $b > 3$ and Hypothesis 5.31 is satisfied. If Z_α, V^α/Z_α and $V_\alpha^{(2)}/V^\alpha$ are FF-modules or trivial modules for $\overline{L_\alpha}$, then $R_\alpha = C_{L_\alpha}(V_\alpha^{(2)})Q_\alpha$.*

Proof Of the configurations described in Theorem C which satisfy $b > 2$, all satisfy $R_\alpha = Q_\alpha$ and so we may assume throughout that G is a minimal counterexample to Theorem C such that $R_\alpha \neq C_{L_\alpha}(V_\alpha^{(2)})Q_\alpha$.

Suppose first that $|V_\beta| \neq q^3$ so that V^α/Z_α contains a non-central chief factor for L_α. Since $L_\alpha/R_\alpha \cong \mathrm{SL}_2(q)$ and $Q_\alpha \in \mathrm{Syl}_p(R_\alpha)$, $|S/Q_\alpha| = q$ and by Lemma 2.41, $L_\alpha/C_{L_\alpha}(V^\alpha/Z_\alpha) \cong L_\alpha/C_{L_\alpha}(V_\alpha^{(2)}/V^\alpha) \cong \mathrm{SL}_2(q)$. Thus, if $C_{L_\alpha}(V^\alpha/Z_\alpha) \neq R_\alpha$, a standard calculation yields that $q \in \{2, 3\}$. Moreover, if $p = 3$ and $R_\alpha C_{L_\alpha}(V^\alpha/Z_\alpha)S < L_\alpha$, then $|R_\alpha C_{L_\alpha}(V^\alpha/Z_\alpha)/R_\alpha| = 2$, $|L_\alpha/C_{L_\alpha}(V^\alpha)Q_\alpha| = 2^4.3$ and Proposition 2.47 (ii) gives a contradiction. Hence, if $C_{L_\alpha}(V^\alpha/Z_\alpha) \neq R_\alpha$ then $L_\alpha = R_\alpha C_{L_\alpha}(V^\alpha/Z_\alpha)S$. But now, $C_{L_\alpha}(V^\alpha/Z_\alpha)$ normalizes $Z_\alpha C_{V_\beta}(O^p(L_\beta))$ and so normalizes $[Z_\alpha C_{V_\beta}(O^p(L_\beta)), Q_\alpha] = Z_\beta$, a contradiction for then $Z_\beta \trianglelefteq L_\alpha$. Thus, $C_{L_\alpha}(V^\alpha/Z_\alpha) = R_\alpha$. Similarly, considering $C_{L_\alpha}(V_\alpha^{(2)}/V^\alpha)$, we have that $V_\beta V^\alpha \trianglelefteq C_{L_\alpha}(V_\alpha^{(2)}/V^\alpha)$ and so $Z_\alpha C_{V_\beta}(O^p(L_\beta)) = Z_\alpha[V_\beta, Q_\alpha] = [V_\beta V^\alpha, Q_\alpha] \trianglelefteq C_{L_\alpha}(V_\alpha^{(2)}/V^\alpha)$. Then $[Z_\alpha C_{V_\beta}(O^p(L_\beta)), Q_\alpha] = Z_\beta$ is normalized by $C_{L_\alpha}(V_\alpha^{(2)}/V^\alpha)$ and, as above, we conclude that $C_{L_\alpha}(V_\alpha^{(2)}/V^\alpha) = R_\alpha$ and the result holds.

Hence, we may assume that $|V_\beta| = q^3$. Since Hypothesis 5.31 is satisfied, $V_\alpha^{(2)}/Z_\alpha$ is an FF-module and $C_{L_\alpha}(V_\alpha^{(2)}/Z_\alpha) \cap R_\alpha = C_{L_\alpha}(V_\alpha^{(2)})Q_\alpha$. Then, $O^p(C_{L_\alpha}(V_\alpha^{(2)}))$ centralizes $Q_\alpha/C_{Q_\alpha}(V_\alpha^{(2)})$ and so normalizes $Q_\alpha \cap Q_\beta > C_{Q_\alpha}(V_\alpha^{(2)})$. We apply Proposition 5.30, taking $\lambda = \alpha$. As $b > 3$ and $V_\alpha^{(2)}/Z_\alpha$ is an FF-module (so admits quadratic action), outcome (a) does not hold. Since $\lambda = \alpha$ outcome (c) does not hold.

Suppose (d) holds. Then, by construction, $\langle V_\beta^{H_\alpha}\rangle = \langle V_\beta^{G_\alpha}\rangle = V_\alpha^{(2)}$ from which it follows that $V_\beta^{(3)} \leq V := \langle Z_\beta^X\rangle$ and the images of both Q_β/C_β and $C_\beta/C_{Q_\beta}(V_\beta^{(3)})$ in $\widetilde{L}_\beta$ contain a non-central chief factor for $\widetilde{L}_\beta$. By Proposition 5.30, $\widetilde{X} \cong \mathrm{G}_2(2)$. It follows from the structure of $\mathrm{G}_2(2)$ that $|Q_\alpha/C_\beta| = 2^2$, $|Q_\alpha/C_{Q_\alpha}(V_\alpha^{(2)})| = 2^4$ and $|C_{Q_\alpha}(V_\alpha^{(2)})| = 2$. Then, $C_S(V) = C_{Q_\beta}(V_\beta^{(3)}) \trianglelefteq X$. By Proposition 2.46 (iii), there are four non-equal subgroups of $L_\alpha/C_{L_\alpha}(V_\alpha^{(2)})Q_\alpha \cong (3 \times 3) : 2$ isomorphic to $\mathrm{Sym}(3)$, and so there is $H_\alpha^* \neq H_\alpha$ such that $S \in H_\alpha^*$, $O^2(H_\alpha^*)$ acts non-trivially on $V_\alpha^{(2)}/Z_\alpha$ and Z_α and $G_\alpha = \langle H_\alpha, H_\alpha^*\rangle$. If H_α^* does not normalize $Q_\alpha \cap Q_\beta$, then setting X^* for the subgroup of G obtained from employing the method in Proposition 5.30 with H_α^* instead of H_α, it follows from the work above that X^* also satisfies outcome (d) and for $V^* := \langle Z_\beta^{X^*}\rangle$, $C_S(V) = C_S(V^*) = C_{Q_\beta}(V_\beta^{(3)}) \trianglelefteq \langle H_\alpha, H_\alpha^*\rangle = G_\alpha$, a contradiction. Hence, H_α^* normalizes $Q_\alpha \cap Q_\beta$. Choose τ in $C_{L_\alpha}(V_\alpha^{(2)}/Z_\alpha) \setminus C_{L_\alpha}(V_\alpha^{(2)})$. Then τ normalizes V_β so normalizes $C_\beta = C_{Q_\alpha}(V_\beta)$, and $G_\alpha = \langle \tau, H_\alpha^*\rangle$. If τ centralizes Q_α/C_β, then τ normalizes $Q_\alpha \cap Q_\beta$ so that G_α normalizes $Q_\alpha \cap Q_\beta$, a contradiction by Proposition 5.27. Thus, τ acts non-trivially on Q_α/C_β and as $|Q_\alpha/C_\beta| = 4$, $C_\beta = (Q_\alpha \cap Q_\beta) \cap (Q_\alpha \cap Q_\beta)^\tau$. Now, $[O^2(H_\alpha^*), \tau] \leq C_{G_\alpha}(V_\alpha^{(2)})$ and as $O^2(H_\alpha^*)$ normalizes $Q_\alpha \cap Q_\beta$, $O^2(H_\alpha^*)$

normalizes $(Q_\alpha \cap Q_\beta)^\tau$. But then H_α^* normalizes $C_\beta = Q_\alpha \cap Q_\beta \cap Q_\beta^\tau$ and so $G_\alpha = \langle \tau, H_\alpha^* \rangle$ normalizes C_β, another contradiction.

Thus, we may assume that outcome (b) of Proposition 5.30 holds so that $p = 3$ and neither R_α nor $C_{L_\alpha}(V_\alpha^{(2)}/Z_\alpha)$ normalizes $Q_\alpha \cap Q_\beta$. Indeed, for the subgroup H_α as constructed in Proposition 5.30, we have that $Q_\alpha \cap Q_\beta \trianglelefteq H_\alpha$. Now, $C_{L_\alpha}(V_\alpha^{(2)}/Z_\alpha)$ normalizes C_β and we may assume that it acts non-trivially on Q_α/C_β for otherwise $Q_\alpha \cap Q_\beta \trianglelefteq G_\alpha = \langle H_\alpha, C_{L_\alpha}(V_\alpha^{(2)}/Z_\alpha)\rangle$, a contradiction by Proposition 5.27. Furthermore, $[O^3(O^{3'}(H_\alpha)), C_{L_\alpha}(V_\alpha^{(2)}/Z_\alpha)] \le C_{L_\alpha}(V_\alpha^{(2)})G_\alpha^{(1)}$ and as H_α normalizes $Q_\alpha \cap Q_\beta$ and $O^3(C_{L_\alpha}(V_\alpha^{(2)}))$ centralizes Q_α/C_β, it follows that for any $r \in C_{L_\alpha}(V_\alpha^{(2)}/Z_\alpha)$ of order coprime to 3 which does not normalize $Q_\alpha \cap Q_\beta$, $O^3(O^{3'}(H_\alpha))$ normalizes $(Q_\alpha \cap Q_\beta)^r$ and H_α normalizes $C_\beta = Q_\alpha \cap Q_\beta \cap Q_\beta^r$, a final contradiction for then $C_\beta \trianglelefteq G_\alpha = \langle H_\alpha, C_{L_\alpha}(V_\alpha^{(2)}/Z_\alpha)\rangle$. Hence, no counterexample exists, and the proof is complete. □

Lemma 5.34 *Suppose that $b > 5$ and Hypothesis 5.31 is satisfied. If $O^p(R_\alpha)$ centralizes $V_\alpha^{(2)}$ and $V_\alpha^{(4)}/V_\alpha^{(2)}$ contains a unique non-central chief factor which, as a* GF$(p)\overline{L_\alpha}$*-module, is an FF-module then $O^p(R_\alpha)$ centralizes $V_\alpha^{(4)}$.*

Proof Since none of the configurations described in Theorem C have $b > 5$, we may assume that G is a minimal counterexample such that $O^p(R_\alpha)$ does not centralize $V_\alpha^{(4)}/V_\alpha^{(2)}$, $V_\alpha^{(4)}/V_\alpha^{(2)}$ contains a unique non-central chief factor and $O^p(R_\alpha)$ centralizes $V_\alpha^{(2)}$. Since $O^p(R_\alpha)$ centralizes $V_\alpha^{(2)}$, an application of the three subgroup lemma implies that $O^p(R_\alpha)$ centralizes $Q_\alpha/C_{Q_\alpha}(V_\alpha^{(2)})$ and as $C_{Q_\alpha}(V_\alpha^{(2)}) \le Q_\alpha \cap Q_\beta$, $Q_\alpha \cap Q_\beta \trianglelefteq R_\alpha$.

We may apply Proposition 5.30 with $\lambda = \alpha$. Since $b > 5$, (a) is not satisfied. Indeed, as $\lambda = \alpha$ and R_α normalizes $Q_\alpha \cap Q_\beta$, we suppose that conclusion (d) is satisfied. For X as constructed in Proposition 5.30, we have that $V_\beta^{(5)} \le V := \langle Z_\beta^X \rangle$ and the images in $\widetilde{L}_\beta$ of Q_β/C_β, $C_\beta/C_{Q_\beta}(V_\beta^{(3)})$ and $C_{Q_\beta}(V_\beta^{(3)})/C_{Q_\beta}(V_\beta^{(5)})$ all contain a non-central chief factor for $\widetilde{L}_\beta$, a contradiction by Proposition 5.30. Hence, no such counterexample exists. □

Lemma 5.35 *Suppose that $b > 3$ and Hypothesis 5.31 is satisfied. If $V_\beta^{(3)}/V_\beta$ contains a unique non-central chief factor which, as a* GF$(p)\overline{L_\beta}$*-module, is an FF-module, then $O^p(R_\beta)$ centralizes $V_\beta^{(3)}$.*

Proof Since the only configuration in Theorem C which satisfies $b > 3$ (where G is parabolic isomorphic to F_3) satisfies $[O^p(R_\beta), V_\beta^{(3)}] = \{1\}$, we may assume that G is a minimal counterexample such that $O^p(R_\beta)$ does not centralize $V_\beta^{(3)}$. Since $O^p(R_\beta)$ centralizes V_β, the three subgroup lemma implies that $O^p(R_\beta)$ centralizes Q_β/C_β so that R_β normalizes $Q_\alpha \cap Q_\beta$. Thus, the hypotheses of Proposition 5.30 are satisfied with $\lambda = \beta$. Since $C_{L_\beta}(V_\beta^{(3)}/V_\beta)$ normalizes $V_\alpha^{(2)}$ and $\lambda = \beta$, conclusions (b) and (c) are not satisfied. As $b > 3$, if outcome (a) is satisfied then $(G_\alpha, H_\beta, G_{\alpha,\beta})$ is parabolic isomorphic to F_3 and $H_\beta/Q_\beta \cong \mathrm{GL}_2(3)$. Then

S is determined up to isomorphism. Indeed, as $V_\beta = \langle Z_\alpha^{G_\beta} \rangle = \langle Z_\alpha^{H_\beta} \rangle = Z_3(S)$, $Q_\beta = C_S(Z_3(S)/Z(S))$ is uniquely determined in S, and so is uniquely determined up to isomorphism. But then one can check (e.g. employing MAGMA [14]) that $\Phi(Q_\beta) = C_\beta$ has index 9 in Q_β, and as $\overline{G_\beta}$ acts faithfully on $Q_\beta/\Phi(Q_\beta)$, $\overline{G_\beta} = \overline{H_\beta} \cong \mathrm{GL}_2(3)$ and $G_\beta = H_\beta$, a contradiction.

Hence, we are left with conclusion (d). But then $V_\alpha^{(4)} \leq V := \langle Z_\beta^X \rangle$ and the images of $Q_\alpha/C_{Q_\alpha}(V_\alpha^{(2)})$ and $C_{Q_\alpha}(V_\alpha^{(2)})/C_{Q_\alpha}(V_\alpha^{(4)})$ in $\widetilde{L_\alpha}$ both contain a non-central chief factor for $\widetilde{L_\alpha}$. Moreover, the images of Q_β/C_β and $C_\beta/C_{Q_\beta}(V_\beta^{(3)})$ also a contain non-central chief factor for $\widetilde{L_\beta}$, and we have a contradiction. Therefore, no counterexample exists and the proof is complete. □

Lemma 5.36 *Suppose that $b > 5$ and Hypothesis 5.31 is satisfied. If $V_\beta^{(5)}/V_\beta^{(3)}$ contains a unique non-central chief factor which, as a $\overline{L_\beta}$-module, is an FF-module and $O^p(R_\beta)$ centralizes $V_\beta^{(3)}$, then $[O^p(R_\beta), V_\beta^{(5)}] = \{1\}$.*

Proof Since none of the configurations in Theorem C satisfy $b > 5$, we may assume the G is a minimal counterexample to Theorem C with $[O^p(R_\beta), V_\beta^{(3)}] = \{1\}$ and $[O^p(R_\beta), V_\beta^{(5)}] \neq \{1\}$. Since $O^p(R_\beta)$ centralizes V_β, $O^p(R_\beta)$ centralizes Q_β/C_β so that R_β normalizes $Q_\alpha \cap Q_\beta$ and we may apply Proposition 5.30 with $\lambda = \beta$. Since $O^p(R_\beta)$ normalizes $V_\alpha^{(2)}$ and $b > 5$, we are in case (d) of Proposition 5.30. Then, $V_\alpha^{(6)} \leq V := \langle Z_\beta^X \rangle$ and the image of $Q_\alpha/C_{Q_\alpha}(V_\alpha^{(6)})$ in $\widetilde{L_\alpha}$ contains at least three non-central chief factors for $\widetilde{L_\alpha}$, a contradiction. Hence, no counterexample exists. □

Chapter 6
$Z_{\alpha'} \not\leq Q_\alpha$

Throughout this chapter, we assume Hypothesis 5.1 and that G is a minimal counterexample. In addition, within this chapter we suppose that $Z_{\alpha'} \not\leq Q_\alpha$ for a chosen critical pair (α, α'). By Lemma 5.16 (iv), this condition is equivalent to $[Z_\alpha, Z_{\alpha'}] \neq \{1\}$. Throughout, we set $S \in \mathrm{Syl}_p(G_{\alpha,\beta})$ and $q_\lambda = |\Omega(Z(T/Q_\lambda))|$ for any $\lambda \in \Gamma$ and $T \in \mathrm{Syl}_p(G_\lambda)$.

Lemma 6.1 *(α', α) is also a critical pair, $Q_{\alpha'} \in \mathrm{Syl}_p(R_{\alpha'})$, $C_{Z_\alpha}(Z_{\alpha'}) = Z_\alpha \cap Q_{\alpha'}$ and $C_{Z_{\alpha'}}(Z_\alpha) = Z_{\alpha'} \cap Q_\alpha$.*

Proof Since $Z_{\alpha'} \not\leq Q_\alpha$ we have that both (α, α') and (α', α) are critical pairs. In particular, all the results we prove in this section hold upon interchanging α and α'. By Lemma 5.17, $C_{Z_\alpha}(Z_{\alpha'}) = Z_\alpha \cap Q_{\alpha'}$. □

Lemma 6.2 *For $\lambda \in \{\alpha, \alpha'\}$, the following hold:*

(i) *$S = Q_\alpha Q_\beta = Z_{\alpha'} Q_\alpha \in \mathrm{Syl}_p(G_{\alpha,\beta})$ and $N_{G_\alpha}(S) = N_{G_\beta}(S) = G_{\alpha,\beta}$;*
(ii) *$Z_\alpha Q_{\alpha'} \in \mathrm{Syl}_p(G_{\alpha',\alpha'-1})$; and*
(iii) *$Z_\lambda/\Omega(Z(L_\lambda))$ is natural $\mathrm{SL}_2(q_\lambda)$-module for $L_\lambda/R_\lambda \cong \mathrm{SL}_2(q_\lambda)$, and $q_\alpha = q_{\alpha'}$.*

Proof Without loss of generality, assume that $|Z_\alpha Q_{\alpha'}/Q_{\alpha'}| \leqslant |Z_{\alpha'} Q_\alpha/Q_\alpha|$. By Lemma 6.1, we have that

$$\begin{aligned}|Z_\alpha/C_{Z_\alpha}(Z_{\alpha'})| &= |Z_\alpha/Z_\alpha \cap Q_{\alpha'}| = |Z_\alpha Q_{\alpha'}/Q_{\alpha'}| \\ &\leqslant |Z_{\alpha'} Q_\alpha/Q_\alpha| = |Z_{\alpha'}/Z_{\alpha'} \cap Q_\alpha| = |Z_{\alpha'}/C_{Z_{\alpha'}}(Z_\alpha)|.\end{aligned}$$

Thus, $Z_{\alpha'}$ is a non-trivial offender on Z_α, and Z_α is an FF-module for L_α/R_α. Since L_α/R_α has a strongly p-embedded subgroup by Lemma 5.20, using Lemma 2.41 we conclude that $L_\alpha/R_\alpha \cong \mathrm{SL}_2(q)$ and $Z_\alpha/\Omega(Z(L_\alpha))$ is a natural $\mathrm{SL}_2(q)$-module.

Since $L_\alpha/R_\alpha \cong \mathrm{SL}_2(q)$ and $Z_\alpha/\Omega(Z(L_\alpha))$ is a natural $\mathrm{SL}_2(q)$-module, we infer that $q = |Z_\alpha/C_{Z_\alpha}(Z_{\alpha'})| \leqslant |Z_{\alpha'}/C_{Z_{\alpha'}}(Z_\alpha)| = |Z_{\alpha'} Q_\alpha/Q_\alpha| \leqslant q$. In

© The Author(s), under exclusive license to Springer Nature Switzerland AG 2024

M. van Beek, *Rank 2 Amalgams and Fusion Systems*, Lecture Notes in Mathematics 2343, https://doi.org/10.1007/978-3-031-54461-3_6

particular, by a symmetric argument, $Z_{\alpha'}/\Omega(Z(L_{\alpha'}))$ is also a natural module for $L_{\alpha'}/R_{\alpha'} \cong \mathrm{SL}_2(q)$. It follows immediately that $Z_{\alpha'}Q_\alpha \in \mathrm{Syl}_p(G_{\alpha,\beta})$ and $Z_\alpha Q_{\alpha'} \in \mathrm{Syl}_p(G_{\alpha',\alpha'-1})$. Since $Z_{\alpha'} \leq Q_\beta$, we have that $S = Q_\alpha Q_\beta$ and since $G_{\alpha,\beta} = G_\alpha \cap G_\beta$ normalizes $Q_\alpha Q_\beta$, the result holds. □

In the following proposition, we divide the analysis of the case $[Z_\alpha, Z_{\alpha'}] \neq \{1\}$ into two subcases. The remainder of this chapter is split into two sections which deal with each of these subcases individually.

Proposition 6.3 *One of the following holds:*

(i) *b is even and $Z_\beta = \Omega(Z(S)) = \Omega(Z(L_\beta))$; or*
(ii) *$Z_\beta \neq \Omega(Z(S))$ and for all $\lambda \in \Gamma$, $Z_\lambda/\Omega(Z(L_\lambda))$ is a natural $\mathrm{SL}_2(q_\lambda)$-module for L_λ/R_λ.*

Proof Notice that if $Z_\beta = \Omega(Z(S))$ then $\{1\} = [Z_\beta, S]^{G_\beta} = [Z_\beta, \langle S^{G_\beta}\rangle] = [Z_\beta, L_\beta]$ so that $Z_\beta = \Omega(Z(L_\beta))$. Since $Z_{\alpha'}$ is not centralized by $Z_\alpha \leq L_{\alpha'}$, it follows immediately in this case that b is even.

Suppose that $Z_\beta \neq \Omega(Z(S))$. If $b = 1$, the result follows immediately from Lemma 6.2 replacing α' by β and so we may assume that $b > 1$. Assume that $V_\alpha \leq Q_{\alpha'-1}$. In particular, $V_\alpha \leq Z_\alpha Q_{\alpha'} \in \mathrm{Syl}_p(L_{\alpha'})$ by Lemma 6.2. Thus, $[V_\alpha, Z_{\alpha'}] \leq [Z_\alpha, Z_{\alpha'}] \leq Z_\alpha$ so that $[V_\alpha, O^p(L_\alpha)] \leq Z_\alpha$ and $Z_\alpha Z_\beta \trianglelefteq L_\alpha$, a contradiction by Proposition 5.27. Hence, there is $\alpha - 1 \in \Delta(\alpha)$ with $Z_{\alpha-1} \not\leq Q_{\alpha'-1}$. Then $(\alpha - 1, \alpha' - 1)$ is a critical pair and since $Z_\alpha \neq \Omega(Z(S)) \neq Z_\beta$, by Lemma 5.16 (ii), we conclude that $[Z_{\alpha-1}, Z_{\alpha'-1}] \neq \{1\}$ and Lemma 6.2 gives the result. □

6.1 $Z_\beta \neq \Omega(Z(S))$

We first consider the case where $[Z_\alpha, Z_{\alpha'}] \neq \{1\}$ and $Z_\beta \neq \Omega(Z(S))$. Under these hypotheses, and using the symmetry in α and α', it is not hard to show that every $\gamma \in \Gamma$ belongs to some critical pair. The main work in this section is then to show that $R_\gamma = Q_\gamma$ and $\overline{L_\gamma} \cong \mathrm{SL}_2(q)$, for then, all examples we obtain arise from weak BN-pairs of rank 2 and G is determined by [27].

As was made apparent in Proposition 5.30, there is a clear distinction between the cases where $p \in \{2, 3\}$ and $p \geqslant 5$ due to the solvability of $\mathrm{SL}_2(p)$ when $p \in \{2, 3\}$. Throughout this section, and the sections to come, this dichotomy will become a prominent theme.

Lemma 6.4 *Suppose that $Z_\beta \neq \Omega(Z(S))$, $b > 1$ and for $\lambda \in \{\alpha, \beta\}$, $Z_\lambda/\Omega(Z(L_\lambda))$ is a natural $\mathrm{SL}_2(q_\lambda)$-module for L_λ/R_λ. Then the following hold:*

(i) *$V_\alpha \not\leq Q_{\alpha'-1}$ and there is a critical pair $(\alpha - 1, \alpha' - 1)$ with $[Z_{\alpha-1}, Z_{\alpha'-1}] \neq \{1\}$ and $V_{\alpha-1} \not\leq Q_{\alpha'-2}$;*
(ii) *$q_\alpha = q_\beta$;*
(iii) *V_λ/Z_λ and Z_λ are FF-modules for $\overline{L_\lambda}$; and*
(iv) *unless $q_\lambda \in \{2, 3\}$, $R_\lambda = C_{L_\lambda}(V_\lambda/Z_\lambda)$ and $L_\lambda/C_{L_\lambda}(V_\lambda)Q_\lambda \cong \mathrm{SL}_2(q_\lambda)$.*

Proof By the proof of Proposition 6.3, we have that $V_\alpha \not\leq Q_{\alpha'-1}$. In particular, there is some $\alpha-1 \in \Delta(\alpha)$ such that $(\alpha-1, \alpha'-1)$ is a critical pair with $[Z_{\alpha-1}, Z_{\alpha'-1}] \neq \{1\}$. By a similar reasoning, $V_{\alpha-1} \not\leq Q_{\alpha'-2}$ else we arrive at a similar contradiction to the above. Hence (i) holds.

Since $(\alpha-1, \alpha'-1)$ is a critical pair satisfying the same hypothesis as (α, α'), we may translate results proved about α and α' to appropriate results about $\alpha-1$ and $\alpha'-1$.

We observe that $V_\alpha \cap Q_{\alpha'-1} = Z_\alpha(V_\alpha \cap Q_{\alpha'})$ has index at most $q_{\alpha'-1}$ in V_α and is centralized, modulo Z_α, by $Z_{\alpha'}$. Furthermore, since $Z_\alpha Z_\beta \ntrianglelefteq L_\alpha$, it follows from Lemma 5.14 (iii) that $Q_\alpha \in \mathrm{Syl}_p(C_{L_\alpha}(V_\alpha/Z_\alpha))$ and by Lemma 2.41, we have that $q_\alpha \leqslant q_{\alpha'-1}$. Since $(\alpha'-1, \alpha-1)$ is also a critical pair, by a similar argument, we have that $q_{\alpha'-1} \leq q_\alpha$. Hence, $q_{\alpha'-1} = q_\alpha$ and for $\lambda \in \{\alpha, \alpha'-1\}$, both V_λ/Z_λ and Z_λ are FF-modules for $\overline{L_\lambda}$.

On the other hand, (α', α) is also a critical pair and a similar argument gives $q_{\alpha'} \leqslant q_\beta$. Finally, we use the critical pair $(\alpha-1, \alpha'-1)$ to see that $q_{\alpha-1} \leqslant q_{\alpha'-2}$ and as $q_{\alpha-1} = q_\beta$ and $q_{\alpha'-2} = q_{\alpha'}$, we deduce that $q_\beta = q_{\alpha'}$. Since $q_\alpha = q_{\alpha'}$ by Lemma 6.2, we have that $q_\alpha = q_\beta$ and (ii) holds. Retracing the proof, we find that $V_{\alpha-1}/Z_{\alpha-1}$ and $Z_{\alpha-1}$ are FF-modules for $\overline{L_{\alpha-1}}$, which substantiates (iii).

It remains to prove (iv). By Lemma 2.41, for all $\lambda \in \Gamma$, $L_\lambda/C_{L_\lambda}(V_\lambda/Z_\lambda) \cong L_\lambda/R_\lambda \cong \mathrm{SL}_2(q_\lambda)$. Suppose that $q_\lambda \notin \{2,3\}$ and assume that $C_{L_\lambda}(V_\lambda/Z_\lambda) \neq R_\lambda$. Since $\{Q_\lambda\} = \mathrm{Syl}_p(C_{L_\lambda}(V_\lambda/Z_\lambda)) = \mathrm{Syl}_p(R_\lambda)$, we infer that $R_\lambda C_{L_\lambda}(V_\lambda/Z_\lambda)$ is a group of order coprime to p and we see immediately that p is odd, $C_{L_\lambda}(V_\lambda/Z_\lambda)R_\lambda/R_\lambda = Z(L_\lambda/R_\lambda)$ and $C_{L_\lambda}(V_\lambda/Z_\lambda)R_\lambda/C_{L_\lambda}(V_\lambda/Z_\lambda) = Z(L_\lambda/C_{L_\lambda}(V_\lambda/Z_\lambda))$. Thus, $L_\lambda/(C_{L_\lambda}(V_\lambda/Z_\lambda) \cap R_\lambda)$ is isomorphic to a central extension of $\mathrm{PSL}_2(q_\lambda)$ by an elementary abelian group of order 4. Since $L_\lambda = O^{p'}(L_\lambda)$ and the 2-part of the Schur multiplier of $\mathrm{PSL}_2(q)$ is of order 2 by Proposition 2.16 (vii) when p is odd, we have a contradiction. Thus, we shown that, unless $q_\lambda \in \{2,3\}$, $C_{L_\lambda}(V_\lambda/Z_\lambda) = R_\lambda$ and (iv) is proved. □

Lemma 6.5 *Suppose that for $Z_\beta \neq \Omega(Z(S))$ and for $\lambda \in \{\alpha, \beta\}$, $Z_\lambda/\Omega(Z(L_\lambda))$ is a natural* $\mathrm{SL}_2(q_\lambda)$*-module for L_λ/R_λ. Then $b \leqslant 2$.*

Proof Assume throughout that $b > 2$ so that V_λ is abelian for all $\lambda \in \Gamma$. For $\delta \in \Gamma$ and $\nu \in \Delta(\delta)$, set $S_{\delta,\nu} \in \mathrm{Syl}_p(G_{\delta,\nu})$ and $Z_{\delta,\nu} := \Omega(Z(S_{\delta,\nu}))$. Choose $\mu \in \Delta(\alpha'-1)$ such that $Z_{\mu,\alpha'-1} \neq Z_{\alpha'-1,\alpha'-2}$. Thus, $Z_{\alpha'-1} = Z_{\mu,\alpha'-1}Z_{\alpha'-1,\alpha'-2}$. Then, using Lemma 6.4 (i), as $V_\alpha \not\leq Q_{\alpha'-1}$ and V_α centralizes $Z_{\alpha'-1,\alpha'-2}$, we have that $L_{\alpha'-1} = \langle Q_\mu, R_{\alpha'-1}, V_\alpha\rangle$.

Set $U_{\alpha'-1,\mu} := \langle Z_\delta \mid Z_{\mu,\alpha'-1} = Z_{\delta,\alpha'-1}, \delta \in \Delta(\alpha'-1)\rangle$. Let $r \in R_{\alpha'-1}Q_\mu$. Since r is an automorphism of the graph, it follows that for Z_δ with $Z_{\mu,\alpha'-1} = Z_{\delta,\alpha'-1}$ and $\delta \in \Delta(\alpha'-1)$, we have that $Z_\delta^r = Z_{\delta\cdot r}$ and $\{\delta, \alpha'-1\}\cdot r = \{\delta\cdot r, \alpha'-1\}$. Since $S_{\delta,\alpha'-1}$ is the unique Sylow p-subgroup of $G_{\delta,\alpha'-1}$, it follows that $Z_{\delta,\alpha'-1}^r = Z_{\delta\cdot r,\alpha'-1}$. Since $R_{\alpha'-1}Q_\mu$ normalizes $Z_{\delta,\alpha'-1}$, we have that $Z_{\delta\cdot r,\alpha'-1} = Z_{\mu,\alpha'-1}$ so that $Z_{\delta\cdot r} \leq U_{\alpha'-1,\mu}$. Thus, $U_{\alpha'-1,\mu} \trianglelefteq R_{\alpha'-1}Q_\mu$.

Suppose that $U_{\alpha'-1,\mu} \leq Q_\alpha$. By Lemma 6.4 (i), there is $\alpha-1 \in \Delta(\alpha)$ such that $Z_{\alpha-1} \not\leq Q_{\alpha'-1}$ and $Z_{\alpha'-1} \not\leq Q_{\alpha-1}$. Moreover, we have that $L_{\alpha'-1} = \langle Q_\mu, R_{\alpha'-1}, Z_{\alpha-1}\rangle$. Then, $U_{\alpha'-1,\mu} = Z_{\alpha'-1}(U_{\alpha'-1,\mu} \cap Q_{\alpha-1})$ is centralized, modulo $Z_{\alpha'-1}$, by $Z_{\alpha-1}$ so that $U_{\alpha'-1,\mu} \trianglelefteq L_{\alpha'-1} = \langle Q_\mu, R_{\alpha'-1}, Z_{\alpha-1}\rangle$. Since $Z_{\alpha-1}$ centralizes $U_{\alpha'-1,\mu}/Z_{\alpha'-1}$, $O^p(L_{\alpha'-1})$ centralizes $U_{\alpha'-1,\mu}/Z_{\alpha'-1}$ and $Z_\mu Z_{\alpha'-1} \trianglelefteq L_{\alpha'-1}$, a contradiction by Proposition 5.27. Thus, $U_{\alpha'-1,\mu} \not\leq Q_\alpha$.

Hence, there is $\delta \in \Delta(\alpha'-1)$ with (α, δ) a critical pair, $L_{\alpha'-1} = \langle Q_\delta, R_{\alpha'-1}, V_\alpha\rangle$ and $Z_{\delta,\alpha'-1} = Z_{\mu,\alpha'-1} \neq Z_{\alpha'-1,\alpha'-2}$. We may as well assume that $\delta = \alpha'$ and $Z_{\alpha',\alpha'-1} \neq Z_{\alpha'-1,\alpha'-2}$. By Lemma 6.1, Lemma 6.4 applies to α' in place of α. Then $V_{\alpha'} \not\leq Q_\beta$ and there is $\alpha'+1 \in \Delta(\alpha')$ with $(\alpha'+1, \beta)$ a critical pair satisfying $Z_{\alpha'+1} \not\leq Q_\beta$ and $Z_\beta \not\leq Q_{\alpha'+1}$. Choose $\mu^* \in \Delta(\alpha')$ such that $Z_{\mu^*,\alpha'} \neq Z_{\alpha',\alpha'-1}$ so that $Z_{\alpha'} = Z_{\mu^*,\alpha'} Z_{\alpha',\alpha'-1}$. Then, as $Z_\alpha \not\leq Q_{\alpha'}$ and Z_α centralizes $Z_{\alpha',\alpha'-1}$, we have that $L_{\alpha'} = \langle Z_\alpha, Q_{\mu^*}, R_{\alpha'}\rangle$. Forming U_{α',μ^*} in an analogous way to $U_{\alpha'-1,\mu}$, we see that $U_{\alpha',\mu^*} \trianglelefteq R_{\alpha'} Q_{\mu^*}$ and $U_{\alpha',\mu^*} \not\leq Q_\beta$. Thus, there is some δ^* with $Z_{\delta^*,\alpha'} \neq Z_{\alpha',\alpha'-1}$, $L_{\alpha'} = \langle Q_{\delta^*}, R_{\alpha'}, Z_\alpha\rangle$ and (β, δ^*) a critical pair. We may as well label $\delta^* = \alpha'+1$ so that $L_{\alpha'} = \langle Z_\alpha, Q_{\alpha'+1}, R_{\alpha'}\rangle$ and $Z_{\alpha'+1,\alpha'} \neq Z_{\alpha',\alpha'-1}$

Now, let $R := [Z_\beta, Z_{\alpha'+1}] \leq Z_\beta \cap Z_{\alpha'+1}$. Then R is centralized by $Z_\beta Q_{\alpha'+1} \in \mathrm{Syl}_p(G_{\alpha'+1,\alpha'})$ so that $R \leq Z_{\alpha'+1,\alpha'}$. Since $b > 1$, Z_α centralizes $R \leq Z_\beta$ and so R is centralized by $L_{\alpha'} = \langle Q_{\alpha'+1}, R_{\alpha'}, Z_\alpha\rangle$ and $R \leq Z(L_{\alpha'}) \leq Z_{\alpha',\alpha'-1}$. But $R \leq Z_\beta \leq V_\alpha$ and since $b > 2$, V_α is abelian so centralizes R. In particular, R is centralized by $L_{\alpha'-1} = \langle V_\alpha, R_{\alpha'-1}, Q_{\alpha'}\rangle$. But then $R \trianglelefteq \langle L_{\alpha'}, L_{\alpha'-1}\rangle$, a final contradiction. Hence, $b \leqslant 2$. □

Proposition 6.6 *Suppose that $Z_\beta \neq \Omega(Z(S))$, $b = 2$ and for $\lambda \in \{\alpha, \beta\}$, $Z_\lambda/\Omega(Z(L_\lambda))$ is a natural $\mathrm{SL}_2(q_\lambda)$-module for $L_\lambda/R_\lambda \cong \mathrm{SL}_2(q_\lambda)$. Then $p = 3$ and G is locally isomorphic to H where $F^*(H) \cong \mathrm{G}_2(3^n)$.*

Proof Since $b > 1$, by Lemma 6.4 (iii), we have that $q_\alpha = q_\beta$ and $V_\alpha \not\leq Q_\beta$. But then $Q_\alpha = V_\alpha(Q_\alpha \cap Q_{\alpha'})$ and it follows that $O^p(L_\alpha)$ centralizes Q_α/V_α. In particular, V_α contains all non-central chief factors for L_α within Q_α, and consequently $C_{L_\alpha}(V_\alpha)$ is a p-group. By Lemma 6.4 (i), there is $\alpha - 1 \in \Delta(\alpha)$ such that $(\alpha-1, \beta)$ is a critical pair with $[Z_{\alpha-1}, Z_\beta] \neq \{1\}$. Applying Lemma 6.4 (ii) again, $C_{L_{\alpha-1}}(V_{\alpha-1})$ is a p-group. By Lemma 6.4 (iv), unless $q_\alpha \in \{2, 3\}$, we conclude that $\overline{L_\alpha} \cong \overline{L_\beta} \cong \mathrm{SL}_2(q_\alpha)$ and G has a weak BN-pair of rank 2. Comparing with [27], the result holds.

Hence, we assume that $q_\alpha = q_\beta \in \{2, 3\}$ and for $\lambda \in \{\alpha, \beta\}$, V_λ/Z_λ and Z_λ are FF-modules for $\overline{L_\lambda}$. Moreover, aiming for a contradiction for some $\delta \in \{\alpha, \beta\}$ we assume that $C_{L_\delta}(V_\delta/Z_\delta) \neq R_\delta$ and $\overline{L_\delta} \not\cong \mathrm{SL}_2(p)$. By Proposition 2.46 (ii) and Proposition 2.47 (ii), $\overline{L_\delta} \cong (3 \times 3) : 2$ or $(Q_8 \times Q_8) : 3$ for $p = 2$ or 3 respectively. Since $O^p(L_\delta)$ centralizes Q_δ/V_δ we have that $C_{L_\delta}(V_\delta/Z_\delta)$ normalizes $Q_\alpha \cap Q_\beta$.

If $p = 2$, by Proposition 2.46 (iii), we may choose $P_\alpha \leq L_\alpha$ such that $\overline{P_\alpha} \cong \mathrm{Sym}(3)$, $\Omega(Z(S)) \ntrianglelefteq P_\alpha$ and $Q_\alpha \cap Q_\beta \ntrianglelefteq P_\alpha$. If $\overline{L_\alpha} \cong \mathrm{Sym}(3)$ then $L_\alpha = P_\alpha$, and if $\overline{L_\alpha} \cong (3 \times 3) : 2$, then as there are two choices for P_α, both are $G_{\alpha,\beta}$-invariant and neither normalizes $Q_\alpha \cap Q_\beta$. For such a P_α, set $H_\alpha = P_\alpha G_{\alpha,\beta}$. We make an analogous choice for $H_\beta \leq G_\beta$ and observe that $P_\lambda = O^{2'}(H_\lambda)$ for $\lambda \in \{\alpha, \beta\}$.

If $p = 3$, by Proposition 2.47 (iii), we may choose $P_\alpha \leq L_\alpha$ such that $\overline{P_\alpha} \cong \mathrm{SL}_2(3)$, $\Omega(Z(S)) \ntrianglelefteq P_\alpha$ and $Q_\alpha \cap Q_\beta \ntrianglelefteq P_\alpha$. If $\overline{L_\alpha} \cong \mathrm{SL}_2(3)$ then $L_\alpha = P_\alpha$, and if $\overline{L_\alpha} \cong (Q_8 \times Q_8) : 3$, then there are three choices for P_α. Since all contain S, there is at least one choice such that P_α is $G_{\alpha,\beta}$-invariant. For this P_α, set $H_\alpha = P_\alpha G_{\alpha,\beta}$ and choose H_β in a similar fashion. Again, observe that $P_\lambda = O^{2'}(H_\lambda)$ for $\lambda \in \{\alpha, \beta\}$.

Set $X := \langle H_\alpha, H_\beta\rangle$ and suppose that there is $\{1\} \neq Q \leq S$ with $Q \trianglelefteq X$. Then $Q \leq O_p(H_\alpha) \cap O_p(H_\beta) = Q_\alpha \cap Q_\beta$. Suppose $\Omega(Z(S)) \not\leq Q$. Then $V_\beta = \langle\langle \Omega(Z(S))^{H_\alpha}\rangle^{H_\beta}\rangle$ centralizes Q and since Q is normal in H_α, $[O^p(P_\alpha), Q] \leq [V_\beta, Q]^{H_\alpha} = \{1\}$. Considering the action of $V_\alpha = \langle\langle \Omega(Z(S))^{H_\beta}\rangle^{H_\alpha}\rangle$ on Q yields $[O^p(P_\beta), Q] = \{1\}$. But $Q \trianglelefteq S$ and so $Q \cap \Omega(Z(S))$ is non-trivial and centralized by $G = \langle H_\alpha, R_\alpha, H_\beta, R_\beta\rangle$, a contradiction. Hence, $\Omega(Z(S)) \leq Q$. But then $Q \geq V_\beta = \langle\langle \Omega(Z(S))^{H_\alpha}\rangle^{H_\beta}\rangle \not\leq Q_\alpha$, a contradiction.

Thus, any subgroup of $G_{\alpha,\beta}$ which is normal in X is a p'-group. Such a subgroup would be contained in H_λ and so would centralize Q_λ for $\lambda \in \{\alpha, \beta\}$. Since $S \leq H_\lambda \leq G_\lambda$, we have that H_λ is of characteristic p, $C_{H_\lambda}(Q_\lambda) \leq Q_\lambda$ and no non-trivial subgroup of $G_{\alpha,\beta}$ is normal in X. Moreover, $\overline{P_\alpha} \cong \overline{P_\alpha} \cong \mathrm{SL}_2(p)$ and X has a weak BN-pair of rank 2. For $\lambda \in \{\alpha, \beta\}$, since Q_λ contains precisely two non-central chief factors for P_λ, and neither P_α nor P_β normalizes $\Omega(Z(S))$, by [27], X is locally isomorphic to $\mathrm{G}_2(3)$ and S is isomorphic to a Sylow 3-subgroup of $\mathrm{G}_2(3)$. Then the pair $\{Q_\alpha, Q_\beta\}$ are uniquely determined in S. Noticing that [63, Lemma 7.8] applies in this situation independent of any fusion system hypothesis, it follows that for $\lambda \in \{\alpha, \beta\}$, $\overline{G_\lambda}$ is isomorphic to a subgroup of $\mathrm{GL}_2(3)$, a contradiction to the assumption that $\overline{L_\delta} \not\cong \mathrm{SL}_2(p)$. Thus, we conclude that G has a weak BN-pair of rank 2 and the result follows upon comparison with [27]. □

Remark The graph automorphism of $\mathrm{G}_2(3^n)$ normalizes $S \in \mathrm{Syl}_3(\mathrm{G}_2(3^n))$ and fuses Q_α and Q_β, and so Hypothesis 5.1 only allows for groups locally isomorphic to $\mathrm{G}_2(3^n)$ decorated by field automorphisms.

Proposition 6.7 *Suppose that $Z_\beta \neq \Omega(Z(S))$ and for $\lambda \in \{\alpha, \beta\}$, $Z_\lambda/\Omega(Z(L_\lambda))$ is a natural $\mathrm{SL}_2(q_\lambda)$-module for L_λ/R_λ. Then G is locally isomorphic to H where $(F^*(H), p)$ is one of $(\mathrm{PSL}_3(p^n), p)$, $(\mathrm{PSp}_4(2^n), 2)$ or $(\mathrm{G}_2(3^n), 3)$.*

Proof By Lemma 6.5 and Proposition 6.6, we may suppose that $b = 1$. Then, $Z_\alpha \not\leq Q_\beta$, $Z_\beta \not\leq Q_\alpha$, $Q_\alpha = Z_\alpha(Q_\alpha \cap Q_\beta)$ and $Q_\beta = Z_\beta(Q_\alpha \cap Q_\beta)$. In particular, $\Phi(Q_\alpha) = \Phi(Q_\alpha \cap Q_\beta) = \Phi(Q_\beta)$ is trivial and so both Q_α and Q_β are elementary abelian. For $\lambda \in \{\alpha, \beta\}$, by coprime action we have that $Q_\lambda = [Q_\lambda, R_\lambda] \times C_{Q_\lambda}(R_\lambda)$ is an S-invariant decomposition. But $\Omega(Z(S)) \leq Z_\lambda \leq C_{Q_\lambda}(R_\lambda)$ and since $[Q_\alpha, R_\lambda] \trianglelefteq S$, we must have that $[Q_\alpha, R_\lambda] = \{1\}$. It follows that R_λ centralizes Q_λ and, as G_λ is of characteristic p, $Q_\lambda = R_\lambda$. Thus, G has a weak BN-pair of rank 2 and is determined by [27], hence the result. □

Remark Similarly to the $\mathrm{G}_2(3^n)$ example, the graph automorphisms for $\mathrm{PSL}_3(p^n)$ and $\mathrm{PSp}_4(2^n)$ fuse Q_α and Q_β and are not permitted by the hypothesis.

6.2 $Z_\beta = \Omega(Z(S))$

Given Proposition 6.3, we assume in this section that b is even and $Z_\beta = \Omega(Z(S))$. The general aim will be to demonstrate that $b = 2$ and $\overline{L_\alpha} \cong \mathrm{SL}_2(q)$ for then, it will quickly follow that the amalgam in question is symplectic and we may apply the classification in [61]. We are able to show that, in all the cases considered, $b = 2$. However, at the end of this section we uncover a configuration where $R_\alpha \neq Q_\alpha$.

Lemma 6.8 *Let $\alpha - 1 \in \Delta(\alpha) \setminus \{\beta\}$ with $Z_{\alpha-1} \neq Z_\beta$. Then $\Omega(Z(L_\alpha)) = \{1\}$, $Z_\alpha = Z_\beta \times Z_{\alpha-1}$ is a natural $\mathrm{SL}_2(q_\alpha)$-module, $Q_\beta \in \mathrm{Syl}_p(R_\beta)$ and $[Z_\alpha, Z_{\alpha'}] = Z_{\alpha'-1} = Z_\alpha \cap Q_{\alpha'} = Z_\beta = [V_\beta, Q_\beta]$.*

Proof Since L_β is transitive on $\Delta(\beta)$ and centralizes $Z_\beta = \Omega(Z(S))$, by Lemma 5.6 (iv), we have that $Z(L_\alpha) = \{1\}$. Then, by Lemma 6.2, Z_α is a natural $\mathrm{SL}_2(q_\alpha)$-module for $L_\alpha/R_\alpha \cong \mathrm{SL}_2(q_\alpha)$.

Now, $[Z_\alpha, S] = [Z_\alpha, Z_{\alpha'}Q_\alpha] = [Z_\alpha, Z_{\alpha'}] = \Omega(Z(S)) = Z_\beta$. Thus, $[V_\beta, Q_\beta] = [\langle Z_\alpha{}^{G_\beta}\rangle, Q_\beta] = Z_\beta \leq C_{V_\beta}(O^p(L_\beta))$ and so $Q_\beta \leq R_\beta$. By Lemma 5.19, we have that $Q_\beta \in \mathrm{Syl}_p(R_\beta)$.

By considering $[Z_{\alpha'}, Z_\alpha Q_{\alpha'}]$ and again employing Lemma 6.2, we deduce that, for $T \in \mathrm{Syl}_p(G_{\alpha',\alpha'-1})$, $[Z_{\alpha'}, Z_\alpha] = \Omega(Z(T)) = Z_{\alpha'-1}$. Then $Z_\beta = Z_{\alpha'-1} \leq Q_{\alpha'}$ and it follows immediately that $Z_\beta = Z_\alpha \cap Q_{\alpha'}$. By properties of natural $\mathrm{SL}_2(q_\alpha)$-modules, $Z_\alpha = Z_\beta \times Z_\beta^x = Z_\beta \times Z_{\beta\cdot x}$ for $x \in L_\alpha \setminus G_{\alpha,\beta}R_\alpha$. In particular, we may choose $\alpha - 1 \in \Delta(\alpha)$ conjugate to β by an element of $L_\alpha \setminus G_{\alpha,\beta}R_\alpha$ so that $Z_\alpha = Z_\beta \times Z_{\alpha-1}$. □

Lemma 6.9 *Suppose that $b > 2$. Then the following hold:*

(i) $[V_\alpha^{(2)}, V_\alpha^{(2)}] = \Phi(V_\alpha^{(2)}) \leq Z_\alpha$;
(ii) $[V_\alpha^{(2)}, V_{\alpha'-1}, V_{\alpha'-1}] \leq Z_\alpha$;
(iii) *V_β/Z_β is a faithful quadratic module for L_β/R_β; and*
(iv) *there is $\alpha - 1 \in \Delta(\alpha)$ such that $Z_{\alpha-1} \neq Z_\beta$, $V_{\alpha-1} \leq Q_{\alpha'-2}$, $V_{\alpha-1} \not\leq Q_{\alpha'-1}$ and $[V_{\alpha'-1} \cap Q_\alpha \cap Q_{\alpha-1}, V_{\alpha-1}] = \{1\}$.*

Proof Notice that

$$[V_\alpha^{(2)}, V_\alpha^{(2)}] = [V_\alpha^{(2)}, V_\beta]^{G_\alpha} \leq [Q_\beta, V_\beta]^{G_\alpha} = \langle Z_\beta^{G_\alpha}\rangle = Z_\alpha.$$

Since $V_\alpha^{(2)}$ is generated by V_λ for $\lambda \in \Delta(\alpha)$ and V_λ is elementary abelian, it follows that $V_\alpha^{(2)}/[V_\alpha^{(2)}, V_\alpha^{(2)}]$ is elementary abelian and (i) holds. Moreover, we have that $[V_\alpha^{(2)}, V_{\alpha'-1}, V_{\alpha'-1}] \leq [V_\alpha^{(2)}, V_{\alpha'-2}^{(2)}, V_{\alpha'-1}] \leq [Q_{\alpha'-1}, V_{\alpha'-1}] = Z_{\alpha'-1} = Z_\beta$, (ii) holds and since $Z_\alpha \leq [V_\alpha^{(2)}, Q_\alpha]$, $V_\alpha^{(2)}/[V_\alpha^{(2)}, Q_\alpha]$ is a quadratic module for $\overline{L_\alpha}$. Furthermore, $[V_{\alpha'-1}, V_\alpha^{(2)} \cap Q_{\alpha'-2}, V_\alpha^{(2)} \cap Q_{\alpha'-2}] \leq V_{\alpha'-1} \cap Z_\alpha = Z_{\alpha'-1}$ and so, provided $V_\alpha^{(2)} \cap Q_{\alpha'-2} \not\leq Q_{\alpha'-1}$, $V_{\alpha'-1}/Z_{\alpha'-1}$ is a quadratic module for $\overline{L_{\alpha'-1}}$, and (iii) holds. We verify that $V_\alpha^{(2)} \cap Q_{\alpha'-2} \not\leq Q_{\alpha'-1}$ in the proof of (iv) which we now task ourselves with.

Set $U := \langle V_\lambda : Z_\lambda = Z_{\alpha-1}, \lambda \in \Delta(\alpha)\rangle$ for a fixed subgroup $Z_{\alpha-1} \neq Z_\beta$. If $U \not\leq Q_{\alpha'-2}$, then up to relabeling, there is some $\alpha-2 \in \Delta^{(2)}(\alpha)$ with $(\alpha-2, \alpha'-2)$ a critical pair and $Z_{\alpha-1} \neq Z_\beta$. But then $Z_\alpha = Z_{\alpha-1} \times Z_\beta \leq V_{\alpha'-1}$, a contradiction since $b \geqslant 4$. Suppose that $U \leq Q_{\alpha'-1}$ so that $[Z_{\alpha'}, U] = [Z_{\alpha'}, Z_\alpha(U \cap Q_{\alpha'})] \leq Z_\alpha \leq U$. Let $r \in R_\alpha Q_{\alpha-1}$. Since r is an automorphism of the graph, it follows that for $V_\delta \leq U$, we have that $V_\delta^r = V_{\delta\cdot r}$. But $Z_{\delta\cdot r} = Z_\delta^r = Z_{\alpha-1}^r = Z_{\alpha-1}$ and so $V_\delta^r \leq U$. Then $U \trianglelefteq L_\alpha = \langle R_\alpha, Z_{\alpha'}, Q_{\alpha-1}\rangle$. Since $Z_{\alpha'}$ centralizes U/Z_α, $[O^p(L_\alpha), V_{\alpha-1}] \leq [O^p(L_\alpha), U] = Z_\alpha \leq V_{\alpha-1}$. In particular, $V_{\alpha-1} \trianglelefteq \langle G_\alpha, G_{\alpha-1}\rangle$, a contradiction. Thus we may assume that $U \not\leq Q_{\alpha'-1}$ and we may choose $V_{\alpha-1} \not\leq Q_{\alpha'-1}$ with $Z_{\alpha-1} \neq Z_\beta$. and (iv) holds. □

Throughout the remainder of this section, whenever $b > 2$, we fix $\alpha - 1 \in \Delta(\alpha)$ with properties described in Lemma 6.9 (iv).

Lemma 6.10 *Suppose that $b > 2$ and $m_p(S/Q_\alpha) > 1$. Then either $L_\beta/R_\beta \cong \mathrm{SL}_2(q) \cong L_\alpha/R_\alpha$ and both Z_α and $V_\beta/C_{V_\beta}(O^p(L_\beta))$ are natural modules; or $q_\beta > q_\alpha$, $[O^p(R_\alpha), V_\alpha^{(2)}] \neq \{1\}$ and $p = 2$.*

Proof Assume that p is an odd prime. Using that $m_p(S/Q_\alpha) \geqslant 2$ and $V_\alpha^{(2)}/Z_\alpha$ admits quadratic action by $V_{\alpha'-1}$, we deduce from Lemma 2.35 (using that $S = Q_\alpha V_{\alpha'-1}$) and a standard argument involving the Schur multiplier of $\mathrm{PSL}_2(q)$, that $R_\alpha = C_{L_\alpha}(V_\alpha^{(2)})Q_\alpha$. Assume now that $p = 2$ and $q_\beta \leqslant q_\alpha$. Then as $V_\alpha^{(2)}/Z_\alpha$ is elementary abelian, $V_\alpha^{(2)} \cap Q_{\alpha'-2} \cap Q_{\alpha'-1}$ has index at most q_α^2 and is centralized, modulo Z_α, by $Z_{\alpha'}$. Applying Lemma 2.55, we deduce in this case also that $R_\alpha = C_{L_\alpha}(V_\alpha^{(2)})Q_\alpha$.

More generally, assume that $R_\alpha = C_{L_\alpha}(V_\alpha^{(2)})Q_\alpha$ with the intention of showing that Z_α and $V_\beta/C_{V_\beta}(O^p(L_\beta))$ are natural modules. An application of Lemma 5.22 to $Z_{\alpha'-1} = Z_\beta$ yields immediately that $b > 4$ so that $V_\alpha^{(2)}$ is elementary abelian. Furthermore, $V_\alpha^{(2)} \leq Q_{\alpha'-2}$ for otherwise there is $\lambda \in \Delta(\alpha)$ with $(\lambda-1, \alpha'-2)$ a critical pair. But then by Lemma 6.8 we deduce that $Z_\lambda = Z_{\alpha'-3}$ and since $Z_\alpha \neq Z_{\alpha'-2}$, we must have that $Z_\lambda = Z_\beta$ and Lemma 5.22 yields that $V_\lambda = V_\beta \leq Q_{\alpha'-2}$, a contradiction.

Now, $V_{\alpha-1} \cap Q_{\alpha'-1} = Z_\alpha(V_{\alpha-1} \cap Q_{\alpha'})$ so that $L_\alpha = \langle Z_{\alpha'}, Q_{\alpha-1}, R_\alpha\rangle$ normalizes $V_{\alpha-1} \cap Q_{\alpha'-1}$. Since L_α acts transitively on the neighbors of α, $V_{\alpha-1} \cap V_\beta \leq V_{\alpha-1} \cap Q_{\alpha'-1} \leq \bigcap_{\lambda\in\Delta(\alpha)} V_\lambda$. By conjugacy, $\bigcap_{\mu\in\Delta(\alpha'-2)} V_\mu$ has index $|V_{\alpha-1}Q_{\alpha'-1}/Q_{\alpha'-1}|$ in $V_{\alpha'-1}$ and is centralized by $V_{\alpha-1}$. By Lemma 2.41 and conjugacy, $V_\beta/C_{V_\beta}(O^p(L_\beta))$ is a natural module for $L_\beta/R_\beta \cong \mathrm{SL}_2(q_\beta)$ and as $Z_\alpha C_{V_\beta}(O^p(L_\beta))/C_{V_\beta}(O^p(L_\beta)) \cong Z_\alpha/Z_\beta$ has order q_α, and $Z_\alpha C_{V_\beta}(O^p(L_\beta))/C_{V_\beta}(O^p(L_\beta))$ is $G_{\alpha,\beta}$-invariant, we deduce that $q_\alpha = q_\beta$ and the proof is complete. □

In the following lemma and proposition, we retain the definition of U from Lemma 6.9. That is, $U := \langle V_\lambda : Z_\lambda = Z_{\alpha-1}, \lambda \in \Delta(\alpha)\rangle$. Furthermore, we define $\mathcal{U} := [U, Q_\alpha; i]Z_\alpha$ with i chosen minimally so that $[U, Q_\alpha; i+1] \leq Z_\alpha$.

Lemma 6.11 *Suppose that $b > 2$. If $V_\beta/C_{V_\beta}(O^p(L_\beta))$ is not a natural module for $L_\beta/R_\beta \cong \mathrm{SL}_2(q_\beta)$ then the following hold:*

(i) $Z(Q_\alpha) = Z_\alpha$;
(ii) $\mathcal{U} = [V_\lambda, Q_\alpha; i] \trianglelefteq G_\alpha$ *for all* $\lambda \in \Delta(\alpha)$; *and*
(iii) $[\mathcal{U}, Q_\alpha] = Z_\alpha$.

Proof Assume first that $Z(Q_\alpha) \not\leq Q_{\alpha'-1}$. In particular, $V_{\alpha'-1} \cap Q_\alpha$ is centralized by $Z(Q_\alpha)$ and $Z(Q_\alpha)$ acts quadratically on $V_{\alpha'-1}$ so that $Z(Q_\alpha)Q_{\alpha'-1}/Q_{\alpha'-1}$ is elementary abelian. We may assume that $m_p(S/Q_\alpha) > 1$ else we have a contradiction to the initial assumptions by Lemma 2.41. If $Z(Q_\alpha) \cap Q_{\alpha'-1}$ has index strictly less than q_α in $Z(Q_\alpha)$ then $O^p(L_\alpha)$ centralizes $Z(Q_\alpha)/Z_\alpha$ and we deduce that S centralizes $Z(Q_\alpha)/Z_\alpha$. In particular, $[V_{\alpha'-1}, Z(Q_\alpha)] \leq Z_\alpha \cap V_{\alpha'-1} = Z_\beta = Z_{\alpha'-1}$, a contradiction since $Z(Q_\alpha) \not\leq Q_{\alpha'-1}$. Hence, $q_\alpha \leqslant |Z(Q_\alpha)Q_{\alpha'-1}/Q_{\alpha'-1}|$ and Lemma 2.41 provides a contradiction to the initial assumption. Therefore, $Z(Q_\alpha) \leq Q_{\alpha'-1}$. Then $O^p(L_\alpha)$ centralizes $Z(Q_\alpha)/Z_\alpha$ and the irreducibility of L_α on Z_α yields that $Z(Q_\alpha)$ is elementary abelian. But then $Z(Q_\alpha) \cap Q_{\alpha'}$ is centralized by $S = Z_{\alpha'}Q_\alpha$ so that $Z(Q_\alpha) \cap Q_{\alpha'} \leq \Omega(Z(S)) = Z_\beta \leq Z_\alpha$. Thus, $Z(Q_\alpha) = Z_\alpha$.

Set $\mathcal{U} := [U, Q_\alpha; i]Z_\alpha$ with i chosen minimally so that $[U, Q_\alpha; i+1] \leq Z_\alpha$. In particular, since $Z(Q_\alpha) = Z_\alpha$, $[\mathcal{U}, Q_\alpha]$ is non-trivial. Let $W := \langle \mathcal{U}^{G_\alpha} \rangle$. Then $[W, Q_\alpha] = [\mathcal{U}, Q_\alpha]^{G_\alpha} = Z_\alpha$. Moreover, if W/Z_α contains no non-central chief factors for L_α, then $W = \mathcal{U} = [V_{\alpha-1}, Q_\alpha; i]$ is normalized by $G_\alpha = O^p(L_\alpha)G_{\alpha,\alpha-1}$ and (ii) and (iii) hold.

Hence, we may assume that W/Z_α contains a non-central chief factor for L_α. Suppose that $|(W \cap Q_{\alpha'-2})Q_{\alpha'-1}/Q_{\alpha'-1}| \leqslant p$. By Lemma 2.41, $|WQ_{\alpha'-2}/Q_{\alpha'-2}| \geqslant q_\alpha/p$. But now, $[V_{\alpha'-2}^{(2)} \cap Q_\alpha, W] \leq Z_\alpha \cap V_{\alpha'-2}^{(2)} = Z_{\alpha'-1}$ so that either $q_\alpha = p$, $W \leq Q_{\alpha'-2}$ and $W \not\leq Q_{\alpha'-1}$; or $V_{\alpha'-2}^{(2)}/Z_{\alpha'-2}$ contains a unique non-central chief factor for $L_{\alpha'-2}$ by Propositions 2.31 and 2.32. In the latter case, by conjugacy and applying Lemma 5.8, we conclude that $\mathcal{U} = U$, $W = V_\alpha^{(2)}$ and $[V_\alpha^{(2)}, Q_\alpha] = Z_\alpha$. Since $q_\alpha \leqslant q_\beta$ by Lemma 6.10 and $[Q_\alpha, V_\beta] = Z_\alpha$, we deduce that V_β/Z_β is dual to an FF-module for $L_\beta/R_\beta \cong \mathrm{SL}_2(q_\beta)$ and by Lemma 2.41, we have a contradiction to the initial hypothesis. If $q_\alpha = p$, then $V_{\alpha'-1} \cap Q_\alpha$ has index p in $V_{\alpha'-1}$ and is centralized, modulo $Z_{\alpha'-1}$, by W and we have a contradiction by Lemma 2.41.

Thus, $|(W \cap Q_{\alpha'-2})Q_{\alpha'-1}/Q_{\alpha'-1}| \geqslant p^2$. As before, we have that $V_{\alpha'-1} \cap Q_\alpha$ has index q_α in $V_{\alpha'-1}$ and is centralized, modulo $Z_{\alpha'-1}$, by W. In particular, Lemma 2.41 yields $q_\alpha > |(W \cap Q_{\alpha'-2})Q_{\alpha'-1}/Q_{\alpha'-1}| \geqslant p^2$ and Lemma 6.10 gives that $q_\beta > q_\alpha$ and $p = 2$. But now, a contradiction is provided by Proposition 2.31 applied to the action of W on $V_{\alpha'-1}/Z_{\alpha'-1}$. □

Proposition 6.12 *Suppose that $b > 2$. Then $L_\beta/R_\beta \cong \mathrm{SL}_2(q) \cong L_\alpha/R_\alpha$ and both Z_α and $V_\beta/C_{V_\beta}(O^p(L_\beta))$ are natural modules.*

Proof We note that if $q_\alpha = p$ and $V_\beta/C_{V_\beta}(O^p(L_\beta))$ is natural module for $L_\beta/R_\beta \cong \mathrm{SL}_2(q_\beta)$ then since $Z_\alpha C_{V_\beta}(O^p(L_\beta))/C_{V_\beta}(O^p(L_\beta))$ is a $G_{\alpha,\beta}$-invariant subgroup of order p, we see that $q_\alpha = q_\beta = p$. Therefore, aiming for a contradiction, assume throughout that $V_\beta/C_{V_\beta}(O^p(L_\beta))$ is not a natural module for $L_\beta/R_\beta \cong \mathrm{SL}_2(q_\beta)$. We may use the results proved in Lemma 6.10 and Lemma 6.11.

Suppose first that $q_\alpha = p$. Since $[V_{\alpha'-1} \cap Q_\alpha \cap Q_{\alpha-1}, V_{\alpha-1}] = \{1\}$, we must have that $V_{\alpha'-1} \cap Q_\alpha \not\leq Q_{\alpha-1}$. Assume that $V_{\alpha'-1} \cap Q_\alpha \cap Q_{\alpha-1}$ has index p^2 in $V_{\alpha'-1}$. Indeed, this is the case if $m_p(S/Q_\beta) = 1$. Then $L_{\alpha'-1}/R_{\alpha'-1}$ and $V_{\alpha'-1}/C_{V_{\alpha'-1}}(O^p(L_{\alpha'-1}))$ are determined by Proposition 2.51. Since $Z_{\alpha'}C_{V_{\alpha'-1}}(O^p(L_{\alpha'-1}))/C_{V_{\alpha'-1}}(O^p(L_{\alpha'-1}))$ has order p and is $G_{\alpha',\alpha'-1}$-invariant, and $V_{\alpha'-1} = \langle Z_{\alpha'}^{L_{\alpha'-1}}\rangle$, by Lemma 2.54 we have that $L_{\alpha'-1}/R_{\alpha'-1} \cong \mathrm{Sz}(2), \mathrm{Dih}(10), (3 \times 3) : 2$ or $(3 \times 3) : 4$. In particular, using coprime action, it follows that for $V := V_{\alpha'-1}/Z_{\alpha'-1}$, $V = [V, O^2(L_{\alpha'-1})] \times C_V(O^2(L_{\alpha'-1}))$ where $[V, O^2(L_{\alpha'-1})]$ is irreducible and $|C_V(O^2(L_{\alpha'-1}))| \leqslant 2$.

Assume that $L_{\alpha'-1}/R_{\alpha'-1} \cong \mathrm{Sz}(2)$ or $(3\times 3) : 4$. Then, by Proposition 2.29 (iii) and Lemma 2.53 (iii), $[V, Q_{\alpha'-2}; 3] \neq \{1\} = [V, Q_{\alpha'-2}; 4]$ and, by conjugacy, we infer that $[V_{\alpha-1}, Q_\alpha; 4] \leq Z_{\alpha-1}$. Then, Lemma 6.11 implies that $\mathcal{U} = [V_{\alpha-1}, Q_\alpha, Q_\alpha]$ and $Z_\alpha = [V_{\alpha-1}, Q_\alpha; 3]$. Moreover, we deduce that $[U, Q_\alpha] \not\leq Q_{\alpha'-1}$, else $[U, Q_\alpha] = Z_\alpha([U, Q_\alpha] \cap Q_{\alpha'})$ is centralized, modulo Z_α, by $Z_{\alpha'}$ from which we have that $[V_{\alpha-1}, Q_\alpha] \trianglelefteq L_\alpha$. But then, by conjugacy, $[V_{\alpha'-1}, Q_{\alpha'-2}] = [V_{\alpha'-3}, Q_{\alpha'-2}]$ is centralized by $V_{\alpha-1}$, contradicting Proposition 2.29 (ii) and Lemma 2.53 (ii). If $[V_\alpha^{(2)}, Q_\alpha] \not\leq Q_{\alpha'-2}$, then as $\Phi(V_\alpha^{(2)}) \leq Z_\alpha \leq Q_{\alpha'-1}$, $V_\alpha^{(2)} \cap Q_{\alpha'-2} = [U, Q_\alpha](V_\alpha^{(2)} \cap Q_{\alpha'-2} \cap Q_{\alpha'-1})$ so that $V_\alpha^{(2)} = [V_\alpha^{(2)}, Q_\alpha](V_\alpha^{(2)} \cap Q_{\alpha'})$ and $V_\alpha^{(2)}/[V_\alpha^{(2)}, Q_\alpha]$ is centralized by $O^p(L_\alpha)$, a contradiction by Lemma 5.8. Thus, as $\Phi(U) \leq \Phi(V_\alpha^{(2)}) \leq Z_\alpha \leq Q_{\alpha'-1}$, $U[V_\alpha^{(2)}, Q_\alpha] = [V_\alpha^{(2)}, Q_\alpha](U[V_\alpha^{(2)}, Q_\alpha] \cap Q_{\alpha'})$ and $U[V_\alpha^{(2)}, Q_\alpha] \trianglelefteq L_\alpha$. In particular, we conclude that $V_\alpha^{(2)} = V_{\alpha-1}[V_\alpha^{(2)}, Q_\alpha]$ from which it follows that $[Q_{\alpha-1}, V_\alpha^{(2)}] \leq [V_\alpha^{(2)}, Q_\alpha]$ and $O^p(L_\alpha)$ centralizes $V_\alpha^{(2)}/[V_\alpha^{(2)}, Q_\alpha]$, and a contradiction is again provided by Lemma 5.8.

Assume that $L_{\alpha'-1}/R_{\alpha'-1} \cong \mathrm{Dih}(10)$ or $(3 \times 3) : 2$. Then, applying Proposition 2.29 (ii) and Proposition 2.46 (v), and using that $P/Q_{\alpha-1} = \Omega(P/Q_{\alpha-1})$ where $P \in \mathrm{Syl}_2(G_{\alpha,\alpha-1})$, $[V_{\alpha-1}, Q_\alpha, Q_\alpha] \leq Z_{\alpha-1}$, $[V_{\alpha-1}, Q_\alpha] \not\leq Z_\alpha$ and Lemma 6.11 gives that $\mathcal{U} = [V_{\alpha-1}, Q_\alpha] \trianglelefteq G_\alpha$. But then $Z_\alpha = [V_{\alpha-1}, Q_\alpha, Q_\alpha] \leq Z_{\alpha-1}$, another contradiction.

Hence, if $q_\alpha = p$ then $|(V_{\alpha'-1} \cap Q_\alpha)Q_{\alpha-1}/Q_{\alpha-1}| \geqslant p^2$. In particular, $m_p(S/Q_\beta) > 1$ and applying Lemma 2.35 when p is odd, since $V_{\alpha'-1} \cap Q_\alpha$ is quadratic on $V_{\alpha-1}$, we conclude that $\overline{L_\beta}/O_{p'}(\overline{L_\beta})$ is isomorphic to one of $\mathrm{PSL}_2(q_\beta)$, $\mathrm{PSU}_3(q_\beta)$ or $\mathrm{Sz}(q_\beta)$ where necessarily $p = 2$ in the last case. In particular, $q_\beta > q_\alpha$. On the other hand, if $q_\alpha > p$, then by Lemma 6.10 we have that $q_\beta > q_\alpha$ and $p = 2$. Hence, in all cases we have that $q_\beta > q_\alpha \geqslant p$.

Assume that $\overline{L_\beta}/O_{p'}(\overline{L_\beta}) \cong \mathrm{Sz}(q_\beta)$ or $\mathrm{PSU}_3(q_\beta)$. Since $V_{\alpha'-1}$ is elementary abelian, $V_{\alpha'-1} \cap Q_\alpha \cap Q_{\alpha-1}$ has index at most $q_\alpha q_\beta$ in $V_{\alpha'-1}$ and is centralized by $V_{\alpha-1}$. Then by Propositions 2.31 and 2.32, $|V_{\alpha-1}Q_{\alpha'-1}/Q_{\alpha'-1}| = p$ and $q_\alpha^3 \geqslant q_\beta$ and $q_\alpha^2 \geqslant q_\beta$ respectively. But now, $[V_{\alpha-1} \cap Q_{\alpha'-1}, x] \leq Z_\alpha$ for some $x \in (V_{\alpha'-1} \cap Q_\alpha) \setminus Q_{\alpha-1}$ from which it follows that x centralizes a subgroup of index at most $pq_\alpha \leq q_\beta$ in $V_{\alpha-1}$. Applying Propositions 2.31 and 2.32, we have a contradiction. Hence, $\overline{L_\beta}/O_{p'}(\overline{L_\beta}) \cong \mathrm{PSL}_2(q_\beta)$.

Now, $[Q_\alpha \cap Q_\beta, \mathcal{U}] \leq Z_\beta$ so that $Q_\alpha \cap Q_\beta \leq C_{Q_\alpha}(\mathcal{U}/Z_\beta) \trianglelefteq R_\alpha G_{\alpha,\beta}$. Since $\overline{L_\beta}/O_{p'}(\overline{L_\beta}) \cong \mathrm{PSL}_2(q_\beta)$, $C_{Q_\alpha}(\mathcal{U}/Z_\beta) \leq Q_\beta$ as $S/Q_\beta \cong Q_\alpha/Q_\alpha \cap Q_\beta$ is irreducible under the action of $G_{\alpha,\beta}$ and Q_α does not centralize $\mathcal{U}/Z_\beta$. More generally, for any $\lambda, \mu \in \Delta(\alpha)$, if $Z_\lambda = Z_\mu$ then $Q_\lambda \cap Q_\alpha = Q_\mu \cap Q_\alpha$. Additionally, we have that $Q_\alpha \cap Q_{\alpha-1} \cap Q_\beta = C_{Q_\alpha}(\mathcal{U}) \trianglelefteq G_\alpha$ so that $[C_{Q_\alpha}(\mathcal{U}), V_\alpha^{(2)}] = [C_{Q_\alpha}(\mathcal{U}), V_\beta]^{G_\alpha} = Z_\alpha$.

Suppose that $[V_\alpha^{(2)}, V_{\alpha'-3}] \neq \{1\}$. Then there is a critical pair $(\lambda + 1, \mu)$ for some $\lambda \in \Delta(\alpha)$ and $\mu \in \Delta(\alpha' - 3)$ and by an argument in Lemma 6.9, we deduce that $Z_\lambda = Z_\beta = Z_{\alpha'-1} = Z_{\alpha'-3}$. But then, $U \leq Q_{\alpha'-3} \cap Q_{\alpha'-2} = Q_{\alpha'-2} \cap Q_{\alpha'-1}$, a contradiction by Lemma 6.9. Indeed, this also proves that $Z_{\alpha'-1} \neq Z_{\alpha'-3}$ and $V_\alpha^{(2)} \leq Q_{\alpha'-2}$. By Lemma 6.11, we have that $[V_{\alpha'-1}, Q_{\alpha'-2}; i] \trianglelefteq G_{\alpha'-2}$ where i is the integer defining $\mathcal{U}$. Moreover, since $Z_{\alpha'-1} \neq Z_{\alpha'-3}$, $C_{Q_{\alpha'-2}}([V_{\alpha'-1}, Q_{\alpha'-2}; i]) = Q_{\alpha'-1} \cap Q_{\alpha'-3}$. But $[V_{\alpha'-1}, Q_{\alpha'-2}; i] \leq V_{\alpha'-3}$ is centralized by $V_\alpha^{(2)}$, a contradiction since $V_\alpha^{(2)} \not\leq Q_{\alpha'-1}$. □

Before continuing, observe that we may now assume that whenever $b > 2$ and $q = p$, both L_α/R_α and L_β/R_β are isomorphic to $\mathrm{SL}_2(p)$. Throughout this section, under these conditions and given a module V on which $\overline{L_\gamma}$ acts, for any $\gamma \in \Gamma$, we will often utilize coprime action. By this, we mean that when $p \geqslant 5$, taking T_γ to be the preimage in $\overline{L_\gamma}$ of $Z(L_\gamma/R_\gamma)$, we have that $V = [V, T] \times C_V(T)$. Indeed, if V is an FF-module for $\overline{L_\gamma}$, then this leads to a splitting $V = [V, \overline{L_\gamma}] \times C_V(\overline{L_\gamma})$. If $p \in \{2, 3\}$, since $\overline{L_\gamma}$ is solvable, we automatically have the conclusion $V = [V, O^p(\overline{L_\gamma})] \times C_V(O^p(\overline{L_\gamma}))$. Without explaining this each time it is used, we will generally just refer to "coprime action" and hope that it is clear in each instance where the conclusions we draw come from.

Lemma 6.13 *Suppose that $b > 2$ and $q = p$. Then $Z_\beta = Z(Q_\beta)$ and $Z_\alpha = Z(Q_\alpha)$.*

Proof By minimality of b, and using that b is even, we infer that $Z(Q_\alpha) \leq Q_\lambda$ for all $\lambda \in \Delta^{(b-2)}(\alpha)$. In particular, $Z(Q_\alpha) \leq Q_{\alpha'-2}$. If $Z(Q_\alpha) \not\leq Q_{\alpha'-1}$ then as $[Z(Q_\alpha), V_{\alpha'-1}, V_{\alpha'-1}] \leq [V_{\alpha'-1}, V_{\alpha'-1}] = \{1\}$ and $[Z(Q_\alpha), V_{\alpha'-1}]$ is centralized by $V_{\alpha'-1}Q_\alpha \in \mathrm{Syl}_p(L_\alpha)$ and has exponent p. Thus, $[Z(Q_\alpha), V_{\alpha'-1}] \leq \Omega(Z(S)) = Z_\beta = Z_{\alpha'-1}$, a contradiction for otherwise $O^p(L_{\alpha'-1})$ centralizes $V_{\alpha'-1}$. Thus, $Z(Q_\alpha) \leq Q_{\alpha'-1}$ so that $Z(Q_\alpha) = Z_\alpha(Z(Q_\alpha) \cap Q_{\alpha'})$, $Z_{\alpha'}$ centralizes $Z(Q_\alpha)/Z_\alpha$ and $O^p(L_\alpha)$ centralizes $Z(Q_\alpha)/Z_\alpha$. Since $Z_\beta \leq Z_\alpha$ an application of coprime action yields $Z(Q_\alpha) = [Z(Q_\alpha), O^p(L_\alpha)] = Z_\alpha$, as desired. As a consequence, using that Q_α is self-centralizing, $Z(S)$ has exponent p.

Let $\alpha - 1 \in \Delta(\alpha)$ such that $Z_{\alpha-1} \neq Z_\beta$, $V_{\alpha-1} \leq Q_{\alpha'-2}$ and $V_{\alpha-1} \not\leq Q_{\alpha'-1}$, as chosen in Proposition 6.12. By minimality of b, and using that b is even, we have that $Z(Q_{\alpha'-1}) \leq Q_\lambda$ for all $\lambda \in \Delta^{(b-1)}(\alpha)$. In particular, $Z(Q_{\alpha'-1}) \leq Q_\alpha$.

If $Z(Q_{\alpha'-1}) \not\leq Q_{\alpha-1}$ then $Z(Q_{\alpha'-1})Q_{\alpha-1} \in \mathrm{Syl}_p(L_{\alpha-1})$. Again, using minimality of b, we infer that $Z(Q_{\alpha-1}) \leq Q_{\alpha'-2}$ so that $[Z(Q_{\alpha'-1}), Z(Q_{\alpha-1})] \leq Z(Q_{\alpha'-1}) \cap Z(Q_{\alpha-1})$. Thus, $[Z(Q_{\alpha'-1}), Z(Q_{\alpha-1})]$ is centralized by $Z(Q_{\alpha'-1})Q_{\alpha-1} \in \mathrm{Syl}_p(L_{\alpha-1})$. Then, $[Z(Q_{\alpha'-1}), Z(Q_{\alpha-1})] \leq Z_{\alpha-1}$ and as $Z_{\alpha-1} \not\leq Z(Q_{\alpha'-1})$,

$[Z(Q_{\alpha'-1}), Z(Q_{\alpha-1})] = \{1\}$ and $Z(Q_{\alpha-1})$ is centralized by $Z(Q_{\alpha'-1})Q_{\alpha-1} \in \mathrm{Syl}_p(L_{\alpha-1})$. But then $Z(Q_{\alpha-1}) = Z_{\alpha-1}$ and by conjugacy, $Z(Q_{\alpha'-1}) = Z_{\alpha'-1} \leq Z_{\alpha'-2} \leq Q_{\alpha-1}$, a contradiction.

Thus, $Z(Q_{\alpha'-1}) \leq Q_{\alpha-1}$ and so, $[Z(Q_{\alpha'-1}), V_{\alpha-1}] \leq Z_{\alpha-1} \cap Z(Q_{\alpha'-1})$. Since $Z_{\alpha-1}$ does not centralize $Z_{\alpha'}$, we deduce that $[Z(Q_{\alpha'-1}), V_{\alpha-1}] = \{1\}$. But then $Z(Q_{\alpha'-1})$ is centralized by $V_{\alpha-1}Q_{\alpha'-1} \in \mathrm{Syl}_p(L_{\alpha'-1})$ and $Z(Q_{\alpha'-1}) = Z_{\alpha'-1}$, as required. □

Combining Proposition 6.12 and Lemma 6.13, we now satisfy Hypothesis 5.31. Thus, whenever b and the non-central chief factors in $V_\lambda^{(n)}$ satisfy the necessary requirements for $\lambda \in \{\alpha, \beta\}$ and various values of n, we may freely apply the results contained between Lemmas 5.32 and 5.36.

Lemma 6.14 *Suppose that $b > 2$. Then $|V_\beta| = q^3$ and $[V_\alpha^{(2)}, Q_\alpha] = Z_\alpha$.*

Proof If $V_\alpha^{(2)} \leq Q_{\alpha'-2}$, then $Z_\alpha(V_\alpha^{(2)} \cap Q_{\alpha'})$ has index q in $V_\alpha^{(2)}$ so that $V_\alpha^{(2)}/Z_\alpha$ has a unique non-central chief factor. Then the result holds by Lemma 5.32. Thus, we suppose that $V_\alpha^{(2)} \not\leq Q_{\alpha'-2}$. Then there is $\alpha - 2$ such that $(\alpha-2, \alpha'-2)$ is a critical pair and by Lemma 6.8, we have that $Z_{\alpha-1} = Z_{\alpha'-3}$. Since $b > 2$ and $Z_\beta Z_{\alpha-1} \leq Z_\alpha \cap Z_{\alpha'-2}$, it follows that $Z_\beta = Z_{\alpha-1} = Z_{\alpha'-3} = Z_{\alpha'-1}$. If $|V_\beta| \neq q^3$, then by Lemma 5.33 since $Z_\alpha(V_\alpha^{(2)} \cap Q_{\alpha'-2} \cap Q_{\alpha'-1})$ has index at most q^2 in $V_\alpha^{(2)}$, $O^p(R_\alpha)$ centralizes $V_\alpha^{(2)}$. By Lemma 5.22, $Z_{\alpha-2} \leq V_{\alpha-1} = V_\beta \leq Q_{\alpha'-2}$, a contradiction. Hence, the lemma holds. □

Lemma 6.15 $b \neq 4$.

Proof Since none of the conclusions of Theorem C satisfy $b = 4$, we may suppose that G is a minimal counterexample with $b = 4$. Suppose that $V_\alpha^{(2)} \leq Q_{\alpha'-2}$. Then $V_\alpha^{(2)} \cap Q_{\alpha'-1} = Z_\alpha(V_\alpha^{(2)} \cap Q_{\alpha'})$ is an index q subgroup of $V_\alpha^{(2)}$ which is centralized, modulo Z_α, by $Z_{\alpha'}$. Thus, $V_\alpha^{(2)}/Z_\alpha$ is an FF-module for $\overline{L_\alpha}$. Then Lemma 5.33 implies that $O^p(R_\alpha)$ centralizes $V_\alpha^{(2)}$ and since $Z_{\alpha'-1} = Z_\beta$, Lemma 5.22 implies that $Z_\alpha \leq V_\beta = V_{\alpha'-1} \leq Q_{\alpha'}$, a contradiction. We have a similar contradiction if $V_\alpha^{(2)} \cap Q_{\alpha'-2} \leq Q_{\alpha'-1}$.

Thus, $V_\alpha^{(2)} \not\leq Q_{\alpha'-2}$ and $V_\alpha^{(2)} \cap Q_{\alpha'-2} \not\leq Q_{\alpha'-1}$. In particular, $V_\alpha^{(2)}$ is non-abelian and $Z_\alpha \leq \Phi(V_\alpha^{(2)})$. Suppose that $r \in L_\alpha$ is of order coprime to p and centralizes $V_\alpha^{(2)}$. Then, by the three subgroup lemma, r centralizes $Q_\alpha/C_{Q_\alpha}(V_\alpha^{(2)})$. Since $C_{Q_\alpha}(V_\alpha^{(2)}) \leq Q_{\alpha'-2}$ and $V_\alpha^{(2)} \cap Q_{\alpha'-2} \not\leq Q_{\alpha'-1}$, we have that $Z_{\alpha'}$ centralizes $C_{Q_\alpha}(V_\alpha^{(2)})V_\alpha^{(2)}/V_\alpha^{(2)}$ so that $O^p(L_\alpha)$ centralizes $C_{Q_\alpha}(V_\alpha^{(2)})V_\alpha^{(2)}/V_\alpha^{(2)}$. By coprime action, r centralizes Q_α, and so $r = 1$. Thus, every p'-element of L_α acts faithfully on $V_\alpha^{(2)}/\Phi(V_\alpha^{(2)})$.

Now, $Z_\alpha(V_\alpha^{(2)} \cap \cdots \cap Q_{\alpha'})$ has index at most q^2 in $V_\alpha^{(2)}$ so that $V_\alpha^{(2)}/Z_\alpha$ is a 2F-module for $\overline{L_\alpha}$. Furthermore,

$$[V_\alpha^{(2)}, V_{\alpha'-1}, V_{\alpha'-1}] \leq [V_\alpha^{(2)}, V_{\alpha'-2}^{(2)}, V_{\alpha'-1}] \leq [Q_{\alpha'-1}, V_{\alpha'-1}] = Z_{\alpha'-1} = Z_\beta$$

and $V_\alpha^{(2)}/Z_\alpha$ is a faithful quadratic 2F-module for $\overline{L_\alpha}$. Then $\overline{L_\alpha}$ is determined by Lemma 2.41, Proposition 2.51 and Lemma 2.55 and since $\overline{L_\alpha}$ has a quotient isomorphic to $\mathrm{SL}_2(q)$, we have that $\overline{L_\alpha} \cong \mathrm{SL}_2(q), \mathrm{SU}_3(2)', (3 \times 3) : 2$ or $(Q_8 \times Q_8) : 3$. Notice that V_β/Z_α is of order q and is not contained in $C_{V_\alpha^{(2)}/Z_\alpha}(O^p(L_\alpha))$. Setting $V := V_\alpha^{(2)}/Z_\alpha$ there is a $G_{\alpha,\beta}$-invariant subgroup of $V/C_V(O^p(L_\alpha))$ of order q which generates V and by Lemma 2.54, we have that $\overline{L_\alpha} \cong (3\times 3) : 2$. Moreover, since $V_\alpha^{(2)}/Z_\alpha$ contains two non-central chief factors for L_α. For $U_\alpha := [V_\alpha^{(2)}, L_\alpha]$, we have that $Z_{\alpha'-2} = Z_{\alpha'-1}[U_\alpha \cap Q_{\alpha'-2}, V_{\alpha'-1}] \leq U_\alpha$ so that $V_\beta \leq U_\alpha$, $V_\alpha^{(2)} = U_\alpha$ and $|V_\alpha^{(2)}/Z_\alpha| = 2^4$.

Let $P_\alpha \leq L_\alpha$ with $S \leq P_\alpha$, $P_\alpha/Q_\alpha \cong \mathrm{Sym}(3)$, $L_\alpha = P_\alpha R_\alpha$ and $O_3(\overline{P_\alpha}) \trianglelefteq \overline{L_\alpha}$. Then P_α is $G_{\alpha,\beta}$-invariant and upon showing that no non-trivial subgroup of S is normalized by both P_α and G_β, then triple $(P_\alpha G_{\alpha,\beta}, G_\beta, G_{\alpha,\beta})$ satisfies Hypothesis 5.1. To this end, suppose that Q is non-trivial subgroup of S normalized by P_α and G_β. Then $Z_\beta \leq Q$ so that $Z_\beta \leq \Omega(Z(Q))$. Taking consecutive normal closure, we deduce that $V_\beta \leq \Omega(Z(Q))$ and $\Omega(Z(Q))/Z_\alpha$ contains some of the non-central L_α-chief factors contained in $V_\alpha^{(2)}/Z_\alpha$. Write W for the preimage in $V_\alpha^{(2)}$ of some non-central chief factor contained in $\Omega(Z(Q)) \cap V_\alpha^{(2)}/Z_\alpha$, noting that by the definition of $V_\alpha^{(2)}$, $W \cap V_\beta = Z_\alpha$. However, $WV_\beta \leq \Omega(Z(Q))$ and $[W, V_\beta] = \{1\}$ so that $W \leq Q_{\alpha'-2}$, $[W, V_{\alpha'-2}] \leq Z_{\alpha'-2} \cap W = Z_\beta = Z_{\alpha'-1}$ and $W = Z_\alpha(W \cap Q_{\alpha'})$. Then W contains no non-central chief factor for L_α, a contradiction. Thus, $Q = \{1\}$ and $(P_\alpha G_{\alpha,\beta}, G_\beta, G_{\alpha,\beta})$ satisfies Hypothesis 5.1. Since G is a minimal counterexample to Theorem C, we conclude that $P_\alpha/Q_\alpha \cong \mathrm{Sym}(3) \cong \overline{L_\beta}$ and $(P_\alpha G_{\alpha,\beta}, G_\beta, G_{\alpha,\beta})$ is a weak BN-pair of rank 2. By Delgado and Stellmacher [27], $|S| \leqslant 2^7$ and since $|V_\alpha^{(2)}| = 2^6$ and $Q_\alpha/V_\alpha^{(2)}$ contains a non-central chief factor for L_α, we have a contradiction. Hence, no such G exists and $b \neq 4$. □

Lemma 6.16 *Suppose that $b > 2$. Then the following hold:*

(i) $V_\alpha^{(2)} \leq Q_{\alpha'-2}$ *but* $V_\alpha^{(2)} \not\leq Q_{\alpha'-1}$*;*
(ii) $[V_\alpha^{(2)}, Q_\alpha] = Z_\alpha$ *and* $|V_\beta| = q^3$*;*
(iii) $O^p(R_\alpha)$ *centralizes* $V_\alpha^{(2)}$ *and* $V_\alpha^{(2)}/Z_\alpha$ *is a faithful FF-module for* $L_\alpha/R_\alpha \cong \mathrm{SL}_2(q)$*;*
(iv) $b \geqslant 8$*; and*
(v) $Z_{\alpha'-2} \leq V_\alpha^{(2)} \leq Z(V_\alpha^{(4)})$.

Proof By Lemma 6.15, we have that $b > 4$ so that $V_\alpha^{(2)}$ is abelian. Moreover, (ii) holds by Lemma 6.14. Suppose first that $V_\alpha^{(2)} \not\leq Q_{\alpha'-2}$ so that there is a critical pair $(\alpha - 2, \alpha' - 2)$ such that $[Z_{\alpha-2}, Z_{\alpha'-2}] = Z_{\alpha-1} = Z_{\alpha'-3}$. Since $b > 2$, we have that $Z_\alpha \neq Z_{\alpha'-2}$ so that $Z_{\alpha-1} = Z_\beta$. Now, $[V_\alpha^{(2)} \cap Q_{\alpha'-2}, V_{\alpha'-1}] \leq Z_{\alpha'-2} \cap V_\alpha^{(2)}$. Since $V_\alpha^{(2)}$ is abelian and $V_\alpha^{(2)} \not\leq Q_{\alpha'-2}$, $Z_{\alpha'-2} \not\leq V_\alpha^{(2)}$. But $Z_{\alpha'-1} \leq V_\alpha^{(2)}$ and so it follows that $[V_\alpha^{(2)} \cap Q_{\alpha'-2}, V_{\alpha'-1}] \leq Z_{\alpha'-1}$ and $V_\alpha^{(2)} \cap Q_{\alpha'-2} \leq Q_{\alpha'-1}$. Then $V_\alpha^{(2)}/Z_\alpha$ is an FF-module and by Lemma 5.33, $O^p(R_\alpha)$ centralizes $V_\alpha^{(2)}$. But then by Lemma 5.22, $Z_{\alpha-2} \leq V_{\alpha-1} = V_\beta \leq Q_{\alpha'-2}$, a contradiction since $(\alpha-2, \alpha'-2)$ is a critical pair.

Thus, $V_\alpha^{(2)} \leq Q_{\alpha'-2}$. If $V_\alpha^{(2)} \leq Q_{\alpha'-1}$, then $V_\alpha^{(2)} = Z_\alpha(V_\alpha^{(2)} \cap Q_{\alpha'})$ and $O^p(L_\alpha)$ would centralize $V_\alpha^{(2)}/Z_\alpha$, a contradiction, and so (i) holds. Now, it follows that $V_\alpha^{(2)}/Z_\alpha$ is an FF-module and by Lemma 5.33, $O^p(R_\alpha)$ centralizes $V_\alpha^{(2)}$ and (iii) holds.

Since $V_\alpha^{(2)} \not\leq Q_{\alpha'-1}$, we infer that $Z_{\alpha'-2} = [V_\alpha^{(2)}, V_{\alpha'-1}]Z_{\alpha'-1} \leq V_\alpha^{(2)}$. If $b \geqslant 8$, then $V_\alpha^{(2)} \leq Z(V_\alpha^{(4)})$ and (v) holds and so to complete the proof, we assume for a contradiction that $b = 6$.

Notice that if $Z_{\alpha'-1} = Z_{\alpha'-3}$, then it follows from Lemma 5.22 that $Z_{\alpha'} \leq V_{\alpha'-1} = V_{\alpha'-3} \leq Q_\alpha$, a contradiction. Since $Z_\beta = Z_{\alpha'-1} \neq Z_{\alpha'-3}$ and $b = 6$, we have that $Z_{\alpha'-2} = Z_{\alpha+2}$. Let $\alpha - 1 \in \Delta(\alpha)$ such that $V_{\alpha-1} \not\leq Q_{\alpha'-1}$ and $Z_{\alpha-1} \neq Z_\beta$, chosen as in Proposition 6.12. We have that $V_{\alpha'-1}^{(3)} \leq Q_{\alpha+2}$ since $V_{\alpha'-1}^{(3)}$ centralizes $Z_{\alpha+2} = Z_{\alpha'-2} \leq V_{\alpha'-1}$. Then $V_{\alpha'-1}^{(3)} \cap Q_\beta = V_{\alpha'-1}(V_{\alpha'-1}^{(3)} \cap Q_\alpha)$ and

$$[V_{\alpha-1}, V_{\alpha'-1}^{(3)} \cap Q_\alpha] \leq [V_\alpha^{(2)}, V_{\alpha'-1}^{(3)} \cap Q_\alpha] \leq Z_\alpha \cap V_{\alpha'-1}^{(3)} = Z_\beta = Z_{\alpha'-1}.$$

In particular, $V_{\alpha'-1}^{(3)}/V_{\alpha'-1}$ contains a unique non-central chief factor $L_{\alpha'-1}$ which, as a $\mathrm{GF}(p)\overline{L_{\alpha'-1}}$-module, is isomorphic to a natural $\mathrm{SL}_2(q)$-module. Thus, we may apply Lemma 5.35 so that $O^p(R_{\alpha'-1})$ acts trivially on $V_{\alpha'-1}^{(3)}$. Since $Z_{\alpha+2} = Z_{\alpha'-2}$, it follows from Lemma 5.22 that $Z_{\alpha'} \leq V_{\alpha'-2}^{(2)} = V_{\alpha+2}^{(2)} \leq Q_\alpha$, an obvious contradiction. Thus, $b \geqslant 8$ and the lemma holds. □

Proposition 6.17 $b = 2$.

Proof To prove the proposition, in pursuit of a contradiction we suppose, by Lemma 6.16, that $b \geqslant 8$. Suppose first that $V_\alpha^{(4)} \not\leq Q_{\alpha'-4}$. Since $Z_{\alpha'-3} \leq Z_{\alpha'-2} \leq Z(V_\alpha^{(4)})$ is centralized by $V_\alpha^{(4)}$, it follows that $Z_{\alpha'-3} = Z_{\alpha'-5}$ and by Lemma 5.22, we have that $V_{\alpha'-3} = V_{\alpha'-5}$. Now, $[V_\alpha^{(4)} \cap Q_{\alpha'-4}, V_{\alpha'-3}] = [V_\alpha^{(4)} \cap Q_{\alpha'-4}, V_{\alpha'-5}] \leq Z_{\alpha'-5} = Z_{\alpha'-3}$ and so $V_\alpha^{(4)} \cap Q_{\alpha'-4} \leq Q_{\alpha'-3}$. Since $V_\alpha^{(4)}$ centralizes $Z_{\alpha'-2}$, we deduce that $V_\alpha^{(4)} \cap Q_{\alpha'-4} = V_\alpha^{(2)}(V_\alpha^{(4)} \cap Q_{\alpha'-4} \cap Q_{\alpha'-1})$ and so $V_\alpha^{(4)}/V_\alpha^{(2)}$ contains a unique non-central chief factor for L_α. Now, by Lemma 5.34 5.34 and Lemma 5.34 5.22, since $Z_{\alpha'-3} = Z_{\alpha'-5}$ we conclude that $Z_{\alpha'} \leq V_{\alpha'-3}^{(3)} = V_{\alpha'-5}^{(3)} \leq Q_\alpha$, a contradiction.

Therefore, we continue assuming that $V_\alpha^{(4)} \leq Q_{\alpha'-4}$. Then $V_\alpha^{(4)} \cap Q_{\alpha'-3}$ centralizes $Z_{\alpha'-2}$ and we may assume that $V_\alpha^{(4)} \not\leq Q_{\alpha'-3}$, else $V_\alpha^{(4)} = V_\alpha^{(2)}(V_\alpha^{(2)} \cap Q_{\alpha'-1})$ and $O^p(L_\alpha)$ centralizes $V_\alpha^{(4)}/V_\alpha^{(2)}$. Since $|V_{\alpha'-3}| = q^3$, $V_\alpha^{(4)} \not\leq Q_{\alpha'-3}$ and $V_\alpha^{(4)}$ centralizes $Z_{\alpha'-2}$, applying Lemma 5.19 we ascertain that $V_{\alpha'-3} \neq Z_{\alpha'-2}Z_{\alpha'-4}$ and so, $Z_{\alpha'-2} = Z_{\alpha'-4}$. If $O^p(R_\beta)$ centralizes $V_\beta^{(3)}$ then applying Lemma 5.22 to $Z_{\alpha'-2} = Z_{\alpha'-4}$ yields $Z_{\alpha'} \leq V_{\alpha'-2}^{(2)} = V_{\alpha'-4}^{(2)} \leq Q_\alpha$, a contradiction. Thus, to obtain a final contradiction, by Lemma 5.35, it suffices to show that $V_{\alpha'-1}^{(3)}/V_{\alpha'-1}$ contains a unique non-central chief factor for $L_{\alpha'-1}$ which, as a $\mathrm{GF}(p)\overline{L_{\alpha'-1}}$-module, is an FF-module.

By the symmetry in the hypothesis of (α, α') and (α', α), we may assume that $Z_{\alpha+2} = Z_{\alpha+4}$. Let $\alpha - 1 \in \Delta(\alpha)$ such that $V_{\alpha-1} \not\leq Q_{\alpha'-1}$ and $Z_{\alpha-1} \neq Z_\beta$, as in Proposition 6.12. Then $V_{\alpha'-1}^{(3)}$ centralizes $Z_{\alpha+2}$ so that $V_{\alpha'-1}^{(3)} \leq Q_{\alpha+2}$, $V_{\alpha'-1}^{(3)} \cap Q_\beta = V_{\alpha'-1}(V_{\alpha'-1}^{(3)} \cap Q_\alpha)$ and

$$[V_{\alpha-1}, V_{\alpha'-1}^{(3)} \cap Q_\alpha] \leq [V_\alpha^{(2)}, V_{\alpha'-1}^{(3)} \cap Q_\alpha] \leq Z_\alpha \cap V_{\alpha'-1}^{(3)} = Z_\beta = Z_{\alpha'-1}.$$

In particular, either $O^p(L_{\alpha'-1})$ centralizes $V_{\alpha'-1}^{(3)}/V_{\alpha'-1}$ or $V_{\alpha'-1}^{(3)}/V_{\alpha'-1}$ contains a unique non-central chief factor for $L_{\alpha'-1}$, and the desired contradiction has been reached. This completes the proof. □

Proposition 6.18 *Suppose $p \geqslant 5$. Then $R_\alpha = Q_\alpha$, G is a symplectic amalgam and one of the following holds:*

(i) *G is locally isomorphic to H where $F^*(H) \cong \mathrm{G}_2(p^n)$;*
(ii) *G is locally isomorphic to H where $F^*(H) \cong {}^3\mathrm{D}_4(p^n)$;*
(iii) *$p = 5$, $|S| = 5^6$, $Q_\beta \cong 5_+^{1+4}$ and $\overline{L_\beta} \cong 2_-^{1+4}.5$;*
(iv) *$p = 5$, $|S| = 5^6$, $Q_\beta \cong 5_+^{1+4}$ and $\overline{L_\beta} \cong 2_-^{1+4}.\mathrm{Alt}(5)$;*
(v) *$p = 5$, $|S| = 5^6$, $Q_\beta \cong 5_+^{1+4}$ and $\overline{L_\beta} \cong 2 \cdot \mathrm{Alt}(6)$; or*
(vi) *$p = 7$, $|S| = 7^6$, $Q_\beta \cong 7_+^{1+4}$ and $\overline{L_\beta} \cong 2 \cdot \mathrm{Alt}(7)$.*

Proof By Proposition 6.17, we have that $b = 2$. Note that $Q_\alpha \cap Q_\beta = Z_\alpha(Q_\alpha \cap Q_\beta \cap Q_{\alpha'})$. Since $Z_{\alpha'} \leq Q_\beta$, it follows that $[Q_\alpha, Z_{\alpha'}, Z_{\alpha'}, Z_{\alpha'}] = \{1\}$. Then by Corollary 2.60 applied to $Q_\alpha/\Phi(Q_\alpha)$, we have that $\overline{L_\alpha} \cong \mathrm{SL}_2(q)$.

We now intend to show that the amalgam is symplectic. We immediately satisfy condition (i) in the definition of a symplectic amalgam. We have that $W := \langle (Q_\alpha \cap Q_\beta)^{L_\alpha} \rangle \not\leq Q_\beta$, for otherwise $W = Q_\alpha \cap Q_\beta \trianglelefteq L_\alpha$, a contradiction by Proposition 5.27. Therefore, by Lemma 5.14 (iii), we have that $G_\beta = \langle W^{L_\beta} \rangle N_{G_\beta}(S)$, satisfying condition (ii). From our hypothesis, we automatically satisfy condition (iii). By Proposition 6.3, we satisfy condition (iv). Since $b = 2$ and $d(\alpha, \beta) = 1$, we have that $Z_\alpha \leq Q_\beta$. Moreover, by hypothesis and the symmetry between α and α' we have that $Z_\alpha \not\leq Q_{\alpha'} = Q_\alpha^x$ for some $x \in G_\beta$. Hence, G is a symplectic amalgam and the result holds by Theorem 4.8. □

Thus, we have reduced to the case where $b = 2$ and $p \in \{2, 3\}$. Since Proposition 6.12 only applied to the cases where $b \geqslant 4$, we have no knowledge of the structure of $\overline{L_\beta}$ or V_β. As intimated earlier, we attempt to show that $R_\alpha = Q_\alpha$ and apply the results in [61].

Lemma 6.19 *Suppose that $b = 2$ and $p \in \{2, 3\}$. Then the following hold:*

(i) *$[V_\beta, Q_\beta] = Z_\beta \leq Z_\alpha \leq \Phi(Q_\alpha)$ and $Q_\alpha/\Phi(Q_\alpha)$ is a faithful quadratic module for $\overline{L_\alpha}$;*

(ii) *if* $q_\alpha > p = 2$ *and* $R_\alpha \neq Q_\alpha$ *then* $q_\beta > q_\alpha^2$ *except perhaps when* $\overline{L_\beta}/O_{2'}(\overline{L_\beta}) \cong \mathrm{PSU}_3(q_\beta)$ *in which case* $q_\beta > q_\alpha$*;*
(iii) *if* $p = 3$ *and* $R_\alpha \neq Q_\alpha$ *then* $q_\alpha = 3$*; and*
(iv) $\Omega(Z(Q_\alpha)) = Z_\alpha$.

Proof That $[V_\beta, Q_\beta] = Z_\beta \leq Z_\alpha$ is contained in Lemma 6.8. Then $[Q_\alpha, V_\beta, V_\beta] \leq Z_\beta \leq Z_\alpha$, and so to prove (i) it suffices to show that $Z_\alpha \leq \Phi(Q_\alpha)$. Indeed, since Z_α is irreducible under the action of L_α, if $\Phi(Q_\alpha)$ is non-trivial, then (i) holds. So assume that Q_α is elementary abelian. If $R_\alpha = Q_\alpha$ then $\mathcal{A}$ is a symplectic amalgam and (i) holds comparing with [61], so we assume that $R_\alpha \neq Q_\alpha$. Then applying coprime action, we have that $Q_\alpha = [Q_\alpha, R_\alpha] \times C_{Q_\alpha}(R_\alpha)$ is an S-invariant decomposition. But $Z_\beta \leq Z_\alpha \leq C_{Q_\alpha}(R_\alpha)$ from which it follows that $Q_\alpha = C_{Q_\alpha}(R_\alpha)$ and $R_\alpha = Q_\alpha$, a contradiction. Hence (i).

Now, if $p = 3$ then since $Q_\alpha/\Phi(Q_\alpha)$ is a quadratic module, with S acting quadratically, applying Lemma 2.35 we deduce that either $R_\alpha = Q_\alpha$ or $|S/Q_\alpha| = 3$ and (iii) holds. Suppose that $p = 2 < q_\alpha$. If $m_2(S/Q_\beta) = 1$, then $\Phi(Q_\alpha)(Q_\alpha \cap Q_\beta)$ has index at most 4 in Q_α and applying Lemma 2.35 we deduce that $\overline{L_\alpha} \cong \mathrm{PSL}_2(4)$. Hence, if $R_\alpha \neq Q_\alpha$, then $m_2(S/Q_\beta) > 1$. If $\overline{L_\beta}/O_{2'}(\overline{L_\beta}) \cong \mathrm{Sz}(q_\beta)$ or $\mathrm{PSL}_2(q_\beta)$, then $\Phi(Q_\alpha)(Q_\alpha \cap Q_\beta)$ has index q_β in Q_α. Applying Lemma 2.55, if $R_\alpha \neq Q_\alpha$, then we have that $q_\beta > q_\alpha^2$. If $\overline{L_\beta}/O_{2'}(\overline{L_\beta}) \cong \mathrm{PSU}_3(q_\beta)$ then $\Phi(Q_\alpha)(Q_\alpha \cap Q_\beta)$ has index q_β^2 in Q_α and applying Lemma 2.55, if $R_\alpha \neq Q_\alpha$ then $q_\beta > q_\alpha$. Hence, (ii) holds.

Finally, ultimately aiming for a contradiction, assume that $Z_\alpha < \Omega(Z(Q_\alpha))$. Further, suppose first that $\Omega(Z(Q_\alpha)) \not\leq Q_\beta$. Since $\Omega(Z(Q_\alpha))$ is $G_{\alpha,\beta}$-invariant and elementary abelian, we have that $[Q_\beta, \Omega(Z(Q_\alpha)), \Omega(Z(Q_\alpha))] = \{1\}$ and applying Lemma 2.35 when p is odd, we deduce in all cases that $|\Omega(Z(Q_\alpha))Q_\beta/Q_\beta| = q_\beta$. But now, by (ii) and (iii), $q_\beta \geqslant q_\alpha$ and since $Q_\beta \cap Q_\alpha$ is a subgroup of Q_β of index q_α which is centralized by $\Omega(Z(Q_\alpha))$, applying Lemma 2.41 the only possibility is that $q_\alpha = q_\beta = p$ and $\overline{L_\beta} \cong \mathrm{SL}_2(p)$. But then, $\Phi(Q_\alpha)(Q_\alpha \cap Q_\beta)$ has index p in Q_α and Lemma 2.41 implies that $\overline{L_\alpha} \cong \mathrm{SL}_2(p)$. Therefore, $\mathcal{A}$ is symplectic and [61, Lemma 5.3] yields a contradiction.

Hence, $\Omega(Z(Q_\alpha)) \leq Q_\beta$ and $\Omega(Z(Q_\alpha)) = Z_\alpha(\Omega(Z(Q_\alpha)) \cap Q_{\alpha'})$. Moreover, $\Omega(Z(Q_\alpha)) \cap Q_{\alpha'}$ is centralized by $S = Z_{\alpha'}Q_\alpha$ so that $\Omega(Z(Q_\alpha)) \cap Q_{\alpha'} = \Omega(Z(S)) = Z_\beta \leq Z_\alpha$. Hence, (iv) holds. □

For the remainder of this section, we set $L := \langle V_\beta, V_\beta^x \rangle Q_\alpha$ with $x \in L_\alpha$ chosen such that $Z_\beta^x \neq Z_\beta$ and $x^2 \in G_{\alpha,\beta}$. In particular, $LR_\alpha = L_\alpha$. We write $\alpha - 1 := \beta^x$.

Note that if $p = 2$ then we may choose $x \in T \setminus Q_\alpha$ for some appropriate $T \in \mathrm{Syl}_2(L_\alpha) \setminus \{S\}$. In particular, we can arrange that $x \in L$. If $p = 3$ then applying Lemma 6.19, we have that $L_\alpha/R_\alpha \cong \mathrm{SL}_2(3)$. Moreover, since $[Q_\alpha, V_\beta, V_\beta] \leq Z_\alpha$, every non-central chief factor U/V within Q_α for L_α has $L_\alpha/C_{L_\alpha}(U/V) \cong \mathrm{SL}_2(3)$. By Proposition 2.47, we deduce that whenever $C_{L_\alpha}(U/V) \neq R_\alpha$, we have that $L_\alpha/C_{L_\alpha}(U/V) \cap R_\alpha \cong \overline{(Q_8 \times Q_3) : 3}$. Extending this argument yields a subgroup $U \leq \overline{L_\alpha}$ with $U \cong Q_8$ and $O^p(\overline{L_\alpha}) \cong U \times \overline{R_\alpha}$. Then for any x of order 4 in the preimage of U in L_α, $x^2 \in G_{\alpha,\beta}$ and $Z_\beta^x \neq Z_\beta$, as required.

Lemma 6.20 *Suppose that $b = 2$ and $p \in \{2, 3\}$. Then the following hold:*

(i) *$Q_\beta \cap Q_\alpha \cap Q_{\alpha-1}$ is elementary abelian, normal in L and satisfies $[O^p(L), Q_\beta \cap Q_\alpha \cap Q_{\alpha-1}] = Z_\alpha$;*
(ii) *$V_\beta \cap Q_\alpha \cap Q_{\alpha-1} = V_{\alpha-1} \cap Q_\alpha \cap Q_\beta$;*
(iii) *$V_\beta \cap Q_\alpha \not\leq Q_{\alpha-1}$ and $V_{\alpha-1} \cap Q_\alpha \not\leq Q_\beta$; and*
(iv) *if $V < V_\beta$ with $V \trianglelefteq G_\beta$ then $V \cap Q_\alpha \cap Q_{\alpha-1} = Z_\beta$.*

Proof Since $[Q_\beta, V_\beta] = Z_\beta \leq Z_\alpha$, $(Q_\beta \cap Q_\alpha \cap Q_{\alpha-1})/Z_\alpha$ is centralized by $O^p(L)$. Since $S = V_\beta Q_\alpha$ normalizes $Q_\beta \cap Q_\alpha \cap Q_{\alpha-1}$, $Q_\beta \cap Q_\alpha \cap Q_{\alpha-1} \trianglelefteq L = SO^p(L)$ and if $Q_\beta \cap Q_\alpha \cap Q_{\alpha-1}$ is not elementary abelian then $Z_\beta \cap \Phi(Q_\beta \cap Q_\alpha \cap Q_{\alpha-1})$ is non-trivial and the construction of L yields that $Z_\alpha \leq \Phi(Q_\beta \cap Q_\alpha \cap Q_{\alpha-1})$, a contradiction. Thus, $Q_\beta \cap Q_\alpha \cap Q_{\alpha-1}$ is elementary abelian, completing the proof of (i).

By the choice of x, we have that $V_\beta^x = V_{\alpha-1}$ and $V_{\alpha-1}^x = V_\beta$. Moreover, $Z_\alpha \leq V_\beta \cap Q_\alpha \cap Q_{\alpha-1}$ and x normalizes $V_\beta \cap Q_\alpha \cap Q_{\alpha-1}$ so that $V_\beta \cap Q_\alpha \cap Q_{\alpha-1} = (V_\beta \cap Q_\alpha \cap Q_{\alpha-1})^x = V_{\alpha-1} \cap Q_\alpha \cap Q_\beta$ and (ii) holds.

Suppose that $V_\beta \cap Q_\alpha \leq Q_{\alpha-1}$. Then, by (ii), $V_\beta \cap Q_\alpha = V_{\alpha-1} \cap Q_\alpha \trianglelefteq L$ and V_β centralizes the L-invariant series $\{1\} \trianglelefteq Z_\alpha \trianglelefteq V_\beta \cap Q_\alpha \trianglelefteq Q_\alpha$. Since $O_p(L) = Q_\alpha$ and $V_\beta \not\leq Q_\alpha$, an application of coprime action yields a contradiction. The action of L implies also that $V_{\alpha-1} \cap Q_\alpha \not\leq Q_\beta$.

Let V be a p-group normal in G_β and strictly contained in V_β. Since $G_{\alpha,\beta}$ acts irreducibly on Z_β, $Z_\beta \leq V$. Moreover, $V \cap Q_\alpha \cap Q_{\alpha-1}$ is elementary abelian, and $(V \cap Q_\alpha \cap Q_{\alpha-1})Z_\alpha \trianglelefteq L$. Then $[Q_\alpha, V \cap Q_\alpha \cap Q_{\alpha-1}] = [Q_\alpha, (V \cap Q_\alpha \cap Q_{\alpha-1})Z_\alpha] \trianglelefteq L$. If $[Q_\alpha, V \cap Q_\alpha \cap Q_{\alpha-1}] \cap Z_\beta$ is non-trivial, then by the construction of L, $Z_\alpha \leq [Q_\alpha, V \cap Q_\alpha \cap Q_{\alpha-1}] \leq V$, a contradiction since $V \trianglelefteq G_\beta$. Thus, $[Q_\alpha, V \cap Q_\alpha \cap Q_{\alpha-1}] = \{1\}$ and $Z_\beta \leq V \cap Q_\alpha \cap Q_{\alpha-1} \leq \Omega(Z(Q_\alpha)) = Z_\alpha$. If $V \cap Z_\alpha > Z_\beta$, then since $G_{\alpha,\beta}$ acts irreducibly on Z_α/Z_β and $V \trianglelefteq G_{\alpha,\beta}$, we have that $Z_\alpha \leq V$ and by the definition of V_β, since $V < V_\beta$ we have a contradiction. Thus, $V \cap Q_\alpha \cap Q_{\alpha-1} = Z_\beta$ and the proof is complete. □

We finally determine some of the structural properties of V_β and its chief factors in a general setting.

Lemma 6.21 *Suppose that $b = 2$ and $p \in \{2, 3\}$. Assume that $R_\alpha \neq Q_\alpha$. Then the following hold:*

(i) *$R_\beta = Q_\beta$ and $O^p(L_\beta)$ centralizes Q_β/V_β;*
(ii) *if $V < V_\beta$ with $V \trianglelefteq G_\beta$ then $V \leq C_{V_\beta}(O^p(L_\beta))$; and*
(iii) *$Z(V_\beta) = C_{V_\beta}(O^p(L_\beta))$ and $V_\beta/Z(V_\beta)$ is $\overline{G_\beta}$-irreducible.*

Proof By Lemma 6.20, we have that $[Q_\beta \cap Q_\alpha, V_{\alpha-1} \cap Q_\alpha] \leq V_{\alpha-1} \cap Q_\alpha \cap Q_\beta \leq V_\beta$ and since $Q_\beta = V_\beta(Q_\beta \cap Q_\alpha)$ and $V_{\alpha-1} \cap Q_\alpha \not\leq Q_\beta$, we infer that $O^p(L_\beta)$ centralizes Q_β/V_β and $R_\beta = Q_\beta$.

Let $V < V_\beta$ with $V \trianglelefteq G_\beta$ and assume that V contains a non-central chief factor for L_β. In particular, for $W := \Phi(V)Z_\beta$, V/W contains a non-central chief factor.

By Lemma 6.20, we have that $V \cap Q_\alpha \cap Q_{\alpha-1} = Z_\beta$ has index at most $q_\alpha r_\beta$ in V, where $r_\beta = |(V_\beta \cap Q_\alpha)Q_{\alpha-1}/Q_{\alpha-1}|$.

Suppose first that $m_p(S/Q_\beta) = 1$. By Lemma 6.19, we have that $q_\alpha = p$ so that W has index at most p^3 in L_β. In particular, unless $p = 2$ and S/Q_β is generalized quaternion, W has index p^2 in V and we deduce by Lemma 5.14 that $L_\beta/C_{L_\beta}(V/W) \cong \mathrm{SL}_2(p)$ and $Q_\beta \in \mathrm{Syl}_p(C_{L_\beta}(V/W))$. Even if $p = 2$ and S/Q_β is generalized quaternion, applying Lemma 5.14, we have that $Q_\beta \in \mathrm{Syl}_p(C_{L_\beta}(V/W))$ and $L_\beta/C_{L_\beta}(V/W)$ is isomorphic to a subgroup of $\mathrm{GL}_3(2)$. In this latter case, S/Q_β is isomorphic to a subgroup of $\mathrm{Dih}(8)$, a contradiction. Thus, $|S/Q_\beta| = p$, $Q_\alpha \cap Q_\beta$ has index p in Q_α and $R_\alpha = Q_\alpha$, a contradiction to the initial assumption.

Suppose now that $m_p(S/Q_\beta) > 1$. Since $R_\alpha \neq Q_\alpha$, we may exploit the bounds in Lemma 6.19. Applying Lemma 5.14, we have that $\overline{C_{L_\beta}(V/W)} \leq O_{p'}(\overline{L_\beta})$ so that $L_\beta/C_{L_\beta}(V/W)$ has a strongly p-embedded subgroup. If $\overline{L_\beta}/O_{2'}(\overline{L_\beta}) \cong \mathrm{Sz}(q_\beta)$, then W has index at most $q_\alpha q_\beta^2$ in V. Since $q_\alpha < q_\beta$ by Proposition 6.19, we deduce that $|V/W| < q_\beta^3$, a contradiction by Proposition 2.31. If $\overline{L_\beta}/O_{3'}(\overline{L_\beta}) \cong \mathrm{Ree}(q_\beta)$ or $\overline{L_\beta}/O_{p'}(\overline{L_\beta}) \cong \mathrm{PSU}_3(q_\beta)$, then W has index at most $q_\alpha q_\beta^3$ in V. Since $q_\alpha < q_\beta$ by Lemma 6.19, we deduce that $|V/W| < q_\beta^5$, and a contradiction is provided by Proposition 2.31 and Proposition 2.32. Finally, if S/Q_β is elementary abelian, then W has index at most $q_\alpha q_\beta$ in V. Using that $q_\alpha < q_\beta$ by Lemma 6.19, we conclude that $|V/W| < q_\beta^2$, and a contradiction is provided by Proposition 2.31 and Proposition 2.32.

To complete the proof, we need only deduce (iii). By (ii), we have that $V_\beta = [V_\beta, O^p(L_\beta)] \leq O^p(L_\beta)$ and $C_{V_\beta}(O^p(L_\beta)) \leq Z(V_\beta)$. Moreover, since V_β is non-abelian, (ii) also implies that $Z(V_\beta) \leq C_{V_\beta}(O^p(L_\beta))$ and there are no proper subgroups of V_β which are normal in G_β and properly contain $C_{V_\beta}(O^p(L_\beta))$. Then (iii) follows immediately from these observations. □

We complete the $b = 2$ case over the next two propositions. The first proposition deals with the case where $m_p(S/Q_\beta) = 1$.

Proposition 6.22 *Suppose that $p \in \{2, 3\}$, $b = 2$ and $m_p(S/Q_\beta) = 1$. Then $R_\alpha = Q_\alpha$, $|S| \leqslant 2^6$, $\mathcal{A} = \mathcal{A}(G_\alpha, G_\beta, G_{\alpha,\beta})$ is a symplectic amalgam and one of the following holds:*

(i) *G has a weak BN-pair of rank 2 and G is locally isomorphic to H where $F^*(H) \cong G_2(2)'$; or*
(ii) *$p = 2$, $|S| = 2^6$, $Q_\beta \cong 2_+^{1+4}$ and $\overline{L_\beta} \cong (3 \times 3) : 2$.*

Proof If $R_\alpha = Q_\alpha$, then $\overline{L_\alpha} \cong \mathrm{SL}_2(q_\alpha)$ and similarly to Proposition 6.18, $\mathcal{A}$ is a symplectic amalgam and the result holds after comparing with the tables listed in [61] and an application of [27] and [28]. Hence, we assume throughout that $\overline{L_\alpha} \not\cong \mathrm{SL}_2(q)$ and $R_\alpha \neq Q_\alpha$, and so we may use the results in Lemmas 6.19–6.21.

If S/Q_β is cyclic then $\Phi(Q_\alpha)(Q_\alpha \cap Q_\beta)$ is an index p subgroup of Q_α and since $V_\beta \not\leq Q_\alpha$ and $[V_\beta, Q_\alpha \cap Q_\beta] \leq \Phi(Q_\alpha)$, it follows that $Q_\alpha/\Phi(Q_\alpha)$ contains a unique non-central chief factor for L_α which is isomorphic to an FF-module for $\overline{L_\alpha} \cong \mathrm{SL}_2(p)$, a contradiction. Hence, we may assume that $p = 2$ and S/Q_β is generalized quaternion. But then, $\Phi(Q_\alpha)(Q_\alpha \cap Q_\beta)$ has index at most 4 in Q_α and we deduce that $L_\alpha/R_\alpha \cong \mathrm{Sym}(3)$ and $Q_\alpha/\Phi(Q_\alpha)$ is a faithful, quadratic 2F-module for $\overline{L_\alpha}$. Applying Proposition 2.51 with the stipulation that $L_\alpha/R_\alpha \cong \mathrm{Sym}(3)$ and $R_\alpha \neq Q_\alpha$, we have that $\overline{L_\alpha} \cong (3 \times 3) : 2$ or $\mathrm{SU}_3(2)'$.

Now, $[V_\beta, V_\beta] = Z_\beta \leq Q_{\alpha-1}$ and so $(V_\beta \cap Q_\alpha)Q_{\alpha-1}/Q_{\alpha-1}$ is abelian and since $m_p(S/Q_\beta) = 1$, $(V_\beta \cap Q_\alpha)Q_{\alpha-1}/Q_{\alpha-1}$ is cyclic. Since $\overline{L_\beta} = \overline{S}O_{2'}(\overline{L_\beta})$. By coprime action, $V_\beta/Z_\beta = [V_\beta/Z_\beta, O^2(L_\beta)] \times C_{V_\beta/Z_\beta}(O^2(L_\beta))$. Since $[V_\beta/Z_\beta, O^2(L_\beta)]$ contains a non-central chief factor for L_β and is normal in G_β, we conclude by Lemma 6.20 that $V_\beta/Z_\beta = [V_\beta/Z_\beta, O^2(L_\beta)]$ and $C_{V_\beta}(O^2(L_\beta)) = Z_\beta$. Similarly, we deduce that $Z_\beta = Z(V_\beta) = \Phi(V_\beta) = [V_\beta, V_\beta]$ and V_β is an extraspecial 2-group. Since $m_2(S/Q_\beta) = 1$, we have that $|(V_\beta \cap Q_\alpha)Q_{\alpha-1}/Q_{\alpha-1}| = 2$ and $V_\beta \cap Q_\alpha \cap V_{\alpha-1}$ is an elementary abelian subgroup of index 4 in V_β. Thus, $|V_\beta| \leqslant 2^5$. Since $R_\beta = Q_\beta$ by Lemma 6.21, $\overline{L_\beta}$ acts faithfully on V_β and has generalized quaternion Sylow 2-subgroups. Comparing with [76], we have a contradiction. □

Proposition 6.23 *Suppose that $p \in \{2, 3\}$, $b = 2$ and $m_p(S/Q_\beta) > 1$. Then one of the following holds:*

(i) *$R_\alpha = Q_\alpha$, $\mathcal{A}$ is a weak BN-pair of rank 2, and either $\mathcal{A}$ is locally isomorphic to H where $(F^*(H), p)$ is $(\mathrm{G}_2(2^n), 2)$ or $({}^3\mathrm{D}_4(p^a), p)$, or $p = 2$ and $\mathcal{A}$ is parabolic isomorphic to J_2 or $\mathrm{Aut}(\mathrm{J}_2)$; or*

(ii) *$p = 2$, $|S| = 2^9$, $\overline{L_\beta} \cong \mathrm{Alt}(5)$, $Q_\beta \cong 2_+^{1+6}$, $V_\beta = O^2(L_\beta)$, V_β/Z_β is a natural $\Omega_4^-(2)$-module for $\overline{L_\beta}$, $\overline{L_\alpha} \cong \mathrm{SU}_3(2)'$, Q_α is a special 2-group of shape 2^{2+6} and Q_α/Z_α is a natural $\mathrm{SU}_3(2)$-module.*

Proof Suppose first that $R_\alpha = Q_\alpha$. Then, as in Proposition 6.18, $\mathcal{A}$ is a symplectic amalgam and the result follows from Theorem 4.8. Indeed, the amalgams presented in [61] satisfying the above hypothesis are either weak BN-pairs of rank 2 (and (i) holds by [27]); or $\mathcal{A}_{42}$ when $p = 2$. In the latter case, $\mathrm{PSp}_6(3)$ is listed as an example completion. But comparing with the list of maximal subgroups in [24], for $G \cong \mathrm{PSp}_6(3)$, $\overline{L_\alpha} \cong 2^{2+6} : \mathrm{SU}_3(2)'$ and from the perspective of this work, $R_\alpha \neq Q_\alpha$. Either way, we assume throughout this proof that

(1) $R_\alpha \neq Q_\alpha$.

The goal is to show that G has "the same" structural properties as $\mathcal{A}_{42}$ in [61] in order to satisfy outcome (ii). We may apply the results in Lemmas 6.19 and 6.21. We set $r_\beta := |(V_\beta \cap Q_\alpha)Q_{\alpha-1}/Q_{\alpha-1}|$.

We aim first to show that $Z(V_\beta) \leq Q_{\alpha-1}$ so that $[V_\beta, V_\beta] = \Phi(V_\beta) = Z(V_\beta) = Z_\beta$ is of order q_α and V_β/Y is an extraspecial group, for Y any maximal subgroup of Z_β. In the language of Beisiegel [12], V_β is a semi-extraspecial group. Towards this goal, we suppose that $Z(V_\beta) \not\leq Q_{\alpha-1}$. Then the action of L implies that $Z(V_{\alpha-1}) \not\leq Q_\beta$. Set $V := V_\beta/Z(V_\beta)$ throughout and let $Z \leq Z(V_{\alpha-1})$ such that $Z_{\alpha-1} \leq Z$, $|Z/Z_{\alpha-1}| = p$ and $Z \not\leq Q_\beta$.

Then $[Z, V_\beta \cap Q_\alpha] \le Z_{\alpha-1}$ and since $|Z_{\alpha-1}| = q_\alpha$, it follows that for $x \in Z \setminus Z_{\alpha-1}$, x centralizes a subgroup of $V_\beta \cap Q_\alpha$ of index at most q_α. Hence, $|V_\beta/C_{V_\beta}(z)| \leqslant q_\alpha^2$.

Assume that $O_{p'}(\overline{L_\beta}) \neq \{1\}$. Then by the irreducibility of V and since $R_\beta = Q_\beta$, we have that $V = [V, O_{p'}(\overline{L_\beta})]$ and by Lemma 6.20, we have that $V_\beta/Z_\beta = [V_\beta/Z_\beta, O_{p'}(\overline{L_\beta})]$. Since $Z(V_\beta)/Z_\beta \le C_{V_\beta/Z_\beta}(O^p(L_\beta)) \le C_{V_\beta/Z_\beta}(O_{p'}(\overline{L_\beta}))$, we infer by coprime action that $Z(V_\beta) = Z_\beta$, as desired. Hence

(2) if $Z(V_\beta) \not\le Q_{\alpha-1}$ then $O_{p'}(\overline{L_\beta}) = \{1\}$.

Suppose that $\overline{L_\beta} \cong \mathrm{M}_{11}$ or $\mathrm{PSL}_3(4)$ and $p = 3$. Then $q_\alpha = 3$ and Z centralizes an index 9 subgroup of V_β. If $\overline{L_\beta} \cong \mathrm{M}_{11}$ then there is $x \in L_\beta$ such that for $J := \langle Z, Z^x, Q_\beta \rangle$, $\overline{J} \cong \mathrm{PSL}_2(11)$ and J centralizes a subgroup of V of index at most 3^4. Since 11 does not divide $|\mathrm{GL}_4(3)|$, J centralizes V, a contradiction since V is a faithful $\overline{L_\beta}$-module. If $\overline{L_\beta} \cong \mathrm{PSL}_3(4)$, then there is $x \in L_\beta$ such that $L_\beta = \langle Z, Z^x, Q_\beta \rangle$ so that $|V| \leqslant 3^4$, a contradiction by Proposition 2.32.

If $p = 3$ and $\overline{L_\beta} \cong \mathrm{Ree}(q_\beta)$ then Z centralizes an index 9 subgroup of V_β and Proposition 2.32 gives a contradiction. If $p = 2$ and $\overline{L_\beta} \cong \mathrm{Sz}(q_\beta)$ then Z centralizes an index q_α^2 subgroup of V_β. Since $q_\alpha^2 < q_\beta$ by Lemma 6.19, Proposition 2.31 gives a contradiction.

If $\overline{L_\beta} \cong \mathrm{PSU}_3(q_\beta)$ then Z centralizes an index q_α^2 subgroup of V_β. Since either $p = 2$ and $q_\alpha < q_\beta$, or $q_\alpha = 3$ by Lemma 6.19, we apply Propositions 2.31 and 2.32 to deduce that either $\overline{L_\beta} \cong \mathrm{SU}_3(3)$, or $p = 2$ and $q_\alpha^2 \geqslant q_\beta^{\frac{3}{2}}$. In the former case, $\overline{L_\beta}$ is generated by only three conjugates of $\overline{Z}$, $|V| \leqslant 3^6$ and V is a natural $\mathrm{SU}_3(3)$-module for $\overline{L_\beta}$. But then, comparing with Proposition 2.28 and using that $V_\beta \cap Q_\alpha$ is a $G_{\alpha,\beta}$-invariant subgroup of index 3, we have a contradiction. In the latter case, writing $U_\alpha := \langle Z(V_\beta)^{L_\alpha} \rangle$, we see that $U_\alpha \trianglelefteq G_\alpha$ and U_α/Z_α is elementary abelian. Since $Z \le Z(V_{\alpha-1}) \le U_\alpha$, we also have that $U_\alpha \not\le Q_\beta$. Thus, $Z_\alpha(U_\alpha \cap Q_{\alpha'})$ has index q_β in U_α. If L_α centralizes U_α/Z_α then $Z(V_{\alpha-1}) \le U_\alpha = Z(V_\beta)Z_\alpha \le Q_\beta$, a contradiction. Since $q_\beta < q_\alpha^2$, an application of Lemma 2.55 reveals that U_α/Z_α contains a unique non-central chief factor for L_α, and this non-central chief factor is a natural $\mathrm{SL}_2(q_\alpha)$-module for L_α/R_α. Then for T_α the preimage in U_α of $C_{U_\alpha/Z_\alpha}(L_\alpha)$ we have that $|U_\alpha/T_\alpha| = q_\alpha^2$ and $T_\alpha Z(V_\beta)$ has index q_α in U_α. Since $q_\alpha < q_\beta$ we must have that $T_\alpha \not\le Q_\beta$ and as T_α is $G_{\alpha,\beta}$ invariant, we see that $U_\alpha = T_\alpha(U_\alpha \cap Q_{\alpha'})$ from which we deduce that L_α centralizes U_α/T_α, a contradiction. Hence, $\overline{L_\beta} \not\cong \mathrm{PSU}_3(q_\beta)$.

Finally, if $\overline{L_\beta} \cong \mathrm{PSL}_2(q_\beta)$ then Z centralizes an index q_α^2 subgroup of V_β. Then, since $q_\alpha^2 < q_\beta$ or $q_\alpha = p$ by Lemma 6.19, and $\overline{L_\beta}$ is generated by three conjugate elements of order p, V is determined by Lemma 2.43. In particular, since Z centralizes an index q_α^2 subgroup of V_β, and $q_\alpha^2 < q_\beta$, we must have that $q_\alpha = p$. Moreover, since $V_\beta \cap Q_\alpha$ is a $G_{\alpha,\beta}$-subgroup of index p we see that V is a natural $\Omega_4^-(p)$-module for $\overline{L_\beta} \cong \mathrm{PSL}_2(p^2)$. Moreover, we have that $|Z(V_\beta)Q_{\alpha-1}/Q_{\alpha-1}| = p$, $|Z(V_\beta)/Z_\beta| = p$ and $|V_\beta/Z_\beta| = p^5$. Since $V_\beta = [V_\beta, L_\beta]$, applying Lemma 2.25, we deduce that $p = 3$ and $[V_\beta/Z_\beta, S, S]$ is 2-

dimensional as a GF(3)-module. But then $Z_\alpha Z(V_\beta) = [V_\beta, S, S] = [V_\beta, V_{\alpha-1} \cap Q_\alpha, V_{\alpha-1} \cap Q_\alpha] \leq V_{\alpha-1}$, a contradiction since $Z(V_\beta) \not\leq Q_{\alpha-1}$. Thus

(3) $Z(V_\beta) \leq Q_{\alpha-1}$, $Z(V_\beta) = Z_\beta = \Phi(V_\beta)$ is of order q_α and V_β is a semi-extraspecial group.

Moreover, $V_\beta \cap Q_\alpha \cap Q_{\alpha-1}$ has index $q_\alpha r_\beta$ in V_β and is elementary abelian. Then for Z a maximal subgroup of Z_β, we set $|V_\beta/Z| = p^{2r+1}$ so that $|V_\beta/Z_\beta| = p^{2r}$. Then $|(V_\beta \cap Q_\alpha \cap Q_\beta)/Z| = p^{2r+1}/q_\alpha r_\beta$ and since the maximal abelian subgroups of V_β/Z have order p^{r+1}, we deduce that $p^{2r+1}/q_\alpha r_\beta \leqslant p^{r+1}$ and $p^r \leqslant q_\alpha r_\beta$.

Suppose that $\overline{L_\beta}/O_{p'}(\overline{L_\beta}) \cong$ (P)SU$_3(q_\beta)$ so that $r_\beta \leqslant q_\beta$ when $p = 2$ and $r_\beta \leqslant q_\beta^2$ when $p = 3$. In particular, $2^r \leqslant q_\alpha q_\beta < q_\beta^3$ when $p = 2$ and $3^r \leqslant 3q_\beta^2$ when $p = 3$. Thus, $|V_\beta/Z_\beta| < q_\beta^6$ when $p = 2$ and $|V_\beta/Z_\beta| \leqslant 3^2 q_\beta^4$ when $p = 3$. Then by Propositions 2.31 and 2.32, we conclude that $q_\beta = 3 = p$, $\overline{L_\beta}/O_{3'}(\overline{L_\beta}) \cong$ (P)SU$_3(3)$ and $|V_\beta/Z_\beta| = 3^6$. A calculation in $\mathrm{Sp}_6(3)$ (e.g. using MAGMA [14]) promises that $\overline{L_\beta} \cong \mathrm{PSU}_3(3)$ and V_β/Z_β is a natural module for $\overline{L_\beta}$. But then $V_\beta \cap Q_\alpha$ is a $G_{\alpha,\beta}$-invariant subgroup of index 3, and we have a contradiction by Proposition 2.28 (iii).

Suppose that $\overline{L_\beta}/O_{3'}(\overline{L_\beta}) \cong \mathrm{Ree}(q_\beta)$. Then we have that $3^r \leqslant r_\beta 3 \leqslant 3q_\beta^2$ and so $|V_\beta/Z_\beta| \leqslant 3^2 q_\beta^4 < \max(q_\beta^6, 3^7)$. Then Proposition 2.32 provides a contradiction. If $\overline{L_\beta}/O_{2'}(\overline{L_\beta}) \cong \mathrm{Sz}(q_\beta)$, then $r_\beta \leq q_\beta$ and so $2^r \leqslant q_\alpha q_\beta$ and $|V_\beta/Z_\beta| \leqslant q_\alpha^2 q_\beta^2 < q_\beta^4$. Then Proposition 2.31 provides a contradiction. Hence, we have that

(4) S/Q_β is elementary abelian of order p^n and $n > 1$.

Then $|V_\beta/Z_\beta| \leqslant q_\alpha^2 q_\beta^2 \leq q_\beta^3$. In particular, it follows from Propositions 2.31 and 2.32 that V_β/Z_β contains a unique non-central chief factor for L_β and so V_β/Z_β is an irreducible $\overline{L_\beta}$-module. If $q_\beta > p^2$ then Lemma 2.44 yields that $\overline{L_\beta} \cong \mathrm{SL}_2(q_\beta)$ or $\mathrm{PSL}_2(q_\beta)$. Furthermore, since $q_\alpha^2 < q_\beta$,using that $V_\beta \cap Q_\alpha$ is $G_{\alpha,\beta}$-invariant, we infer that V_β/Z_β is a triality module and $q_\alpha^3 = q_\beta$. Then for K_α a critical subgroup of Q_α, we have that $[V_\beta, K_\alpha, K_\alpha, K_\alpha] = \{1\}$. By Proposition 2.26, this yields $K_\alpha \leq Q_\beta$ so that $K_\alpha = Z_\alpha(K_\alpha \cap Q_{\alpha'})$, $[O^p(L_\alpha), K_\alpha] = Z_\alpha$ and $R_\alpha = Q_\alpha$, a contradiction.

Hence, we may suppose that S/Q_β is elementary abelian of order p^2 so that $q_\alpha = p$, V_β is an extraspecial group and $|V_\beta/Z_\beta| \in \{p^4, p^6\}$. In particular, applying [76], if $p = 3$ then $\overline{L_\beta}$ is isomorphic to a subgroup of $\mathrm{Sp}_6(3)$ and if $p = 2$, then $\overline{L_\beta}$ is isomorphic to a subgroup of $\mathrm{PSL}_4(2)$ or $\mathrm{PSU}_4(2)$. We deduce in both cases that

(5) $\overline{L_\beta} \cong \mathrm{SL}_2(p^2)$ or $\mathrm{PSL}_2(p^2)$, and V_β/Z_β is described by Lemma 2.43.

Since $V_\beta \cap Q_\alpha$ is a $G_{\alpha\beta}$-invariant subgroup of index p containing $[V_\beta, S]$ Lemma 2.43 implies that V_β/Z_β is a natural $\Omega_4^-(p)$-module and $|V_\beta| = p^5$. Now, as L_β/C_β embeds in the automorphism group of V_β, we infer that $Q_\beta = V_\beta C_\beta$. Moreover, using [76], if $p = 2$ then $\overline{L_\beta} \cong \mathrm{Out}(V_\beta) \cong \Omega_4^-(2)$ and $V_\beta \cong Q_8 \circ D_8 \cong 2_-^{1+4}$; and if $p = 3$ then V_β has exponent 3.

Suppose that $p = 3$ and let $K \in \mathrm{Syl}_2(L_\beta)$. Since $\overline{L_\beta} \cong \mathrm{PSL}_2(9)$, $K \cong \mathrm{Dih}(8)$. Letting $i \in Z(K)^\#$, we have that $|C_{V_\beta/Z_\beta}(i)| = 9$ and by coprime action $V_\beta = C_{V_\beta}(i)[V_\beta, i]$. Since $[V_\beta, V_\beta] \le C_{V_\beta}(i)$ it follows from the three subgroup lemma that $[[V_\beta, i], C_{V_\beta}(i)] = \{1\}$ and since $|[V_\beta, i]| \leqslant 3^3$, it follows that $Z_\beta = C_{V_\beta}(i) \cap [V_\beta, i]$ and $C_{V_\beta}(i) \cong [V_\beta, i] \cong 3_+^{1+2}$. Since $i \le Z(K)$, K normalizes $[V_\beta, i]$ and since $Z_\beta = Z(L_\beta)$, K acts trivially on $Z_\beta = Z([V_\beta, i])$ and by Winter [76], K embeds into $\mathrm{Sp}_2(3) \cong \mathrm{SL}_2(3)$. But $\mathrm{SL}_2(3)$ has quaternion Sylow 2-subgroups, a contradiction. Thus, we have shown that

(6) $p = 2$.

Now, $Z_\alpha \not\le C_\beta$ and so $Z_\beta = C_\beta \cap Q_{\alpha-1}$ has index at most 4 in C_β and $|C_\beta| \leqslant 8$. Since $Z(C_\beta)$ is centralized by $L_\beta = O^2(L_\beta)C_\beta$ and Q_α is self centralizing, $Z(C_\beta) \le Z(Q_\alpha) = Z_\alpha$. Thus, $Z(C_\beta) = Z_\beta$ and as $|C_\beta| \leqslant 8$, either $C_\beta = Z_\beta$, or $C_\beta \cong Q_8$ or $\mathrm{Dih}(8)$. If $C_\beta = Z_\beta$ then we have that $Q_\beta = V_\beta \cong 2_-^{1+4}$, $|S| = 2^7$ and $|Q_\alpha| = 2^6$. Since $Z_\alpha \le \Phi(Q_\alpha)$ and $R_\alpha \ne Q_\alpha$, we have that $Z_\alpha = \Phi(Q_\alpha)$ and Q_α/Z_α is a faithful quadratic 2F-module for $\overline{L_\alpha}$. As $L_\alpha/R_\alpha \cong \mathrm{Sym}(3)$, it follows that $\overline{L_\alpha} \cong (3 \times 3) : 2$. Now, for every subgroup Z of Z_α of order 2, is easy to check that Q_α/Z is an extraspecial group. In the language of Beisiegel [12], Q_α is an ultraspecial 2-group of order 2^6. Checking in MAGMA [14] utilizing the Small Groups library, the automorphism groups of all such groups have 3-part at most 9. Since there is $r \in (L_\beta \cap G_{\alpha,\beta})$ a 3-element centralizing Z_α by Proposition 2.24 (v), $r \in G_\alpha \setminus L_\alpha$ and a Sylow 3-subgroup of $\overline{G_\alpha}$ has order at least 27, and as $\overline{G_\alpha}$ acts faithfully on Q_α, we have a contradiction.

Thus, C_β is non-abelian of order 8. Furthermore, $|S| = 2^9$ and if Q_α/Z_α is a natural $\mathrm{SU}_3(2)$-module for $\overline{L_\alpha} \cong \mathrm{SU}_3(2)'$, then since C_β is $G_{\alpha,\beta}$-invariant, there is a 3-element in $L_\alpha \cap G_{\alpha,\beta}$ which acts non-trivially on C_β so that $C_\beta \cong Q_8$ and $Q_\beta = 2_+^{1+6}$. Thus, to complete the proof, it suffices to show that Q_α/Z_α is a natural $\mathrm{SU}_3(2)$-module. Now, $Q_\alpha \cap Q_\beta = Z_\alpha(Q_\alpha \cap Q_{\alpha'})$ has index 4 in Q_α and, modulo Z_α, is centralized by $Z_{\alpha'}$. It is clear that $Z_{\alpha'}$ acts quadratically on Q_α/Z_α and, since $Z_\alpha \le \Phi(Q_\alpha)$ and $R_\alpha \ne Q_\alpha$, $\overline{L_\alpha}$ is determined by Proposition 2.51. Since $L_\alpha/R_\alpha \cong \mathrm{Sym}(3)$, we need only rule out the case where $\overline{L_\alpha} \cong (3 \times 3) : 2$.

Assume that $\overline{L_\alpha} \cong (3\times 3) : 2$ and $|C_\beta| = 8$. Observe that $Q_\alpha = (Q_\alpha \cap Q_\beta)(Q_\alpha \cap Q_{\alpha-1}) = (V_\beta \cap Q_\alpha)(V_{\alpha-1} \cap Q_\beta)(Q_\beta \cap Q_\alpha \cap Q_{\alpha-1})$. Then, $V_\beta \cap Q_\alpha \cap Q_{\alpha-1} = V_{\alpha-1} \cap Q_\alpha \cap Q_\beta = Z_\alpha$, and it follows that $Z_\alpha = \Phi(Q_\alpha)$. By coprime action, we have that $Q_\alpha/Z_\alpha = [Q_\alpha/Z_\alpha, O^2(L_\alpha)] \times C_{Q_\alpha/Z_\alpha}(O^2(L_\alpha))$ where $|[Q_\alpha/Z_\alpha, O^2(L_\alpha)]| = 2^4$. Taking Q_α^* to be the preimage in Q_α of $[Q_\alpha/Z_\alpha, O^2(L_\alpha)]$, form $S^* = V_\beta Q_\alpha^*$ and $L_\lambda^* = \langle (S^*)^{L_\lambda} \rangle$ for $\lambda \in \{\alpha, \beta\}$. It is clear that $S^* \in \mathrm{Syl}_2(L_\lambda^*)$, $V_\beta = O_2(L_\beta^*)$ and $Q_\alpha^* = O_2(L_\alpha^*)$, and $L_\lambda^*/O_2(L_\lambda^*) \cong \overline{L_\lambda}$ for $\lambda \in \{\alpha, \beta\}$. Then for K a Hall $2'$-subgroup of $G_{\alpha,\beta}$, we conclude that $(L_\alpha^* K, L_\beta^* K, S^* K)$ satisfies Hypothesis 5.1 and since G is a minimal counterexample, comparing with Theorem C, we have a contradiction. Hence, $\overline{L_\alpha} \cong \mathrm{SU}_3(2)'$ and the result holds. □

In summary, in this chapter we have demonstrated that a minimal counterexample to Theorem C has $[Z_\alpha, Z_{\alpha'}] = \{1\}$. After this second case has been dispelled (which we do in the following chapter) we will have proved the following:

Theorem 6.24 *Suppose that $\mathcal{A} = \mathcal{A}(G_\alpha, G_\beta, G_{\alpha,\beta})$ is an amalgam satisfying Hypothesis 5.1. If $Z_{\alpha'} \not\leq Q_\alpha$, then one of the following holds:*

(i) *$\mathcal{A}$ is a weak BN-pair of rank 2;*
(ii) *$\mathcal{A}$ is a symplectic amalgam; or*
(iii) *$p = 2$, $|S| = 2^9$, $\overline{L_\beta} \cong \mathrm{PSL}_2(4)$, $Q_\beta \cong 2_-^{1+6}$, $V_\beta = O^2(L_\beta)$, V_β/Z_β is a natural $\Omega_4^-(2)$-module for $\overline{L_\beta}$, $\overline{L_\alpha} \cong \mathrm{SU}_3(2)'$, Q_α is a special 2-group of shape 2^{2+6} and Q_α/Z_α is a natural $\mathrm{SU}_3(2)$-module.*

Chapter 7
$Z_{\alpha'} \le Q_\alpha$

We now begin the second half of our analysis, where $Z_{\alpha'} \le Q_\alpha$ so that $[Z_\alpha, Z_{\alpha'}] = \{1\}$. Again, we assume Hypothesis 5.1 and that G is a minimal counterexample to Theorem C.

Throughout, we fix $S \in \mathrm{Syl}_p(G_{\alpha,\beta})$ and $q_\lambda = |\Omega(Z(T/Q_\lambda))|$ for any $\lambda \in \Gamma$ and $T \in \mathrm{Syl}_p(G_\lambda)$.

Lemma 7.1 *The following hold:*

(i) $Z_\beta = \Omega(Z(S)) = \Omega(Z(L_\beta))$ *and* b *is odd; and*
(ii) $Z(L_\alpha) = \{1\}$.

Proof Since $Z_{\alpha'} \le Q_\alpha$ we have that $\{1\} = [Z_\alpha, Z_{\alpha'}]$. Then, for $T \in \mathrm{Syl}_p(G_{\alpha',\alpha'-1})$, since $Z_\alpha \not\le Q_{\alpha'}$, we deduce that $Q_{\alpha'} < C_T(Z_{\alpha'})$ and by Lemma 5.16 (ii), we get that $Z_{\alpha'} = \Omega(Z(T)) = \Omega(Z(L_{\alpha'})$. By Lemma 5.6 (iii), $Z_\alpha \not\le \Omega(Z(L_\alpha))$ and so α and α' are not conjugate. Thus, α' is conjugate to β, b is odd and $Z_\beta = \Omega(Z(S)) = \Omega(Z(L_\beta)))$. Since L_β acts transitively on $\Delta(\beta)$, by Lemma 5.6 (iv), we conclude that $Z(L_\alpha) = \{1\}$. □

Lemma 7.2 *Suppose that* $b > 1$. *Then* V_β *is abelian,* $\{1\} \ne [V_\beta, V_{\alpha'}] \le V_{\alpha'} \cap V_\beta$ *and* V_β *acts quadratically on* $V_{\alpha'}$.

Proof By Proposition 5.7, both V_β and $V_{\alpha'}$ are abelian. Since $Z_\alpha \le V_\beta$ and $Z_\alpha \not\le Q_{\alpha'}$ it follows that $V_\beta \not\le C_{L_{\alpha'}}(V_{\alpha'})$. By minimality of b, $V_\beta \le Q_{\alpha'-1} \le L_{\alpha'}$ and so $\{1\} \ne [V_\beta, V_{\alpha'}] \le V_{\alpha'}$. Again, by minimality of b, $V_{\alpha'} \le Q_{\alpha+2} \le L_\beta$ and so $[V_\beta, V_{\alpha'}] \le V_{\alpha'} \cap V_\beta$. Since V_β is abelian, $[V_{\alpha'}, V_\beta, V_\beta] = \{1\}$, completing the proof. □

Lemma 7.3 *Suppose that* $b > 1$ *and* p *is an odd prime. Then either*

(i) $L_\beta/R_\beta Q_\beta \cong \mathrm{SL}_2(p^n)$ *where* p *is any odd prime;*
(ii) $L_\beta/R_\beta Q_\beta \cong \mathrm{(P)SU}_3(p^n)$ *where* p *is any odd prime;*

© The Author(s), under exclusive license to Springer Nature Switzerland AG 2024

M. van Beek, *Rank 2 Amalgams and Fusion Systems*, Lecture Notes in Mathematics 2343, https://doi.org/10.1007/978-3-031-54461-3_7

(iii) $L_\beta/C_{L_\beta}(U/V) \cong 4 \circ 2_-^{1+4}.\mathrm{Alt}(6)$ *for some non-trivial irreducible composition factor* U/V *of* $V_\beta/C_{V_\beta}(O^p(L_\beta))$, $p = |V_{\alpha'}Q_\beta/Q_\beta| = |V_\beta Q_{\alpha'}/Q_{\alpha'}| = 3$ *and* $|S/Q_\beta| = 9$;
(iv) $L_\beta/C_{L_\beta}(U/V) \cong 2 \cdot \mathrm{Alt}(5)$ *or* $2_-^{1+4}.\mathrm{Alt}(5)$ *for every non-trivial irreducible composition factor* U/V *of* $V_\beta/C_{V_\beta}(O^p(L_\beta))$ *and* $p = |S/Q_\beta| = 3$; *or*
(v) $L_\beta/C_{L_\beta}(U/V) \cong \mathrm{SL}_2(3)$ *for every non-trivial irreducible composition factor* U/V *of* $V_\beta/C_{V_\beta}(O^p(L_\beta))$ *and* $p = |S/Q_\beta| = 3$.

Proof Since $[V_{\alpha'}, V_\beta, V_\beta] = \{1\}$, this follows applying Lemma 2.35 since β is conjugate to α'. □

The following lemma is essentially part of the proof of the qrc lemma, which we recreate for the purposes of this work.

Lemma 7.4 *Suppose that* $b > 1$ *and* $V_\beta/C_{V_\beta}(O^p(L_\beta))$ *is an irreducible module for* $O^p(L_\beta)R_\beta/R_\beta$. *Then* Z_α *is a natural module for* $L_\alpha/R_\alpha \cong \mathrm{SL}_2(q_\alpha)$.

Proof Let $Q := Q_\beta \cap O^p(L_\beta)$ so that, by Proposition 5.27, $Q \not\leq Q_\alpha$. In particular, Q acts non-trivially on Z_α and centralizes $C_{V_\beta}(O^p(L_\beta))$. Since $V_\beta/C_{V_\beta}(O^p(L_\beta))$ is irreducible, $C_{Z_\alpha}(Q) = Z_\alpha \cap C_{V_\beta}(O^p(L_\beta))$. Moreover, $[Z_\alpha, Q, Q] = \{1\}$.

Let $z \in Z_\alpha \setminus (Z_\alpha \cap C_{V_\beta}(O^p(L_\beta)))$ and let $W_\beta := \langle z^{O^p(L_\beta)} \rangle$. Since $V_\beta/C_{V_\beta}(O^p(L_\beta))$ is irreducible, we have that $V_\beta = W_\beta C_{V_\beta}(O^p(L_\beta))$ so that $[Z_\alpha, Q] \leq [V_\beta, Q] = [W_\beta, Q]$. Moreover, since $[z, Q] \leq C_{V_\beta}(O^p(L_\beta))$, we have that $[z, Q] = [z, Q]^{O^p(L_\beta)} = [W_\beta, Q]$ from which we conclude that $[Z_\alpha, Q] \leq [W_\beta, Q] = [z, Q] \leq [Z_\alpha, Q]$.

Now, there is a surjective homomorphism $\theta : Q \to [z, Q]$ and with kernel $C_Q(z) \geqslant C_Q(Z_\alpha)$. In particular,

$$|[Z_\alpha, Q]| = |[z, Q]| = |Q/C_Q(z)| \leqslant |Q/C_Q(Z_\alpha)|.$$

Indeed, Z_α is dual to an FF-module for L_α/R_α. Hence, Lemma 2.41 yields that Z_α is a natural module for $L_\alpha/R_\alpha \cong \mathrm{SL}_2(q_\alpha)$, as desired. □

We provide some generic results in the case $C_{V_\beta}(V_{\alpha'}) = V_\beta \cap Q_{\alpha'}$. These will also be useful for certain inductive arguments in the case $C_{V_\beta}(V_{\alpha'}) < V_\beta \cap Q_{\alpha'}$ later.

Lemma 7.5 *Suppose that* $b > 1$, $C_{V_\beta}(V_{\alpha'}) = V_\beta \cap Q_{\alpha'}$ *and* $V_{\alpha'} \leq Q_\beta$. *Then* $Q_\beta \in \mathrm{Syl}_p(R_\beta)$, $L_\alpha/R_\alpha \cong \mathrm{SL}_2(p^n) \cong L_\beta/R_\beta$, *both* Z_α *and* $V_\beta/C_{V_\beta}(O^p(L_\beta))$ *are natural* $\mathrm{SL}_2(p^n)$*-modules and* $Z_\alpha Q_{\alpha'} \in \mathrm{Syl}_p(L_{\alpha'})$. *Moreover,* $[Q_\beta, V_\beta] = Z_\beta = [V_{\alpha'}, V_\beta] \leq V_{\alpha'} \cap V_\beta$.

Proof Suppose that $C_{V_\beta}(V_{\alpha'}) = V_\beta \cap Q_{\alpha'}$ and $V_{\alpha'} \leq Q_\beta$. Note, that if $V_{\alpha'} \leq Q_\alpha$, then $[Z_\alpha, V_{\alpha'}] = \{1\}$ and $Z_\alpha \leq Q_{\alpha'}$, a contradiction. Additionally, $[Z_\alpha, V_{\alpha'}, V_{\alpha'}] \leq [V_\beta, V_{\alpha'}, V_{\alpha'}] = \{1\}$ and it follows that both Z_α and $V_{\alpha'}$ admit quadratic action. Notice that $Z_\alpha \cap Q_{\alpha'} = C_{Z_\alpha}(V_{\alpha'})$ and that $V_{\alpha'} \cap Q_\alpha = C_{V_{\alpha'}}(Z_\alpha)$, and set $r_\alpha = |Z_\alpha Q_{\alpha'}/Q_{\alpha'}|$ and $r_{\alpha'} = |V_{\alpha'}Q_\alpha/Q_\alpha|$.

Assume first that $r_{\alpha'} < r_\alpha$. Then for $V := V_{\alpha'}/[V_{\alpha'}, Q_{\alpha'}]$, V contains a non-central chief factor for $L_{\alpha'}$ by Lemma 5.8 so that $Q_{\alpha'} \in \mathrm{Syl}_p(C_{L_{\alpha'}}(V))$. Now, we have that $|V/C_V(Z_\alpha)| \leqslant r_{\alpha'} < r_\alpha = |Z_\alpha/C_{Z_\alpha}(V)|$ and we obtain a contradiction by Lemma 2.41.

Suppose now that $r_\alpha \leqslant r_{\alpha'}$. Then $|Z_\alpha/C_{Z_\alpha}(V_{\alpha'})| = r_\alpha \leqslant r_{\alpha'} = |V_{\alpha'}/C_{V_{\alpha'}}(Z_\alpha)|$ and Z_α is an FF-module for $L_\alpha/R_\alpha \cong \mathrm{SL}_2(r_\alpha)$, and applying Lemma 2.41 acknowledging that $Z(L_\alpha) = \{1\}$, Z_α is a natural module. But then $r_{\alpha'} \leqslant r_\alpha \leqslant r_{\alpha'}$ so that $r_\alpha = r_{\alpha'}$, $[Z_{\alpha'-1}, Q_{\alpha'}] = [V_{\alpha'}, Q_{\alpha'}] = Z_{\alpha'}$ is of order r_α and $V_{\alpha'}/Z_{\alpha'}$ is an FF-module for $L_{\alpha'}/R_{\alpha'} \cong \mathrm{SL}_2(r_{\alpha'})$. In particular, $q_{\alpha'} = q_\beta = q_\alpha$. By Lemma 5.19, $Q_\beta \in \mathrm{Syl}_p(R_\beta)$ and as $\{1\} \neq [V_{\alpha'}, Z_\alpha] \leq [V_{\alpha'}, V_\beta] \leq [Q_\beta, V_\beta]$, we conclude that $Z_\beta = [V_{\alpha'}, V_\beta] \leq V_{\alpha'} \cap V_\beta$. Finally applying Lemma 2.41, the result holds. □

Lemma 7.6 *Suppose that* $b > 1$, $C_{V_\beta}(V_{\alpha'}) = V_\beta \cap Q_{\alpha'}$ *and* $V_{\alpha'} \not\leq Q_\beta$. *Then* $Q_\beta \in \mathrm{Syl}_p(R_\beta)$, $L_\alpha/R_\alpha \cong \mathrm{SL}_2(p^n) \cong L_\beta/R_\beta$, *both* Z_α *and* $V_\beta/C_{V_\beta}(O^p(L_\beta))$ *are natural* $\mathrm{SL}_2(p^n)$*-modules,* $Z_\alpha Q_{\alpha'} \in \mathrm{Syl}_p(L_{\alpha'})$ *and* $V_{\alpha'}Q_\beta \in \mathrm{Syl}_p(L_\beta)$.

Proof Assume throughout that $C_{V_\beta}(V_{\alpha'}) = V_\beta \cap Q_{\alpha'}$ and $V_{\alpha'} \not\leq Q_\beta$. Suppose first that $|V_\beta/C_{V_\beta}(V_{\alpha'})| = |V_\beta Q_{\alpha'}/Q_{\alpha'}| = p$. Then by Lemma 2.41, $V_\beta/C_{V_\beta}(O^p(L_\beta))$ is a natural $\mathrm{SL}_2(p)$-module for $L_\beta/R_\beta \cong \mathrm{SL}_2(p)$. Since $Q_\alpha \cap Q_\beta \ntrianglelefteq L_\beta$ by Proposition 5.27, $Q_\beta \cap O^p(L_\beta) \not\leq Q_\alpha$ and $Z_\alpha \cap C_{V_\beta}(O^p(L_\beta))$ is centralized by $Q_\beta \cap O^p(L_\beta)$. Now, $V_\beta \neq Z_\alpha C_{V_\beta}(O^p(L_\beta))$, for otherwise Q_α centralizes $V_\beta/C_{V_\beta}(O^p(L_\beta))$ and $O^p(L_\beta)$ centralizes V_β, and so $Z_\alpha \cap C_{V_\beta}(O^p(L_\beta))$ has index p in Z_α. Thus, Z_α is an FF-module and by Lemma 2.41, using that $Z(L_\alpha) = \{1\}$, Z_α is a natural $\mathrm{SL}_2(p)$-module for $L_\alpha/R_\alpha \cong \mathrm{SL}_2(p)$. Then, $[Q_\beta, V_\beta] = [Q_\beta, Z_\alpha]^{G_\beta} = Z_\beta \leq C_{V_\beta}(O^p(L_\beta))$ and by Lemma 5.19, $Q_\beta \in \mathrm{Syl}_p(R_\beta)$ and the result holds. Thus, we assume that

(1) $|V_\beta Q_{\alpha'}/Q_{\alpha'}| \geqslant p^2$.

As V_β is elementary abelian, $m_p(S/Q_\beta) \geqslant 2$. Since V_β acts quadratically on $V_{\alpha'}$ we infer, using Proposition 2.15 when $p = 2$, and Lemma 7.3 and Proposition 2.36 when p is odd, that $\overline{L_\beta}/O_{p'}(\overline{L_\beta})$ is a rank 1 group of Lie type but not a Ree group, $V_\beta Q_{\alpha'}/Q_{\alpha'} \leq \Omega(Z(T/Q_{\alpha'}))$ where $T \in \mathrm{Syl}_p(G_{\alpha',\alpha'-1})$, and $q_\beta \geqslant p^2$.

Assume first that $|V_{\alpha'}Q_\beta/Q_\beta| = p$. Then a subgroup of index at most pr_α is centralized by Z_α, where $r_\alpha := |(V_{\alpha'} \cap Q_\beta)Q_\alpha/Q_\alpha| \neq \{1\}$. Applying Lemma 2.41 we have that if $|Z_\alpha Q_{\alpha'}/Q_{\alpha'}| \geqslant pr_\alpha$ then $|Z_\alpha Q_{\alpha'}/Q_{\alpha'}| = pr_\alpha$, $V_{\alpha'}/C_{V_{\alpha'}}(O^p(L_{\alpha'}))$ is natural module for $L_{\alpha'}/R_{\alpha'} \cong \mathrm{SL}_2(q'_\alpha)$ and the results holds upon applying Lemma 7.4. Hence, we may assume that

(2) if $|V_{\alpha'}Q_\beta/Q_\beta| = p$ then $|Z_\alpha Q_{\alpha'}/Q_{\alpha'}| \leqslant r_\alpha$.

Since $Z_\alpha \cap Q_{\alpha'}$ is centralized by $V_{\alpha'} \cap Q_\beta$, applying Lemma 2.41 we deduce that $r_\alpha = |Z_\alpha Q_{\alpha'}/Q_{\alpha'}| \leqslant q_\beta$ and Z_α is a natural module for $L_\alpha/R_\alpha \cong \mathrm{SL}_2(r_\alpha)$. If $r_\alpha = p$ then Z_α is a natural module for $L_\alpha/R_\alpha \cong \mathrm{SL}_2(p)$. Furthermore, if $V_{\alpha'}$ contains more than two non-central chief factors for $L_{\alpha'}$ then each is an FF-module for $\overline{L_{\alpha'}}$ from which it follows that $m_p(S/Q_\beta) = 1$, a contradiction. Hence, $V_{\alpha'}$ contains a unique non-central chief factor and Lemma 5.8 reveals that $[V_{\alpha'}, Q_{\alpha'}] \leq C_{V_{\alpha'}}(O^p(L_{\alpha'}))$ and $Q_{\alpha'} \in \mathrm{Syl}_p(R_{\alpha'})$. Moreover, $V_{\alpha'}/C_{V_{\alpha'}}(O^p(L_{\alpha'}))$

is a 2F-module determined by Proposition 2.51. Since $|V_\beta Q_{\alpha'}/Q_{\alpha'}| \geqslant p^2$ and V_β acts quadratically on $V_{\alpha'}$, the only possibility is that $V_{\alpha'}/C_{V_{\alpha'}}(O^p(L_{\alpha'}))$ is a natural $\mathrm{SL}_2(p^2)$-module, a contradiction by Proposition 2.21.

Suppose now that $|V_{\alpha'} Q_\beta/Q_\beta| = p$ and $r_\alpha > p$. If $r_\alpha = p^2$ then a subgroup of index at most p^3 of $V_{\alpha'}$ is centralized by Z_α and by Lemma 2.41, Propositions 2.31 and 2.32, we deduce that $V_{\alpha'}$ contains a unique non-central chief factor for $L_{\alpha'}$ and $\overline{L_{\alpha'}}/O_{p'}(\overline{L_{\alpha'}}) \cong \mathrm{PSL}_2(q_\beta)$ where $q_\beta \in \{p^2, p^3\}$. But then, Lemma 2.55 yields that $V_{\alpha'}/C_{V_{\alpha'}}(O^p(L_{\alpha'}))$ is a natural $\mathrm{SL}_2(q_\beta)$-module for $L_{\alpha'}/R_{\alpha'} \cong \mathrm{SL}_2(q_\beta)$. Since $Z_\alpha C_{V_\beta}(O^p(L_\beta))/C_{V_\beta}(O^p(L_\beta))$ is $G_{\alpha,\beta}$-invariant subgroup of order $r_\alpha = p^2$, we see by Proposition 2.21 that $q_\alpha = q_\beta = p^2$, and the result holds. Hence, we assume

(3) if $|V_{\alpha'} Q_\beta/Q_\beta| = p$ then $r_\alpha \geqslant p^3$.

Let $A \le Z_\alpha$ of index p such that $Z_\alpha \cap Q_{\alpha'} < A$. Since $r_\alpha \geqslant p^3$, there is also $B \le Z_\alpha$ of index p in A strictly containing $Z_\alpha \cap Q_{\alpha'}$. Furthermore, since Z_α is a natural $\mathrm{SL}_2(r_\alpha)$-module, $C_{V_{\alpha'} \cap Q_\beta}(B) = V_{\alpha'} \cap Q_\beta \cap Q_\alpha$. Since $|V_{\alpha'} Q_\beta/Q_\beta| = p$, without loss of generality and writing $V := V_{\alpha'}/Z_{\alpha'}$, either $C_V(A) = C_V(Z_\alpha)$, or $C_V(A) = C_V(B)$ for every $Z_\alpha \cap Q_{\alpha'} < B \le A$ with $[A : B] = p$. We apply Proposition 2.8 so that $O_{p'}(\overline{L_{\alpha'}})$ is generated by centralizers of subgroups similar to A, or subgroups similar to B for a fixed subgroup A. In particular, by the previous observations, $O_{p'}(\overline{L_{\alpha'}})$ normalizes $C_V(A)$ for some appropriate A. Then writing $H := \langle (AQ_{\alpha'}/Q_{\alpha'})^{O_{p'}(\overline{L_{\alpha'}})} \rangle$, we have that $[H, C_V(A)] = \{1\}$. Moreover, $V = [V, O_{p'}(H)] \times C_V(O_{p'}(H))$ is an $AQ_{\alpha'}/Q_{\alpha'}$-invariant decomposition and we conclude that $O_{p'}(H)$ centralizes $V_{\alpha'}$. Thus, $O_{p'}(L_{\alpha'}/R_{\alpha'})$ normalizes $AQ_{\alpha'}/Q_{\alpha'}$ and $L_{\alpha'}/R_{\alpha'}$ is a central extension of a Lie type group.

Since $V_{\alpha'} \cap Q_\beta \cap Q_\alpha$ has index at most pq_β in $V_{\alpha'}$ and is centralized by Z_α, we apply Proposition 2.36 to deduce that $L_{\alpha'}/R_{\alpha'} \cong \mathrm{SL}_2(q_{\alpha'})$. In particular, since $L_{\alpha'} = \langle Z_\alpha, Z_\alpha^x, R_{\alpha'} \rangle$ for some $x \in L_{\alpha'}$, we deduce that $C_{V_{\alpha'}}(O^p(L_{\alpha'}))$ has index at most $p^2 r_\alpha^2 \leqslant p^2 q_\beta^2 \leqslant q_\beta^3$. Since $q_\beta/p \leqslant r_\alpha \geqslant p^2$, appealing to Lemma 2.43, we deduce that $V_{\alpha'}/C_{V_{\alpha'}}(O^p(L_{\alpha'}))$ is a natural module for $L_{\alpha'}/R_{\alpha'} \cong \mathrm{SL}_2(q_{\alpha'})$. Then Proposition 2.21 yields that $q'_\alpha = q_\beta = q_\alpha$ and the result holds. Hence, we may assume for the remainder of the proof that

(4) $|V_{\alpha'} Q_\beta/Q_\beta| \geqslant p^2$.

Since $[V_\beta \cap Q_{\alpha'}, V_{\alpha'}] = \{1\}$, Propositions 2.31 and 2.32 yield $\overline{L_\beta}/O_{p'}(\overline{L_\beta}) \cong \mathrm{PSL}_2(q_\beta)$ and $V_\beta Q_{\alpha'} \in \mathrm{Syl}_p(L_{\alpha'})$. An application of Lemma 2.55 reveals that either $L_\beta/R_\beta \cong \mathrm{SL}_2(q_\beta)$, $V_\beta/C_{V_\beta}(O^p(L_\beta))$ is a natural module, and the result holds; or $q_\beta \geqslant p^4$.

Assume that $q_\beta \geqslant p^4$ and set $A \le V_{\alpha'}$ such that $V_{\alpha'} \cap Q_\beta < A$ and $|AQ_\beta/Q_\beta|p = |V_{\alpha'} Q_\beta/Q_\beta|$. Write $V := V_\beta/Z_\beta$. Suppose that $C_V(A) > (V_\beta \cap Q_{\alpha'})/Z_\beta$ so that by Proposition 2.31 and Proposition 2.32, $|AQ_\beta/Q_\beta| = p$. Hence, $|V_{\alpha'} Q_\beta/Q_\beta| = p^2$ and Z_α centralizes an index $p^2 r_\alpha$ subgroup of $V_{\alpha'}$. If $r_\alpha \geqslant p^3$, then Lemma 2.55 gives $V_{\alpha'}/C_{V_{\alpha'}}(O^p(L_{\alpha'}))$ is a natural module for $L_{\alpha'}/R_{\alpha'} \cong \mathrm{SL}_2(q_\beta)$, and the result holds.

If $r_\alpha = p^2$ then Propositions 2.31 and 2.32 imply that $q_\beta = p^4$. By Proposition 2.51, $C_V(A)$ has index p^3 in V. By conjugacy, there is $x \in Z_\alpha \setminus (Z_\alpha \cap Q_{\alpha'})$

such that $C_{V_{\alpha'}/Z_{\alpha'}}(x)$ has index p^3 in $V_{\alpha'}/Z_{\alpha'}$. Let $P_{\alpha'} = \langle Z_\alpha, Z_\alpha^x, R_{\alpha'}\rangle$ with $x \in L_{\alpha'}$ chosen such that $\overline{L_{\alpha'}} = \overline{P_{\alpha'}}O_{p'}(\overline{L_{\alpha'}})$. Then $|V_{\alpha'}/C_{V_{\alpha'}}(P_{\alpha'})| \leqslant q_\beta^2$ so that by Lemma 2.44, we have that $P_{\alpha'}R_{\alpha'}/R_{\alpha'} \cong \mathrm{SL}_2(p^4)$ and $V_{\alpha'}/C_{V_{\alpha'}}(P_{\alpha'})$ is described by Lemma 2.43. Since V_β acts quadratically on $V_{\alpha'}$ and $C_{V_{\alpha'}/Z_{\alpha'}}(x)$ has index p^3 in $V_{\alpha'}/Z_{\alpha'}$, we have a contradiction.

Finally, if $r_\alpha = p$ then for $x \in V_\beta \setminus (V_\beta \cap Q_{\alpha'})$, with $[x, A] \leq Z_\beta$, then the commutation homomorphism $\phi : A/Z_{\alpha'} \to A/Z_{\alpha'}$ such that $aZ_{\alpha'}\phi = [x, a]Z_{\alpha'}$ has image contained in $Z_\beta Z_{\alpha'}/Z_{\alpha'}$ of order at most p from which it follows that x centralizes a subgroup of $V_{\alpha'}/Z_{\alpha'}$ of order at most p^2. Hence, $V_{\alpha'}/Z_{\alpha'}$ is determined by Proposition 2.51 and since $q_\beta \geqslant p^4$, we have a contradiction. Thus,

(5) $C_{V_\beta/Z_\beta}(A) = (V_\beta \cap Q_{\alpha'})/Z_\beta$ for all $A \leq V_{\alpha'}$ of index p such that $V_{\alpha'} \cap Q_\beta < A$.

We again apply the coprime action argument from Proposition 2.8 in a similar manner as before so that $O_{p'}(L_\beta/R_\beta)$ normalizes $V_{\alpha'}R_\beta/R_\beta$ and $L_\beta/R_\beta \cong \mathrm{SL}_2(q_\beta)$. Since V_β acts quadratically on $V_{\alpha'}$, applying Lemma 2.42, we deduce that $V_\beta/C_{V_\beta}(O^p(L_\beta))$ is a natural module for L_β/R_β. Then Proposition 2.21 yields $q_\alpha = q_\beta$, as desired. This completes the proof. □

As in Section 6.2, throughout this chapter, we intend to control the action of $O^p(R_\alpha)$ and $O^p(R_\beta)$ using the methods in Lemmas 5.32–5.36 in the expectation of applying Lemma 5.22 or [27] to force contradictions. In the following lemmas, we demonstrate that we satisfy Hypothesis 5.31, required for the application of these lemmas. Also, as in Section 6.2, whenever $L_\alpha/R_\alpha \cong L_\beta/R_\beta \cong \mathrm{SL}_2(p)$, we will often make a generic appeal to coprime action, utilizing that L_λ is solvable when $p = 2$ for $\lambda \in \{\alpha, \beta\}$, and that there is a central involution $t_\lambda \in L_\lambda/R_\lambda$ which acts fixed point freely on natural modules when p is odd.

Lemma 7.7 *Suppose that $C_{V_\beta}(V_{\alpha'}) = V_\beta \cap Q_{\alpha'}$, $V_{\alpha'} \leq Q_\beta$ and $q_\alpha = q_\beta = p$. Then $Z_\alpha = Z(Q_\alpha)$ and $Z_\beta = Z(Q_\beta)$.*

Proof We aim to show that if the conclusion of the lemma fails to hold then $R = Z_\beta = Z_{\alpha'}$ for then, as $V_\beta \not\leq Q_{\alpha'}$, $O^p(L_{\alpha'})$ centralizes $V_{\alpha'}$, a contradiction.

Suppose that $V_{\alpha'} \leq Q_\beta$ and $Z_\alpha \neq Z(Q_\alpha)$. By minimality of b, and using that b is odd, we have that $Z_\lambda \leq Q_\alpha$ and $Z(Q_\alpha) \leq Q_\lambda$ for all $\lambda \in \Delta^{(b-1)}(\alpha)$. In particular, $Z(Q_\alpha) \leq Q_{\alpha'-1}$ and $Z(Q_\alpha) = Z_\alpha(Z(Q_\alpha) \cap Q_{\alpha'})$. If $[V_{\alpha'}, Z(Q_\alpha) \cap Q_{\alpha'}] = \{1\}$, it follows that $O^p(L_\alpha)$ centralizes $Z(Q_\alpha)/Z_\alpha$ and an application of coprime action, observing that $Z_\beta \leq Z_\alpha = [Z(Q_\alpha), O^p(L_\alpha)]$, gives a contradiction. If $[V_{\alpha'}, Z(Q_\alpha) \cap Q_{\alpha'}] \neq \{1\}$, then $Z_{\alpha'} = [V_{\alpha'}, Z(Q_\alpha) \cap Q_{\alpha'}] \leq Z(Q_\alpha)$ and so $Z_{\alpha'}$ is centralized by $V_{\alpha'}Q_\alpha \in \mathrm{Syl}_p(L_\alpha)$ from which it follows that $Z_{\alpha'} = Z_\beta$, a contradiction. Thus, $Z_\alpha = Z(Q_\alpha)$. Since $Z(S) \leq Z(Q_\alpha)$ we conclude that $Z(S) = \Omega(Z(S)) = Z_\beta$ is of exponent p.

Since $Z_\lambda \leq Q_{\alpha'}$ for all $\lambda \in \Delta^{(b-2)}(\alpha')$, again using the minimality of b and that b is odd, we argue that $Z(Q_{\alpha'}) \leq Q_{\alpha+2}$. If $Z(Q_{\alpha'}) \not\leq Q_\beta$ then, as $Z(S) = Z_\beta$, $\{1\} \neq [Z(Q_{\alpha'}), Z(Q_\beta)] \leq Z(Q_{\alpha'}) \cap Z(Q_\beta)$, for otherwise $Z(Q_\beta)$ is centralized by $Z(Q_{\alpha'})Q_\beta \in \mathrm{Syl}_p(L_\beta)$ and the result holds. Then, $[Z(Q_{\alpha'}), Z(Q_\beta)]$ is centralized by $Z(Q_{\alpha'})Q_\beta \in \mathrm{Syl}_p(L_\beta)$ and since $Z(S) = Z_\beta$, $[Z(Q_{\alpha'}), Z(Q_\beta)] = Z_\beta$.

Moreover, since $[Z(Q_{\alpha'}), Z(Q_\beta)] \neq \{1\}$, $Z(Q_\beta) \not\leq Q_{\alpha'}$, and by a similar reasoning, $[Z(Q_{\alpha'}), Z(Q_\beta)] = Z_{\alpha'}$. But then $Z_\beta = Z_{\alpha'}$, a contradiction. Hence, $Z(Q_{\alpha'}) \leq Q_\beta$.

Observe that $Z(Q_{\alpha'}) \not\leq Q_\alpha$, else $Z(Q_{\alpha'})$ is centralized by $Z_\alpha Q_{\alpha'} \in \mathrm{Syl}_p(L_{\alpha'})$ and $Z(Q_{\alpha'}) = Z_{\alpha'}$, as desired. Then $Z_\beta = [Z(Q_{\alpha'}), Z_\alpha] \leq \Omega(Z(Q_{\alpha'}))$ so that Z_β is centralized by $Z_\alpha Q_{\alpha'} \in \mathrm{Syl}_p(L_{\alpha'})$ and $Z_\beta = Z_{\alpha'}$, again a contradiction. Therefore, if $V_{\alpha'} \leq Q_\beta$, we have shown that $Z(Q_\beta) = Z_\beta$. □

Lemma 7.8 *Suppose that $C_{V_\beta}(V_{\alpha'}) = V_\beta \cap Q_{\alpha'}$, $V_{\alpha'} \not\leq Q_\beta$, $q_\alpha = q_\beta = p$ and Z_α is a natural module for $L_\alpha/R_\alpha \cong \mathrm{SL}_2(p)$. Then $Z_\alpha = Z(Q_\alpha)$ and $Z_\beta = Z(Q_\beta)$.*

Proof Suppose that $V_{\alpha'} \not\leq Q_\beta$. Set $Y^\beta := \langle Z(Q_\lambda) : Z_\lambda = Z_\alpha, \lambda \in \Delta(\beta)\rangle$ and let $r \in R_\beta Q_\alpha$. Since r is a graph automorphism, for $\lambda \in \Delta(\beta)$ such that $Z_\lambda = Z_\alpha$, $Z(Q_\lambda)^r = Z(Q_{\lambda\cdot r})$. But now, $Z_{\lambda\cdot r} = Z_\lambda^r = Z_\alpha^r = Z_\alpha$ and so $Z(Q_\lambda)^r \leq Y^\beta$. Thus, $Y^\beta \trianglelefteq R_\beta Q_\alpha$. Now, observe that by minimality of b, and using that b is odd, $V_\delta \leq Q_\lambda$ and $Z(Q_\lambda) \leq Q_\delta$ for all $\lambda \in \Delta(\beta)$ with $Z_\lambda = Z_\alpha$ and $\delta \in \Delta^{(b-2)}(\lambda)$ by Lemma 5.19. In particular, $Z(Q_\alpha) \leq Y^\beta \leq Q_{\alpha'-1}$. Thus, $Z(Q_\alpha) = Z_\alpha(Z(Q_\alpha) \cap Q_{\alpha'})$ and $Y^\beta = Z_\alpha(Y^\beta \cap Q_{\alpha'})$.

Since $Z(Q_\alpha) \cap Q_{\alpha'}$ is a maximal subgroup of $Z(Q_\alpha)$ not containing Z_α, we must have that $Z_\alpha \not\leq \Phi(Z(Q_\alpha))$. But then, by the irreducibility of Z_α under the action of G_α, $Z_\beta \cap \Phi(Z(Q_\alpha)) = \Omega(Z(S)) \cap \Phi(Z(Q_\alpha)) = \{1\}$ so that $\Phi(Z(Q_\alpha)) = \{1\}$ and $Z(Q_\alpha) = \Omega(Z(Q_\alpha))$ is elementary abelian.

Assume first that $[Y^\beta \cap Q_{\alpha'}, V_{\alpha'}] = Z_{\alpha'}$ so that $Y^\beta \not\leq V_\beta$ and there is some $\alpha'+1 \in \Delta(\alpha')$ with $Y^\beta \cap Q_{\alpha'} \not\leq Q_{\alpha'+1}$. Again, using the minimality of b and that b is odd, we deduce that $Z(Q_{\alpha'+1}) \leq Q_{\alpha+2}$. Write $Y_\beta = \langle Z(Q_\alpha)^{G_\beta}\rangle$ so that $Y^\beta \leq Y_\beta \trianglelefteq G_\beta$ and, as $b > 2$, Y_β is abelian. Then $Z(Q_{\alpha'+1})$ normalizes Y_β, $[Z(Q_{\alpha'+1}), Y^\beta \cap Q_{\alpha'}, Y^\beta \cap Q_{\alpha'}] \leq [Z(Q_{\alpha'+1}), Y_\beta, Y_\beta] = \{1\}$ and $Z(Q_{\alpha'+1})$ is quadratic module for $\overline{L_{\alpha'+1}}$. Moreover, by coprime action, $Z(Q_{\alpha'+1}) = [Z(Q_{\alpha'+1}), R_{\alpha'+1}] \times C_{Z(Q_{\alpha'+1})}(R_{\alpha'+1})$ is invariant under $T \in \mathrm{Syl}_p(G_{\alpha',\alpha'+1})$ and as $Z_{\alpha'} \leq Z_{\alpha'+1} \leq C_{Z(Q_{\alpha'+1})}(R_{\alpha'+1})$, we infer that $Z(Q_{\alpha'+1}) = C_{Z(Q_{\alpha'+1})}(R_{\alpha'+1})$ and $Z(Q_{\alpha'+1})$ is a faithful module for $L_{\alpha'+1}/R_{\alpha'+1} \cong \mathrm{SL}_2(p)$. But then by Lemma 2.42, $Z(Q_{\alpha'+1})$ is a direct sum of natural $\mathrm{SL}_2(p)$-modules. Now, since $[Z(Q_{\alpha'+1}), Y^\beta \cap Q_{\alpha'}]$ is of exponent p and centralized by $(Y^\beta \cap Q_{\alpha'})Q_{\alpha'+1} \in \mathrm{Syl}_p(G_{\alpha',\alpha'+1})$, we have that $[Z(Q_{\alpha'+1}), Y^\beta \cap Q_{\alpha'}] = Z_{\alpha'}$ is of order p from which it follows that $Z(Q_{\alpha'+1})$ contains a unique summand. Hence, $Z(Q_{\alpha'+1}) = Z_{\alpha'+1}$ and by conjugacy, $Z_\alpha = Z(Q_\alpha)$. But then $Y^\beta \leq V_\beta$, and we have a contradiction.

Suppose now that $[Y^\beta \cap Q_{\alpha'}, V_{\alpha'}] = \{1\}$. Then $[V_{\alpha'}, Y^\beta] \leq V_\beta$ and, as $Z_\alpha \neq Z_{\alpha+2}$, we conclude that $Y^\beta V_\beta \trianglelefteq L_\beta = \langle V_{\alpha'}, R_\beta, Q_\alpha\rangle$. But $V_{\alpha'}$ centralizes $Y^\beta V_\beta/V_\beta$ so that $O^p(L_\beta)$ centralizes $Y^\beta V_\beta/V_\beta$ and it follows that $Y^\beta V_\beta = Z(Q_\alpha)V_\beta \trianglelefteq L_\beta$. Then $[Z(Q_\alpha), Q_\beta] \trianglelefteq L_\beta$ and since $Q_\alpha \cap Q_\beta$ centralizes $[Z(Q_\alpha), Q_\beta]$ and $Q_\alpha \cap Q_\beta \ntrianglelefteq L_\beta$ by Proposition 5.27, we must have that $[Z(Q_\alpha), Q_\beta] \leq \Omega(Z(S)) = Z_\beta$ and $[Z(Q_\alpha), Q_\beta, L_\beta] = \{1\}$. Now, $[O^p(L_\beta), Z(Q_\alpha), Q_\beta] \leq [V_\beta, Q_\beta] = Z_\beta$ and by the three subgroup lemma $[Q_\beta, O^p(L_\beta), Z(Q_\alpha)] \leq Z_\beta \leq Z_\alpha$. Since $[Q_\beta, O^p(L_\beta)] \not\leq Q_\alpha$, it follows that $O^p(L_\alpha)$ centralizes $Z(Q_\alpha)/Z_\alpha$ and coprime action yields $Z(Q_\alpha) =$

$[Z(Q_\alpha), O^p(L_\alpha)] \times C_{Z(Q_\alpha)}(O^p(L_\alpha))$. But $Z_\beta \leq Z_\alpha = [Z(Q_\alpha), O^p(L_\alpha)]$ and $Z(Q_\alpha) = Z_\alpha$. In particular, $Z(S) = Z_\beta$. It remains to show that $Z(Q_\beta)$ is centralized by S (or equivalently that $Z(Q_{\alpha'})$ is centralized by $T \in \mathrm{Syl}_p(G_{\alpha',\alpha'-1})$).

To this end, assume that $Z(Q_\beta) \not\leq Q_{\alpha'}$. Then $[Z(Q_\beta), Z(Q_{\alpha'})] = [Z(Q_\beta), V_{\alpha'}Q_\beta]$ is non-trivial, of exponent p and centralized by both $Z(Q_\beta)Q_{\alpha'} \in \mathrm{Syl}_p(L_{\alpha'})$ and $V_{\alpha'}Q_\beta \in \mathrm{Syl}_p(L_\beta)$. It follows that $Z_\beta = [Z(Q_\beta), Z(Q_{\alpha'})] = Z_{\alpha'}$. But then $[Z(Q_\beta), V_{\alpha'}] = Z_{\alpha'}$, a contradiction. Hence, $Z(Q_\beta) \leq Q_{\alpha'}$ and likewise $Z(Q_{\alpha'}) \leq Q_\beta$. Again, $\{1\} \neq [Z(Q_\beta), V_{\alpha'}] \leq Z_{\alpha'} \cap Z_\beta$ so $Z_\beta = Z_{\alpha'}$ and we have that $Z(Q_{\alpha'}) \not\leq Q_\alpha$.

Set $Y^\alpha := \langle Z(Q_\delta) : \delta \in \Delta(\alpha), Z_\delta = Z_{\alpha-1}\rangle$ where $\alpha - 1 \in \Delta(\alpha)$ is such that $Z_{\alpha-1} \neq Z_\beta$. Since $Z_{\alpha'-1} \leq Q_\alpha$, we have that if $Y^\alpha \not\leq Q_{\alpha'-1}$ then $Z_{\alpha-1} = [Y^\alpha, Z_{\alpha'-1}] = Z_{\alpha'-2}$ so that $Z_\alpha = Z_{\alpha-1} \times Z_\beta = Z_{\alpha'-2} \times Z_{\alpha'} \leq Q_{\alpha'}$, a contradiction. Hence, $[Y^\alpha, Z(Q_{\alpha'})] = Z_{\alpha'} \leq Z_\alpha$. Furthermore, in a similar manner as with Y^β above, one can show that $Y^\alpha \trianglelefteq R_\alpha Q_{\alpha-1}$ so that $Y^\alpha \trianglelefteq L_\alpha = \langle Z(Q_{\alpha'}), Q_{\alpha-1}, R_\alpha\rangle$. Since $[Z(Q_{\alpha'}), Y^\alpha] \leq Z_\alpha$, we see that L_α centralizes Y^α/Z_α and $Z(Q_{\alpha-1})Z_\alpha \trianglelefteq L_\alpha$. But then $Z_{\alpha-1} = [Z(Q_{\alpha-1}), Q_\alpha] \trianglelefteq L_\alpha$, a final contradiction. Hence, $Z(Q_{\alpha'}) \leq Q_\alpha$ so that $Z(Q_{\alpha'})$ is centralized by $Z_\alpha Q_{\alpha'} \in \mathrm{Syl}_p(L_{\alpha'})$ so that $Z(Q_{\alpha'}) = Z_{\alpha'}$. This completes the proof. □

Thus, throughout this chapter, whenever we assume the necessary values of b, we are able to apply Lemma 5.32 through Lemma 5.36. That the hypotheses of these lemmas are satisfied will often be left implicit in proofs.

We now prove the "converse" statement to Lemmas 7.5 and 7.6.

Lemma 7.9 *If $b > 1$ and $V_\beta/C_{V_\beta}(O^p(L_\beta))$ is a natural module for $L_\beta/R_\beta \cong \mathrm{SL}_2(q_\beta)$, then $C_{V_\beta}(V_{\alpha'}) = V_\beta \cap Q_{\alpha'}$.*

Proof Applying Lemma 7.4, we have that Z_α is a natural module for $L_\alpha/R_\alpha \cong \mathrm{SL}_2(q_\alpha)$ and $q_\alpha = q_\beta$. Throughout, aiming for a contradiction, we assume that $C_{V_\beta}(V_{\alpha'}) < V_\beta \cap Q_{\alpha'}$ so that $[V_\beta \cap Q_{\alpha'}, V_{\alpha'}] = Z_{\alpha'}$. We may suppose that $C_{V_\beta}(O^p(L_\beta))$ acts non-trivially on $V_{\alpha'}$ for otherwise $V_\beta \cap Q_{\alpha'} = C_{V_\beta}(O^p(L_\beta))Z_{\alpha+2}$ centralizes $V_{\alpha'}$. As in the proof of Lemma 5.32, we have that

(1) $\{1\} \neq [V_{\alpha'}, C_{V_\beta}(O^p(L_\beta))] \leq [L_\beta, C_{V_\beta}(O^p(L_\beta))] \leq Z_\beta$.

Let $\alpha' + 1 \in \Delta(\alpha')$ with $[V_\beta \cap Q_{\alpha'}, Z_{\alpha'+1}] \neq \{1\}$. Indeed, we have that $[V_\beta \cap Q_{\alpha'}, Z_{\alpha'+1}] = Z_{\alpha'}$. We set $V^{\alpha'+1} := \langle C_{V_{\alpha'}}(O^p(L_{\alpha'}))^{G_{\alpha'+1}}\rangle$ throughout. By Lemma 5.32

(2) both $V^{(2)}_{\alpha'+1}/V^{\alpha'+1}$ and $V^{\alpha'+1}/Z_{\alpha'+1}$ contains non-central chief factors for $L_{\alpha'+1}$.

Assume first that $|Z_{\alpha'+1}Q_\beta/Q_\beta| < q_\beta$. Then, $Z_\beta = [V_\beta \cap Q_{\alpha'}, Z_{\alpha'+1} \cap Q_\beta] = Z_{\alpha'}$. Indeed, since $V_\beta \not\leq Q_{\alpha'}$, $|V_{\alpha'}Q_\beta/Q_\beta| = q_\beta$. But then, since $Z_{\alpha'-1} \leq Q_\beta$, we must have that $C_{V_{\alpha'}}(O^p(L_{\alpha'})) \not\leq Q_\beta$, and $[C_{V_{\alpha'}}(O^p(L_{\alpha'})), V_\beta] \leq Z_{\alpha'} = Z_\beta$, a contradiction. Hence

(3) $Z_{\alpha'+1}Q_\beta \in \mathrm{Syl}_p(L_\beta)$.

Note that if $C_{V_\beta}(O^p(L_\beta)) \not\leq Q_{\alpha'}$, then $[V_{\alpha'}, C_{V_\beta}(O^p(L_\beta))] \leq Z_\beta$ and as $|[V_{\alpha'}, C_{V_\beta}(O^p(L_\beta))]| \geqslant q_\beta$, we deduce that $[V_{\alpha'}, C_{V_\beta}(O^p(L_\beta))] = Z_\beta \leq V_{\alpha'}$ and $Z_{\alpha'} \cap Z_\beta = \{1\}$. If $C_{V_\beta}(O^p(L_\beta)) \leq Q_{\alpha'}$ then as $C_{V_\beta}(O^p(L_\beta))$ is non-trivial on $V_{\alpha'}$, $Z_{\alpha'} = [C_{V_\beta}(O^p(L_\beta)), Z_{\alpha'+1}] = Z_\beta$.

Suppose that $Z_{\alpha+3} \neq Z_\beta$. If $b = 3$, then $Z_{\alpha'} \cap Z_\beta = \{1\}$. Then $[V_{\alpha'}, C_{V_\beta}(O^p(L_\beta))] \leq Z_\beta \leq Z_{\alpha'-1}$. Moreover, $[Q_{\alpha'-1}Q_{\alpha'}, V_{\alpha'}] = [\langle C_{V_\beta}(O^p(L_\beta))^{G_{\alpha',\alpha'-1}}\rangle Q_{\alpha'}, V_{\alpha'}] \leq Z_{\alpha'-1}$ so that $Z_{\alpha'-1}Z_{\alpha'-1}^g$ is of order q_β^3 and normalized by $L_{\alpha'} = \langle Q_{\alpha'-1}, Q_{\alpha'-1}^g, R_{\alpha'}\rangle$ for some appropriately chosen $g \in L_{\alpha'}$. But then, by definition, $|V_{\alpha'}| = |V_\beta| = q_\beta^3$ and $V_\beta \cap Q_{\alpha'} = Z_{\alpha'-1}$ centralizes $V_{\alpha'}$, a contradiction.

Hence, $b > 3$ so that $V_{\alpha'}^{(3)}$ centralizes $Z_\beta \leq V_{\alpha'}$. Then $V_{\alpha'}^{(3)}$ centralizes $Z_{\alpha+2} = Z_{\alpha+3}Z_\beta$ and either $V_{\alpha'}^{(3)} = V_{\alpha'}(V_{\alpha'}^{(3)} \cap Q_\beta)$, or $V_{\alpha'}^{(3)} \not\leq Q_{\alpha+3}$ and $Z_{\alpha+4}C_{V_{\alpha+3}}(O^p(L_{\alpha+3})) = Z_{\alpha+2}C_{V_{\alpha+3}}(O^p(L_{\alpha+3}))$. The former case yields a contradiction since $V_\beta \not\leq Q_{\alpha'}$ and $[V_\beta, V_{\alpha'}^{(3)} \cap Q_\beta] \leq Z_\beta \leq V_{\alpha'}$. In the latter case, we still have that $[V_\beta, V_{\alpha'}^{(3)} \cap Q_{\alpha+3}] \leq V_{\alpha'}$ so that $V_{\alpha'}^{(3)}/V_{\alpha'}$ contains a unique non-central chief factor for $L_{\alpha'}$, which as $\overline{L_{\alpha'}}$-module, is an FF-module. By Lemma 5.32 we have that $Z_{\alpha+2} = Z_{\alpha+4}$ and Lemma 5.35 with Lemma 5.22 implies that $Z_\alpha \leq V_{\alpha+2}^{(2)} = V_{\alpha+4}^{(2)} \leq Q_{\alpha'}$, a contradiction. Therefore, for the remainder of this proof, we may assume that

(4) $Z_{\alpha+3} = Z_\beta$.

Assume first that $Z_{\alpha'} = Z_\beta$. Then $V_{\alpha'+1}^{(2)} \cap Q_{\alpha+3} \cap Q_{\alpha+2}$ is centralized, modulo $Z_{\alpha'+1}$, by $V_\beta \cap Q_{\alpha'}$. If $b = 3$ and $V^{\alpha'+1} \not\leq Q_{\alpha'}$, then since $V^{\alpha'+1}$ is $G_{\alpha'+1,\alpha'}$ invariant, $V^{\alpha'+1}Q_{\alpha'} \in \mathrm{Syl}_p(G_{\alpha'+1,\alpha'})$. But then, $[V^{\alpha'+1}, V_{\alpha'}] \leq Z_{\alpha'+1}$ so that $Z_{\alpha'+1}Z_{\alpha'+1}^g$ is normalized by $L_{\alpha'} = \langle Q_{\alpha'+1}, Q_{\alpha'+1}^g, R_{\alpha'}\rangle$ for some appropriate $g \in L_{\alpha'}$, and $|V_{\alpha'}| = |Z_{\alpha'+1}Z_{\alpha'+1}^g| = q_\beta^3$, a contradiction. Hence, $V^{\alpha'+1} \leq Q_{\alpha'}$. Then $V^{\alpha'+1} \cap Q_{\alpha'-1}$ is centralized, modulo $Z_{\alpha'+1}$, by $V_\beta \cap Q_{\alpha'}$. But then adapting Lemma 5.33, $O^p(R_{\alpha'+1})$ centralizes $V^{\alpha'+1}$ and an extension of Lemma 5.22 to $Z_{\alpha'} = Z_\beta$ yields that $Z_{\alpha'-1}C_{V_{\alpha'}}(O^p(L_{\alpha'})) = Z_{\alpha'+1}C_{V_\beta}(O^p(L_\beta)) = V_\beta \cap Q_{\alpha'}$, a contradiction. Hence, $b > 3$.

Note that $V_{\alpha+3} \leq Q_{\alpha'+1}$ so that $[V_{\alpha+3}, V^{\alpha'+1}] \leq Z_{\alpha'+1} \cap V_{\alpha+3} = Z_{\alpha'} = Z_{\alpha+3}$. In particular, $V^{\alpha'+1} \leq Q_{\alpha+3}$. Assume that $C_{Z_{\alpha+2}}(V^{\alpha'+1}) > Z_{\alpha+3}$. Then $V^{\alpha'+1} \leq Q_{\alpha+2}$ so that $V_\beta \cap Q_{\alpha'}$ centralizes $V^{\alpha'+1}/Z_{\alpha'+1}$, a contradiction. Since $b > 3$ and $V_{\alpha'+1}^{(2)}$ is abelian, $Z_{\alpha+2} \cap V_{\alpha'+1}^{(2)} = Z_{\alpha+3} = Z_{\alpha'}$. Set $V^{\alpha+2} = \langle Z_{\alpha+2}C_{V_{\alpha+3}}(O^p(L_{\alpha+3}))^{G_{\alpha+2}}\rangle$. Then $[V_{\alpha'}, V^{\alpha+2}] \leq V_{\alpha'} \cap Z_{\alpha+2} = Z_{\alpha'}$ and $V^{\alpha+2} \leq Q_{\alpha'}$. Now, $V^{\alpha'+1} \cap Q_{\alpha+2}$ is centralized, modulo $Z_{\alpha'}$, by $V^{\alpha+2}$ and $V^{\alpha+2} \cap Q_{\alpha'+1}$ is centralized, modulo $Z_{\alpha+3}$, by $V^{\alpha'+1}$. it follows from Lemma 2.41, using that $\alpha' + 1, \alpha + 2 \in \alpha^G$, that V^α/Z_α is an FF-module for $\overline{L_\alpha}$ so that $V^{\alpha'+1}Q_{\alpha+2} \in \mathrm{Syl}_p(L_{\alpha+2})$ and $V^{\alpha+2}Q_{\alpha'+1} \in \mathrm{Syl}_p(L_{\alpha'+1})$. But now $V_{\alpha'+1}^{(2)} \cap Q_{\alpha'}$ is centralized, modulo $V^{\alpha'+1}$, by $V^{\alpha+2}$ so that by Lemma 2.41, $V_\alpha^{(2)}/V^\alpha$ is also an FF-module for $\overline{L_\alpha}$. By Lemma 5.33, $O^p(R_\alpha)$ centralizes $V_\alpha^{(2)}$ and Lemma 5.22 applied to $Z_\beta = Z_{\alpha+3}$ yields that $V_\beta = V_{\alpha+3} \le Q_{\alpha'}$, a contradiction. Thus

(5) $Z_{\alpha'} \cap Z_\beta = \{1\}$ so that $b > 3$ and $C_{V_\beta}(O^p(L_\beta)) \not\le Q_{\alpha'}$.

In particular, since $V^{\alpha'+1} \cap V_{\alpha'} = Z_{\alpha'+1}C_{V_{\alpha'}}(O^p(L_{\alpha'}))$, it follows that $Z_\beta \cap V^{\alpha'+1} = \{1\}$. Since $V_{\alpha+3} \le Q_{\alpha'+1}$, we have that $[V_{\alpha'+1}^{(2)} \cap Q_{\alpha+3}, V_{\alpha+3}] \le V^{\alpha'+1} \cap Z_{\alpha+3} = \{1\}$ and $V_{\alpha'+1}^{(2)} \cap Q_{\alpha+3} \le Q_{\alpha+2}$. But now, $V^{\alpha'+1} \cap Q_{\alpha+3}$ is centralized, modulo $Z_{\alpha'+1}$, by $V_\beta \cap Q_{\alpha'}$ and we have that $V^{\alpha'+1}Q_{\alpha+3} \in \mathrm{Syl}_p(L_{\alpha+3})$. Moreover, $[V^{\alpha'+1}, V_{\alpha+3}] \le V_{\alpha+3} \cap Z_{\alpha'+1} = Z_{\alpha'}$. Then, $Z_{\alpha'}Z_{\alpha'}^g Z_{\alpha+3}$ is normalized by $L_{\alpha+3} = \langle V_{\alpha'+1}, V_{\alpha'+1}^g, R_{\alpha+3}\rangle$ for some appropriate $g \in L_{\alpha+3}$. Hence, $|[V_{\alpha+3}, O^p(L_{\alpha+3})]| = q_\beta^3$ and $V_{\alpha+3} = [V_{\alpha+3}, O^p(L_\alpha)]Z_{\alpha+2}$ has order either q_β^3 or q_β^4. Either way, we have that $[V_{\alpha'}, V_\beta]Z_{\alpha+2}$ has index q_β in V_β, from which it follows that $V_\beta \cap Q_{\alpha'} = [V_{\alpha'}, V_\beta]Z_{\alpha+2}$ centralizes $V_{\alpha'}$, a final contradiction. Hence, the lemma holds. □

Proposition 7.10 *Suppose that* $b > 1$, $C_{V_\beta}(V_{\alpha'}) = V_\beta \cap Q_{\alpha'}$ *and* $V_{\alpha'} \not\le Q_\beta$. *Then* $C_{V_{\alpha'}}(V_\beta) = V_{\alpha'} \cap Q_\beta$.

Proof Since $V_{\alpha'} \not\le Q_\beta$, there is $\alpha'+1 \in \Delta(\alpha')$ with $Z_{\alpha'+1} \not\le Q_\beta$ and $(\alpha'+1, \beta)$ a critical pair. Moreover, by Lemma 7.6 and since $\alpha' \in \beta^G$, $V_{\alpha'}/C_{V_{\alpha'}}(O^p(L_{\alpha'}))$ is a natural module for $L_{\alpha'}/R_{\alpha'} \cong \mathrm{SL}_2(q_{\alpha'})$. Then Lemma 7.9 applied to $(\alpha'+1, \alpha')$ in place of (α, α') gives the result. □

7.1 $C_{V_\beta}(V_{\alpha'}) < V_\beta \cap Q_{\alpha'}$

The hypothesis for this section is $b > 1$ and $C_{V_\beta}(V_{\alpha'}) < V_\beta \cap Q_{\alpha'}$. Notice by Lemma 7.2 this condition is equivalent to $[V_\beta \cap Q_{\alpha'}, V_{\alpha'}] \ne \{1\}$. By Lemmas 7.5, 7.6, and 7.9, throughout this section, whenever $(\alpha^*, \alpha^{*\prime})$ is a critical pair, we have that $[V_{\alpha^*+1} \cap Q_{\alpha^{*\prime}}, V_{\alpha^{*\prime}}] \ne \{1\}$ and $V_{\alpha^{*\prime}}/C_{\alpha^{*\prime}}(O^p(L_{\alpha^{*\prime}}))$ is never a natural $\mathrm{SL}_2(q_{\alpha^{*\prime}})$-module.

The aim of this section will be to recognize amalgams of type ${}^2\mathrm{F}_4(2^n)$ and ${}^2\mathrm{F}_4(2)'$ via the identifications provided in [27].

Proposition 7.11 *Suppose that* $b > 1$ *and* $C_{V_\beta}(V_{\alpha'}) < V_\beta \cap Q_{\alpha'}$. *Then* $V_{\alpha'} \not\le Q_\beta$.

Proof Aiming for a contradiction, suppose that $V_{\alpha'} \le Q_\beta$. Note that if Z_α is a natural $\mathrm{SL}_2(q_\alpha)$-module for L_α/R_α, then $[V_\beta, Q_\beta] = [Z_\alpha, Q_\beta]^{L_\beta} = Z_\beta$. Moreover, for $\lambda \in \Delta(\alpha')$ with $V_\beta \cap Q_{\alpha'} \not\le Q_\lambda$, $Z_{\alpha'} = [V_\beta \cap Q_{\alpha'}, Z_\lambda] \le [V_\beta, Q_\beta] = Z_\beta$ so that $Z_{\alpha'} = Z_\beta = [V_{\alpha'}, V_\beta]$, a contradiction since $V_\beta \not\le Q_{\alpha'}$. Hence, by Lemma 7.4, $V_\beta/C_{V_\beta}(O^p(L_\beta))$ contains at least two non-central chief factors.

Let $\alpha' + 1 \in \Delta(\alpha')$ with $Z_{\alpha'+1} \not\leq Q_\alpha$. Since neither $Z_{\alpha'+1}$ nor Z_α are FF-modules, we have that

$$|(Z_\alpha \cap Q_{\alpha'})Q_{\alpha'+1}/Q_{\alpha'+1}| < |Z_{\alpha'+1}Q_\alpha/Q_\alpha| < |Z_\alpha/(Z_\alpha \cap Q_{\alpha'} \cap Q_{\alpha'+1})|.$$

In particular, $|Z_\alpha Q_{\alpha'}/Q_{\alpha'}| > p$ and as Z_α acts quadratically on $V_{\alpha'}$, by conjugacy and applying Lemma 7.3, $\overline{L_\beta}/O_{p'}(\overline{L_\beta}) \cong \mathrm{PSL}_2(q_\beta)$, $\mathrm{PSU}_3(q_\beta)$ or $\mathrm{Sz}(q_\beta)$. If $p = 2$, then $Z_\alpha Q_{\alpha'}/Q_{\alpha'} \leq \Omega(Z(T/Q_{\alpha'}))$ and if p is odd, then applying Lemma 7.3 and Proposition 2.36, we again conclude that $Z_\alpha Q_{\alpha'}/Q_{\alpha'} \leq \Omega(Z(T/Q_{\alpha'}))$ for $T \in \mathrm{Syl}_p(G_{\alpha',\alpha'-1})$. Hence $p < |Z_\alpha Q_{\alpha'}/Q_{\alpha'}| \leqslant q_\beta$ and by Lemmas 2.41 and 7.9, we have that $p < |Z_\alpha Q_{\alpha'}/Q_{\alpha'}| < |V_{\alpha'}Q_\alpha/Q_\alpha|$. Arguing as before, $p < |V_{\alpha'}Q_\alpha/Q_\alpha| \leqslant q_\alpha$ and $\overline{L_\alpha}/O_{p'}(\overline{L_\alpha}) \cong \mathrm{PSL}_2(q_\alpha)$, $\mathrm{PSU}_3(q_\alpha)$ or $\mathrm{Sz}(q_\alpha)$.

Since Z_α centralizes a subgroup of $V_{\alpha'}$ of index at most q_α and $V_{\alpha'}$ contains at least two non-central chief factors for $L_{\alpha'}$, applying Propositions 2.31 and 2.32, we deduce that $q_\alpha \geqslant q_\beta^2$. By a similar reasoning, if $|(Z_\alpha \cap Q_{\alpha'})Q_{\alpha'+1}/Q_{\alpha'+1}| = p$, then Propositions 2.31 and 2.32 yields $q_\beta p \geqslant q_\alpha$ so that $q_\alpha p^2 \geqslant q_\beta^2 p^2 \geqslant q_\alpha^2$ and $p^2 < q_\beta^2 \leqslant q_\alpha \leqslant p^2$, a contradiction. Then $|(Z_\alpha \cap Q_{\alpha'})Q_{\alpha'+1}/Q_{\alpha'+1}| > p$ and applying Propositions 2.31 and 2.32 we conclude that $|Z_{\alpha'+1}Q_\alpha/Q_\alpha| = q_\alpha$, $\overline{L_\alpha}/O_{p'}(\overline{L_\alpha}) \cong \mathrm{PSL}_2(q_\alpha)$ and $S = Z_{\alpha'+1}Q_\alpha$. Then Z_β has index at most $|Z_\alpha/(Z_\alpha \cap Q_{\alpha'} \cap Q_{\alpha'+1})| \leqslant q_\alpha q_\beta \leqslant q_\alpha^{\frac{3}{2}}$ in Z_α. Applying Lemma 2.55, we deduce that Z_α is a natural module for $L_\alpha/R_\alpha \cong \mathrm{SL}_2(q_\alpha)$, a contradiction. Hence, $V_{\alpha'} \not\leq Q_\beta$. □

For the remainder of this section, we fix $\alpha' + 1 \in \Delta(\alpha')$ such that $V_\beta \cap Q_{\alpha'} \not\leq Q_{\alpha'+1}$. Note also that $[Z_{\alpha'+1}, V_\beta \cap Q_{\alpha'}, V_\beta \cap Q_{\alpha'}] = \{1\}$ so that both $Z_{\alpha'+1}$ and $V_{\alpha'}$ admit non-trivial quadratic action. Throughout, we set $R := [V_\beta \cap Q_{\alpha'}, V_{\alpha'}]$.

The following lemma, along with its proof, appeared earlier as Proposition 2.51 and Lemma 2.54 where the necessary additional hypothesis there follow from Hypothesis 5.1. We recall it here as it will be applied liberally throughout this section.

Proposition 7.12 *For $\gamma \in \Gamma$, $G := \overline{L_\gamma}$ and $S \in \mathrm{Syl}_p(G)$, assume that V is a faithful* $\mathrm{GF}(p)G$*-module with $C_V(O^p(G)) = \{1\}$ and $V = \langle C_V(S)^G \rangle$. If there is a p-element $x \in G^\#$ such that $[V, x, x] = \{1\}$ and $|V/C_V(x)| = p^2$ then, setting $L := \langle x^G \rangle$, one of the following holds:*

(i) *p is odd, $G = L \cong$ (P)$\mathrm{SU}_3(p)$ and V is the natural module;*
(ii) *p is arbitrary, $G \cong \mathrm{SL}_2(p^2)$ and V is the natural module;*
(iii) *$p = 2$, $G = L \cong \mathrm{PSL}_2(4)$ and V is a natural $\Omega_4^-(2)$-module;*
(iv) *$p = 3$, $G = L \cong 2 \cdot \mathrm{Alt}(5)$ or $2_-^{1+4}.\mathrm{Alt}(5)$ and V is the unique irreducible quadratic $2F$-module of dimension 4;*
(v) *p is arbitrary, $G = L \cong \mathrm{SL}_2(p)$ and V is the direct sum of two natural* $\mathrm{SL}_2(p)$*-modules;*
(vi) *$p = 2$, $L \cong \mathrm{SU}_3(2)'$, G is isomorphic to a subgroup of* $\mathrm{SU}_3(2)$ *which contains* $\mathrm{SU}_3(2)'$ *and V is a natural* $\mathrm{SU}_3(2)$*-module viewed as an irreducible* $\mathrm{GF}(2)G$*-module by restriction;*

(vii) $p = 2$, $L \cong \mathrm{Dih}(10)$, $G \cong \mathrm{Dih}(10)$ *or* $\mathrm{Sz}(2)$ *and V is a natural* $\mathrm{Sz}(2)$*-module viewed as an irreducible* $\mathrm{GF}(2)G$*-module by restriction;*
(viii) $p = 3$, $G = L \cong (Q_8 \times Q_8) : 3$ *and* $V = V_1 \times V_2$ *where V_i is a natural* $\mathrm{SL}_2(3)$*-module for* $G/C_G(V_i) \cong \mathrm{SL}_2(3)$*;*
(ix) $p = 2$, $G = L \cong (3 \times 3) : 2$ *and* $V = V_1 \times V_2$ *where V_i is a natural* $\mathrm{SL}_2(2)$*-module for* $G/C_G(V_i) \cong \mathrm{Sym}(3)$*; or*
(x) $p = 2$, $L \cong (3\times 3) : 2$, $G \cong (3\times 3) : 4$, *V is irreducible as a* $\mathrm{GF}(2)G$*-module and* $V|_L = V_1 \times V_2$ *where V_i is a natural* $\mathrm{SL}_2(2)$*-module for* $L/C_L(V_i) \cong \mathrm{Sym}(3)$.

Moreover, if V is generated by a $N_G(S)$-invariant subspace of order p then (G, V) satisfies outcome (iii), (vii) (ix) or (x).

We now begin the task of restricting the structure of the amalgam, with the ultimate aim of demonstrating that it is a weak BN-pair of rank 2.

Lemma 7.13 *Suppose that $b > 1$, $C_{V_\beta}(V_{\alpha'}) < V_\beta \cap Q_{\alpha'}$ and Z_α is not a natural* $\mathrm{SL}_2(q_\alpha)$*-module. Then $Z_{\alpha'+1} \not\leq Q_\beta$ and either:*

(i) $|Z_{\alpha'+1}Q_\beta/Q_\beta| > p$*; or*
(ii) *L_β/R_β is isomorphic to one of* $\mathrm{SL}_2(p)$, $(3\times 3) : 2$, $(3\times 3) : 4$ *or* $(Q_8\times Q_8) : 3$, $Q_\beta \in \mathrm{Syl}_p(R_\beta)$, $[V_\beta, Q_\beta] \leq C_{V_\beta}(O^p(L_\beta))$, *$V_\beta/C_{V_\beta}(O^p(L_\beta))$ is described in Proposition 7.12 and $q_\alpha = p$.*

Proof Note that if $m_p(S/Q_\beta) > 1$ then as $V_{\alpha'}$ acts quadratically on V_β and vice versa, we have that $\overline{L_\beta}/O_{p'}(\overline{L_\beta}) \cong \mathrm{PSL}_2(q_\beta), \mathrm{PSU}_3(q_\beta)$ or $\mathrm{Sz}(q_\beta)$ by Proposition 2.15 and Lemma 7.3. As in Proposition 7.11, we conclude that $|V_{\alpha'}Q_\beta/Q_\beta| \leqslant q_\beta$. We assume throughout that Z_α is not a natural module for $L_\alpha/R_\alpha \cong \mathrm{SL}_2(q_\alpha)$ so that Lemma 7.4 implies that $V_\beta/C_{V_\beta}(O^p(L_\beta))$ contains at least two non-central chief factors for L_β.

Assume first that $V_{\alpha'} \cap Q_\beta \leq Q_\alpha$. Applying Lemmas 2.41 and 7.9, we deduce that $p < |V_{\alpha'}Q_\beta/Q_\beta| \leqslant q_\beta$ and $m_p(S/Q_\beta) > 1$. Applying Propositions 2.31 and 2.32, since $V_{\alpha'}$ contains at least two non-central chief factors for $L_{\alpha'}$, we deduce that $q_\beta \geqslant q_\beta^{\frac{4}{3}}$, a contradiction. Hence

(1) $V_{\alpha'} \cap Q_\beta \not\leq Q_\alpha$ and $q_\alpha \geqslant q_\beta^{\frac{1}{3}}$.

Taking Propositions 2.31 and 2.32 further whenever $|Z_\alpha Q_{\alpha'}/Q_{\alpha'}| > p$, we have that $q_\alpha \geqslant |(V_{\alpha'} \cap Q_\beta)Q_\alpha/Q_\alpha| \geqslant q_\beta$. Hence, with any condition on the order of $|Z_\alpha Q_{\alpha'}/Q_{\alpha'}|$, we have by Lemma 2.41 that there is $\lambda \in \Delta(\alpha')$ with $Z_\alpha \cap Q_{\alpha'} \not\leq Q_{\alpha'+1}$. We may relabel such that $\lambda = \alpha' + 1$. Define $r_\beta := |Z_{\alpha'+1}Q_\beta/Q_\beta|$, $r_\alpha := |(Z_{\alpha'+1} \cap Q_\beta)Q_\alpha/Q_\alpha|$, $r_{\alpha'} = |Z_\alpha Q_{\alpha'}/Q_{\alpha'}|$ and $r_{\alpha'+1} := |(Z_\alpha \cap Q_{\alpha'})Q_{\alpha'+1}/Q_{\alpha'+1}|$ so that $r_\lambda \leq q_\lambda$ for $\lambda \in \{\alpha, \beta, \alpha', \alpha'+1\}$. We note that outcome (i) is exactly $r_\beta > p$ and so for the remainder of the proof we assume that

(2) $r_\beta \leqslant p$.

By Lemma 2.41, since Z_α is not a natural $\mathrm{SL}_2(p)$-module, we have that $p \leqslant r_\alpha < r_{\alpha'} r_{\alpha'+1}$ and $r_{\alpha'+1} < r_\alpha r_\beta$.

Suppose that $r_\alpha > p$ so that by Proposition 2.15 and Lemma 7.3, $\overline{L_\alpha}/O_{p'}(\overline{L_\alpha}) \cong \mathrm{PSL}_2(q_\alpha), \mathrm{PSU}_3(q_\alpha)$ or $\mathrm{Sz}(q_\alpha)$. Assume first that $r_{\alpha'+1} = p$. Applying Lemma 2.41, we have that $p < r_{\alpha'} \leqslant q_\beta$ so that $q_\beta \leqslant q_\alpha$. Applying Propositions 2.31 and 2.32, we conclude that $\overline{L_\alpha}/O_{p'}(\overline{L_\alpha}) \cong \mathrm{PSL}_2(q_\alpha)$ and $q_\alpha/p \leqslant r_{\alpha'} \leqslant q_\beta \leqslant q_\alpha$. If $q_\alpha = p^2$ then applying Lemma 2.55 to Z_α yields that Z_α is a natural module for $L_\alpha/R_\alpha \cong \mathrm{SL}_2(q_\alpha)$. Hence, we either have that $q_\alpha = q_\beta > p^2$, or $p^2 \leqslant q_\alpha/p = r_{\alpha'} = q_\beta$. Applying Lemma 2.55 to non-central chief factors contained in $V_{\alpha'}$, we deduce that there are exactly two non-central chief factors for $L_{\alpha'}$ within $V_{\alpha'}$ and both, as $\overline{L_{\alpha'}}$-modules, are FF-modules. Hence, applying Lemmas 2.42 and 2.55, if $[V_\beta, Q_\beta]$ contains no non-central chief factors for L_β, then $V_\beta/C_{V_\beta}(O^p(L_\beta))$ is a direct sum of two natural modules for $L_\beta/R_\beta \cong \mathrm{SL}_2(q_\beta)$. But then, $Z_\beta = Z_\alpha \cap C_{V_\beta}(O^p(L_\beta))$ has index $q_\beta^2 \leqslant q_\alpha^2$ in Z_α and Lemma 2.55 implies that $q_\beta = q_\alpha$ and $L_\alpha/R_\alpha \cong \mathrm{SL}_2(q_\alpha)$ and Z_α is a direct sum of two natural modules, a contradiction since an index $pr_\alpha < q_\alpha^2$ subgroup of $Z_{\alpha'+1}$ is centralized by $Z_\alpha \cap Q_{\alpha'} \not\leq Q_{\alpha'+1}$. Hence, $[V_\beta, Q_\beta]$ contains a non-central chief factor for L_β and Q_β is not quadratic on Z_α. Since $(V_{\alpha'} \cap Q_\beta)Q_\alpha$ acts quadratically on Z_α and $Z_\beta = \Omega(Z(S))$, we must have that $(V_{\alpha'} \cap Q_\beta)Q_\alpha < S$ and $r'_\alpha = q_\beta = q_\alpha/p$. Applying Lemma 2.55, we deduce that $|Z_\alpha/C_{Z_\alpha}(V_{\alpha'} \cap Q_\beta)| > q_\alpha^2/p^2$. But then we may form $L^\alpha = \langle Z_{\alpha'+1} \cap Q_\beta, (Z_{\alpha'+1} \cap Q_\beta)^x, Q_\alpha \rangle$ with $x \in L_\alpha$ chosen such that $\overline{L^\alpha}/O_{p'}(\overline{L^\alpha}) \cong \overline{L_\alpha}/O_{p'}(\overline{L_\alpha})$. Since $r'_\alpha p = q_\alpha$, we have that $|Z_\alpha/C_{Z_\alpha}(L^\alpha)| \leqslant q_\alpha^2$. Note that $V_{\alpha'} \cap Q_\beta \leq L^\alpha$ and $[Z_\alpha, V_{\alpha'} \cap Q_\beta] \leq C_{Z_\alpha}(V_{\alpha'} \cap Q_\beta)$ from which it follows that $|[Z_\alpha, V_{\alpha'} \cap Q_\beta]C_{Z_\alpha}(L^\alpha)/C_{Z_\alpha}(L^\alpha)| < p^2$. Hence, $|Z_\alpha/C_{Z_\alpha}(L^\alpha)|$ is dual to an FF-module for $\overline{L^\alpha}$ and applying Lemma 2.41, since $q_\alpha > p$, we have a contradiction. Thus

(3) if $r_\alpha > p$, then $r_{\alpha'+1} > p$.

Then applying Propositions 2.31 and 2.32, we infer have that $q_\alpha/p \leqslant r_\alpha$, $\overline{L_\alpha}/O_{p'}(\overline{L_\alpha}) \cong \mathrm{PSL}_2(q_\alpha)$ and Z_α is irreducible under the action of L_α. Since $r_\alpha r_\beta \leqslant r_\alpha p \leqslant q_\alpha p$ and $r_{\alpha'+1} \geqslant p^2$, Lemma 2.55 applied to the action of $Z_\alpha \cap Q_{\alpha'}$ on $Z_{\alpha'+1}$ yields that $q_\alpha > p^2$. Hence, $r_\alpha > q_\alpha^{\frac{1}{2}}$ and Lemma 2.55 yields $r_\alpha^2 \leqslant r_{\alpha'+1} r_{\alpha'}$. If $r_{\alpha'+1} \leqslant q_\alpha^{\frac{1}{2}}$ then as $r_{\alpha'+1} \geqslant p^2$, it follows that $r_{\alpha'} = q_\alpha = p^4$, $r_\alpha = q_\alpha/p$ and $r_{\alpha'+1} = p^2$. But then, Lemma 2.55 on the action of $Z_{\alpha'+1} \cap Q_\beta$ on Z_α implies that Z_α is direct sum of two natural $\mathrm{SL}_2(p^4)$-modules. Since $r_\alpha r_\beta \leqslant q_\alpha$, this is a contradiction. Hence, $r_{\alpha'+1} > q_\alpha^{\frac{1}{2}}$ so that $q_\alpha p < r_{\alpha'+1}^2$. However, then Lemma 2.55 delivers, another contradiction. Hence, we deduce that

(4) $r_\alpha = p$.

Again applying Lemma 2.41, $r_\beta = p$, $Z_{\alpha'+1} \not\leq Q_\beta$ and $Z_\alpha \cap Q_{\alpha'}$ has a quadratic 2F-action $Z_{\alpha'+1}$. Applying Proposition 7.12, we conclude that $q_\alpha = p$ or $L_\alpha/R_\alpha \cong \mathrm{PSL}_2(4)$ and Z_α is a natural $\Omega_4^-(2)$-module for $L_\alpha/R_\alpha \cong \mathrm{PSL}_2(4)$. Since $V_{\alpha'} \cap Q_\beta$

acts quadratically on Z_α, in either case we deduce that $|(V_{\alpha'} \cap Q_\beta)Q_\alpha/Q_\alpha| = p$. But then Z_α centralizes a subgroup of index at most $q_\beta p$ in $V_{\alpha'}$, and since $V_{\alpha'}$ contains at least two non-central chief factors for $L_{\alpha'}$, we have that $q_\beta p \geqslant q_\beta^{\frac{4}{3}}$ so that $q_\beta \leqslant p^3$ and $V_{\alpha'}$ contains exactly two non-central chief factors. If $q_\beta = p^3$, then it follows by Lemma 2.41 that both non-central chief factors are quadratic 2F-modules, each with an index p^2 subgroup centralized, a contradiction by Proposition 7.12 since $q_\beta > p^2$. If $q_\beta = p^2$, then an index p subgroup of one of the non-central chief factors is centralized and by Lemma 2.41 we have a contradiction. Hence

(5) $q_\beta = p$

We note that $r_{\alpha'+1} < r_\alpha r_\beta$ so that $r_{\alpha'+1} = p$. Indeed, $Z_{\alpha'+1}$ is a quadratic 2F-module and so is determined by Proposition 7.12.

Suppose that $[V_\beta, Q_\beta]$ contains a non-central chief factor for L_β. Then $[Z_\alpha, Q_\beta] \not\leq Z_\beta$ and since $V_{\alpha'} \cap Q_\beta$ centralizes $[Z_\alpha, V_{\alpha'} \cap Q_\beta]$, we have that $|S/Q_\alpha| > p$. Moreover, by Lemma 5.8, $V_\beta/[V_\beta, Q_\beta]$ also contains a non-central chief factor and by Lemma 2.41, both $V_\beta/[V_\beta, Q_\beta]C_{V_\beta}(O^p(L_\beta))$ and $[V_\beta, Q_\beta]C_{V_\beta}(O^p(L_\beta))/C_{V_\beta}(O^p(L_\beta))$ are natural $\mathrm{SL}_2(p)$-modules. In particular, by Propositions 2.46, 2.47, and Lemma 7.3 when $p \geqslant 5$, we have that $L_\beta/R_\beta Q_\beta \cong \mathrm{SL}_2(p), (3 \times 3) : 2$ or $(Q_8 \times Q_8) : 3$ so that $S = Q_\alpha Q_\beta$. Furthermore, $(Q_\beta \cap O^p(L_\beta))Q_\alpha$ centralizes an index p subgroup of $[Z_\alpha, S]$.

If $L_\alpha/R_\alpha \cong \mathrm{PSL}_2(4)$ and Z_α is a natural $\Omega_4^-(2)$-module then $S = Q_\beta \cap O^p(L_\beta))Q_\alpha$, impossible for then S centralizes a subgroup of order p^2. If Z_α is a natural $\mathrm{SU}_3(p)$-module for L_α/R_α which is isomorphic to a subgroup $\mathrm{SU}_3(p)$ containing $\mathrm{SU}_3(p)'$, then $Z_\alpha[V_\beta, Q_\beta]C_{V_\beta}(O^p(L_\beta))/[V_\beta, Q_\beta]C_{V_\beta}(O^p(L_\beta))$ has order p since $V_\beta/[V_\beta, Q_\beta]C_{V_\beta}(O^p(L_\beta))$ is a natural $\mathrm{SL}_2(p)$-module. Thus, there is $G_{\alpha,\beta}$-invariant subgroup of Z_α of index p, a contradiction.

If $L_\alpha/R_\alpha \cong (3\times3):4$ or $\mathrm{Sz}(2)$, then we observe that $Q_\alpha \cap Q_\beta$ has index 4 in Q_β and centralizes $[Z_\alpha, Q_\beta, Q_\beta] \trianglelefteq L_\beta$. Since $Q_\alpha \cap Q_\beta \ntrianglelefteq L_\beta$, and $[Z_\alpha, Q_\beta, Q_\beta] \not\leq Z(S)$, we must have that $Q := \langle (Q_\alpha \cap Q_\beta)^{G_\beta} \rangle$ has index 2 in Q_β. Moreover, for $L_\alpha^* := \langle Q^{L_\alpha} \rangle Q_\alpha$, we have that $L_\alpha^* R_\alpha/R_\alpha \cong (3 \times 3) : 2$ or $\mathrm{Dih}(10)$ respectively. It follows that $Q = O_2(L_\beta^*)$ where $L_\beta^* := \langle Q_\alpha^{L_\beta} \rangle$. Let K be a Hall $2'$-subgroup of G_α and set $G_\lambda^* := L_\lambda^* K$ so that G_λ^* has index 2 in G_λ for $\lambda \in \{\alpha, \beta\}$. Then any subgroup of S normalized by G_α^* and G_β^* is normalized by G and checking against the required constraints, we see that the triple $(G_\alpha^*, G_\beta^*, KQ_\alpha Q)$ satisfies Hypothesis 5.1. Since G is a minimal counterexample, $(G_\alpha^*, G_\beta^*, KQ_\alpha Q)$ is determined in Theorem C. Checking against the possible outcomes, we have a contradiction.

Appealing to Proposition 7.12 for the structure of Z_α, there are no other groups and modules satisfying our requirements. Hence, we deduce that

(6) $[V_\beta, Q_\beta] \leq C_{V_\beta}(O^p(L_\beta))$, $Q_\beta \in \mathrm{Syl}_p(R_\beta)$ and $V_\beta/C_{V_\beta}(O^p(L_\beta))$ is described in Proposition 7.12

Since $V_\beta/C_{V_\beta}(O^p(L_\beta))$ contains two non-central chief factors for $O^p(L_\beta)$, we have that $L_\beta/R_\beta \cong \mathrm{SL}_2(p), (3 \times 3) : 2, (3 \times 3) : 4$ or $(Q_8 \times Q_8) : 3$. Applying

Lemma 2.41, we have that $Z_\alpha \cap C_{V_\beta}(O^p(L_\beta))$ is centralized by $Q_\beta \cap O^p(L_\beta)$ and has index p^2 in Z_α so that $L_\alpha/R_\alpha \not\cong \mathrm{PSL}_2(4)$ and $q_\alpha = p$, as desired. □

Lemma 7.14 *Suppose that $b > 1$, $C_{V_\beta}(V_{\alpha'}) < V_\beta \cap Q_{\alpha'}$ and Z_α is not a natural $\mathrm{SL}_2(q_\alpha)$-module. If $|Z_{\alpha'+1}Q_\beta/Q_\beta| > p$ then $L_\beta/R_\beta \cong \mathrm{SL}_2(q_\alpha) \cong L_\alpha/R_\alpha$, $q_\alpha > p$ and both $V_\beta/C_{V_\beta}(O^p(L_\beta))$ and Z_α are a direct sum of two natural modules.*

Proof As in the proof of Lemma 7.13, we choose $\alpha'+1 \in \Delta(\alpha')$ with $[Z_{\alpha'+1}, Z_\alpha \cap Q_{\alpha'}] \neq \{1\}$, and set $r_\beta := |Z_{\alpha'+1}Q_\beta/Q_\beta|$, $r_\alpha := |(Z_{\alpha'+1} \cap Q_\beta)Q_\alpha/Q_\alpha|$, $r_{\alpha'} = |Z_\alpha Q_{\alpha'}/Q_{\alpha'}|$ and $r_{\alpha'+1} := |(Z_\alpha \cap Q_{\alpha'})Q_{\alpha'+1}/Q_{\alpha'+1}|$ so that $r_\lambda \leq q_\lambda$ for $\lambda \in \{\alpha, \beta, \alpha', \alpha'+1\}$. Since $r_\beta > p$, we have that

(1) $m_p(S/Q_\beta) > 1$ and $\overline{L_\beta}/O_{p'}(\overline{L_\beta}) \cong \mathrm{PSL}_2(q_\beta), \mathrm{PSU}_3(q_\beta)$ or $\mathrm{Sz}(q_\beta)$ and $|V_\beta/V_\beta \cap Q_{\alpha'} \cap Q_{\alpha'+1}| \leqslant q_\alpha q_\beta$.

Moreover, V_β contains at least two non-central chief factors for L_β by Lemma 7.4. As intimated in Lemma 7.13, since $r_{\alpha'} > p$, we have that $q_\alpha \geqslant q_\beta$. We begin by assuming

(⋆) if V_β contains exactly two non-central chief factors then, as $\overline{L_\beta}$-modules, both chief factors are not FF-modules.

Note that if $r_\beta > q_\beta^{\frac{1}{2}}$ then as a subgroup of V_β of index $q_\alpha q_\beta$ is centralized by $Z_{\alpha'+1}$ and V_β contains at least two non-central chief factors, applying Propositions 2.31 and 2.32, a subgroup of a non-central L_β-chief factor of index at most q_α is centralized by $Z_{\alpha'+1}$. Applying Lemma 2.55, by (⋆) we infer that $r_\beta^2 \leqslant q_\alpha$ with equality yielding $|(V_\beta \cap Q_{\alpha'})Q_{\alpha'+1}| = q_\alpha$. Likewise, if $r_\beta \leqslant q_\beta^{\frac{1}{2}}$ then as $q_\beta \leqslant q_\alpha$, $r_\beta^2 \leqslant q_\alpha$. Hence, $p^4 \leqslant r_\beta^2 \leqslant q_\alpha$. In a similar manner, we also have that $r_{\alpha'}^2 \leqslant q_\alpha$. Thus

(2) if (⋆) then $r_\beta^2 \leqslant q_\alpha \geqslant r_{\alpha'}^2$.

We observe first that if $r_\alpha = 1$ then a subgroup of $Z_{\alpha'+1}$ of index at most $q_\alpha^{\frac{1}{2}}$ is centralized by $Z_\alpha \cap Q_{\alpha'}$ and we obtain a contradiction by Propositions 2.31 and 2.32. Hence, $r_\alpha > 1$.

If $r_{\alpha'+1} > p$, then Propositions 2.31 and 2.32 imply that $r_\alpha r_\beta \geqslant q_\alpha$ and $\overline{L_{\alpha'+1}}/O_{p'}(\overline{L_{\alpha'+1}}) \cong \mathrm{PSL}_2(q_\alpha)$. Moreover, we also have that $r_\alpha > p$ and $r_{\alpha'}r_{\alpha'+1} \geqslant q_\alpha$ in this case. Instead assuming that $r_\alpha > p$ yields a similar result. In fact, by Lemma 2.55, using that Z_α is not a natural module, we have that $r_{\alpha'} = r_{\alpha'+1} = r_\beta = r_\alpha = q_\alpha^{\frac{1}{2}}$ or $r_\alpha > q_\alpha^{\frac{1}{2}} < r_{\alpha'+1}$. In the latter case, since Z_α is not a natural $\mathrm{SL}_2(q_\alpha)$-module, applying Lemma 2.55, Propositions 2.31 and 2.32, we deduce that $r_\alpha^2 \leqslant r_{\alpha'}r_{\alpha'+1}$ and $r_{\alpha'+1}^2 \leqslant r_\alpha r_\beta$. Then $r_\alpha^4 \leqslant r_{\alpha'}^2 r_{\alpha'+1}^2 \leqslant r_{\alpha'}^2 r_\alpha r_\beta$ so that $r_\alpha^3 \leqslant r_{\alpha'}^2 r_\beta$. But then, $q_\alpha^{\frac{3}{2}} < r_\alpha^3 \leqslant r_{\alpha'}^2 r_\beta \leqslant q_\alpha^{\frac{3}{2}}$, a contradiction. Hence, $r_{\alpha'} = r_{\alpha'+1} = r_\beta = r_\alpha = q_\alpha^{\frac{1}{2}}$. Note that by Propositions 2.31 and 2.32, if

$q_\alpha = q_\beta$, then ($\star$) gives that $S = (V_{\alpha'} \cap Q_\beta)Q_\alpha$ acts quadratically on Z_α. If $q_\alpha > q_\beta$, then since $r_\beta = q_\alpha^{\frac{1}{2}}$, $q_\beta^2 \geqslant q_\alpha$. As so a subgroup of index at most q_β^3 in V_β is centralized by $Z_{\alpha'+1}$. Applying Lemma 2.55 since $r_\beta > q_\beta^{\frac{1}{2}}$, and using ($\star$) we see that $r_\beta = q_\beta = q_\alpha^2$ and $S = (V_{\alpha'} \cap Q_\beta)Q_\alpha$. Once again, we have that $S = (V_{\alpha'} \cap Q_\beta)Q_\alpha$ acts quadratically on Z_α. Summed up, we have

(3) if ($\star$) and $r_{\alpha'+1} > p$ then $\overline{L_\alpha}/O_{p'}(\overline{L_\alpha}) \cong \mathrm{PSL}_2(q_\alpha)$, $r_{\alpha'} = r_{\alpha'+1} = r_\beta = r_\alpha = q_\alpha^{\frac{1}{2}}$, S is quadratic on Z_α and either $q_\alpha = q_\beta$ or $q_\alpha = q_\beta^2$.

Assume that $r_\alpha = r_{\alpha'+1} = p$. Then an index $r_\beta p \leqslant q_\alpha^{\frac{1}{2}} p$ subgroup of Z_α is centralized by $Z_{\alpha'+1} \cap Q_\beta$, and we deduce by Propositions 2.31 and 2.32 that $\overline{L_\alpha}/O_{p'}(\overline{L_\alpha}) \cong \mathrm{PSL}_2(q_\alpha)$ and $q_\alpha \leqslant p^6$. Note that if $r_\beta^2 < q_\alpha$ then $r_\beta^2 p \leqslant q_\alpha$ and since $q_\alpha^{2/3} \leqslant r_\beta p$ by Propositions 2.31 and 2.32, we have that $q_\alpha^{2/3} r_\beta \leqslant r_\beta^2 p \leqslant q_\alpha$ so that $r_\beta \leqslant q_\alpha^{1/3}$. Since $r_\beta > p$, we have that $r_\beta = p^2$ and $q_\alpha = p^6$. But then a subgroup of index $r_\beta p = p^3$ of $Z_{\alpha'+1}$ is centralized by $Z_\alpha \cap Q_{\alpha'}$ and we have a contradiction by Propositions 2.31 and 2.32. Hence, $r_\beta^2 = q_\alpha$. If $q_\beta < q_\alpha$ then $r_\beta > q_\beta^{\frac{1}{2}}$ and by an earlier observation, $S = (V_{\alpha'} \cap Q_\beta)Q_\alpha$ acts quadratically on Z_α. If $q_\beta = q_\alpha$ then $r_\beta > p$ and by ($\star$), applying Proposition 2.31 and Proposition 2.32, $(V_\beta \cap Q_{\alpha'})Q_{\alpha'+1} \in \mathrm{Syl}_p(L_{\alpha'+1})$ and conjugating to α, S acts quadratically on Z_α. Hence,

(4) if ($\star$) and $r_{\alpha'+1} = p$ then $\overline{L_\alpha}/O_{p'}(\overline{L_\alpha}) \cong \mathrm{PSL}_2(q_\alpha)$, $r_{\alpha'} = r_{\alpha'+1} = p, r_\beta^2 = q_\alpha$ and S is quadratic on Z_α.

In all cases, $[Z_\alpha, Q_\beta] \leqslant [Z_\alpha, S] \le Z_\beta$ and $\overline{L_\alpha}/O_{p'}(\overline{L_\alpha}) \cong \mathrm{PSL}_2(q_\alpha)$. Forming L^α from Q_α and d conjugates of $Z_{\alpha'+1} \cap Q_\beta$ where $d = 2$ when $r_\alpha > p$ and $d = 3$ when $r_\alpha = p$, we have that $\overline{L^\alpha}/O_{p'}(\overline{L^\alpha}) \cong \overline{L_\alpha}/O_{p'}(\overline{L_\alpha})$, $S \in \mathrm{Syl}_p(L^\alpha)$ and $|Z_\alpha/C_{Z_\alpha}(L^\alpha)| \leqslant q_\alpha^2$ when $d = 2$ and $|Z_\alpha/C_{Z_\alpha}(L^\alpha)| \leqslant r_\beta^3 p^3$ when $d = 3$. Since $O^p(L^\alpha)$ acts non-trivially on Z_α, $C_{Z_\alpha}(L^\alpha) < C_{Z_\alpha}(L^\alpha)[Z_\alpha, Q_\beta] \le Z_\beta$ and applying Lemma 2.41, we have that $|[Z_\alpha, Q_\beta]C_{Z_\alpha}(L^\alpha)/C_{Z_\alpha}(L^\alpha)| \geqslant q_\alpha$. But then Z_β has index at most q_α when $d = 2$ and index $r_\beta^3 p^3/q_\alpha < q_\alpha^2$ when $d = 3$ and by Lemma 2.55, we deduce that Z_α is a natural $\mathrm{SL}_2(q_\alpha)$-module for $L_\alpha/R_\alpha \cong \mathrm{SL}_2(q_\alpha)$, a contradiction. Therefore,

(5) V_β contains exactly two non-central chief factors for L_β and both, as $\overline{L_\beta}$-modules, are FF-modules.

In particular, we have that $L_\beta/R_\beta Q_\beta \cong \mathrm{SL}_2(q_\beta)$. By Lemma 5.8, we infer that $[V_\beta, Q_\beta, Q_\beta]$ contains no non-central chief factors for L_β and that $[Z_\alpha, Q_\beta, Q_\beta] \le C_{V_\beta}(O^p(L_\beta))$.

Assume that $[V_\beta, Q_\beta]$ contains a non-central chief factor for L_β. Note that $[Z_\alpha, V_{\alpha'} \cap Q_\beta]Z_\beta$ is normalized by $L_\beta = \langle V_{\alpha'}, Q_\alpha, R_\beta\rangle$ and we deduce that $[Z_\alpha, V_{\alpha'} \cap Q_\beta]Z_\beta \le C_{V_\beta}(O^p(L_\beta))$. Indeed, we ascertain that $[Z_\alpha, V_{\alpha'} \cap Q_\beta, Q_\beta \cap O^p(L_\beta)] = \{1\}$ and if $\overline{L_\alpha}/O_{p'}(\overline{L_\alpha}) \cong \mathrm{PSL}_2(q_\alpha)$, then taking the closure of this commutator under the action of $G_{\alpha,\beta}$ yields $[Z_\alpha, S, S] = \{1\}$ and $[Z_\alpha, Q_\beta] \leq Z_\beta$, a contradiction. Hence,

(6) if $[V_\beta, Q_\beta]$ contains a non-central chief factor for L_β then $\overline{L_\alpha}/O_{p'}(\overline{L_\alpha}) \not\cong \mathrm{PSL}_2(q_\alpha)$.

Without loss of generality, we may assume that $r_\alpha \leqslant r_{\alpha'+1}$ and an index $q_\beta r_\alpha$ subgroup of $Z_{\alpha'+1}$ is centralized by $Z_\alpha \cap Q_{\alpha'}$. Applying Propositions 2.31 and 2.32, we infer that $r_\alpha = r_{\alpha'+1} = p$ and $q_\alpha = q_\beta = r_\beta = r_{\alpha'}$; or $q_\alpha = q_\beta = r_\alpha = r_\beta = r_{\alpha'} = r_{\alpha'+1}$.

In the former case, we have that an index $pq_\beta < q_\alpha^2$ subgroup of Z_α is centralized by $X := \langle Z_{\alpha'+1}, Q_\alpha, O^p(R_\beta)\rangle$ and $|(X \cap Q_\beta)Q_\alpha/Q_\alpha| > p$. Then applying Proposition 2.31 and Proposition 2.32 yields a contradiction.

In the latter case, we have by Lemma 2.55 that Z_α is a natural module for $L_\alpha/R_\alpha \cong \mathrm{SU}_3(q_\alpha)$ or $\mathrm{Sz}(q_\alpha)$. If Z_α is a natural $\mathrm{SU}_3(q_\alpha)$-module then $|Z_\alpha/[Z_\alpha, Q_\beta]| = q_\alpha^2$ and since $V_\beta/[V_\beta, Q_\beta]$ contains a unique non-central chief factor which is a natural $\mathrm{SL}_2(q_\beta)$-module, we have a contradiction. Hence, Z_α is a natural $\mathrm{Sz}(q_\alpha)$-module. Using a Frattini argument, we can choose an element $k \in G_{\alpha,\beta} \cap L_\beta$ such that kR_β/R_β has order $q_\beta - 1$. Then k acts on L_α/R_α and since the outerautomorphism group of $\mathrm{Sz}(q_\alpha)$ consists of field automorphisms, and using that k centralizes $[Z_\alpha, S, S]$, we see that some power of k centralizes L_α/R_α and so centralizes S/Q_α. Choose such an element and call it x. Since x centralizes $Z_\beta = C_{Z_\alpha}(S)$, apply the A×B-lemma implies that x centralizes Z_α, a contradiction since the non-central chief factors with V_β are natural $\mathrm{SL}_2(q_\beta)$-modules, on which x acts fixed point freely by Proposition 2.21 (v). Hence

(7) $[V_\beta, Q_\beta]$ contains no non-central chief factors for L_β, $[V_\beta, Q_\beta] \leq C_{V_\beta}(O^p(L_\beta))$ and $Q_\beta \in \mathrm{Syl}_p(R_\beta)$.

Then $Z_\alpha \cap C_{V_\beta}(O^p(L_\beta))$ has index q_β^2 in Z_α and is centralized by $Q_\beta \cap O^p(L_\beta)$. Since $q_\alpha \geqslant q_\beta$, and Z_α is not a natural module, we infer that $q_\alpha = q_\beta$.

Moreover, it follows that $Z_\alpha Q_{\alpha'} \in \mathrm{Syl}_p(L_{\alpha'})$ and $(V_{\alpha'} \cap Q_\beta)Q_\alpha$ acts quadratically on Z_α. Using Lemma 2.55 and observing that $[Z_\alpha, Q_\beta, Q_\beta \cap O^p(L_\beta)] \neq \{1\}$ when Z_α is a natural $\mathrm{Sz}(q_\alpha)$-module, to complete the proof we may assume that Z_α is a natural $\mathrm{SU}_3(q_\alpha)$-module. But then, $[Z_\alpha \cap Q_{\alpha'}, Z_{\alpha'+1} \cap Q_\beta]$ has order q_α^2 and is contained in $Z_\beta \cap Z_{\alpha'}$ from which it follows that $Z_\beta = Z_{\alpha'}$. But then $[V_\beta \cap Q_{\alpha'}, V_{\alpha'}] = Z_{\alpha'} = Z_\beta$, a contradiction by Lemmas 2.41 and 7.9. This completes the proof. □

Lemma 7.15 *Suppose that $b > 1$ and $C_{V_\beta}(V_{\alpha'}) < V_\beta \cap Q_{\alpha'}$. If Z_α is a natural module for $L_\alpha/R_\alpha \cong \mathrm{SL}_2(q_\alpha)$ and $q_\alpha \neq q_\beta$, then we may assume that $Z_{\alpha'} \cap Z_\beta = \{1\}$.*

Proof We first observe that if $Z_\beta = Z_{\alpha'}$ then $[V_\beta \cap Q_{\alpha'}, V_{\alpha'}] \leq Z_{\alpha'} = Z_\beta \geq [V_{\alpha'} \cap Q_\beta, V_{\alpha'}]$ and we conclude that V_β/Z_β is an FF-module for $L_\beta/R_\beta \cong \mathrm{SL}_2(q_\beta)$. This is a contradiction by Lemma 7.9. We further note that every critical pair satisfies the same initial hypothesis as (α, α'). We aim to observe that for at least one critical pair $(\alpha^*, (\alpha^*)')$ we have that $Z_{\beta^*} \cap Z_{(\alpha^*)'} = \{1\}$. We assume throughout, aiming for a contradiction, that $\{1\} \neq H := Z_\beta \cap Z_{\alpha'} < Z_\beta$. Note that immediately we have that $q_\alpha > p$.

If $b = 3$, then since $Z_{\alpha+2}$ is a natural module, we deduce that $H \trianglelefteq \langle R_{\alpha+2}, Q_\beta, Q_{\alpha'}\rangle = L_{\alpha+2}$, a contradiction since $Z(L_{\alpha+2}) = \{1\}$. Hence, we have shown that

(1) if $\{1\} \neq Z_\beta \cap Z_{\alpha'} < Z_\beta$ then $b \geqslant 5$.

We note that if $V_\beta^{(3)} \not\leq Q_{\alpha'-2}$ then $(\beta-3, \alpha'-2)$ is critical pair. Similarly, if $V_{\beta-2}^{(3)} \not\leq Q_{\alpha'-4}$, then $((\beta-2)-3, \alpha'-4)$ is also a critical pair. Observe that if every critical pair $(\alpha^*, (\alpha^*)')$ has $V_{\beta^*}^{(3)} \not\leq Q_{(\alpha^*)'-2}$ and $Z_{(\alpha^*)'} = Z_{(\alpha^*)'-2}$ then iterating backwards throughout critical pairs, we have that $Z_{\alpha'} = Z_\beta$. Hence, relabeling if necessary, we may assume that the critical pair (α, α') satisfies either $V_\beta^{(3)} \leq Q_{\alpha'-2}$ or $Z_{\alpha'} \neq Z_{\alpha'-2}$.

Assume first that $Z_{\alpha'} = Z_{\alpha'-2}$ so that $V_\beta^{(3)} \leq Q_{\alpha'-2}$. If $b = 5$, then $H \leq Z_\beta \cap Z_{\alpha'-2}$ and so we conclude that $Z_{\alpha'} = Z_{\alpha'-2} = Z_\beta$, a contradiction. Therefore, we may proceed under the added assumption that $b \geqslant 7$. Suppose that $V_\beta^{(3)} \not\leq Q_{\alpha'-1}$ and so there is $\beta - 2 \in \Delta^{(2)}(\beta)$ with $V_{\beta-2} \not\leq Q_{\alpha'-1}$. Since $V_{\beta-2} \leq V_\beta^{(3)} \leq Q_{\alpha'-2}$, $(\alpha'-1, \beta-2)$ is not a critical pair by Proposition 7.11. Hence, $Z_{\alpha'} = Z_{\alpha'-2} = [V_{\beta-2}, Z_{\alpha'-1}] = Z_{\beta-2}$. Then $\{1\} \neq H = Z_{\beta-2} \cap Z_\beta$, a contradiction.

Thus $V_\beta^{(3)} \leq Q_{\alpha'-1}$. Since $b \geqslant 7$, $[V_{\alpha'}, V_\beta^{(3)}, V_\beta^{(3)}] = \{1\}$ and so we have that $|V_\beta^{(3)} Q_{\alpha'}/Q_{\alpha'}| \leqslant q_\beta$. Indeed, $V_\beta(V_\beta \cap Q_{\alpha'})$ has index strictly less than q_β in $V_\beta^{(3)}$ and is centralized, modulo V_β, by $V_{\alpha'}$. Applying Propositions 2.31 and 2.32, we conclude that $|V_{\alpha'} Q_\beta/Q_\beta| = p$ and $\overline{L_\beta}/O_{p'}(\overline{L_\beta}) \cong \mathrm{PSL}_2(q_\beta)$ where $q_\beta > p$. Let $X \leq L_\beta$ be such that X is generated by Q_β and three conjugates of $V_{\alpha'}$ with the property that $\overline{X} O_{p'}(\overline{L_\beta}) = \overline{L_\beta}$. Note that $[V_\beta^{(3)}, V_{\alpha'}, V_{\alpha'}] = [V_\beta^{(3)}, V_{\alpha'-2}^{(3)}, V_{\alpha'-2}^{(3)}] = \{1\}$ as $b \geqslant 7$, and so applying Lemmas 2.44 and 2.43, and using that a subgroup of $V_\beta^{(3)}/V_\beta$ of index strictly less than q_β, we conclude that $O^p(X)$ centralizes $V_\beta^{(3)}/V_\beta$. It then follows that $O^p(L_\beta)$ centralizes $V_\beta^{(3)}/V_\beta$, a contradiction. Thus,

(2) if $\{1\} \neq Z_\beta \cap Z_{\alpha'} < Z_\beta$ then $b \geqslant 5$ and $Z_{\alpha'} \neq Z_{\alpha'-2}$.

Now, $V_\beta^{(3)} \cap Q_{\alpha'-2}$ centralizes $HZ_{\alpha'-2} \leq Z_{\alpha'-1}$ and so we must have that $V_\beta^{(3)} \cap Q_{\alpha'-2} \leq Q_{\alpha'-1}$. By a similar argument as before, we deduce that $V_\beta^{(3)} \not\leq Q_{\alpha'-2}$. Indeed, this condition holds for any critical pair. If $V_\alpha^{(2)} \not\leq Q_{\alpha'-2}$, then $(\alpha-2, \alpha'-2)$ is also a critical pair, and by the standing assumption, we have that $\{1\} \neq Z_{\alpha-1} \cap Z_{\alpha'-2} \leq Z_\alpha \cap Z_{\alpha'-1}$. It follows that $Z_{\alpha-1} \cap Z_{\alpha'-2} \leq Z_\beta$ so that $Z_{\alpha-1} = Z_\beta$. Again, iterating backwards through critical pairs we will eventually reach a situation where we have a critical pair $(\alpha^*, (\alpha^*)')$ with $Z_{\beta^*} = Z_{(\alpha^*)'}$, a contradiction. Hence, again up to relabeling, we may assume that $V_\alpha^{(2)} \leq Q_{\alpha'-1}$.

Now, if $q_\alpha > q_\beta$, then since Z_α acts quadratically on $V_{\alpha'}$, we deduce that $Z_\alpha \cap Q_{\alpha'} \not\leq Q_{\alpha'+1}$. Similarly, $Z_{\alpha'+1} \cap Q_\beta \not\leq Q_\alpha$ and so $Z_\alpha \cap Q_{\alpha'} \cap Q_{\alpha'+1} = Z_\beta$.

Hence $Z_\alpha(V_\alpha^{(2)} \cap Q_{\alpha'+1})$ has index q_β in $V_\alpha^{(2)}$ and is centralized by $Z_{\alpha'+1} \cap Q_\beta$. By Proposition 2.31 and Proposition 2.32, we have that $|(Z_{\alpha'+1} \cap Q_\beta)Q_\alpha/Q_\alpha| = p$ and $\overline{L_\alpha}/O_{p'}(\overline{L_\alpha}) \cong \mathrm{PSL}_2(q_\alpha)$. Set $X \le L_\alpha$ generated by Q_α and three conjugates of $Z_{\alpha'+1} \cap Q_\beta$ so that $\overline{X}O_{p'}(\overline{L_\alpha}) = \overline{L_\alpha}$ and $O^p(X)$ does not centralize $V_\alpha^{(2)}/Z_\alpha$. Then $O^p(X)$ centralizes a subgroup of $V_\alpha^{(2)}/Z_\alpha$ of index at most q_β^3. By Lemma 2.44, using that $V_{\alpha'} \cap Q_\beta$ is quadratic on $V_\alpha^{(2)}$, we deduce that $O^p(R_\alpha)$ centralizes $V_\alpha^{(2)}/Z_\alpha$. Then by Lemma 2.43, using that an index q_β subgroup of $V_\alpha^{(2)}/Z_\alpha$ is centralized by $Z_{\alpha'+1} \cap Q_\beta$ and that $q_\beta < q_\alpha$, we have a contradiction. Hence, we have that

(3) $V_\alpha^{(2)} \le Q_{\alpha'-1}$, $V_\beta^{(3)} \not\le Q_{\alpha'-2}$, $V_\beta^{(3)} \cap Q_{\alpha'-2} \le Q_{\alpha'-1}$ and $q_\alpha < q_\beta$.

Assume first that $|Z_{\alpha'+1}Q_\beta/Q_\beta| = p$. Then $Z_{\alpha'+1} \cap Q_\beta$ is a maximal subgroup of $Z_{\alpha'+1}$ and $C_{Z_{\alpha'+1} \cap Q_\beta}(V_\beta \cap Q_{\alpha'}) = Z_{\alpha'}$. It follows that $[Z_{\alpha'+1} \cap Q_\beta, V_\beta \cap Q_{\alpha'}] = H$ has index p in Z_β, and index p in $Z_{\alpha'}$. But now, $[V_\beta \cap Q_{\alpha'}, V_{\alpha'}] \le Z_{\alpha'}$ and since $V_\beta \cap Q_{\alpha'}$ has index at most q_β in V_β, we have that for some $x \in V_{\alpha'} \setminus (V_{\alpha'} \cap Q_\beta)$, $C_{V_\beta/Z_\beta}(x)$ has index at most pq_β in V_β/Z_β. Since $p < q_\alpha < q_\beta$, by Propositions 2.31 and 2.32, we deduce that $\overline{L_\beta}/O_{p'}(\overline{L_\beta}) \cong \mathrm{PSL}_2(q_\beta)$; or $q_\beta = 8$, $q_\alpha = 4$ and $\overline{L_\beta}/O_{2'}(\overline{L_\beta}) \cong \mathrm{Sz}(8)$. In the latter case, we necessarily have $V_\beta \cap Q_{\alpha'}$ has index exactly q_β in V_β.

More generally, assume that $|V_\beta Q_{\alpha'}/Q_{\alpha'}| = q_\beta$. Since $V_\alpha^{(2)} \le Q_{\alpha'-1}$ and $V_\alpha^{(2)}$ is elementary abelian, we conclude that $V_\alpha^{(2)} = V_\beta(V_\alpha^{(2)} \cap Q_{\alpha'})$. But then $[V_\alpha^{(2)}, V_{\alpha'} \cap Q_\beta] = [V_\beta(V_\alpha^{(2)} \cap Q_{\alpha'}), V_{\alpha'} \cap Q_\beta] \le Z_{\alpha'}Z_\beta$. Since $|Z_{\alpha'}Z_\alpha/Z_\alpha| = p$ and $V_\alpha^{(2)}/Z_\alpha$ contains a non-central chief factor for $\overline{L_\alpha}$, Lemma 2.41 yields a contradiction as $q_\alpha > p$.

Hence, a subgroup of V_β/Z_β of index at most q_β is centralized by $x \in V_{\alpha'} \setminus (V_{\alpha'} \cap Q_\beta)$ and $\overline{L_\beta}/O_{p'}(\overline{L_\beta}) \cong \mathrm{PSL}_2(q_\beta)$. Forming $X \le L_\beta$ generated by Q_β and three conjugates of x with the property that $\overline{X}O_{p'}(\overline{L_\beta}) \cong \overline{L_\beta}$, we have that V_β/Z_β contains a non-central chief factor for X and X centralizes a subgroup of V_β/Z_β of index at most q_β^3. Since $V_{\alpha'}$ is quadratic on V_β, we deduce by Lemmas 2.43 and 2.44, that the non-central X-chief factor with V_β/Z_β is either a natural $\mathrm{SL}_2(q_\beta)$-module; or an $\Omega_4^-(q_\beta^{\frac{1}{2}})$-module. Either way, we deduce that x centralizes a subgroup of V_β/Z_β of index exactly q_β and so $|V_\beta Q_{\alpha'}/Q_{\alpha'}| = q_\beta/p$. Since V_β acts quadratically on $V_{\alpha'}$, if the non-central chief factor is an $\Omega_4^-(q_\beta^{\frac{1}{2}})$-module then we must have that $q_\beta = p^2$, a contradiction since $p < q_\alpha < q_\beta$.

If the non-central chief factor is a natural module, then we deduce that $[V_\beta/Z_\beta, P, P] = \{1\}$ for $P \in \mathrm{Syl}_p(X) \subseteq \mathrm{Syl}_p(L_\beta)$ so that $[V_\beta/Z_\beta, P]C_{V_\beta/Z_\beta}(X) \le C_{V_\beta/Z_\beta}(P)$ and $[V_\beta/Z_\beta, P]C_{V_\beta/Z_\beta}(X)$ has index at most q_β^2 in V_β/Z_β. Hence, V_β/Z_β is determined by Lemma 2.55 and since an element of $Z_{\alpha'+1} \setminus (Z_{\alpha'+1} \cap Q_\beta)$ centralizes a subgroup of V_β/Z_β of index strictly less than q_β^2, using that $V_\beta = \langle Z_\alpha^{L_\beta} \rangle$ and Lemma 2.30, we conclude that $V_\beta/C_{V_\beta}(O^p(L_\beta))$ is a natural module for $L_\beta/R_\beta \cong \mathrm{SL}_2(q_\beta)$. Since Z_α is $G_{\alpha,\beta}$-invariant, we must have that

$Z_\beta = Z_\alpha \cap C_{V_\beta}(O^p(L_\beta))$ has index q_β in Z_α, impossible since Z_β has index q_α in Z_α and $q_\alpha < q_\beta$.

Thus, we have that

(4) $|Z_{\alpha'+1}Q_\beta/Q_\beta| > p$.

Since $q_\alpha < q_\beta$ and $|Z_{\alpha'+1}Q_\beta/Q_\beta| \geqslant p^2$, Propositions 2.31 and 2.32 yield that $\overline{L_\beta}/O_{p'}(\overline{L_\beta}) \cong \mathrm{PSL}_2(q_\beta)$.

Suppose that $b = 5$. Then $Z_{\alpha'} = [V_\beta \cap Q_{\alpha'}, V_{\alpha'}] \leq [V_{\alpha'-2}^{(3)}, V_{\alpha'-2}^{(3)}]$. Since $[V_{\alpha'-2}^{(3)}, V_{\alpha'-2}^{(3)}]$ is normalized by $G_{\alpha'-2}$, we deduce that $V_{\alpha'-2} \leq [V_{\alpha'-2}^{(3)}, V_{\alpha'-2}^{(3)}]$. Assume that $[V_\beta^{(3)}, Q_\beta]/V_\beta$ has no non-central chief factor for L_β. Then $[V_\beta^{(3)}, Q_\beta] = [V_\alpha^{(2)}, Q_\beta]V_\beta$ is centralized by $V_\alpha^{(2)}$ and it follows that $[V_\beta^{(3)}, Q_\beta, V_\beta^{(3)}] = \{1\}$. By the three subgroups lemma, we deduce that $V_\beta \leq [V_\beta^{(3)}, V_\beta^{(3)}] \leq Z(Q_\beta)$, a contradiction. Thus, applying Lemma 5.8, $V_\beta^{(3)}$ contains at least two non-central chief factors for $\overline{L_\beta}$. Now, $V_{\alpha'}$ centralizes $V_\beta(V_\beta^{(3)} \cap Q_{\alpha'})$, modulo V_β, which has index strictly less than q_β^2 in $V_\beta^{(3)}$. An application of Propositions 2.31 and 2.32 provide a contradiction. Therefore

(5) $b \geqslant 7$.

Now, since $V_\beta^{(3)}$ acts quadratically on $V_{\alpha'-2}$ and $V_{\alpha'}$ (using that $b \geqslant 7$), we deduce that $V_\beta^{(3)} \cap \dots \cap Q_{\alpha'+1}$ has index at $q_\beta^2 q_\alpha$ in $V_\beta^{(3)}$. Moreover, both $V_\beta^{(3)}/V_\beta$ and V_β/V_β contain a non-central chief factor for $\overline{L_\beta}$. Our first aim will be to demonstrate that $Z_{\alpha'+1}$ centralizes a subgroup of index at most $q_\beta^{\frac{3}{2}}$ in V_β/Z_β.

Assume that a subgroup of $V_\beta^{(3)}/V_\beta$ of index at most $q_\beta^{\frac{3}{2}}$ is centralized by $Z_{\alpha'+1}$. Form $X \leq L_\beta$ from Q_β and two conjugates of $Z_{\alpha'+1}$ so that $\overline{X}O_{p'}(\overline{L_\beta}) \cong \overline{L_\beta}$ and $O^p(X)$ centralizes a subgroup of $V_\beta^{(3)}/V_\beta$ of index at most q_β^3. Since $V_{\alpha'-2}^{(3)}$ acts quadratically on $V_\beta^{(3)}$, applying Lemmas 2.44 and 2.43 we deduce that $V_\beta^{(3)}/V_\beta$ contains a unique non-central chief factor for $\overline{X}$ and then quadratic action reveals that this non-central chief factor is either a natural $\mathrm{SL}_2(q_\beta)$-module, or has the structure of an $\Omega_4^-(q_\beta^{\frac{1}{2}})$-module.

In the latter case, since V_β acts quadratically on $V_{\alpha'}^{(3)}$ we conclude that $V_\beta \cap Q_{\alpha'} \cap Q_{\alpha'+1}$ has index at most $q_\beta^{\frac{1}{2}} q_\alpha < q_\beta^{\frac{3}{2}}$.

In the former case, writing $V := V_\beta^{(3)}/[V_\beta^{(3)}, Q_\beta]$, we have that V contains the unique non-central X-chief factor within $V_\beta^{(3)}/V_\beta$ by Lemma 5.8. Then we deduce that $[V, P, P] = \{1\}$ for $P \in \mathrm{Syl}_p(X) \subseteq \mathrm{Syl}_p(L_\beta)$ so that $[V, P]C_V(X) \leq C_V(P)$ and $[V, P]C_V(X)$ has index at most q_β^2 in V. Hence, V is determined by Lemma 2.55 and since an element of $Z_{\alpha'} \setminus (Z_{\alpha'+1} \cap Q_\beta)$ centralizes a subgroup of V of index strictly less than q_β^2, V contains a unique non-central chief factor for L_β, which has the structure of a natural $\mathrm{SL}_2(q_\beta)$-module. Now, writing $W := V/C_V(L_\beta)$, we may have that $[W, L_\beta] < W$ so that $p = 2$.

Assume that $r := |W/[W, L_\beta]| \neq 1$ so that for $x \in Z_{\alpha'+1} \setminus (Z_{\alpha'+1} \cap Q_\beta)$, we have that $|C_W(x)| = rq_\beta$. Then, for some $g \in L_\beta$, we have that $M := \langle x, x^g\rangle C_{L_\beta}(W)/C_{L_\beta}(W) \cong \mathrm{Dih}(2k)$ for some odd k, $L_\beta = \langle x, x^g, Z_{\alpha'+1}\rangle C_{L_\beta}(W)$ and $|C_W(M)| \geqslant r$. Moreover, $W = C_W(M)[W, L_\beta]$. Indeed, $C_W(x) = C_W(M)C_W(Z_{\alpha'+1})$ so that $q_\beta = |C_W(P)| \leqslant |C_W(Z_{\alpha'+1})| = |C_W(x)/C_W(M)| \leqslant q_\beta$ for some appropriate $P \in \mathrm{Syl}_2(L_\beta)$ and we deduce that $|C_W(Z_{\alpha'+1})| = q_\beta$. But $Z_{\alpha'+1}$ centralizes a subgroup of $V_\beta^{(3)}$ of index at most $q_\beta^{\frac{3}{2}}$ in $V_\beta^{(3)}$. Hence, $|W| \leqslant q_\beta^{\frac{5}{2}}$.

Whether $r = 1$ or not, writing U_β for the preimage in $V_\beta^{(3)}$ of $C_V(L_\beta)$, we have that $[U_\beta, O^p(L_\beta)] \leq V_\beta$ and $V_\alpha^{(2)} \cap U_\beta$ has index at most $q_\beta^{\frac{3}{2}}$ in $V_\alpha^{(2)}$. Indeed, for an appropriately defined $U_{\alpha+3}$, we have that $V_{\alpha+2}^{(2)} \cap U_{\alpha+3} = V_{\alpha+4}^{(2)} \cap U_{\alpha+3} \leq Q_{\alpha'+1}$ and so a subgroup of V_β of index at most $q_\beta^{\frac{3}{2}}$ is centralized by $Z_{\alpha'+1}$.

Since $q_\alpha < q_\beta$, we have shown in all cases that

(6) $Z_{\alpha'+1}$ centralizes a subgroup of index at most $q_\beta^{\frac{3}{2}}$ in V_β/Z_β.

We form X as before, and applying Lemmas 2.43 and 2.44, this time recognizing that V_β/Z_β contains a unique non-central chief factor for $\overline{X}$ which is either a natural $\mathrm{SL}_2(q_\beta)$-module; or has the structure of an $\Omega_4^-(q_\beta^{\frac{1}{2}})$-module.

In the former case, we deduce that $[V_\beta/Z_\beta, P, P] = \{1\}$ for $P \in \mathrm{Syl}_p(X) \subseteq \mathrm{Syl}_p(L_\beta)$ so that $[V_\beta/Z_\beta, P]C_{V_\beta/Z_\beta}(X) \leq C_{V_\beta/Z_\beta}(P)$ and $[V_\beta/Z_\beta, P]C_{V_\beta/Z_\beta}(X)$ has index at most q_β^2 in V_β/Z_β. Hence, V_β/Z_β is determined by Lemma 2.55 and since an element of $Z_{\alpha'+1} \setminus (Z_{\alpha'+1} \cap Q_\beta)$ centralizes a subgroup of V_β/Z_β of index strictly less than q_β^2, using that $V_\beta = \langle Z_\alpha^{L_\beta}\rangle$ and Lemma 2.30, we conclude that $V_\beta/C_{V_\beta}(O^p(L_\beta))$ is a natural module for $L_\beta/R_\beta \cong \mathrm{SL}_2(q_\beta)$. Since Z_α is $G_{\alpha,\beta}$-invariant, we must have that $Z_\beta = Z_\alpha \cap C_{V_\beta}(O^p(L_\beta))$ has index q_β in Z_α, impossible since Z_β has index q_α in Z_α and $q_\alpha < q_\beta$.

In the latter case, since $V_\beta^{(3)}$ acts quadratically on $V_{\alpha'-2}$, and $V_\beta^{(3)} \cap Q_{\alpha'-1}$ acts quadratically on $V_{\alpha'}$, we have that $V_\beta^{(3)} \cap \ldots Q_{\alpha'+1}$ has index at most $q_\alpha q_\beta$ in $V_\beta^{(3)}$. Since $q_\alpha < q_\beta$, and $V_\beta^{(3)}$ contains at least two non-central chief factors, this provides a final contradiction. □

Lemma 7.16 *Suppose that $b > 1$ and $C_{V_\beta}(V_{\alpha'}) < V_\beta \cap Q_{\alpha'}$. Then $Z_{\alpha'+1} \not\leq Q_\beta$, $Q_\beta \in \mathrm{Syl}_p(R_\beta)$, $[V_\beta, Q_\beta] \leq C_{V_\beta}(O^p(L_\beta))$ and either:*

(i) *Z_α is a natural module for $L_\alpha/R_\alpha \cong \mathrm{SL}_2(q_\alpha)$ and $q_\alpha = q_\beta$;*
(ii) *$L_\beta/R_\beta \cong \mathrm{SL}_2(q_\alpha) \cong L_\alpha/R_\alpha$, $q_\alpha > p$ and both $V_\beta/C_{V_\beta}(O^p(L_\beta))$ and Z_α are a direct sum of two natural modules; or*
(iii) *$L_\beta/R_\beta \cong \mathrm{SL}_2(p)$, $(3 \times 3) : 2$, $(3 \times 3) : 4$ or $(Q_8 \times Q_8) : 3$, $V_\beta/C_{V_\beta}(O^p(L_\beta))$ is described in Proposition 7.12 and $q_\alpha = p$.*

Proof By Lemmas 7.13 and 7.14, we may assume that Z_α is a natural $\mathrm{SL}_2(q_\alpha)$-module for L_α/R_α and so it remains to show that $q_\alpha = q_\beta$.

Assume that $Z_{\alpha'} < Z_{\alpha'+1} \cap Q_\beta \not\leq Z_{\alpha'}$. Then $\{1\} \neq [Z_{\alpha'+1} \cap Q_\beta, V_\beta \cap Q_{\alpha'}] \leq Z_{\alpha'} \cap Z_\beta$, a contradiction by Lemma 7.15. More generally, we deduce in all cases that

(1) $Z_{\alpha'} \cap Z_\beta = \{1\}$, $Z_\beta = Z_\alpha \cap Q_\beta$ and $Z_{\alpha'} = Z_{\alpha'+1} \cap Q_\beta$.

Hence, $|Z_\alpha Q_{\alpha'}/Q_{\alpha'}| = |Z_{\alpha'+1}Q_\beta/Q_\beta| = q_\alpha \leqslant q_\beta$. To complete the proof we assume, aiming for a contradiction, that $q_\alpha < q_\beta$.

Since Z_α centralizes a subgroup of index at most $q_\beta q_\alpha$ in $V_{\alpha'}$, we deduce by Propositions 2.31 and 2.32 that either $\overline{L_{\alpha'}}/O_{p'}(\overline{L_{\alpha'}}) \cong \mathrm{PSL}_2(q_{\alpha'})$; or $q_\alpha = p$ and $\overline{L_{\alpha'}}/O_{p'}(\overline{L_{\alpha'}}) \cong \mathrm{PSU}_3(p^2)$ or $\mathrm{Sz}(8)$.

Assume that $q_\alpha = p$. Suppose further that $|V_{\alpha'}Q_\beta/Q_\beta| \geqslant p^2$. Then we may choose $\lambda \in \Delta(\alpha') \setminus \{\alpha' + 1\}$ such that $|Z_{\alpha'+1}Z_\lambda Q_\beta/Q_\beta| = p^2$ and $Z_{\alpha'+1}Z_\lambda$ centralizes a subgroup of index at most $q_\beta p^2$ in V_β. By Propositions 2.31 and 2.32, if $\overline{L_\beta}/O_{p'}(\overline{L_\beta}) \not\cong \mathrm{PSL}_2(q_\beta)$ then $q_\beta = p^2$ and by Lemma 2.55 we conclude that $L_\beta/R_\beta \cong \mathrm{SU}_3(p^2)$ and $V_\beta/C_{V_\beta}(O^p(L_\beta))$ is a natural module. But $Z_\alpha C_{V_\beta}(O^p(L_\beta))/C_{V_\beta}(O^p(L_\beta))$ is a $G_{\alpha,\beta}$-invariant subgroup of order $q_\alpha = p$, a contradiction. Hence, $\overline{L_\beta}/O_{p'}(\overline{L_\beta}) \cong \mathrm{PSL}_2(q_\beta)$. Set $X \leq L_\beta$ generated by Q_β and two conjugates of $Z_{\alpha'+1}Z_\lambda$ so that $\overline{X}O_{p'}(\overline{L_\beta}) = \overline{L_\beta}$, X centralizes a subgroup of V_β/Z_β of index at most $q_\beta^2 p^4$ and V_β/Z_β contains a non-central chief factor for X.

If $q_\beta = p^2$ then Lemma 2.55 reveals that $V_\beta/C_{V_\beta}(O^p(L_\beta))$ is a direct sum of natural modules for $L_\beta/R_\beta \cong \mathrm{SL}_2(q_\beta)$, impossible since $Z_\alpha C_{V_\beta}(O^p(L_\beta))/C_{V_\beta}(O^p(L_\beta))$ is a $G_{\alpha,\beta}$-invariant subgroup of order p. If $q_\beta = p^3$, then either $V_\beta Q_{\alpha'} \in \mathrm{Syl}_p(L_{\alpha'})$ in which case there is $\mu, \gamma \in \Delta(\beta)$ with $V_\beta \leq Z_\alpha Z_\gamma Z_\mu Q_{\alpha'}$ so that V_β centralizes a subgroup of $V_{\alpha'}/Z_{\alpha'}$ of index at most $q_\beta p^3 = q_\beta$; or $Z_{\alpha'+1}Z_\lambda$ centralizes a subgroup of V_β/Z_β of index at most p^4. In either case, an application of Lemma 2.55 provides a contradiction.

Hence, we may assume that $q_\beta \geqslant p^4$ so that X centralizes a subgroup of V_β/Z_β of index at most q_β^3. Then applying Lemmas 2.43 and 2.44, using that $V_{\alpha'}$ acts quadratically on V_β, we conclude that the unique non-central chief factor for X in V_β/Z_β is either a natural module or an $\Omega_4^-(q_\beta^{\frac{1}{2}})$-module. Using that $|Z_\alpha/Z_\beta| = p$ implies a contradiction in either case. Thus, we have shown that $|V_{\alpha'}Q_\beta/Q_\beta| = p$. Since $q_\beta > p$ and $|Z_\alpha/Z_\beta| = p$, appealing to Proposition 2.51, we conclude that $q_\alpha = 2$, $q_\beta = 4$, $L_\beta/R_\beta \cong \mathrm{PSL}_2(4)$ and $V_\beta/C_{V_\beta}(L_\beta)$ is an $\Omega_4^-(2)$-module.

Suppose that $q_\alpha = p$ and $b \geqslant 7$. Then $V_\beta^{(3)}$ acts quadratically on $V_{\alpha'-2}$ and $V_\beta^{(3)} \cap Q_{\alpha'-2} \cap Q_{\alpha'-1}$ acts quadratically on $V_{\alpha'}$ so that $V_\beta(V_\beta^{(3)} \cap Q_{\alpha'})$ is centralized, modulo V_β, by $V_{\alpha'}$ and has index 4 in $V_\beta^{(3)}$. By Lemma 2.41, we have that $V_\beta^{(3)} \not\leq Q_{\alpha'-2}$ and $V_\beta^{(3)} \cap Q_{\alpha'-2} \not\leq Q_{\alpha'-1}$. But $V_\beta^{(3)}$ centralizes $Z_{\alpha'}$ and so we deduce that $Z_{\alpha'} = Z_{\alpha'-2}$. Since $V_\beta^{(3)} \not\leq Q_{\alpha'-2}$, there is $\beta-3 \in \Delta^{(3)}(\beta)$ such that $(\beta-3, \alpha'-2)$ is a critical pair. By Lemmas 7.5 and 7.6, $(\beta-3, \alpha'-2)$ satisfies the same hypothesis as (α, α') and so we conclude that $Z_{\alpha'} = Z_{\alpha'-2} = Z_{\alpha'-4}$ and there are more critical pairs to iterate through. We conclude that $Z_{\alpha'} = \cdots = Z_\beta$, a contradiction by Lemma 7.15. Hence, we have proved that

(2) if $q_\alpha = p$ then $q_\alpha = 2$, $q_\beta = 4$, $L_\beta/R_\beta \cong \mathrm{PSL}_2(4)$, $V_\beta/C_{V_\beta}(L_\beta)$ is an $\Omega_4^-(2)$-module and $b \leqslant 5$.

Assume that $b \geqslant 7$ so that $q_\alpha \geqslant p^2$, $|V_{\alpha'}Q_\beta/Q_\beta| \geqslant q_\alpha \leqslant |V_\beta Q_{\alpha'}/Q_{\alpha'}|$ and $\overline{L_\beta}/O_{p'}(\overline{L_\beta}) \cong \mathrm{PSL}_2(q_\beta)$. Suppose that $V_\beta^{(3)} \leq Q_{\alpha'-2}$. Then applying Propositions 2.31 and 2.32 via the action of $V_{\alpha'}$ on $V_\beta^{(3)}/V_\beta$, using that $Z_{\alpha'} \leq V_\beta$, we must have that $q_\alpha q_\beta/|V_\beta Q_{\alpha'}/Q_{\alpha'}| \geqslant q_\beta$. But then, $q_\alpha \leqslant |V_\beta Q_{\alpha'}/Q_{\alpha'}| \leqslant q_\alpha$. Applying Propositions 2.31 and 2.32 via the action of $Z_{\alpha'+1}$ on V_β, we have that $q_\alpha^2 \geqslant q_\beta > q_\alpha$. Let P_β be a p-minimal subgroup of L_β containing $Q_{\alpha+2}$ such that $\overline{L_\beta} = \overline{P_\beta}O_{p'}(\overline{L_\beta})$. Set $V^P := \langle Z_\alpha^{P_\beta}\rangle/Z_\beta$ so that P_β non-trivially on V^P. Still, $Z_{\alpha'+1}$ is a quadratic 2F-offender on the module $V^P/C_{V^P}(P_\beta)$ and applying [22, Theorem 1], using that $|Z_\alpha C_{V^P}(P_\beta)/C_{V^P}(P_\beta)| = q_\alpha < q_\beta$, we have that $V^P/C_{V^P}(P_\beta)$ is a natural $\Omega_4^-(q_\alpha)$-module for $\overline{P_\beta} \cong \mathrm{PSL}_2(q_\beta)$ and $q_\beta = q_\alpha^2$. By conjugacy, and using that $V_\beta^{(3)} \cap Q_{\alpha'-1}$ acts quadratically on $V_{\alpha'}$, we have that $V_\beta(V_\beta^{(3)} \cap Q_{\alpha'-1})$ has index q_α in $V_\beta^{(3)}$ and is centralized, modulo V_β, by $V_{\alpha'}$, a contradiction by Propositions 2.31 and 2.32. Thus,

(3) if $b \geqslant 7$ then $V_\beta^{(3)} \not\leq Q_{\alpha'-2}$.

Note that $(\beta-3, \alpha'-2)$ is also a critical pair which satisfies the same conditions as the critical pair (α, α'). Indeed, iterating backward through critical pair, upon relabeling we may assume that $Z_{\alpha'} \neq Z_{\alpha'-2}$, for otherwise $Z_{\alpha'} = \cdots = Z_\beta$, yielding a contradiction by Lemma 7.15. Since $Z_{\alpha'} \leq V_\beta$, we see that $V_\beta^{(3)}$ centralizes $Z_{\alpha'-1}$. Hence,

(4) if $b \geqslant 7$ then $Z_{\alpha'} \cap Z_{\alpha'-2} = \{1\}$, $V_\beta^{(3)} \not\leq Q_{\alpha'-2}$ and $V_\beta^{(3)} \cap Q_{\alpha'-2} \leq Q_{\alpha'-1}$.

Assume that $L_\beta/R_\beta \not\cong \mathrm{SL}_2(q_\beta)$ so that $q_\alpha > p$. Since Z_α acts quadratically on $V_{\alpha'}$, if p is odd then we arrive at an immediate contradiction by Lemma 2.35. Hence, we proceed under the additional constraint that $p = 2$. Note that, for $Z_{\alpha'} < A \leq Z_{\alpha'+1}$, $\{1\} \neq [V_\beta \cap Q_{\alpha'}, A] \leq Z_{\alpha'}$. Since $Z_\beta \cap Z_{\alpha'} = \{1\}$ by Lemma 7.15, for $V := V_\beta/Z_\beta$ and C_A the preimage in V_β of $C_V(A)$, we deduce that $C_A \cap Q_{\alpha'} \leq Q_{\alpha'+1}$.

If there is $Z_{\alpha'} < B \leq Z_{\alpha'+1}$ with $C_V(Q) = C_V(B)$ for all $Z_{\alpha'} < Q < B$ with $[B : Q] = 2$, then applying coprime action, we have that $O_{2'}(\overline{L_\beta})$ normalizes $C_V(B)$. Then writing $H := \langle (BQ_\beta)/Q_\beta)^{O_{2'}(\overline{L_\beta})}\rangle$, we have that $[H, C_V(B)] = \{1\}$. Moreover, $V = [V, O_{2'}(H)] \times C_V(O_{2'}(H))$ is a BQ_β/Q_β-invariant decomposition and we conclude that $O_{2'}(H)$ centralizes V. Thus, $O_{2'}(L_\beta/R_\beta)$ normalizes BQ_β/Q_β and L_β/R_β is a central extension of $\mathrm{PSL}_2(q_\beta)$, a contradiction. Hence, we deduce that

(5) if $b \geqslant 7$ and $L_\beta/R_\beta \not\cong \mathrm{SL}_2(q_\beta)$ then $p = 2$ and for every $Z_{\alpha'} < Q < B \leq Z_{\alpha'+1}$ with $|Q/Z_{\alpha'}| = 2 = |B/Q|$ we have that $C_{V_\beta/Z_\beta}(Q)$ has index at most

$|V_\beta Q_{\alpha'}/Q_{\alpha'}|2$ in V_β/Z_β and $C_{V_\beta/Z_\beta}(B)$ has index at most $|V_\beta Q_{\alpha'}/Q_{\alpha'}|4$ in V_β/Z_β.

We set $r_{\alpha'} := |V_\beta Q_{\alpha'}/Q_{\alpha'}|$. Applying Proposition 2.31, we deduce that $r_{\alpha'} p^2 \geqslant q_\beta$. Form $X := \langle B, B^g, Q_\beta\rangle$ for such a B and some appropriately chosen $g \in L_\beta$ with $\overline{X} O_{2'}(\overline{L_\beta}) = \overline{L_\beta}$. Then $|V/C_V(X)| \leqslant r_{\alpha'}^2 2^4$ and applying Lemma 2.44, we deduce that either $XR_\beta/R_\beta \cong \mathrm{PSL}_2(q_\beta)$ and the unique non-central X-chief factor within V is determined by Lemma 2.43, or $q_\beta = r_{\alpha'} = 8$ and $q_\alpha = 4$. In the first instance, since $V_\beta^{(3)}$ acts quadratically on $V_{\alpha'-2}$, $V_\beta^{(3)} \cap Q_{\alpha'-2}$ acts quadratically on $V_{\alpha'}$, and $V_\beta^{(3)}$ contains a non-central chief factor for L_β; we deduce that the unique non-central X-chief factor within $V_\beta^{(3)}/V_\beta$ is a natural $\mathrm{SL}_2(q_\beta)$-module. Then for $T \in \mathrm{Syl}_2(X)$, we have that $[V, x]C_V(X) = [V, y]C_V(X)$ for all $x, y \in T \setminus Q_\beta$ and we conclude that $C_V(Q) = C_V(B)$ where Q and B are subgroups of $Z_{\alpha'+1}$ defined previously, a contradiction.

Hence, $q_\beta = r_{\alpha'} = 8$ and $q_\alpha = 4$. Then $V_\beta = Z_\alpha Z_\lambda(V_\beta \cap Q_{\alpha'})$ for some $\lambda \in \Delta(\beta)$ with $Z_\lambda \cap Z_\alpha = Z_\beta$. But then $|Z_\alpha Z_\lambda| = 2^6$ so that $|Z_\alpha Z_\lambda \cap Q_{\alpha'}| = 2^3$. Then $X := Z_\lambda(Z_\alpha Z_\lambda \cap Q_{\alpha'})$ has order 2^5 from which it follows that X is a maximal subgroup of $Z_\alpha Z_\lambda$. Indeed, $X \cap Z_\alpha$ is a maximal subgroup of Z_α. Now, $[V_{\alpha'} \cap Q_\beta, Z_\alpha Z_\lambda \cap Q_{\alpha'}] \leq Z_\beta \cap Z_{\alpha'} = \{1\}$ from which it follows that $[V_{\alpha'} \cap Q_\beta \cap Q_\lambda, X] = \{1\}$. But then $V_{\alpha'} \cap Q_\beta \cap Q_\lambda$ centralizes a maximal subgroup of Z_α and we deduce that $V_{\alpha'} \cap Q_\beta \cap Q_\lambda$ has index 2^5 in $V_{\alpha'}$ and centralizes $Z_\alpha Z_\lambda$. Applying Lemma 2.55, we conclude that $V_{\alpha'}/Z_{\alpha'}$ contains a unique non-central chief factor which is a natural module for $L_{\alpha'}/R_{\alpha'} \cong \mathrm{SL}_2(q_{\alpha'})$, a contradiction to our hypothesis. Hence, we conclude that

(6) if $b \geqslant 7$ then $L_\beta/R_\beta \cong \mathrm{SL}_2(q_\beta)$.

Assume that $r_{\alpha'}^2 q_\alpha^2 \leqslant q_\beta^3$ so that $V_\beta/C_{V_\beta}(O^p(L_\beta))$ is determined by Lemma 2.43. Since V_β acts quadratically on $V_{\alpha'}$, by Lemma 7.9, it follows that $r_{\alpha'} \leqslant q_\beta^{\frac{1}{2}}$. But $q_\alpha \leqslant r_{\alpha'}$ from which it follows that $q_\alpha = r_{\alpha'} = q_\beta^{\frac{1}{2}}$ and $V_\beta/C_{V_\beta}(O^2(L_\beta))$ is a natural $\Omega_4(q_\alpha)$-module. But then, since $V_\beta^{(3)}$ acts quadratically on $V_{\alpha'-2}$ and $V_\beta^{(3)} \cap Q_{\alpha'-2}$ is quadratic on $V_{\alpha'}$, we have that $V_\beta(V_\beta^{(3)} \cap Q_{\alpha'})$ has index q_α in $V_\beta^{(3)}$ and is centralized, modulo V_β, by $V_{\alpha'}$ and Proposition 2.31 yields a contradiction. Hence

(7) if $b \geqslant 7$ then $r_{\alpha'}^2 q_\alpha^2 > q_\beta^3$.

In particular, $r_{\alpha'} > q_\beta^{\frac{1}{2}}$ and writing $P_\beta = \langle V_{\alpha'}, V_{\alpha'}^g, C_{L_\beta}(V_\beta^{(3)}/V_\beta), Q_\beta\rangle$, we have that P_β has an $\mathrm{SL}_2(q_\beta)$ quotient and centralizes a subgroup of $V_\beta^{(3)}/[V_\beta^{(3)}, Q_\beta]V_\beta$ of index strictly less than q_β^3. Then Lemma 2.44 implies that $P_\beta/C_{P_\beta}(V_\beta^{(3)})Q_\beta \cong \mathrm{SL}_2(q_\beta)$ and Lemma 2.43 yields that $V_\beta^{(3)}/V_\beta$ contains a unique non-central chief factor which, as a $\overline{P_\beta}$-module, is a natural $\mathrm{SL}_2(q_\beta)$-module. It follows that all elements of order p in S/Q_β have the same centralizer on $V_\beta^{(3)}/[V_\beta^{(3)}, Q_\beta]V_\beta$ and applying coprime action and arguing as above, we get that $L_\beta/C_{L_\beta}(V_\beta^{(3)})Q_\beta \cong \mathrm{SL}_2(q_\beta)$. Since $V_\beta^{(3)} \not\leq Q_{\alpha'-2}$ yet centralizes $Z_{\alpha'-1}$, we deduce that $Z_{\alpha'-1} = Z_{\alpha'-3}$ and application of Lemma 5.22 implies that $V_{\alpha'} \leq V_{\alpha'-1}^{(2)} = V_{\alpha'-3}^{(2)} \leq Q_\beta$, a contradiction. Therefore

(8) $b \leqslant 5$.

Suppose that $b = 5$. Since $Z_{\alpha'} \neq Z_\beta$, without loss of generality, we have that $Z_\beta \neq Z_{\alpha'-2}$. But $Z_\beta = [V_\beta, V_{\alpha'} \cap Q_\beta] \leq [V_{\alpha'-2}^{(3)}, V_{\alpha'-2}^{(3)}]$ from which we deduce that $Z_{\alpha+2} \leq [V_{\alpha'-2}^{(3)}, V_{\alpha'-2}^{(3)}]$ and ultimately $V_{\alpha'-2} \leq [V_{\alpha'-2}^{(3)}, V_{\alpha'-2}^{(3)}] \leq \Phi(V_{\alpha'-2}^{(3)})$. By conjugation, $V_\beta \leq [V_\beta^{(3)}, V_\beta^{(3)}]$. If $[V_\beta^{(3)}, Q_\beta]/V_\beta$ does not contain a non-central chief factor for L_β, then $[V_\beta^{(3)}, Q_\beta] = [V_\lambda^{(2)}, Q_\beta]V_\beta$ for all $\lambda \in \Delta(\beta)$ and as $V_\lambda^{(2)}$ is elementary abelian, $[V_\beta^{(3)}, Q_\beta] \leq Z(V_\beta^{(3)})$. But then, by the three subgroups lemma, $[V_\beta^{(3)}, V_\beta^{(3)}, Q_\beta] = \{1\}$, a contradiction since $[V_\beta, Q_\beta] \neq \{1\}$. Hence, by Lemma 5.8, $V_\beta^{(3)}/V_\beta$ contains at least two non-central chief factors for L_β.

Note that if $|V_\beta Q_{\alpha'}/Q_{\alpha'}| = p$, then $V_\lambda/C_{V_\lambda}(O^2(L_\lambda))$ is isomorphic to a natural $\Omega_4^-(2)$-module for all $\lambda \in \beta^G$. Since $V_{\alpha'}^{(3)}$ is quadratic on $V_{\alpha'-2}$, we get that $V_{\alpha'}(V_{\alpha'}^{(3)} \cap Q_\beta)$ has index at most 8 in $V_{\alpha'}^{(3)}$ and is centralized, modulo $V_{\alpha'}$, by V_β. Applying Lemma 2.41 since $q_\beta = 4$ and $V_\beta^{(3)}/V_\beta$ contains at least two non-central chief factors for L_β, this is a contradiction. Hence, $|V_\beta Q_{\alpha'}/Q_{\alpha'}| \geqslant p^2$ and by a similar argument, $|V_{\alpha'} Q_\beta/Q_\beta| \geqslant p^2$. We observe that as $Z_\beta \leq V_{\alpha'} \leq Z(V_{\alpha'}^{(3)})$, $V_{\alpha'}^{(3)} \cap Q_{\alpha'-2} \leq Q_{\alpha+2}$. We deduce that V_β centralizes a subgroup of $V_{\alpha'}^{(3)}/V_{\alpha'}$ of index strictly less than q_β^2 and applying Proposition 2.31 and Proposition 2.32, and using that $V_{\alpha'}^{(3)}$ contains at least two non-central chief factors, we have a contradiction. So

(9) $b = 3$.

We assume that $|V_{\alpha'} Q_\beta/Q_\beta| \geqslant p^2$. Then $q_\alpha > p$ and $\overline{L_\beta}/O_{p'}(\overline{L_\beta}) \cong \mathrm{PSL}_2(q_\beta)$. Now, $Z_{\alpha'+1}$ centralizes a subgroup of Q_β of index at most $q_\alpha^2 q_\beta$ and as Q_β/C_β and V_β contain non-central chief factors for L_β, we deduce that $q_\alpha^2 \geqslant q_\beta$ and $O^p(L_\beta)$ centralizes C_β/V_β. Furthermore, for any $r \in L_\beta$ of order coprime to p, if $[r, Q_\beta, V_\beta] = \{1\}$, then $[r, V_\beta] \leq Z(Q_\beta)$ by the three subgroups lemma, from which it follows that $[r, Q_\beta] = \{1\}$, $r = 1$ and $C_{L_\beta}(Q_\beta/C_\beta) \leq Q_\beta$. Similarly, if $[r, V_\beta] = \{1\}$ then $[r, Q_\beta, V_\beta] = \{1\}$, $[r, Q_\beta] = \{1\}$ and $R_\beta = Q_\beta$.

(10) if $|V_{\alpha'} Q_\beta/Q_\beta| \geqslant p^2$ then $q_\alpha > p$, $\overline{L_\beta}/O_{p'}(\overline{L_\beta}) \cong \mathrm{PSL}_2(q_\beta)$, $R_\beta = Q_\beta$ and both Q_β/C_β and $V_\beta/Z_{|beta}$ are faithful modules for $\overline{L_\beta}$.

Assume that $|V_{\alpha'} Q_\beta/Q_\beta| \geqslant p^2$ and $Z_{\alpha'+1}$ centralizes a subgroup of index strictly less than q_α^2 in Q_β/C_β or V_β. If $q_\alpha^2 \leqslant q_\beta$, then we have a contradiction by Propositions 2.31 and 2.32. Hence, $q_\alpha^2 > q_\beta$. Applying Lemma 2.55, we infer that $\overline{L_\beta} \cong \mathrm{SL}_2(q_\beta)$ and both Q_β/C_β and V_β contain a unique non-central chief factor for $\overline{L_\beta}$, which has the structure of a natural module. Since $Z_\alpha C_{V_\beta}(O^p(L_\beta))/C_{V_\beta}(O^p(L_\beta))$ is $G_{\alpha,\beta}$-invariant of order q_α, and $Q_\beta \cap Q_\alpha$ is $G_{\alpha,\beta}$ and of index q_α, using that $q_\alpha < q_\beta$, we have forced another contradiction. We have proved

(11) if $|V_{\alpha'}Q_\beta/Q_\beta| \geqslant p^2$ then $C_\beta(Q_\beta \cap \cdots \cap Q_{\alpha'+1})$ has index at least q_α^2 in Q_β and $V_\beta \cap \cdots \cap Q_{\alpha'+1}$ has index at least q_α^2 in V_β.

We continue under the assumption that $|V_{\alpha'}Q_\beta/Q_\beta| \geqslant p^2$. Since $Z_{\alpha'+1}$ centralizes a subgroup of Q_β of index at most $q_\alpha^2 q_\beta$ we deduce that $q_\alpha^2 = q_\beta$. Indeed, since $[V_{\alpha'}, Q_\beta \cap Q_{\alpha'}] \leq V_\beta$, we must have that $|V_{\alpha'}Q_\beta/Q_\beta| = |V_\beta Q_{\alpha'}/Q_{\alpha'}| = q_\alpha$ and $(V_\beta \cap Q_{\alpha'})Q_{\alpha'+1} \in \mathrm{Syl}_p(L_{\alpha'+1})$.

Assume that $q_\alpha = p^2$ and $C_{V_\beta/Z_\beta}(x) > (V_\beta \cap Q_{\alpha'+1})/Z_\beta$ for all $x \in Z_{\alpha'+1} \setminus Z_{\alpha'}$. Then for C_x the preimage in V_β of $C_{V_\beta/Z_\beta}(x)$, we must have that $[C_x \cap Q_{\alpha'}, Z_{\alpha'}\langle x\rangle] \leq Z_\beta \cap Z_{\alpha'} = \{1\}$ so that $C_x \cap Q_{\alpha'} \leq Q_{\alpha'+1}$. If $V_\beta = C_x(V_\beta \cap Q_{\alpha'})$ then using $q_\beta = p^4$ and applying Proposition 7.12, we have a contradiction. Thus, $|C_x Q_{\alpha'}/Q_{\alpha'}| = p$ and x centralizes a subgroup of index p^3 in V_β/Z_β. Let P_β be a p-minimal subgroup of L_β containing $Q_{\alpha'-1}$ such that $\overline{L_\beta} = \overline{P_\beta} O_{p'}(\overline{L_\beta})$. Set $V^P := \langle Z_\alpha^{P_\beta}\rangle/Z_\beta$ so that P_β non-trivially on V^P. Still, $Z_{\alpha'+1}$ is a 2F-offender on the module $V^P/C_{V^P}(P_\beta)$ and applying [22, Theorem 1] and using that x centralizes a subgroup of index at most $p^3 < q_\beta$ in V^P, we obtain a contradiction.

If $q_\alpha > p^2$ then by Propositions 2.31 and 2.32 we must have that $C_{V_\beta/Z_\beta}(A) = (V_\beta \cap Q_{\alpha'+1})/Z_\beta$ for all $A \leq Z_{\alpha'+1}$ such that $Z_{\alpha'} \leq A$ and $|Z_{\alpha'+1}/A| = p$. Hence, for $q_\alpha \geqslant p^2$ and $C := V_\beta \cap Q_{\alpha'} \cap Q_{\alpha'+1}$, applying Proposition 2.8, we deduce that C is normalized by $\overline{Z_{\alpha'+1}} O_{p'}(\overline{L_\beta})$. Then writing $H := \langle (\overline{Z_{\alpha'+1}})^{O_{p'}(\overline{L_\beta})}\rangle$, we have that $[H, C] \leq Z_\beta$. Moreover, $V_\beta/C_{V_\beta}(O^p(L_\beta)) = [V_\beta/C_{V_\beta}(O^p(L_\beta)), O_{p'}(H)] \times C_{V_\beta/C_{V_\beta}(O^p(L_\beta))}(O_{p'}(H))$ is a $\overline{Z_{\alpha'+1}} O_{p'}(\overline{L_\beta})$-invariant decomposition and we conclude that $O_{p'}(H)$ centralizes V_β. Thus, $O_{p'}(L_\beta)$ normalizes $\overline{Z_{\alpha'+1}}$ and $\overline{L_\beta}$ is a central extension of $\mathrm{PSL}_2(q_\beta)$ with a faithful quadratic module. Since $\overline{L_\beta}$ is generated by two conjugates of $\overline{Z_{\alpha'+1}}$, we deduce that $|V_\beta/C_{V_\beta}(O^p(L_\beta))| \leqslant q_\beta^2$. Applying Lemma 2.43, we have that $V_\beta/C_{V_\beta}(O^p(L_\beta))$ is a natural $\Omega_4^-(q_\alpha)$-module for $\overline{L_\beta} \cong \mathrm{PSL}_2(q_\beta)$ and $|V_\beta Q_{\alpha'}/Q_{\alpha'}| = q_\alpha$. Since V_β admits quadratic action, we deduce that $p = 2$. Thus, including the situation where $|V_{\alpha'}Q_\beta/Q_\beta| = p$, we have reduced to the case where

(12) $b = 3$, $p = 2$, $q_\beta = q_\alpha^2$, $\overline{L_\beta} \cong \mathrm{PSL}_2(q_\beta)$ and $V_\beta/C_{V_\beta}(O^2(L_\beta))$ is a natural $\Omega_4^-(q_\alpha)$-module.

We note that first that as S/Q_β is irreducible under the action of $G_{\alpha,\beta}$ and $V_\alpha^{(2)} \not\leq Q_\beta$, we must have that $[V_\alpha^{(2)}, V_\beta] = [Q_\alpha, V_\beta]$ from which it follows that $[V_\alpha^{(2)}, V_\alpha^{(2)}] = [V_\alpha^{(2)}, Q_\alpha]$. Moreover, we have that $[V_{\alpha'}, Q_{\alpha'-1}] \leq [Q_{\alpha'-1}, Q_{\alpha'-1}] \leq Q_\beta$ so that $V_{\alpha'} \cap Q_\beta = [V_{\alpha'}, Q_{\alpha'-1}]C_{V_{\alpha'}}(O^2(L_{\alpha'}))$. Now,

$V_{\alpha'} \cap Q_\beta$ is $G_{\alpha,\alpha'-1}$-invariant and so $[Z_\alpha Q_{\alpha'}, V_{\alpha'} \cap Q_\beta] = Z_{\alpha'-1}$ from which it follows that $[Q_{\alpha'-1} Q_{\alpha'}, V_{\alpha'} \cap Q_\beta] = Z_{\alpha'-1}$.

Let K_α be a critical subgroup of Q_α. Then $[K_\alpha, Q_\alpha, K_\alpha] = \{1\}$ and by the three subgroups lemma, $[K_\alpha, K_\alpha] \leq Z(Q_\alpha) \leq C_\beta$. Since the unique non-central chief factor within Q_β/C_β is not a natural $\mathrm{SL}_2(q_\beta)$-module, we deduce that $K_\alpha \leq Q_\beta$. Since $\overline{L_\beta} \cong \mathrm{PSL}_2(q_\beta)$ is 2-transitive on the set $\{Z_\lambda : \lambda \in \Delta(\beta)\}$, we have that $V_\beta = Z_\alpha \langle Z_{\alpha'-1}^S \rangle$. Hence, if $K_\alpha \cap Z_{\alpha'} \neq \{1\}$, then we deduce that $V_\beta \leq K_\alpha$ so that $V_\alpha^{(2)} \leq K_\alpha \leq Q_\beta$, a clear contradiction since $V_{\alpha'} \leq V_{\alpha'-1}^{(2)} \not\leq Q_\beta$. Hence, $[K_\alpha \cap Q_{\alpha'-1}, V_{\alpha'} \cap Q_\beta] \leq Z_{\alpha'-1} \cap K_\alpha = Z_\beta$. Hence, by Lemma 2.41, K_α/Z_α contains at most one non-central chief factor for $\overline{L_\alpha}$. Applying an argument on the Schur multiplier of $\mathrm{PSL}_2(q_\beta)$, we conclude that $q_\alpha = 2$ and $q_\beta = 4$, for otherwise G has a weak BN-pair of rank 2, and a contradiction is retrieved from [27]. But now, by Lemma 2.25, and using that $[V_{\alpha'} \cap Q_\beta, Q_{\alpha'-1}] = Z_{\alpha'-1}$, we deduce that $V_{\alpha'}/Z_{\alpha'}$ is irreducible.

Now, $[V_\alpha^{(2)}, Q_\alpha] \leq Q_\beta$ and $[[V_\alpha^{(2)}, Q_\alpha] \cap Q_{\alpha'-1}, V_{\alpha'} \cap Q_\beta] \leq Z_{\alpha'-1}$. Since $V_\beta \not\leq [V_\alpha^{(2)}, Q_\alpha]$ and $V_\beta = Z_\alpha \langle Z_{\alpha'-1}^S \rangle$, we deduce that $V_{\alpha'} \cap Q_\beta$ centralizes an index 2 subgroup of $[V_\alpha^{(2)}, Q_\alpha]/Z_\alpha$ and writing $V := [V_\alpha^{(2)}, Q_\alpha]/Z_\alpha$ and by Lemma 2.41, $V/C_V(L_\alpha)$ is a natural $\mathrm{SL}_2(2)$-module. Set C^α the preimage in $V_\alpha^{(2)}$ of $C_V(L_\alpha)$. It follows that either $[V_\alpha^{(2)}, Q_\alpha] = [V_\beta, Q_\alpha] C^\alpha$ or $[V_\beta, Q_\alpha] \leq C^\alpha$. In the former case, we have that Q_β centralizes $[V_\alpha^{(2)}, Q_\alpha]/C^\alpha$, an obvious contradiction; whereas in the latter case, conjugating from α to $\alpha+2$, we have that $[V_\beta, Q_{\alpha+2}] = [V_{\alpha'}, Q_{\alpha+2}]$ is centralized by V_β, a contradiction to the structure of $V_{\alpha'}/C_{V_{\alpha'}}(O^p(L_{\alpha'-1}))$ by Proposition 2.24. This final contradiction completes the proof. □

Given our claims that the conclusion of this section is that G is parabolic isomorphic to ${}^2\mathrm{F}_4(2)'$ or ${}^2\mathrm{F}_4(2^n)$, we must first rid ourselves of the situation where Z_α is *not* a natural $\mathrm{SL}_2(q_\alpha)$-module. The following lemma investigates the consequences of this assumption.

Lemma 7.17 *Suppose that $C_{V_\beta}(V_{\alpha'}) < V_\beta \cap Q_{\alpha'}$, $q_\alpha = p$ and Z_α is not a natural module for $L_\alpha/R_\alpha \cong \mathrm{SL}_2(q_\alpha)$. Then the following hold:*

- (i) $S = Q_\alpha Q_\beta$;
- (ii) $|S/Q_\alpha| = p$;
- (iii) $L_\alpha/R_\alpha \in \{\mathrm{SL}_2(p), \mathrm{SU}_3(2)', \mathrm{Dih}(10), (3 \times 3) : 2, (Q_8 \times Q_8) : 3, 2 \cdot \mathrm{Alt}(5), 2_-^{1+4}.\mathrm{Alt}(5)\}$;
- (iv) $R = [Q_{\alpha'}, V_{\alpha'}] \leq Z_{\alpha'}$;
- (v) $Q_\beta \in \mathrm{Syl}_p(R_\beta)$;
- (vi) $|Z_\alpha/Z_\beta| = p^2$; *and*
- (vii) *unless $L_\alpha/R_\alpha \cong \mathrm{SU}_3(2)'$ and $R < Z_{\alpha'}$, we have that $R = Z_{\alpha'}$ and $Z_\alpha = Z_\beta \times Z_{\alpha-1}$ for some $\alpha - 1 \in \Delta(\alpha)$.*

Proof Since this result holds in all the relevant cases in Theorem C, we may assume that G is a minimal counterexample to the lemma. We assume throughout that Z_α is not a natural module for $L_\alpha/R_\alpha \cong \mathrm{SL}_2(q_\alpha)$ and so $q_\alpha = p$ and Z_α is a quadratic 2F-module determined by Proposition 7.12. By Lemma 7.16, we have that $L_\beta/R_\beta \cong \mathrm{SL}_2(p), (3\times 3):2, (3\times 3):4$ or $(Q_8\times Q_8):3$ and $V_\beta/C_{V_\beta}(O^p(L_\beta))$ is the associated 2F-module described in Proposition 7.12.

Suppose first that $S \neq Q_\alpha Q_\beta$ so that $L_\beta/R_\beta \cong (3\times 3):4$. Moreover, since Z_α is also a quadratic 2F-module determined by Proposition 7.12, $L_\alpha/R_\alpha \cong (3\times 3):4, \mathrm{Sz}(2), \mathrm{SU}_3(2)'.2$ or $\mathrm{SU}_3(2)$. For $\mu \in \{\alpha,\beta\}$, let O_μ be the preimage in L_μ of $O_{2'}(\overline{L_\mu})$ and set $L_\mu^* := O_\mu Q_\alpha Q_\beta$. Then $L_\mu^* \trianglelefteq L_\mu$ and L_μ^* has index 2 in L_μ. Set K to be a Hall $2'$-subgroup of $G_{\alpha,\beta}$ and set $G_\mu^* := L_\mu^* K$. Then G_μ^* has index 2 in G_μ, and is normal in G_μ. Moreover, for $X = \langle G_\alpha^*, G_\beta^*\rangle$, X is normalized by $G_{\alpha,\beta}$ and $G = \langle X, G_{\alpha,\beta}\rangle$. Thus, the largest subgroup of S which is normal in X is also normal in G and so is trivial. Hence, any subgroup of $G_{\alpha,\beta}\cap X$ which is normal in X is a $2'$-group and we can arrange that it is contained in $K \le G_\mu^*$, a contradiction since G_μ^* is of characteristic 2. Thus, the amalgam $(G_\alpha^*, G_\beta^*, KQ_\alpha Q_\beta)$ satisfies Hypothesis 5.1. Since G_α^* and G_β^* are solvable, by minimality, $(G_\alpha^*, G_\beta^*, KQ_\alpha Q_\beta)$ is a weak BN-pair; or X is a symplectic amalgam with $|S| = 2^6$. In all cases, for some $\mu \in \{\alpha,\beta\}$, we infer that $\overline{L_\mu^*} \cong \mathrm{Sym}(3)$. But then, it follows that $\overline{L_\mu} \cong \mathrm{Sym}(3)\times R$, where R is a 2-group, a contradiction since $m_p(S/Q_\mu) = 1$. Hence, $S = Q_\alpha Q_\beta$ and (i) is proved.

Suppose that $|S/Q_\alpha| > p$. Since $S = Q_\alpha Q_\beta$ and $Q_\beta \cap O^p(L_\beta)$ centralizes $[Z_\alpha, Q_\beta]$ by Lemma 7.16, $L_\alpha/R_\alpha \not\cong \mathrm{Sz}(2)$ or $(3\times 3):4$. Set $Q_\beta^* := \langle (Q_\alpha \cap Q_\beta)^{G_\beta}\rangle$. Then Q_β^* centralizes $[Z_\alpha, Q_\beta]$. If $S = Q_\beta^* Q_\alpha$ then S centralizes $[Z_\alpha, Q_\beta]$. However, since $|S/Q_\alpha| > p$, comparing with the list in Proposition 7.12, we have a contradiction. So $Q_\beta^* < Q_\beta$ and $Q_\beta^* Q_\alpha < S$. Then, $Q_\beta^* Q_\alpha$ is a proper $G_{\alpha,\beta}$-invariant subgroup of S/Q_α, from which it follows that $L_\alpha/R_\alpha \cong \mathrm{(P)SU}_3(p)$ or $\mathrm{SU}_3(2)'.2$. Since Q_β^* centralizes $[Z_\alpha, Q_\beta]$, $|Q_\beta^* Q_\alpha/Q_\alpha| = p$.

If $p = 2$, then as $m_2(S/Q_\beta) = 1$, L_β is solvable. Set $L_\beta^* := C_{L_\beta}([V_\beta, Q_\beta])$. Then, $Q_\beta^* \le O_2(L_\beta^*)$ and since $O_2(L_\beta^*)$ is $G_{\alpha,\beta}$-invariant and centralizes $[Z_\alpha, Q_\beta]$, $|O_2(L_\beta^*)Q_\alpha/Q_\alpha| = 2$ and $Q_\beta^* = O_2(L_\beta^*)$. Moreover, $S^* := Q_\alpha Q_\beta^* = S \cap L_\beta^* \in \mathrm{Syl}_2(L_\beta^*)$. Setting $L_\alpha^* := \langle (S^*)^{G_\alpha}\rangle$, we have that $L_\alpha^* \trianglelefteq G_\alpha$ and $S^* \in \mathrm{Syl}_2(L_\alpha^*)$. For $\mu \in \{\alpha,\beta\}$, set $G_\mu^* := L_\mu^* K$, where K is a Hall $2'$-subgroup of $G_{\alpha,\beta}$. Then the amalgam $X := (G_\alpha^*, G_\beta^*, S^*K)$ satisfies Hypothesis 5.1 and by the minimality of G, we have a contradiction.

Thus, p is odd and $L_\alpha/R_\alpha \cong \mathrm{(P)SU}_3(p)$ and $L_\beta/R_\beta \cong \mathrm{SL}_2(p)$ or $(Q_8\times Q_8):3$. Suppose first that $b > 3$. Then $V_\alpha^{(2)}$ is elementary abelian so that $V_\alpha^{(2)} \cap Q_{\alpha'}$ acts quadratically on $Z_{\alpha'+1}\cap Q_\beta$. In particular, $V_\alpha^{(2)} \cap Q_{\alpha'-2} \cap Q_{\alpha'-1} = Z_\alpha(V_\alpha^{(2)} \cap Q_{\alpha'+1})$, $V_\alpha^{(2)} \not\le Q_{\alpha'-2}$ and $V_\alpha^{(2)} \cap Q_{\alpha'-2} \not\le Q_{\alpha'-1}$ so that $V_\alpha^{(2)}/Z_\alpha$ contains a unique non-central chief factor for L_α which is also a natural $\mathrm{SU}_3(p)$-module. Then an argument on the Schur multiplier of $\mathrm{SU}_3(p)$ yields that $O^p(R_\alpha)$ centralizes $V_\alpha^{(2)}$. Indeed, $Z_{\alpha'} \neq Z_{\alpha'-2}$ else by Lemma 5.22 $V_{\alpha'} = V_{\alpha'-2} \le Q_\beta$. But now, $L_{\alpha'-1} = \langle V_\alpha^{(2)} \cap Q_{\alpha'-2}, Q_{\alpha'}, R_{\alpha'-1}\rangle$ centralizes $Z_{\alpha'} = [V_\beta \cap Q_{\alpha'}, Z_{\alpha'+1}] \le V_\beta$, so that $Z_{\alpha'} \le Z(L_{\alpha'+1}) = \{1\}$, a contradiction.

Suppose now that $b = 3$. Note that $Z_\beta = Z(R_{\alpha'-1}Q_\beta)$ and so either $Z_{\alpha'} = Z_\beta$ and $R_{\alpha'-1}Q_\beta = R_{\alpha'-1}Q_{\alpha'}$, or $Z_{\alpha'} \cap Z_\beta$ is centralized by $L_{\alpha'-1} =$

$\langle Q_{\alpha'}, Q_\beta, R_{\alpha'-1}\rangle$. In the former case, we have that $[V_\beta \cap Q_{\alpha'}, Z_{\alpha'+1}] = Z_{\alpha'} = Z_\beta$, a contradiction by Lemma 7.9. In the latter case, $Z_{\alpha'} \cap Z_\beta = \{1\}$ so that $[Z_{\alpha'+1} \cap Q_\beta, V_\beta \cap Q_{\alpha'}] \le Z_{\alpha'} \cap Z_\beta = \{1\}$ and an index p subgroup of $Z_{\alpha'+1}$ is centralized by $V_\beta \cap Q_{\alpha'}$. Hence, Z_α is an FF-module, a contradiction.

Thus, $|S/Q_\alpha| = p$ and, as $Q_\alpha \cap Q_\beta \ntrianglelefteq L_\beta$ by Proposition 5.27, $Q_\beta = (Q_\alpha \cap Q_\beta)(Q_\gamma \cap Q_\beta)$ for some $\gamma \in \Delta(\beta)$. Thus, $L_\beta = \langle Q_\gamma : \gamma \in \Delta(\beta)\rangle$ centralizes $[V_\beta, Q_\beta]$ and $[V_\beta, Q_\beta] \le Z_\beta$. The remaining properties follow from Proposition 7.12 and may be checked in MAGMA [14]. □

Lemma 7.18 *Suppose that $C_{V_\beta}(V_{\alpha'}) < V_\beta \cap Q_{\alpha'}$ and $V_\alpha^{(2)}/Z_\alpha$ contains a unique non-central chief factor U/V for L_α. Then U/V is not an FF-module for $\overline{L_\alpha}$.*

Proof Suppose that U/V is an FF-module for $\overline{L_\alpha}$. By Lemma 7.16, $q_\alpha = q_\beta$. By Lemma 5.8, $V_\alpha^{(2)}/[V_\alpha^{(2)}, Q_\alpha]$ contains a non-central chief factor for L_α. Set C to be the preimage in $V_\alpha^{(2)}$ of $C_{V_\alpha^{(2)}/Z_\alpha}(O^p(L_\alpha))$. Then $[V_\alpha^{(2)}, Q_\alpha] \le C$ and since U/V is an FF-module, by Lemma 2.41, $V_\alpha^{(2)}/C$ is isomorphic to a natural $\mathrm{SL}_2(q_\alpha)$-module. In particular, as $V_\beta \not\le C$, $V_\beta C/C$ is of order q_α for otherwise Q_β centralizes $V_\alpha^{(2)}/C$. But now, $V_\beta \cap C$ has index q_α in V_β and is normalized by L_α. By conjugacy, an index q_α subgroup of V_β is normalized by $L_{\alpha+2}$, and by transitivity, this subgroup is contained in $V_{\alpha+3}$ so that $V_\beta \cap V_{\alpha+3}$ is of index q_α in V_β. But then, as $V_\beta \not\le Q_{\alpha'}$, $V_\beta \cap Q_{\alpha'} = V_\beta \cap V_{\alpha+3} = V_\beta \cap C_{\alpha'}$ and $[V_\beta \cap Q_{\alpha'}, V_{\alpha'}] = \{1\}$, contradicting the initial assumption. □

Proposition 7.19 *Suppose that $C_{V_\beta}(V_{\alpha'}) < V_\beta \cap Q_{\alpha'}$. Then Z_α is a natural $\mathrm{SL}_2(q_\alpha)$-module.*

Proof Throughout, aiming for a contradiction, we assume that Z_α is not a natural module and so we are in the case (ii) or (iii) or Lemma 7.16.

Suppose that $b > 3$ and Z_α is not a natural $\mathrm{SL}_2(q_\alpha)$-module. By Lemma 7.16, we have that $q_\alpha = q_\beta$. If $q_\alpha = p$ then Z_α is as described in Lemma 7.17, whereas if $q_\alpha > p$ then Z_α is a direct sum of two natural $\mathrm{SL}_2(q_\alpha)$-modules. Since $V_\alpha^{(2)}$ is elementary abelian and $Z_\alpha \cap Q_{\alpha'} \not\le Q_{\alpha'+1}$, $V_\alpha^{(2)} \cap Q_{\alpha'-2} \cap Q_{\alpha'-1} = Z_\alpha(V_\alpha^{(2)} \cap Q_{\alpha'+1})$ is centralized, modulo Z_α, by $Z_{\alpha'+1} \cap Q_\beta \not\le Q_\alpha$. By Lemma 7.18, we conclude that $V_\alpha^{(2)} \not\le Q_{\alpha'-2}$ and $V_\alpha^{(2)} \cap Q_{\alpha'-2} \not\le Q_{\alpha'-1}$.

Suppose first that $L_\alpha/R_\alpha \cong \mathrm{SU}_3(2)'$ so that $q_\alpha = q_\beta = p$. Then $R = [V_\beta \cap Q_{\alpha'}, V_{\alpha'}] = [V_{\alpha'}, Q_{\alpha'}]$ is of order 4 and strictly contained in $Z_{\alpha'}$. Moreover, since $b > 3$, R is centralized by $X_{\alpha'-1} := \langle V_\alpha^{(2)} \cap Q_{\alpha'-2}, R_{\alpha'-1}, Q_{\alpha'}\rangle$ and so either $Q_{\alpha'}Q_{\alpha'-1}$ is conjugate to $Q_{\alpha'}Q_{\alpha'-2}$ by an element of $R_{\alpha'-1}$; or $X_{\alpha'-1}/C_{X_{\alpha'-1}}(Z_{\alpha'-1}) \cong \mathrm{Sym}(3)$. In the latter case, it follows that R is invariant under the action of a subgroup of index 3 in $L_{\alpha'-1}$, a contradiction to structure of $Z_{\alpha'-1}$. In the former case, it follows that $[V_{\alpha'}, Q_{\alpha'}] = [V_{\alpha'-2}, Q_{\alpha'-2}]$ and since $V_\alpha^{(2)} \not\le Q_{\alpha'-2}$, we may iterate backwards through critical pairs $(\alpha - 2k, \alpha' - 2k)$ for $k \geqslant 0$ so that $R = [V_{\alpha'}, Q_{\alpha'}] = [V_\beta, Q_\beta] \le Z_\beta$ and so an index p subgroup of V_β/Z_β is centralized by $V_{\alpha'}$. We have a contradiction by Lemma 7.9.

Now, for all values of q_α, we have that $Z_{\alpha'} = H \le V_\beta$. Then $V_\alpha^{(2)} \cap Q_{\alpha'-2}$ is not contained in $Q_{\alpha'-1}$ and centralizes $Z_{\alpha'}Z_{\alpha'-2}$. It follows that $Z_{\alpha'} = Z_{\alpha'-2}$. Moreover, since $V_\alpha^{(2)} \not\le Q_{\alpha'-2}$ there is some $\alpha-2$, with $(\alpha-2, \alpha'-2)$ a critical pair. By Lemma 7.9, we may assume that $(\alpha-2, \alpha'-2)$ satisfies the same hypothesis as (α, α'). Iterating through critical pairs, we conclude that $Z_{\alpha'} = \cdots = Z_\beta$. But then $R = [V_\beta \cap Q_{\alpha'}, V_{\alpha'}] = Z_\beta$ and $V_\beta/C_{V_\beta}(O^p(L_\beta))$ is a natural $\mathrm{SL}_2(q_\alpha)$-module for L_β/R_β, a contradiction by Lemma 7.9. Hence if Z_α is not a natural $\mathrm{SL}_2(q_\alpha)$-module then $b = 3$.

Suppose that $L_\alpha/R_\alpha \cong \mathrm{SU}_3(2)'$, $b = 3$ and Z_α is the restriction of a natural $\mathrm{SU}_3(2)$-module. Since Q_α is non-abelian, by the irreducibility of Z_α, $Z_\alpha \le \langle (Z_\beta \cap \Phi(Q_\alpha))^{G_\alpha} \rangle \le \Phi(Q_\alpha)$. If $|S/Q_\beta| = 2$, then $Q_\alpha \cap Q_\beta \cap Q_{\alpha'-1} = Z_\alpha(Q_\alpha \cap \cdots \cap Q_{\alpha'+1})$ and since $Q_\alpha/\Phi(Q_\alpha)$ is not an FF-module, $\overline{L_\alpha} \cong \mathrm{SU}_3(2)'$ and $Q_\alpha/\Phi(Q_\alpha)$ contains a unique non-central chief factor, U/V say. Moreover, U/V is isomorphic as an $\overline{L_\alpha}$-module to Z_α and $U \not\le Q_\beta$. But $U \cap Q_\beta$ is $G_{\alpha,\beta}$-invariant subgroup of index 2 in U, a contradiction. Hence, $|S/Q_\beta| > 2$.

Applying Proposition 7.12, we see that $L_\beta/R_\beta \cong (3 \times 3) : 4$. Now $V_\beta(Q_\beta \cap Q_{\alpha'-1} \cap Q_{\alpha'})$ has index at most 4 in Q_β and since $|S/Q_\beta| \ne 2$, no non-central chief factor is an FF-module for $\overline{L_\beta}$ and so Q_β/V_β contains a unique non-central chief factor for $\overline{L_\beta}$, and this chief factor lies in Q_β/C_β. Then, an application of the three subgroup lemma implies that $R_\beta = Q_\beta$. Thus, $\overline{L_\beta} \cong (3 \times 3) : 4$. However, from the structure of Z_α, we conclude that $Z_\beta = Z_\alpha \cap C_{V_\beta}(O^p(L_\beta))$ has index 4 in Z_α so that a subgroup of order 4 of $V_\beta/C_{V_\beta}(O^2(L_\beta))$ is centralized by $S = Q_\alpha Q_\beta$, contradicting the structure of the 2F-module associated to $(3\times 3) : 4$. Hence, $L_\alpha/R_\alpha \not\cong \mathrm{SU}_3(2)'$.

Therefore, using Lemma 7.17 when $q_\alpha = p$, we may assume that $b = 3$ and $Z_\alpha = Z_\beta \times Z_{\alpha-1}$ for some $\alpha - 1 \in \Delta(\alpha)$. Then $[V_\beta \cap Q_{\alpha'}, V_{\alpha'} \cap Q_\beta] \le Z_{\alpha'} \cap Z_\beta$. But $Z_{\alpha'}Z_\beta \le Z_{\alpha'-1}$ and by Lemma 7.17, either $Z_{\alpha'} = Z_\beta$, or $Z_{\alpha'} \cap Z_\beta = \{1\}$. If $Z_{\alpha'} = Z_\beta$, then $R = [V_\beta \cap Q_{\alpha'}, V_{\alpha'}] = Z_\beta$ and so $V_\beta/C_{V_\beta}(O^p(L_\beta))$ is a natural $\mathrm{SL}_2(q_\alpha)$-module, a contradiction by Lemma 7.9. Hence, $[Z_\alpha \cap Q_{\alpha'}, Z_{\alpha'+1} \cap Q_\beta] \le [V_\beta \cap Q_{\alpha'}, V_{\alpha'} \cap Q_\beta] \le Z_{\alpha'} \cap Z_\beta = \{1\}$ and by Lemma 2.41, Z_α is an FF-module, a final contradiction. Thus, Z_α is a natural module. □

Lemma 7.20 *Suppose that $C_{V_\beta}(V_{\alpha'}) < V_\beta \cap Q_{\alpha'}$. Then $p = 2$, $q_\alpha = q_\beta$ and for $V := V_\beta/C_{V_\beta}(O^p(L_\beta))$ either:*

(i) *V is a natural $\mathrm{Sz}(q_\beta)$-module for $L_\beta/R_\beta \cong \mathrm{Sz}(q_\beta)$;*
(ii) *V is a natural $\mathrm{Sz}(2)$-module for $L_\beta/R_\beta \cong \mathrm{Dih}(10)$; or*
(iii) *V is a 2F-module for $L_\beta/R_\beta \cong (3 \times 3) : 2$ or $(3 \times 3) : 4$.*

Moreover, for $T \in \mathrm{Syl}_p(G_{\alpha',\alpha-1})$ and $P \in \mathrm{Syl}_p(G_{\beta,\alpha+2})$, we have that $V_\beta Q_{\alpha'}/Q_{\alpha'} = \Omega(T/Q_{\alpha'})$, $V_{\alpha'}Q_\beta/Q_\beta = \Omega(P/Q_\beta)$, $(V_\beta \cap Q_{\alpha'})Q_{\alpha'+1} \in \mathrm{Syl}_p(L_{\alpha'+1})$ and $S = (V_{\alpha'} \cap Q_\beta)Q_\alpha$.

Proof Suppose first that $q_\beta = p$. By Lemma 7.9, V is not an FF-module and so, as V is a quadratic 2F-module, the structure of V and L_β/R_β follows from Proposition 7.12. Since $Z_\alpha C_{V_\beta}(O^p(L_\beta))/C_{V_\beta}(O^p(L_\beta))$ is of order p, $G_{\alpha,\beta}$-

invariant and $V_\beta = \langle Z_\alpha^{L_\beta} \rangle$, by Proposition 7.12, we conclude that $L_\beta/R_\beta \cong$ Sz(2), Dih(10), $(3 \times 3) : 2$ or $(3 \times 3) : 4$ and V is determined. Thus, the lemma holds in this case.

Hence, we assume that $q_\alpha = q_\beta > p$ so that $\overline{L_\beta}/O_{p'}(\overline{L_\beta}) \cong \mathrm{PSL}_2(q_\beta)$, $\mathrm{PSU}_3(q_\beta)$ or $\mathrm{Sz}(q_\beta)$. Then $V_\beta \cap Q_\beta \cap Q_\alpha$ has index at most q_β^2 and is centralized by Z_α. Applying Lemmas 2.55 and 7.9, we deduce that $L_{\alpha'}/R_{\alpha'} \cong \mathrm{SL}_2(q_\beta)$, (P)$\mathrm{SU}_3(q_\beta)$ or $\mathrm{Sz}(q_\beta)$ and $V_\beta/C_{V_\beta}(O^2(L_\beta))$ is either a direct sum of two natural $\mathrm{SL}_2(q_\beta)$-modules, or the associated natural module in the latter two cases. Moreover, $Z_\alpha C_{V_\beta}(O^2(L_\beta))/C_{V_\beta}(O^2(L_\beta)) \cong Z_\alpha/Z_\beta$ has order $q_\alpha = q_\beta$ in $V_\beta/C_{V_\beta}(O^2(L_\beta))$ and is $G_{\alpha,\beta}$-invariant. Then Lemma 2.22 and Proposition 2.28 imply that $L_\beta/R_\beta \cong \mathrm{Sz}(q_\beta)$ and $V_\beta/C_{V_\beta}(O^2(L_\beta))$ is a natural module.

In all cases, we see a subgroup of index exactly q_β^2 in V_β resp. $V_{\alpha'}$ is centralized by $Z_{\alpha'+1}$ resp. Z_α, and so the remaining claims follow. □

Proposition 7.21 *Suppose that $C_{V_\beta}(V_{\alpha'}) < V_\beta \cap Q_{\alpha'}$. Then $b = 3$.*

Proof Aiming for a contradiction, suppose that $b > 3$. Since V_β is centralized by $V_\beta^{(3)}$ we deduce that $[V_{\alpha'}, V_\beta, V_\beta^{(3)}] = \{1\}$. We observe that as $V_\beta^{(3)}$ is generated by elementary abelian subgroups, $V_\beta^{(3)} Q_{\alpha'-2}/Q_{\alpha'-2} \leq \Omega(P/Q_{\alpha'-2})$ where $P \in \mathrm{Syl}_p(L_{\alpha'-2})$. Heavily using the structure of the modules in involved, and that $V_\beta^{(3)}$ centralizes $[V_{\alpha'}, V_\beta]$, we must have that $V_\beta^{(3)} \cap Q_{\alpha'-2} \cap Q_{\alpha'-1} = V_\beta(V_\beta^{(3)} \cap \cdots \cap Q_{\alpha'})$ and $[V_\beta^{(3)} \cap \cdots \cap Q_{\alpha'}, V_{\alpha'}] = [V_{\alpha'}, Q_{\alpha'}] = Z_{\alpha'} = R \leq V_\beta$.

Assume that $|S/Q_\beta| \neq 2$. Since $V_\beta^{(3)} \cap Q_{\alpha'-2} \cap Q_{\alpha'-1}$ has index at most q_β^2 in $V_\beta^{(3)}$, applying Lemma 2.55 when $q_\beta > 2$ and Lemma 2.41 when $|S/Q_\beta| \neq 2$, we conclude that $V_\beta^{(3)} \not\leq Q_{\alpha'-2}$ and $(V_\beta^{(3)} \cap Q_{\alpha'-2})Q_{\alpha'-1} \in \mathrm{Syl}_p(L_{\alpha'-1})$. Then $Z_{\alpha'}$ is centralized by $\langle Q_{\alpha'}, R_{\alpha'-1}, (V_\beta^{(3)} \cap Q_{\alpha'-2})\rangle$ and we conclude that $R_{\alpha'-1}Q_{\alpha'-2} = R_{\alpha'-1}Q_{\alpha'}$ and $Z_{\alpha'} = Z_{\alpha'-2}$. Since $V_\beta^{(3)} \not\leq Q_{\alpha'-2}$ there is a critical pair $(\beta - 3, \alpha' - 2)$ satisfying the same hypothesis as (α, α') by Lemmas 7.5 and 7.6, and iterating back through critical pairs, we conclude that $Z_{\alpha'} = \cdots = Z_\beta$, $[V_\beta \cap Q_{\alpha'}, V_{\alpha'}] = R = Z_{\alpha'} = Z_\beta$ and $V_\beta/C_{V_\beta}(O^2(L_\beta))$ is a natural $\mathrm{SL}_2(q_\beta)$-module, a contradiction by Lemma 7.9.

Hence, $|S/Q_\beta| = 2$ and we may assume by the above argument that either $V_\beta^{(3)} \leq Q_{\alpha'-2}$ and $Z_{\alpha'} = Z_{\alpha'-2}$, or $V_\beta^{(3)} \cap Q_{\alpha'-2} \leq Q_{\alpha'-1}$. Repeating the proof of Lemma 7.18, we deduce that if $V_\alpha^{(2)}/Z_\alpha$ contains a unique non-central chief factor for L_α, then it is not an FF-module for $\overline{L_\alpha}$, and so we have that either $V_\alpha^{(2)} \leq Q_{\alpha'-2}$, $V_\alpha^{(2)} \not\leq Q_{\alpha'-1}$ and $Z_{\alpha'} = Z_{\alpha'-2}$; or $V_\alpha^{(2)} \not\leq Q_{\alpha'-2}$ and $V_\alpha^{(2)} \cap Q_{\alpha'-2} \leq Q_{\alpha'-1}$.

In the former case, there is $\alpha - 1 \in \Delta(\alpha)$ with $V_{\alpha-1} \leq Q_{\alpha'-2}$ and $V_{\alpha-1} \not\leq Q_{\alpha'-1}$. Hence, an index 2 subgroup of $V_{\alpha-1}$ is centralized by $Z_{\alpha'-1}$ and Lemma 2.41 yields that $Z_{\alpha'-1} \leq Q_{\alpha-1}$. But then $Z_{\alpha'-2} = [Z_{\alpha'-1}, V_{\alpha-1}] = Z_{\alpha-1}$ and since $b > 3$ we have that $Z_{\alpha'} = Z_{\alpha'-2} = Z_{\alpha-1} = Z_\beta$. But then, as above, Lemma 7.9 gives a contradiction. Hence, $V_\alpha^{(2)} \not\leq Q_{\alpha'-2}$ and so there is a critical pair $(\alpha - 2, \alpha' - 2)$.

Applying Lemmas 7.5 and 7.6, $(\alpha - 2, \alpha' - 2)$ satisfies the same properties as (α, α') and we can iterate far back enough through critical pairs so that we find a value m such that for all $k \geqslant m$, $(\alpha - 2k, \alpha' - 2k)$ is a critical pair and $(\alpha' - 3 - 2k, \alpha' - 2k + b - 3)$ is a critical pair. In particular, $Z_{\alpha'-1-2k} \neq Z_{\alpha'-3-2k}$ so that $L_{\alpha'-2-2k} = \langle V_{\alpha-2k}^{(2)}, Q_{\alpha'-1-2k}, R_{\alpha'-2-2k}\rangle$. Since $Z_{\alpha'-2k}Z_{\alpha'-2-2k}$ is normalized by $L_{\alpha'-2-2k}$, we deduce that $Z_{\alpha'-2k} = Z_{\alpha'-2-2k}$ for all $k \geqslant m$ and iterating even further back through the critical pairs, we have that $Z_{\alpha'-2k} = Z_{\beta-2k}$ and again, Lemma 7.9 provides a contradiction. □

Lemma 7.22 *Suppose that $C_{V_\beta}(V_{\alpha'}) < V_\beta \cap Q_{\alpha'}$ and $b = 3$. Then $p = 2$, $q_\alpha = q_\beta$, $L_\alpha/R_\alpha \cong \mathrm{SL}_2(q_\alpha)$, Z_α is natural $\mathrm{SL}_2(q_\alpha)$-module, $O^2(L_\beta)$ centralizes C_β/V_β and one of the following holds:*

(i) *$\overline{L_\beta} \cong \mathrm{Sz}(q_\beta)$ and V_β/Z_β is a natural module $\mathrm{Sz}(q_\beta)$-module; or*
(ii) *$\overline{L_\beta} \cong (3 \times 3) : 4$ and V_β/Z_β is an irreducible 2F-module.*

In particular, L_β is 2-minimal in either case.

Proof Suppose that $|S/Q_\beta| = 2$ so that $L_\beta/R_\beta \cong \mathrm{Dih}(10)$ or $(3 \times 3) : 2$. Then $C_\beta \leq Q_{\alpha'-1}$ and $C_\beta = V_\beta(C_\beta \cap Q_{\alpha'})$. Since $[V_{\alpha'}, Q_{\alpha'}] = Z_{\alpha'} \leq V_\beta$, we deduce that $O^2(L_\beta)$ centralizes C_β/V_β. Then for $r \in R_\beta$ of odd order, if $[r, Q_\beta, V_\beta] = \{1\}$ then $[r, V_\beta, Q_\beta] = \{1\}$ by the three subgroup lemma, and so r centralizes Q_β. But now, $Q_\beta \cap Q_{\alpha'-1} = V_\beta(Q_\beta \cap Q_{\alpha'})$, and so Q_β/V_β contains a unique non-central chief factor for L_β, which is a faithful FF-module for $\overline{L_\beta}$, and $\overline{L_\beta} \cong \mathrm{Sym}(3)$ by Lemma 2.41, a contradiction.

Thus, $|S/Q_\beta| = 4$ or $q_\beta > 2$ and by Lemma 2.41, no non-central chief factor within Q_β is an FF-module for $\overline{L_\beta}$. Since $C_\beta \leq Q_{\alpha'-1}$, $V_\beta(C_\beta \cap Q_{\alpha'})$ has index at most q_β in C_β and since $[Q_{\alpha'}, V_{\alpha'}] = Z_{\alpha'} \leq V_\beta$, $V_{\alpha'}$ centralizes C_β/V_β so that $O^2(L_\beta)$ centralizes C_β/V_β. Now, applying the three subgroup lemma, any odd order element of R_β centralizes Q_β/C_β and V_β so centralizes Q_β, and we deduce that $R_\beta = Q_\beta$. By Lemma 7.20, $\overline{L_\beta} \cong \mathrm{Sz}(q_\beta)$ or $(3 \times 3) : 4$ and $V_\beta/C_{V_\beta}(O^2(L_\beta))$ is a natural module or an irreducible 2F-module described in Proposition 7.12.

Assume that $q_\beta = 2$ so that L_β is solvable. Applying coprime action, we have that $V_\beta/Z_\beta = [V_\beta/Z_\beta, O^2(L_\beta)] \times C_{V_\beta/Z_\beta}(O^2(L_\beta))$ where $[V_\beta/Z_\beta, O^2(L_\beta)]$ is irreducible of dimension 4. Letting V^β be the preimage in V_β of $[V_\beta/Z_\beta, O^2(L_\beta)]$, we must have that $[V^\beta \cap Q_{\alpha'}, V_{\alpha'}] = Z_{\alpha'} \leq V^\beta$ so that $Z_{\alpha'-1} = Z_{\alpha'} \times Z_\beta \leq V^\beta$. But then, by definition, $V^\beta = V_\beta$ and V_β/Z_β is irreducible of dimension 4. Since Q_β/C_β is dual to V_β/Z_β, Q_β/C_β is also irreducible of dimension 4.

Assume that $q_\beta > 2$. Note that $V_\beta(Q_\beta \cap Q_{\alpha'-1} \cap Q_{\alpha'})$ has index q_β^2 in Q_β and centralized, modulo V_β, by $V_{\alpha'}$. Since $L_\beta = \langle V_{\alpha'}, V_{\alpha'}^g, Q_\beta\rangle$ for some appropriately chosen $g \in L_\beta$, we deduce that $|Q_\beta/\Phi(Q_\beta)V_\beta| \leqslant q_\beta^4$. Furthermore, since $\Phi(Q_\beta)V_\beta \leq Q_{\alpha'-1} \cap Q_\beta \ntrianglelefteq L_\beta$ by Proposition 5.27, $Q_\beta/\Phi(Q_\beta)V_\beta$ is irreducible. But now, $\Phi(Q_\beta)V_\beta \leq Q_\beta \cap \bigcap_{\lambda \in \Delta(\beta)} Q_\lambda \leq C_\beta$ so that $|Q_\beta/C_\beta| = q_\beta^4$, $Q_\beta = (Q_\beta \cap O^2(L_\beta))C_\beta$ and L_β centralizes $C_{V_\beta}(O^2(L_\beta))$. Hence, V_β/Z_β is a natural module and the result holds. □

We are now ready to complete this section, proving that $C_{V_\beta}(V_{\alpha'}) < V_\beta \cap Q_{\alpha'}$ yields only amalgams locally isomorphic to ${}^2\mathrm{F}_4(2)'$ or ${}^2\mathrm{F}_4(q)$. We achieve this over the following two propositions.

Proposition 7.23 *Suppose that $C_{V_\beta}(V_{\alpha'}) < V_\beta \cap Q_{\alpha'}$ and $b > 1$. If $q_\beta > 2$ then G is locally isomorphic to ${}^2\mathrm{F}_4(q_\beta)$.*

Proof We have that $b = 3$, $L_\alpha/R_\alpha \cong \mathrm{SL}_2(q_\alpha)$, Z_α is natural $\mathrm{SL}_2(q_\alpha)$-module, $\overline{L_\beta} \cong \mathrm{Sz}(q_\beta)$ and V_β/Z_β is a natural module $\mathrm{Sz}(q_\beta)$-module. Our aim will be to demonstrate that $R_\alpha = Q_\alpha$. Assume that $Z_{\alpha'-1} \cap [V_\beta, Q_\alpha] > Z_\beta$. Then, by the 2-transitivity of $\overline{L_\beta}$ on the neighbours of β, $Z_\alpha \cap [V_\beta, Q_{\alpha'-1}] > Z_\beta$. Since $[V_\beta, Q_{\alpha'-1}] \leq Q_{\alpha'}$, this is a contradiction. Hence, $V_\beta = Z_{\alpha'} \times [V_\beta, Q_\alpha]$ so that $[V_\alpha^{(2)}, Q_\alpha] \cap Z_{\alpha'} = \{1\}$.

We note that as $V_\alpha^{(2)}$ is generated by elementary abelian subgroups, $V_\alpha^{(2)} Q_\beta/Q_\beta \leq \Omega(S/Q_\beta)$. Hence, $[V_\alpha^{(2)}, Q_\alpha] \leq Q_\beta$. For the remainder of this proof, set $V := [V_\alpha^{(2)}, Q_\alpha]$ and $C := C_{Q_\alpha}(V)$. Then C centralizes $[V_\beta, Q_\alpha]$ so that $C \leq Q_\beta$. Moreover, $C \cap Q_{\alpha'-1}$ centralizes $V_\beta \leq Z_{\alpha'} V$ so that $C \cap Q_{\alpha'-1} = C \cap C_\beta$. From the structure of Q_β/C_β as an $\overline{L_\beta}$-module, we infer that $C_\beta = V_\beta(C_\beta \cap Q_{\alpha'})$. Hence, $C \cap Q_{\alpha'-1} = Z_\alpha(C \cap Q_{\alpha'})$.

Now, $V \geq [V_\alpha^{(2)}, Q_\alpha \cap Q_\beta] \not\leq C_\beta$. In particular, $V \not\leq Q_{\alpha'-1}$, else $V \leq Q_\alpha \cap Q_\beta \cap Q_{\alpha'-1}$ and as $\overline{L_\beta}$ acts 2-transitively on neighbours of β, this would imply that $V \leq Q_\beta \cap \bigcap_{\lambda \in \Delta(\beta)} Q_\lambda = C_\beta$, a contradiction. Hence, $Z_{\alpha'} \cap C = Z_{\alpha'} \cap V = \{1\}$ and applying Lemma 2.41, we have that C/Z_α contains a unique non-central chief factor which is an FF-module for $\overline{L_\alpha}$ so that $O^2(R_\alpha)$ centralizes C as $q_\beta > 2$. Assume that $|VC_\beta/C_\beta| > q_\beta$. Then $[V, Q_\alpha] \not\leq C_\beta$ and as $[V_\beta, Q_\alpha, Q_\alpha] \leq C_\beta$, $[V, Q_\alpha]/Z_\alpha$ also contains a non-central chief factor for $\overline{L_\alpha}$. Then both $V/[V, Q_\alpha]$ and $[V, Q_\alpha]/Z_\alpha$ are FF-modules for $\overline{L_\alpha}$ and $O^2(R_\alpha)$ centralizes V. Assume that $|VC_\beta/C_\beta| = q_\beta$ so that $V \cap Q_{\alpha'-1} = V \cap C_\beta$ and we deduce that $V \cap Q_{\alpha'-1} = Z_\alpha(V \cap Q_{\alpha'})$ and V/Z_α contains at most one non-central chief factor and, if it exists, it is an FF-module for $\overline{L_\alpha}$ so that $O^2(R_\alpha)$ centralizes V. Hence, $O^2(R_\alpha)$ centralizes CV.

Finally, by the three subgroups lemma $[O^2(R_\alpha), Q_\alpha, V] = \{1\}$ so that $[O^2(R_\alpha), Q_\alpha] \leq C$ and coprime action gives $[O^2(R_\alpha), Q_\alpha] = \{1\}$ and $R_\alpha = Q_\alpha$. Then G has a weak BN-pair of rank 2 and comparing with [27], the result holds. □

Proposition 7.24 *Suppose that $C_{V_\beta}(V_{\alpha'}) < V_\beta \cap Q_{\alpha'}$ and $b > 1$. Then G is locally isomorphic to ${}^2\mathrm{F}_4(2^n)$ or ${}^2\mathrm{F}_4(2)'$.*

Proof By Propositions 7.23, 7.21, and Lemma 7.22, we have that $b = 3$, $L_\alpha/R_\alpha \cong \mathrm{Sym}(3)$, Z_α is natural $\mathrm{SL}_2(2)$-module and either $\overline{L_\beta} \cong \mathrm{Sz}(2)$ or $\overline{L_\beta} \cong (3 \times 3) : 4$. Suppose first that L_α is also a 2-minimal group. Then the amalgam is determined in [41], G has a weak BN-pair of rank 2 and the result follows by [27] and [28]. Hence, to complete the proof, we assume that L_α is not 2-minimal and derive a contradiction. We may choose $P_\alpha < L_\alpha$ such that P_α is 2-minimal. Better, by McBride's lemma (Proposition 2.5), we may choose P_α such that $P_\alpha \not\leq R_\alpha$ and $L_\alpha = P_\alpha R_\alpha$. Moreover, we may assume that G is a minimal counterexample to

Theorem C. Form $X := \langle P_\alpha, L_\beta(G_{\alpha,\beta} \cap P_\alpha)\rangle$ and let K be the largest subgroup of S which is normal in X.

If $K = \{1\}$, then it follows that any non-trivial normal subgroup of X which is contained in $G_{\alpha,\beta} \cap P_\alpha$ is a $2'$-group, a contradiction for then Q_λ is not self centralizing in G_λ, where $\lambda \in \{\alpha, \beta\}$. Thus, no non-trivial normal subgroup of $G_{\alpha,\beta} \cap P_\alpha$ is normal in X and the triple $(P_\alpha, L_\beta(G_{\alpha,\beta} \cap P_\alpha), G_{\alpha,\beta} \cap P_\alpha)$ satisfies Hypothesis 5.1. Then, by minimality and comparing with the list of amalgams in Theorem C, it follows that X is locally isomorphic to ${}^2\mathrm{F}_4(2)$ or ${}^2\mathrm{F}_4(2)'$. In particular, $P_\alpha/Q_\alpha \cong \mathrm{Sym}(3)$, $G_\beta/Q_\beta \cong \mathrm{Sz}(2)$ and S is isomorphic to a Sylow 2-subgroup of ${}^2\mathrm{F}_4(2)$ or ${}^2\mathrm{F}_4(2)'$. But then $2^2 \leqslant |Q_\alpha/\Phi(Q_\alpha)| \leqslant 2^3$ and so, $\overline{L_\alpha}$ is isomorphic to a subgroup of $\mathrm{GL}_3(2)$ which has a strongly 2-embedded subgroup. An elementary calculation, which may be performed in MAGMA [14], yields $\overline{L_\alpha} \cong \overline{P_\alpha} \cong \mathrm{Sym}(3)$ and L_α is 2-minimal, a contradiction.

Thus, $K \neq \{1\}$ and since P_α does not centralize Z_β and $K \trianglelefteq S$, we deduce that $Z_\alpha \leq K$ and so $V_\beta \leq K$. Moreover, since $K \leq Q_\alpha \cap Q_\beta$ and $K \trianglelefteq L_\beta$, $K \leq C_\beta$. If $\Phi(K) \neq \{1\}$ then $Z_\beta \leq \Phi(K)$ and arguing as above, $V_\beta \leq \Phi(K)$. But then $O^2(L_\beta)$ centralizes $K/\Phi(K)$, a contradiction. Thus, K is elementary abelian and since $C_S(K) \leq C_\beta$, $C_S(K) = C_{Q_\alpha}(K) = C_{Q_\beta}(K) \trianglelefteq X$ and $C_S(K) = K$.

Suppose that there is $r \in P_\alpha$ of odd order such that $[r, Q_\alpha] \leq K$. If r centralizes $C_K(Q_\alpha)$, then by the A$\times$B-lemma, r centralizes K. But then r centralizes Q_α, and so r is trivial. Now, since Q_α is self centralizing in S, $C_K(Q_\alpha) \leq Z(Q_\alpha)$. But $V_{\alpha'} \cap Q_\alpha$ is of index 4 in $V_{\alpha'}$, contains $Z_{\alpha'-1}$ and is centralized by $Z(Q_\alpha)$ from which it follows that $Z(Q_\alpha) = Z_\alpha(Z(Q_\alpha) \cap Q_{\alpha'})$. Since $Z_{\alpha'} \not\leq Z(Q_\alpha)$, otherwise $Z_{\alpha'-1} = Z_{\alpha'} \times Z_\beta$ would be normalized by $L_\beta = \langle Q_\beta, Q_\alpha, Q_{\alpha'-1}\rangle$, it follows that $V_{\alpha'} \cap Q_\beta$ centralizes $Z(Q_\alpha)/Z_\alpha$ and so $O^2(L_\alpha)$ centralizes $Z(Q_\alpha)/Z_\alpha$. Since $Z_\beta \leq Z_\alpha = [Z(Q_\alpha), O^2(L_\alpha)]$, it follows from coprime action that $Z(Q_\alpha) = Z_\alpha$. Hence, for r of odd order such that $[r, Q_\alpha] \leq K$, we have that $\langle r\rangle \cap R_\alpha = \{1\}$. It follows that r is of order 3 and $\langle r\rangle K/K = O_{2'}(P_\alpha/K)$. Then, by coprime action and as r acts non-trivially on Z_α, we have that $K = [K, r]$. But now, as K is elementary abelian and contains V_β, it $K \cap Q_{\alpha'} \cap Q_{\alpha'+1}$ is has index 4 in K and is centralized by $Z_{\alpha'+1} \cap Q_\beta \not\leq Q_\alpha$. In particular, K contains at most two non-central chief factors for P_α, and we may arrange that one of these non-central chief factors is Z_α. Then the other non-central chief factor is also an FF-module and so has order 4. But then since $K = [K, r]$, by coprime action, we now have that $K/Z_\alpha = [K/Z_\alpha, r]$ has order 4 and $|K| = 2^4 < 2^5 = |V_\beta|$, absurd as $V_\beta \leq K$. Thus, no such r exists and P_α/K is of characteristic 2.

Suppose that there is $s \in L_\beta(P_\alpha \cap G_{\alpha,\beta})$ such that $[s, Q_\beta] \leq K$. Since $L_\beta/Q_\beta \cong$ $\mathrm{Sz}(2)$ it follows that $L_\beta(P_\alpha \cap G_{\alpha,\beta})/Q_\beta = L_\beta/Q_\beta \times O_{2'}((P_\alpha \cap G_{\alpha,\beta})/Q_\beta)$. Since $K \leq C_\beta$ and Q_β/C_β is an irreducible module for $\overline{L_\beta}$, $s \not\leq L_\beta$. Hence, s centralizes S/Q_β and so centralizes S/K. Then $s \in P_\alpha$ and centralizes Q_α/K, and by the previous paragraph, $s = 1$. Thus, $L_\beta(P_\alpha \cap G_{\alpha,\beta})/K$ is of characteristic 2. Moreover, no subgroup of S properly containing K is normal in X and since P_α/K is of characteristic 2, it follows that no non-trivial subgroup of $(G_{\alpha,\beta} \cap P_\alpha)/K$ is normal in X/K. Then the triple $(P_\alpha/K, (L_\beta(G_{\alpha,\beta} \cap P_\alpha))/K, (G_{\alpha,\beta} \cap P_\alpha)/K)$ satisfies Hypothesis 5.1. By minimality and since $L_\beta/Q_\beta \cong \mathrm{Sz}(2)$, X/K is locally isomorphic to ${}^2\mathrm{F}_4(2)$ or ${}^2\mathrm{F}_4(2)'$. But there is only one non-central chief factor in Q_β/K for L_β, and we have a final contradiction. □

7.2 $C_{V_\beta}(V_{\alpha'}) = V_\beta \cap Q_{\alpha'}$

We continue with the analysis of the case $[Z_\alpha, Z_{\alpha'}] = \{1\}$, this time with the additional assumptions that $b > 1$ and $[V_\beta \cap Q_{\alpha'}, V_{\alpha'}] = \{1\}$. Recall from Lemmas 7.5 and 7.6 that this hypothesis implies that $q := q_\alpha = q_\beta$, $L_\alpha/R_\alpha \cong L_\beta/R_\beta \cong \mathrm{SL}_2(q)$ and Z_α and $V_\beta/C_{V_\beta}(O^p(L_\beta))$ are natural $\mathrm{SL}_2(q)$-modules.

Throughout this subsection, we fix the notation $V^\lambda := \langle (C_{V_\mu}(O^p(L_\mu)))^{G_\lambda} \rangle$ whenever $\lambda \in \alpha^G$, $\mu \in \Delta(\lambda)$ and $|V_\beta| \neq q^3$, and we remark that when $|V_\beta| \neq q^3$ and $b > 3$, for $\gamma \in \beta^G$ and some fixed $\delta \in \Delta(\gamma)$, the subgroup $\langle V^\eta : Z_\eta = Z_\delta, \eta \in \Delta(\gamma)\rangle$ is normal in $R_\gamma Q_\delta$ by essentially the same argument as Lemma 5.23.

We set $R := [V_{\alpha'}, V_\beta]$ so that $R \leq Z_{\alpha+2} C_{V_\beta}(O^p(L_\beta)) \cap Z_{\alpha'-1} C_{V_{\alpha'}}(O^p(L_{\alpha'})) \leq V_\beta \cap V_{\alpha'}$. In particular, if $|V_\beta| = q^3$ then $R \leq Z_{\alpha+2} \cap Z_{\alpha'-1}$. By Lemma 7.9, we may assume in this section that every critical pair (α, α') satisfies the condition $C_{V_\beta}(V_{\alpha'}) = V_\beta \cap Q_{\alpha'}$. We reiterate that whenever we assume the necessary values of b, we are able to apply Lemma 5.32 through Lemma 5.36. That the hypotheses of these lemmas are satisfied will often be left implicit in proofs.

The first goal in the analysis of the case $C_{V_\beta}(V_{\alpha'}) = V_\beta \cap Q_{\alpha'}$ will be to show that $b \leqslant 5$. Then the methods for $b = 5$ differ slightly from the techniques employed for larger values of b and so, for the most part, we treat the case when $b = 5$ independently from the other cases. The case when $b = 3$ is different again and so this case is also treated separately.

The following lemma is also valid whenever $b = 3$ but, as mentioned above, since the techniques we apply when $b = 3$ are somewhat disparate from the rest of this subsection, we only prove it here whenever $b > 3$.

Lemma 7.25 *Suppose that $C_{V_\beta}(V_{\alpha'}) = V_\beta \cap Q_{\alpha'}$ and $b > 3$. If $V_\alpha^{(2)} \leq Q_{\alpha'-2}$ and $V_{\alpha'} \leq Q_\beta$ then $R = Z_\beta \leq Z_{\alpha'-1}$, $|V_\beta| = q^3$, $V_\alpha^{(2)}/Z_\alpha$ is an FF-module for $\overline{L_\alpha}$ and one of the following holds:*

(i) *$V_\alpha^{(2)} \leq Q_{\alpha'-1}$ and $[V_\alpha^{(2)} \cap Q_{\alpha'}, V_{\alpha'}] = Z_{\alpha'} \leq V_\alpha^{(2)}$; or*
(ii) *$V_\alpha^{(2)} \not\leq Q_{\alpha'-1}$, $[V_\alpha^{(2)} \cap Q_{\alpha'}, V_{\alpha'}] = \{1\}$, $R = Z_\beta = Z_{\alpha'-2}$ and $Z_{\alpha'-1} = Z_{\alpha'} \times Z_\beta$.*

Proof By Lemma 7.5, $Z_\beta = [V_\beta, Q_\beta]$ and $Z_\alpha Q_{\alpha'} \in \mathrm{Syl}_p(L_{\alpha'})$. Suppose first that $V_\alpha^{(2)} \leq Q_{\alpha'-1}$. Then $V_\alpha^{(2)} = Z_\alpha(V_\alpha^{(2)} \cap Q_{\alpha'})$ and since $V_\alpha^{(2)}/Z_\alpha$ contains a non-central chief factor for L_α, $[V_\alpha^{(2)} \cap Q_{\alpha'}, V_{\alpha'}] = Z_{\alpha'} \not\leq Z_\alpha$. Then, for $\alpha'+1 \in \Delta(\alpha')$ with $Z_{\alpha'+1} \not\leq Q_\alpha$ it follows that $[Z_{\alpha'+1}, V_\alpha^{(2)} \cap Q_{\alpha'} \cap Q_{\alpha'+1}] = \{1\}$ and $V_\alpha^{(2)}/Z_\alpha$ contains a unique non-central chief factor which is an FF-module for $\overline{L_\alpha}$. Then by Lemma 5.32, $|V_\beta| = q^3$, $[V_\alpha^{(2)}, Q_\alpha] = Z_\alpha$ and $Z_\beta = R \leq Z_{\alpha'-1} \cap Z_{\alpha+2}$.

Suppose now that $V_\alpha^{(2)} \not\leq Q_{\alpha'-1}$ and $|V_\beta| > q^3$. Then by Lemma 5.32, both V^α/Z_α and $V_\alpha^{(2)}/V^\alpha$ contain a non-central chief factor for L_α. If $V^\alpha \not\leq Q_{\alpha'-1}$, then $V^\alpha(V_\alpha^{(2)} \cap Q_{\alpha'})$ has index strictly less than q in $V_\alpha^{(2)}$ and so, we have that $[V_{\alpha'}, V_\alpha^{(2)} \cap Q_{\alpha'}] = Z_{\alpha'} \leq V_\alpha^{(2)}$ and $Z_{\alpha'} \not\leq V^\alpha$. In particular, $[V^\alpha \cap Q_{\alpha'}, V_{\alpha'}] = \{1\}$ and since V^α/Z_α contains a non-central chief factor, we deduce that $V^\alpha Q_{\alpha'-1} \in \mathrm{Syl}_p(L_{\alpha'-1})$. Since $b > 3$, $V_\alpha^{(2)}$ is elementary abelian and $V_\alpha^{(2)} \not\leq Q_{\alpha'-1}$, we deduce that $Z_{\alpha'} = C_{Z_{\alpha'-1}}(V_\alpha^{(2)}) = Z_{\alpha'-2} = [V^\alpha, Z_{\alpha'-1}] \leq V^\alpha$, a contradiction.

Thus, if $V_\alpha^{(2)} \not\leq Q_{\alpha'-1}$ and $|V_\beta| > q^3$ then $V^\alpha \leq Q_{\alpha'-1}$ and since V^α/Z_α contains a non-central chief factor, it follows that $[V^\alpha \cap Q_{\alpha'}, V_{\alpha'}] = Z_{\alpha'} \leq V^\alpha$, $(V^\alpha \cap Q_{\alpha'})Q_{\alpha'+1} \in \mathrm{Syl}_p(L_{\alpha'+1})$ for any $\alpha'+1 \in \Delta(\alpha')$ with $Z_{\alpha'+1} \not\leq Q_\alpha$ and V^α/Z_α is an FF-module for $\overline{L_\alpha}$. Since $V_\alpha^{(2)} \not\leq Q_{\alpha'-1}$ and $V_\alpha^{(2)}$ is abelian, $Z_{\alpha'} = C_{Z_{\alpha'-1}}(V_\alpha^{(2)}) = Z_{\alpha'-2}$, $(V_\alpha^{(2)} \cap Q_{\alpha'-1})/V^\alpha$ is centralized by $V_{\alpha'}$ and $V_\alpha^{(2)}/V^\alpha$ is also an FF-module for $\overline{L_\alpha}$. Then, applying Lemmas 5.33 and 5.22 to $Z_{\alpha'} = Z_{\alpha'-2}$, we conclude that $V_{\alpha'} = V_{\alpha'-2} \leq Q_\alpha$, a contradiction.

Thus, we assume now that $|V_\beta| = q^3$ and $V_\alpha^{(2)} \leq Q_{\alpha'-1}$ so that $Z_\beta = R \leq Z_{\alpha'-1} \cap Z_{\alpha+2}$. Indeed, $Z_\beta = C_{Z_{\alpha'-1}}(V_\alpha^{(2)}) = Z_{\alpha'-2}$ intersects $Z_{\alpha'}$ trivially, and $Z_{\alpha'-1} = Z_{\alpha'} \times Z_\beta$. If $[V_\alpha^{(2)} \cap Q_{\alpha'}, V_{\alpha'}] = Z_{\alpha'} \leq V_\alpha^{(2)}$, then $Z_{\alpha'-1} = Z_\beta \times Z_{\alpha'}$ is centralized by $V_\alpha^{(2)}$ and $V_\alpha^{(2)} \leq Q_{\alpha'-1}$, a contradiction. Thus, $[V_\alpha^{(2)} \cap Q_{\alpha'}, V_{\alpha'}] = \{1\}$, $(V_\alpha^{(2)} \cap Q_{\alpha'-1})/Z_\alpha$ is centralized by $V_{\alpha'}$ and $V_\alpha^{(2)}/Z_\alpha$ is an FF-module for $\overline{L_\alpha}$. □

Lemma 7.26 *Suppose that $C_{V_\beta}(V_{\alpha'}) = V_\beta \cap Q_{\alpha'}$ and $b > 5$. If $V_{\alpha'} \not\leq Q_\beta$ and $V_\alpha^{(2)} \leq Q_{\alpha'-2}$, then $|V_\beta| = q^3$.*

Proof Aiming for a contradiction, suppose throughout that $|V_\beta| \neq q^3$ so that both V^α/Z_α and $V_\alpha^{(2)}/V^\alpha$ contain a non-central chief factor for L_α by Lemma 5.32. Choose $\alpha'+1 \in \Delta(\alpha')$ with $Z_{\alpha'+1} \not\leq Q_\beta$. In particular, $(\alpha'+1, \beta)$ is a critical pair and we may assume that $C_{V_{\alpha'}}(V_\beta) = V_{\alpha'} \cap Q_\beta$. Set $U^\beta := \langle V^\lambda : \lambda \in \Delta(\beta), Z_\lambda = Z_\alpha \rangle$ so that $R_\beta Q_\alpha$ normalizes U^β by Lemma 5.23. Setting $U^{\alpha'} := \langle V^\mu : \mu \in \Delta(\alpha'), Z_\mu = Z_{\alpha'+1} \rangle$, it follows similarly that $U^{\alpha'} \trianglelefteq R_{\alpha'} Q_{\alpha'+1}$. Throughout, for $\mu \in \beta^G$, we set $U_\mu := \langle (V^{\mu+1})^{L_\mu} \rangle$ where $\mu+1 \in \Delta(\mu)$. In particular, $U^\beta \leq U_\beta \trianglelefteq L_\beta$. Note that for any $\lambda \in \Delta(\beta)$, whenever $Z_\alpha = Z_\lambda$ we have that $Q_\alpha R_\beta = Q_\lambda R_\beta$. Furthermore, $Q_\alpha O^p(R_\beta) = C_{L_\beta}(Z_\alpha) = C_{L_\beta}(Z_\lambda) = Q_\lambda O^p(R_\beta)$, $Q_\alpha \in \mathrm{Syl}_p(Q_\lambda O^p(R_\beta))$, Q_α is conjugate to Q_λ by an element of $O^p(R_\beta)$ and since $O^p(R_\beta)$ centralizes Q_β/C_β, $Q_\alpha \cap Q_\beta = Q_\lambda \cap Q_\beta$. In particular, $[Q_\alpha \cap Q_\beta, U^\beta] = Z_\alpha$.

We will commonly use that if $R \leq Z_{\alpha'-1}$, then $Z_{\alpha'-1} Z_{\alpha'-1}^g$ is normalized by $L_{\alpha'} = \langle V_\beta, (V_\beta)^g, R_{\alpha'} \rangle$ for some suitable $g \in L_{\alpha'}$. Then, from the definition of $V_{\alpha'}$, we conclude that $V_{\alpha'} = Z_{\alpha'-1} Z_{\alpha'-1}^g$ is of order q^3, forcing the contradiction. A similar conclusion follows if $R \leq Z_{\alpha+2}$.

Suppose first that $U^\beta \not\leq Q_{\alpha'-2}$ and so there is some $\lambda \in \Delta(\beta)$ with $V^\lambda \not\leq Q_{\alpha'-2}$ and $Z_\lambda = Z_\alpha$. In particular, since $V_{\alpha'-2} \leq Q_\lambda$ and $Z_\alpha \cap Q_{\alpha'} = Z_\beta$, we deduce that $[V_{\alpha'-2}, V^\lambda] = Z_\beta \leq V_{\alpha'-2}$ and $Z_{\alpha'-2} \cap Z_\beta = \{1\}$. Moreover, there is $\lambda - 1 \in \Delta(\lambda)$ such that $C_{V_{\lambda-1}}(O^p(L_{\lambda-1})) \not\leq Q_{\alpha'-2}$. In particular, $V_{\lambda-1} \not\leq Q_{\alpha'-2}$ and there is a critical pair $(\lambda - 2, \alpha' - 2)$. Since $C_{V_{\lambda-1}}(O^p(L_{\lambda-1}))(V_{\lambda-1} \cap Q_{\alpha'-2})$ has index strictly less than q in $V_{\lambda-1}$, we must have that $V_{\alpha'-2} \leq Q_{\lambda-1}$. Then, $[V_{\alpha'-2}, V_{\lambda-1}] = Z_{\lambda-1} \leq V_{\alpha'-2} \cap Z_\lambda$. Hence, $Z_{\lambda-1} = Z_\beta$.

We aim to show that $C_{L_{\lambda-2}}(V^{\lambda-2}/Z_{\lambda-2}) \cong \mathrm{SL}_2(q)$ for then, applying techniques from Lemma 5.33 reveals that $C_{L_{\lambda-2}}(V^{\lambda-2}/Z_{\lambda-2}) = R_{\lambda-2}$ and Lemma 5.22 and conjugacy yields that $C_{V_{\lambda-1}}(O^p(L_{\lambda-1}))Z_\lambda = C_{V_\beta}(O^p(L_\beta))Z_\lambda \leq Q_{\alpha'-2}$, a contradiction. Now, $[V^{\lambda-2} \cap Q_{\alpha'-4}, V_{\alpha'-4}] \leq Z_{\alpha'-4} \cap Z_{\lambda-2}$. If this is non-trivial, it follows that $Z_{\alpha'-4} = Z_{\lambda-1} = Z_\beta$, impossible for then $[V_{\alpha'-2}, V^\lambda] = Z_\beta \leq Z_{\alpha'-3}$ and we infer that $|V_{\alpha'-2}| = q^3$. Hence, $V^{\lambda-2} \cap Q_{\alpha'-4} = Z_{\lambda-2}(V^{\lambda-2} \cap Q_{\alpha'-2})$. By Lemma 2.41, we must have that $V^{\lambda-2} \not\leq Q_{\alpha'-4}$ and $Z_{\alpha'-2} \leq V^{\lambda-2}$. Since $V_{\lambda-2}^{(2)}/V^{\lambda-2}$ contains a non-central chief factor, and $V_{\lambda-2}^{(2)}$ centralizes $V^{\lambda-2}$, we ascertain that $Z_{\alpha'-2} = Z_{\alpha'-4}$.

Now, $V^{\lambda-2}/Z_{\lambda-2}$ is a quadratic 2F-module generated by $C_{V_{\lambda-1}}(O^p(L_{\lambda-1}))Z_{\lambda-2}/Z_{\lambda-2}$ and applying Proposition 2.51, Lemmas 2.54 and 2.55, we must have that $q = 2$ and $L_{\lambda-2}/C_{L_{\lambda-2}}(V^{\lambda-2}/Z_{\lambda-2})$ is isomorphic to Dih(10) or $(3 \times 3) : 2$. Arguing as in Lemma 5.33, since $Z_\lambda - 1 \ntrianglelefteq L_{\lambda-2}$, we must have that $C_{L_{\lambda-2}}(V^{\lambda-2}/Z_{\lambda-2}) < R_{\lambda-2}$. Therefore, $\overline{L_\alpha}/C_{\overline{L_\alpha}}(V_\alpha^{(2)}) \cong (3 \times 3) : 2$ and V^α/Z_α contains two non-central chief factors. Now, $V_\beta V^\alpha \trianglelefteq R_\alpha S$. Then, $[V^\beta V^\alpha, Q_\alpha] = Z_\alpha C_{V_\beta}(O^2(L_\beta)) \trianglelefteq R_\alpha S$. But one can verify (e.g. using MAGMA [14]) that R_α centralizes one of the non-central chief factors in V^α/Z_α and that $C_{V_\beta}(O^2(L_\beta))$ projects in exactly one of the summands in $[V^\alpha, O^2(L_\alpha)]/Z_\alpha$. But V^α is the normal closure in L_α of $C_{V_\beta}(O^2(L_\beta))$ and we have a contradiction.

Thus, $U^\beta \leq Q_{\alpha'-2}$. Notice that the hypothesis $V_\alpha^{(2)} \leq Q_{\alpha'-2}$ is not involved in the above arguments and so we may repeat the above arguments to conclude also that $U^{\alpha'} \leq Q_{\alpha+3}$. Hence, we have demonstrated that

(1) $U^\beta \leq Q_{\alpha'-2}$ and $U^{\alpha'} \leq Q_{\alpha+3}$.

Assume that $U^\beta \not\leq Q_{\alpha'-1}$. Then, as $Z_{\alpha'-1} \leq Q_\alpha$, it follows by Lemma 5.32 that $Z_{\alpha'-2} = [U^\beta, Z_{\alpha'-1}] \leq Z_\alpha$ and $Z_{\alpha'-2} = Z_\beta = Z_\alpha \cap Q_{\alpha'}$. Then $[V^{\alpha'-1}, V_\beta] \leq Z_{\alpha'-1} \cap V_\beta$ and since $V_\beta U^\beta \leq V_\beta^{(3)}$ is abelian, it follows that $[V^{\alpha'-1}, V_\beta] \leq Z_{\alpha'-2} = Z_\beta$ and $V^{\alpha'-1} \leq Q_\beta$. If $V^{\alpha'-1} \leq Q_\lambda$ for some $\lambda \in \Delta(\beta)$ with $Z_\lambda = Z_\alpha$ and $V^\lambda \not\leq Q_{\alpha'-1}$, then $[V^{\alpha'-1}, V^\lambda] \leq Z_\lambda \cap Q_{\alpha'} = Z_\beta = Z_{\alpha'-2} \leq Z_{\alpha'-1}$, a contradiction since $V^\lambda \not\leq Q_{\alpha'-1}$. Therefore $V^{\alpha'-1} \not\leq Q_\lambda$ and as

$$[V^\lambda \cap Q_{\alpha'-1}, V^{\alpha'-1}] \leq Z_{\alpha'-1} \cap V^\lambda = C_{Z_{\alpha'-1}}(U^\beta) = Z_{\alpha'-2} = Z_\beta \leq Z_\alpha = Z_\lambda,$$

V^λ/Z_λ is an FF-module for $\overline{L_\lambda}$ and $V^\lambda Q_{\alpha'-1} \in \mathrm{Syl}_p(L_{\alpha'-1})$. Moreover, $V_\lambda^{(2)} \cap Q_{\alpha'-2} = V^\lambda(V_\lambda^{(2)} \cap Q_{\alpha'-1})$, $V_\lambda^{(2)}/V^\lambda$ is also an FF-module for $\overline{L_\lambda}$ and $V_\lambda^{(2)} Q_{\alpha'-2} \in \mathrm{Syl}_p(L_{\alpha'-2})$. Then Lemma 5.33 implies that $O^p(R_\lambda)$ centralizes $V_\lambda^{(2)}$.

By Lemma 5.22, $Z_{\alpha+3} \neq Z_\beta = Z_{\alpha'-2}$ and so $V_{\alpha'}^{(3)} \cap Q_{\alpha+3}$ centralizes $Z_{\alpha+2} = Z_{\alpha+3} \times Z_{\alpha'-2}$ and $V_{\alpha'}^{(3)} \cap Q_{\alpha+3} = V_{\alpha'}(V_{\alpha'}^{(3)} \cap Q_\beta)$. Since $Z_\beta \leq V_{\alpha'}$, have that $V_{\alpha'}^{(3)} Q_{\alpha+3} \in \mathrm{Syl}_p(L_{\alpha+3})$, $V_{\alpha'}^{(3)}/V_{\alpha'}$ contains a unique non-central chief factor for $L_{\alpha'}$ which, as a $\overline{L_{\alpha'}}$-module, is itself an FF-module and, by Lemma 5.35 and conjugacy, $O^p(R_\beta)$ centralizes $V_\beta^{(3)}$. By Lemma 5.22, $Z_\alpha = Z_\lambda$ implies that $V^\alpha = V^\lambda = U^\beta$ and $V_\alpha^{(2)} = V_\lambda^{(2)}$. But $V_\alpha^{(2)} \leq Q_{\alpha'-2}$, and this is a contradiction. Hence, we have proved

(2) $U^\beta \leq Q_{\alpha'-1}$.

If $[U^\beta \cap Q_{\alpha'}, V_{\alpha'}] \leq V_\beta U^\beta$, then $V_{\alpha'}$ normalizes $V_\beta U^\beta$ and so $U_\beta = V_\beta U^\beta \trianglelefteq L_\beta = \langle V_{\alpha'}, R_\beta, Q_\alpha \rangle$. Since U_β / V_β contains a non-central chief factor for L_β, we have that $Z_{\alpha'} \leq U_\beta$, $Z_{\alpha'} \not\leq V_\beta$ and $U^\beta \cap Q_{\alpha'} \not\leq Q_{\alpha'+1}$.

Assume first that $Z_{\alpha'} = Z_{\alpha'-2}$ and $q = p$. Then, $(U^\beta \cap Q_{\alpha'})Q_{\alpha'+1} \in \mathrm{Syl}_p(L_{\alpha'+1})$. Now, $U^{\alpha'} \leq Q_{\alpha+3}$ and if $U^{\alpha'} \not\leq Q_{\alpha+2}$, then $Z_{\alpha+3} = [U^{\alpha'}, Z_{\alpha+2}] \leq Z_{\alpha'+1}$ from which we deduce that $Z_{\alpha'} = Z_{\alpha+3} \leq V_\beta$. But then, $Z_{\alpha'+1}$ centralizes $U^\beta V_\beta / V_\beta$ and we deduce that $V^\alpha V_\beta \trianglelefteq L_\beta$, a contradiction by Lemma 5.32. Hence, $U^{\alpha'} \leq Q_{\alpha+2}$ so that $V^{\alpha'+1} = Z_{\alpha'+1}(V^{\alpha'+1} \cap Q_\beta)$.

If $V^{\alpha'+1} \cap Q_\beta \leq Q_\alpha$, then $[V^{\alpha'+1} \cap Q_\beta, U^\beta \cap Q_{\alpha'}] \leq Z_\alpha \cap V^{\alpha'+1}$ and since $V^{\alpha+1}/Z_{\alpha'+1}$ contains a non-central chief factor we must have that $Z_\beta \leq V^{\alpha'+1}$ and $V^{\alpha'+1}/Z_{\alpha'+1}$ is (dual to) an FF-module. But then $U^{\alpha'} V_{\alpha'} \trianglelefteq L_{\alpha'} = \langle V_\beta, Q_{\alpha'+1}, R_{\alpha'} \rangle$. Since $Q_{\alpha'} \cap Q_{\alpha'+1} \leq C_{Q_{\alpha'}}(U^{\alpha'} V_{\alpha'}/V_{\alpha'})$, we infer by Proposition 5.27 that $[Q_{\alpha'}, U_{\alpha'}] = [Q_{\alpha'}, U^{\alpha'} V_{\alpha'}] \leq V_{\alpha'}$. But then $[V^{\alpha'+1} \cap Q_\beta, U^\beta \cap Q_{\alpha'}] \leq Z_\alpha \cap V_{\alpha'}$ so that $Z_\beta \leq V_{\alpha'}$ and V_β centralizes $U^{\alpha'} V_{\alpha'}/V_{\alpha'}$, a contradiction by Lemma 5.32.

Hence, $V^{\alpha'+1} \cap Q_\beta \not\leq Q_\alpha$. In particular, $Z_\beta = [Z_\alpha, V^{\alpha'+1} \cap Q_\beta] \leq U_{\alpha'}$. Since V^α / Z_α contains a non-central chief factor and $[V^\alpha \cap Q_{\alpha'+1}, V^{\alpha'+1} \cap Q_\beta] \leq Z_{\alpha'} \cap V^{\alpha'+1} = \{1\}$, we must have that $V^\alpha \cap Q_{\alpha'} \not\leq Q_{\alpha'+1}$. Furthermore, since $Z_{\alpha'} = Z_{\alpha'-2} \not\leq V^\alpha$, we have that $[V_\alpha^{(2)}, Z_{\alpha'-1}] = \{1\}$ and $V_\alpha^{(2)} = V^{\alpha'+1}(V_\alpha^{(2)} \cap Q_{\alpha'+1})$. Hence, $Z_{\alpha'} \leq V_\alpha^{(2)}$, $V_\alpha^{(2)}/V^\alpha$ is an FF-module for $\overline{L_\alpha}$ and by Lemma 5.33, $O^p(R_\alpha)$ centralizes $V_\alpha^{(2)}$. Then, Lemma 5.22 applied to $Z_{\alpha'} = Z_{\alpha'-2}$ yields $V_{\alpha'} = V_{\alpha'-2} \leq Q_\beta$, a contradiction. Thus

(3) if $Z_{\alpha'} = Z_{\alpha'-2}$ then $q > p$.

Continue assuming that $Z_{\alpha'} = Z_{\alpha'-2}$. Since $Z_{\alpha'} \not\leq V_\beta$ and $[U_\beta, C_\beta] \leq V_\beta$, we have that $U_\beta \cap Q_{\alpha'-2} \leq Q_{\alpha'-1}$ and as $Z_{\alpha'+1}$ acts quadratically on U_β, we deduce by Lemma 2.55 that $O^p(R_\beta)$ centralizes U_β. Since $[V_\beta^{(3)} \cap Q_{\alpha'}, V_{\alpha'}] \leq Z_{\alpha'} \leq U_\beta$, we also deduce by Lemma 2.55 that $O^p(R_\beta)$ centralizes $V_\beta^{(3)}$. In particular, $U^\beta = V^\alpha$ and by Lemmas 5.22 and 5.32, $Z_{\alpha+2} C_{V_{\alpha+3}}(O^p(L_{\alpha+3})) \neq Z_{\alpha+4} C_{V_{\alpha+3}}(O^p(L_{\alpha+3}))$.

Note that $[V_\alpha^{(2)}, Z_{\alpha'-1}] \leq Z_{\alpha'} \cap V^\alpha = \{1\}$ so that $V_\alpha^{(2)} = V_\beta(V_\alpha^{(2)} \cap Q_{\alpha'})$. Since $Z_{\alpha'} \leq U_\beta$, we deduce that $V_\beta^{(3)} = V_\alpha^{(2)} U_\beta \trianglelefteq \langle V_{\alpha'}, Q_\alpha, R_\beta \rangle$. In particular, by conjugation, $V_{\alpha'-2}^{(3)} = U_{\alpha'-2}V_{\alpha'-3}^{(2)}$ so that $R = [V_\beta, V_{\alpha'}] \leq [V_\beta, U_{\alpha'-2}]$. Hence, there is $\delta \in \Delta(\alpha'-2)$ such that $Z_\delta \neq Z_{\alpha'-3}$ and $[V_\beta, V^\delta] \leq V_\beta \cap Z_\delta \not\leq Z_{\alpha'-2}$.

Since U_β centralizes $Z_{\alpha'-3}(V_\beta \cap Z_\delta)$ we infer that $V_\beta^{(3)} \leq Q_{\alpha'-2}$ and since $U_\beta \leq Q_{\alpha'-1}$, we have that $V_\beta^{(3)} = V_\alpha^{(2)}U_\beta = V_\beta(V_\beta^{(3)} \cap Q_{\alpha'})$. Then U_β/V_β contains a unique non-central chief factor for L_β, $Z_{\alpha'} \leq U_\beta$ and $Z_{\alpha'} \cap V_\beta = \{1\}$. Since U_β/V_β contains a unique non-central chief factor for L_β, writing C for the preimage in U_β of $C_{U_\beta/V_\beta}(O^p(L_\beta))$, we have that $U_\beta = CV^\alpha V^{\alpha+2}$, $Z_{\alpha'} \cap C = \{1\}$ and $V^\alpha \cap V^{\alpha+2} \leq C$.

Note that $[C \cap Q_{\alpha'}, Z_{\alpha'+1}] = \{1\}$ and $[V^{\alpha+2} \cap Q_{\alpha'}, Z_{\alpha'+1}] \leq Z_{\alpha+2} \cap Z_{\alpha'} = \{1\}$ from which we deduce that $U_\beta = V_\beta V^\alpha(U_\beta \cap Q_{\alpha'+1})$ and $(V^\alpha \cap Q_{\alpha'})Q_{\alpha'+1} \in \mathrm{Syl}_p(L_{\alpha'+1})$. Then $V_\alpha^{(2)} = V^\alpha(V_\alpha^{(2)} \cap V^{\alpha'+1})$ and we also infer by a similar reasoning that $S = (V^{\alpha'+1} \cap Q_\beta)Q_\alpha$. Since both $V_\alpha^{(2)}/V^\alpha$ and V^α/Z_α contain a non-central chief factor for L_α, we deduce that both $V_\alpha^{(2)}/V^\alpha$ and V^α/Z_α are FF-modules for $\overline{L_\alpha}$ and applying Lemma 5.33, $O^p(R_\alpha)$ centralizes $V_\alpha^{(2)}$. Then Lemma 5.22 applied to $Z_{\alpha'} = Z_{\alpha'-2}$ provides a contradiction as in the $q = p$ case. Hence, we have shown that if $|V_\beta| > q^3$ then

(4) $Z_{\alpha'} \neq Z_{\alpha'-2}$.

Note that if $R \leq V_{\alpha'-2}$, then as $R \not\leq Z_{\alpha'-1}$, we have that $RZ_{\alpha'-1} \leq V_{\alpha'} \cap V_{\alpha'-2} \leq Z_{\alpha'-1}C_{V_{\alpha'}}(O^p(L_{\alpha'}))$. Then $[R, Q_{\alpha'}] \leq Z_{\alpha'}$ and $[R, Q_{\alpha'-2}] \leq Z_{\alpha'-2}$ and so $RZ_{\alpha'-1} \trianglelefteq L := \langle Q_{\alpha'}, Q_{\alpha'-2}\rangle$. But now, $[R, Q_{\alpha'-1} \cap Q_{\alpha'}] = [R, V_\beta(Q_{\alpha'-1} \cap Q_{\alpha'})] = [RZ_{\alpha'-1}, Q_{\alpha'-1}] \trianglelefteq L$ and since $R \leq V_{\alpha'}$, $[RZ_{\alpha'-1}, Q_{\alpha'-1}] \leq Z_{\alpha'}$ and $R \leq \Omega(Z(Q_{\alpha'-1})) \cap V_{\alpha'}$. If $\Omega(Z(Q_{\alpha'-1})) \cap C_{V_{\alpha'}}(O^p(L_{\alpha'})) > Z_{\alpha'}$ then $Q_{\alpha'-1} \cap Q_{\alpha'} \leq C_{Q_{\alpha'}}(\Omega(Z(Q_{\alpha'-1})) \cap C_{V_{\alpha'}}(O^p(L_{\alpha'}))) \trianglelefteq G_{\alpha'}$ and Proposition 5.27 yields a contradiction. Hence, $\Omega(Z(Q_{\alpha'-1})) \cap V_{\alpha'} = Z_{\alpha'-1}$, a contradiction since $R \not\leq Z_{\alpha'-1}$. Therefore

(5) $R \not\leq V_{\alpha'-2}$.

Now, if $q > p$ then $V_\beta^{(3)} \cap Q_{\alpha'-2}$ centralizes $Z_{\alpha'-1}$ so that $V_\beta(V_\beta^{(3)} \cap Q_{\alpha'+1})$ has index at most q^2 in $V_\beta^{(3)}$ and is centralized, modulo V_β, by $Z_{\alpha'+1}$. Furthermore, $Z_{\alpha'+1}$ acts quadratically on $V_\beta^{(3)}$ and by Lemma 2.55 we deduce that $O^p(R_\beta)$ centralizes $V_\beta^{(3)}$. Then by Lemma 5.22, $Z_{\alpha'-1} \neq Z_{\alpha'-3}$ for otherwise $V_{\alpha'} \leq V_{\alpha'-1}^{(2)} = V_{\alpha'-3}^{(2)} \leq Q_{\alpha'}$. Hence, by Lemma 5.32, $Z_{\alpha'-1}C_{V_{\alpha'-2}}(O^p(L_{\alpha'-2})) \neq Z_{\alpha'-3}C_{V_{\alpha'-2}}(O^p(L_{\alpha'-2}))$ and since $V_\beta^{(3)}$ centralizes $Z_{\alpha'-1}$, $V_\beta^{(3)} = V_\beta(V_\beta^{(3)} \cap Q_{\alpha'})$ and $V_\beta^{(3)}/V_\beta$ contains a unique non-central chief factor for L_β which, as a $\overline{L_\beta}$-module, is an FF-module. Since $V^\alpha V_\beta \not\trianglelefteq L_\beta$, we deduce that U_β/V_β contains the non-central chief factor and $V_\beta^{(3)} = U_\beta V_\alpha^{(2)}$. But then, by conjugacy, $V_{\alpha'} \leq V_{\alpha'-2}^{(3)} = V_{\alpha'-3}^{(2)}U_{\alpha'-2}$ and since V_β centralizes $V_{\alpha'-3}^{(2)}$, $R = [V_\beta, V_{\alpha'}] \leq [V_\beta, U_{\alpha'-2}] \leq V_{\alpha'-2}$, a contradiction.

If $q = p$, then the above observations yield that $V_\beta^{(3)}/U_\beta$ contains a non-central chief factor for L_β. We deduce that $U_\beta \leq Q_{\alpha'-2}$ and as $Z_{\alpha'-1} = Z_{\alpha'} \times Z_{\alpha'-2}$ is centralized by U_β, $U_\beta \leq Q_{\alpha'-1}$. Then, as $V^\alpha V_\beta \ntrianglelefteq L_\beta$ by Lemma 5.32, U_β/V_β contains a unique non-central chief factor and $(U_\beta \cap Q_{\alpha'})Q_{\alpha'+1} \in \mathrm{Syl}_p(L_{\alpha'+1})$. Moreover, by a similar argument, $V_\beta^{(3)} \cap Q_{\alpha'-2} \leq Q_{\alpha'-1}$, $V_\beta^{(3)} Q_{\alpha'-2} \in \mathrm{Syl}_p(L_{\alpha'-2})$ and as $V_\beta^{(3)}$ centralizes $Z_{\alpha'-1}$, by Lemma 5.32 we have

(6) $q = p$ and $Z_{\alpha'-1} = Z_{\alpha'-3}$

To force a contradiction via Lemma 5.22 as before, we need only show that $O^p(R_\beta)$ centralizes $V_\beta^{(3)}$. Note that both $V_\beta^{(3)}/U_\beta$ and U_β/V_β contain exactly one non-central chief factor for L_β and in both cases, the non-central chief factor is an FF-module for $\overline{L_\beta}$. We set $R_1 := C_{L_\beta}(U_\beta/V_\beta)$ and $R_2 := C_{L_\beta}(V_\beta^{(3)}/U_\beta)$ and, aiming for a contradiction, assume that at least one of R_1 or R_2 is not equal to R_β.

Since the non-central chief factor within $V_\beta^{(3)}/U_\beta$ is an FF-module, it follows that either $R_2 Q_\beta = R_\beta$; or $L_\beta = \langle R_2, R_\beta, S\rangle$ and $q \in \{2, 3\}$ by Proposition 2.46 (ii), (iii) and Proposition 2.47 (ii), (iii). In the former case, since $V_\alpha^{(2)} \leq Q_{\alpha'-1}$, we see that $V_\beta^{(3)} = V_\alpha^{(2)} U_\beta \trianglelefteq L_\beta = \langle V_{\alpha'}, R_\beta, Q_\alpha\rangle$ and $V_\beta^{(3)} \leq Q_{\alpha'-2}$, a contradiction. In the latter case, $V_\alpha^{(2)} U_\beta \trianglelefteq R_2 S$ and if $[C_\beta, V_\alpha^{(2)} U_\beta] \leq V_\beta$, then $[C_\beta, V_\alpha^{(2)} U_\beta]$ is centralized by $O^p(R_\beta)$ and so $[C_\beta, V_\alpha^{(2)} U_\beta] \trianglelefteq L_\beta = \langle R_2, R_\beta, S\rangle$. Thus, $[C_\beta, V_\beta^{(3)}] = [C_\beta, V_\alpha^{(2)} U_\beta] \leq V_\beta$ and by conjugacy, $R \leq [V_{\alpha'-2}^{(3)}, V_\beta] \leq [V_{\alpha'-2}^{(3)}, C_{\alpha'-2}] \leq V_{\alpha'-2}$, a contradiction. Thus, $[C_\beta, V_\alpha^{(2)}] \leq V^\alpha$ but $[C_\beta, V_\alpha^{(2)}] \not\leq V_\beta$. If $R_1 Q_\beta = R_2 Q_\beta$ then, assuming that G is a minimal counterexample to Theorem C, we may apply Proposition 5.30 with $\lambda = \beta$. Since $b > 5$, $R_1 Q_\beta$ normalizes $V_\alpha^{(2)}$ and $\lambda = \beta$, conclusion (d) holds. Then, $V_\alpha^{(4)} \leq V := \langle Z_\beta^X\rangle$ and the images of $Q_\alpha/C_{Q_\alpha}(V_\alpha^{(2)})$ and $C_{Q_\alpha}(V_\alpha^{(2)})/C_{Q_\alpha}(V_\alpha^{(4)})$ resp. Q_β/C_β and $C_\beta/C_{Q_\beta}(V_\beta^{(3)})$ contain a non-central chief factor for $\widetilde{L}_\alpha$ resp. $\widetilde{L}_\beta$, and we have a contradiction. Hence, we have that

(7) $R_1 Q_\beta \neq R_2 Q_\beta \neq R_\beta$.

By Proposition 2.46 (iii) and Proposition 2.47 (iii), we deduce that $L_\beta = \langle R_1, R_2, S\rangle$. Then $V_\alpha^{(2)} U_\beta \trianglelefteq R_2 S$ so that $V^\alpha V_\beta \geq [C_\beta, V_\alpha^{(2)} U_\beta] V_\beta \trianglelefteq R_2 S$. Furthermore, as $O^p(R_1)$ centralizes U_β/V_β, we have that $[C_\beta, V_\alpha^{(2)} U_\beta] V_\beta \trianglelefteq R_1 S$ so that $[C_\beta, V_\alpha^{(2)} U_\beta] V_\beta \trianglelefteq L_\beta$. Since $V^\alpha V_\beta \ntrianglelefteq L_\beta$, we may assume that $[C_\beta, V_\alpha^{(2)}] V_\beta < V^\alpha V_\beta$. Note that $[V^{\alpha'+1} \cap Q_{\alpha+3}, Z_{\alpha+2}] \leq Z_{\alpha+3} \cap Z_{\alpha'} = \{1\}$ since $Z_{\alpha'} \not\leq V_\beta$. Then, if $V^{\alpha'+1} \cap Q_\beta \leq Q_\alpha$, we have that $[V^{\alpha'+1} \cap Q_\beta, U^\beta \cap Q_{\alpha'}] \leq Z_\beta \cap V^{\alpha'+1}$. By Lemma 2.41 (and by conjugacy), either V^α/Z_α is an FF-module, or $V^{\alpha'+1} \not\leq Q_{\alpha+3}$ and $Z_\beta \leq V^{\alpha'+1}$. In the latter case, $V_{\alpha'+1}^{(2)} = V^{\alpha'+1}(V_{\alpha'+1}^{(2)} \cap Q_\alpha)$ and as $Z_\beta \leq V^{\alpha'+1}$, we have a contradiction since $V_{\alpha'+1}^{(2)}/V^{\alpha'+1}$ contains a non-central chief factor. Now, V^α/Z_α is an FF-module generated by $C_{V_\beta}(O^p(L_\beta))/Z_\alpha$ of order p so that by Lemma 2.41, $p^2 \leqslant |V^\alpha/Z_\alpha| \leqslant p^3$ and $p^4 \leqslant |V^\alpha| \leqslant p^5$.

Hence, $p^5 \leqslant |V^\alpha V_\beta| \leqslant p^6$, accordingly. But now, as $[C_\beta, V_\alpha^{(2)} U_\beta]V_\beta > V_\beta$, $|[C_\beta, V_\alpha^{(2)} U_\beta]V_\beta| \leqslant p^5$ and as $[C_\beta, V_\alpha^{(2)}]V_\beta < V^\alpha V_\beta$, we get that

(8) $|V^\alpha| = p^5, |V^\alpha V_\beta| = p^6$ and $[Q_\beta, V^\alpha] \not\leq Z_\alpha C_{V_\beta}(O^p(L_\beta))$.

Writing C^α for the preimage in V^α of $C_{V^\alpha/Z_\alpha}(O^p(L_\alpha))$, we have that $|C^\alpha| = p^3, C^\alpha \cap V_\beta = Z_\alpha, |Q_\alpha/C_{Q_\alpha}(C^\alpha)| \leqslant p^2$ and a calculation using the three subgroup lemma yields $[R_\alpha, Q_\alpha] \leq C_{Q_\alpha}(C^\alpha)$. Since $Z(Q_\alpha) = Z_\alpha$, calculating in $\mathrm{GL}_3(p)$, we infer that $Q_\alpha/C_{Q_\alpha}(C^\alpha)$ is a non-central chief factor of order p^2 for L_α. Hence, $Q_\alpha/C_{Q_\alpha}(C^\alpha)$ is a natural $\mathrm{SL}_2(p)$ module for L_α/R_α.

Now, by Lemma 5.8, $U_\beta/([U_\beta, Q_\beta]V_\beta)$ contains the unique non-central chief factor within U_β/V_β and so $O^p(L_\beta)$ centralizes $[U_\beta, Q_\beta]V_\beta/V_\beta$. Thus, $[V^\alpha, Q_\beta]V_\beta \trianglelefteq L_\beta$ from which it follows that $Z_\alpha \geq [V^\alpha, Q_\beta, Q_\beta] \trianglelefteq L_\beta$ and $[V^\alpha, Q_\beta, Q_\beta] = Z_\beta$. But $C^\alpha \leq Z_\alpha C_{V_\beta}(O^p(L_\beta))[V^\alpha, Q_\beta]$ so that $[Q_\beta, C^\alpha] = Z_\beta$. In particular, $C_{Q_\alpha}(C^\alpha) \leq Q_\beta$ for otherwise $Z_\beta = [C^\alpha, Q_\alpha \cap Q_\beta] = [C^\alpha, Q_\alpha] \trianglelefteq L_\alpha$, a contradiction.

If $V^{\alpha'-1} \not\leq Q_\beta$, then $RZ_\beta \leq [V^{\alpha'-1}, V_\beta]Z_\beta \leq Z_{\alpha'-1}Z_\beta$. Then, as $R \not\leq Z_{\alpha'-1}$, we get that $Z_\beta \leq RZ_{\alpha'-1} \leq V_{\alpha'}$. If $V^{\alpha'-1} \leq Q_\beta$ but $V_\beta \not\leq C_\beta$, we deduce that $Z_\beta = [V^{\alpha'-1}, V_\beta] \leq Z_{\alpha'-1}$. In either case, since $O^p(R_\alpha)$ centralizes $V_\alpha^{(2)}$, by Lemma 5.22, $Z_\beta \neq Z_{\alpha+3}$ and so $V_{\alpha'}^{(3)}$ centralizes $Z_{\alpha+2} = Z_\beta Z_{\alpha+3}$. But then $V_{\alpha'}^{(3)} \cap Q_{\alpha+3} = V_{\alpha'}(V_{\alpha'}^{(3)} \cap Q_\beta)$ and since $Z_\beta \leq Z_{\alpha'-1} \leq V_{\alpha'}$, we conclude that $V_{\alpha'}^{(3)}/V_{\alpha'}$ contains a unique non-central chief factor, a contradiction. Thus, $[V_\beta, V^{\alpha'-1}] = \{1\}$ and $V_\beta \leq C_{Q_{\alpha'-1}}(C^{\alpha'-1}) \leq Q_{\alpha'}$, and we have a final contradiction. This completes the proof. □

Lemma 7.27 *Suppose that $C_{V_\beta}(V_{\alpha'}) = V_\beta \cap Q_{\alpha'}$ and $b > 5$. If $V_{\alpha'} \not\leq Q_\beta$ and $V_\alpha^{(2)} \leq Q_{\alpha'-2}$ then $V_\alpha^{(2)} \leq Q_{\alpha'-1}$. Moreover, if $V_\beta^{(3)} \cap Q_{\alpha'-2} \leq Q_{\alpha'-1}$, then $V_\beta^{(3)} \leq Q_{\alpha'-1}$.*

Proof By Lemma 7.26, $|V_\beta| = q^3$ so that $R = [V_\beta, V_{\alpha'}] \leq Z_{\alpha'-1} \cap Z_{\alpha+2}$. If $V_\alpha^{(2)} \not\leq Q_{\alpha'-1}$, then $Z_{\alpha'-2} = [V_\alpha^{(2)}, V_{\alpha'-2}] \leq Z_\alpha \cap Q_{\alpha'-2}$, so that $Z_\beta = Z_{\alpha'-2}$. But $Z_\beta \cap R = \{1\}$ and $Z_{\alpha'-1} = R \times Z_\beta \leq V_\beta$, a contradiction since $V_\alpha^{(2)}$ is abelian. Thus, $V_\alpha^{(2)} \leq Q_{\alpha'-1}$, as desired.

Aiming for a contradiction, we suppose throughout that $V_\beta^{(3)} \not\leq Q_{\alpha'-2}$ but $V_\beta^{(3)} \cap Q_{\alpha'-2} \leq Q_{\alpha'-1}$. Since $Z_{\alpha'-1} = Z_{\alpha'}R \leq V_\beta^{(3)}$, $b > 5$ and $V_\beta^{(3)}$ does not centralize $V_{\alpha'-2}$ we deduce that

(1) $Z_{\alpha'-1} = Z_{\alpha'-3}$.

By Lemma 5.22, $O^p(R_\beta)$ does not centralize $V_\beta^{(3)}$ and so by Lemma 5.35, either $V_\beta^{(3)}/V_\beta$ contains more than one non-central chief factor, or a non-central chief factor within $V_\beta^{(3)}/V_\beta$ is not an FF-module. Hence, we infer that $Z_{\alpha'-1} = [V_\beta^{(3)} \cap Q_{\alpha'-2}, V_{\alpha'}] \not\leq V_\beta$. Moreover, since $b > 5$, $[V_\beta^{(3)}, Z_{\alpha'+1}, Z_{\alpha'+1}] \leq$

$[V_\beta^{(3)}, V_{\alpha'-2}^{(3)}, V_{\alpha'-2}^{(3)}] = \{1\}$ and $V_\beta^{(3)}$ admits quadratic action. In particular, if $p \geqslant 5$ then the Hall–Higman theorem implies that $O^p(R_\beta)$ centralizes $V_\beta^{(3)}$ and so $p \in \{2, 3\}$. Moreover, we may apply Lemma 7.3 when $p = 3$ and Lemma 2.55 when $p = 2$ to deduce that

(2) $q = p \in \{2, 3\}$.

Notice that $Z_{\alpha'-1} = Z_{\alpha'-3} \leq V_\beta^{(3)} \leq Z(V_\beta^{b-4})$. If $b = 7$, then $Z_{\alpha'-1} = Z_{\alpha'-3} = Z_{\alpha+4}$ by definition. Suppose that $b > 7$ and let $n \leqslant \frac{b-5}{2}$ be chosen minimally such that $V_\beta^{(2n+1)} \leq Q_{\alpha'-2n}$. Since $V_\beta^{(3)} \not\leq Q_{\alpha'-2}$, if such an n exists then $n \geqslant 2$. Notice $V_\beta^{(5)}$ centralizes $Z_{\alpha'-3} \leq V_\beta^{(3)}$ so that either $Z_{\alpha'-3} = Z_{\alpha'-5} \leq V_\beta^{(3)}$ or $V_\beta^{(5)} \leq Q_{\alpha'-4}$ and $n = 2$. Extending through larger subgroups, it is clear that for a minimally chosen n, $Z_{\alpha'-1} = Z_{\alpha'-3} = \cdots = Z_{\alpha'-2n+1} \leq V_\beta^{(3)}$ is centralized by $V_\beta^{(2n+1)}$ so that $V_\beta^{(2n+1)} \leq Q_{\alpha'-2n+1}$. Then $V_\beta^{(2n+1)} = V_\beta^{(2(n-1)+1)}(V_\beta^{(2n+1)} \cap Q_{\alpha'-2n+2})$. Moreover, $Z_{\alpha'-1} = \cdots = Z_{\alpha'-2n+1}$, $V_\beta^{(2n+1)} \cap Q_{\alpha'-2a} \leq Q_{\alpha'-2a+1}$ and $V_\beta^{(2n+1)} \cap Q_{\alpha'-2a} = V_\beta^{(2(a-2)+1)}(V_\beta^{(2n+1)} \cap Q_{\alpha'-2a+2})$ from which it follows that $V_\beta^{(2n+1)} = V_\beta^{(2(n-1)+1)}(V_\beta^{(2n+1)} \cap Q_{\alpha'})$ so that $O^p(L_\beta)$ centralizes $V_\beta^{(2n+1)}/V_\beta^{(2(n-1)+1)}$, a contradiction. Thus, no such n exists for $n \leqslant \frac{b-5}{2}$ and it follows that $V_\beta^{(b-4)} \not\leq Q_{\alpha'-b+5} = Q_{\alpha+5}$ and (for any $b \geqslant 7$) that

(3) $Z_{\alpha'-1} = \cdots = Z_{\alpha+6} = Z_{\alpha+4}$.

Since $Z_{\alpha'-1} \not\leq V_\beta$, to obtain a contradiction, we need only show that $Z_{\alpha+2} = Z_{\alpha+4}$. If Z_β is centralized by $V_{\alpha'}^{(3)}$, then $V_{\alpha'}^{(3)}$ centralizes $Z_{\alpha+2} = R \times Z_\beta$ and if $Z_{\alpha+2} \neq Z_{\alpha+4}$, then $V_{\alpha'}^{(3)}$ centralizes $V_{\alpha+3}$ and $V_{\alpha'}^{(3)} = V_{\alpha'}(V_{\alpha'}^{(3)} \cap Q_\beta)$ so that $V_{\alpha'}^{(3)}/V_{\alpha'}$ contains a unique non-central chief factor which is an FF-module and by Lemma 5.35, $O^p(R_{\alpha'})$ centralizes $V_{\alpha'}^{(3)}$. By conjugacy, $O^p(R_\beta)$ centralizes $V_\beta^{(3)}$, a contradiction. Thus, $V_{\alpha'}^{(3)}$ does not centralize Z_β. Since $V_{\alpha'}^{(3)}$ centralizes $Z_{\alpha+3} \times R \leq Z_{\alpha+2}$, we deduce that $R = Z_{\alpha+3}$. Furthermore, since $b > 5$ and $V_{\alpha'}^{(3)}$ is abelian, $V_{\alpha'}^{(3)} \cap Q_{\alpha+3} \cap Q_{\alpha+2} \cap Q_\beta \leq C_\beta$.

Now, $V_\beta \leq C_{\alpha'-2}$ and since $[Q_\lambda, V_\lambda^{(2)}] = Z_\lambda$ for all $\lambda \in \Delta(\alpha'-2)$, we have that $R \leq [V_\beta, V_{\alpha'-2}^{(3)}] \leq Z_{\alpha+2} \cap V_{\alpha'-2}$. If $Z_{\alpha+2} \leq V_{\alpha'-2}$, then $Z_{\alpha+2} = Z_{\alpha'-3} = Z_{\alpha'-1} \leq V_\beta$, a contradiction. Thus, $[V_\beta, V_{\alpha'-2}^{(3)}] = R$ and $[V_\beta, V_{\alpha'-2}^{(3)} \cap Q_\beta] = R \cap Z_\beta = \{1\}$. Then $V_{\alpha'-2}^{(3)} \cap Q_\beta \leq C_\beta$ so that $[V_\beta^{(3)}, V_{\alpha'-2}^{(3)} \cap Q_\beta] \leq V_\beta \cap V_{\alpha'-2}^{(3)}$. Since $b > 5$, $V_\beta \not\leq V_{\alpha'-2}^{(3)}$ and since $R \leq V_{\alpha'-2}$, $Z_{\alpha+2} \leq V_{\alpha'-2}^{(3)}$ and $Z_\beta \leq V_{\alpha'-2}^{(3)}$ but $Z_\beta \not\leq V_{\alpha'-2}$. If $b > 7$, $V_{\alpha'}^{(3)}$ centralizes Z_β, a contradiction by the above. Thus, we shown that

(4) $b = 7$, $V_\beta^{(3)} \not\leq Q_{\alpha'-2}$, $V_\beta^{(3)} \cap Q_{\alpha'-2} \leq Q_{\alpha'-1}$, $Z_{\alpha'-1} = Z_{\alpha'-3} \neq Z_{\alpha+2}$ and $[Z_\beta, V_{\alpha'}^{(3)}] \neq \{1\}$.

Set $W^\beta = \langle V_\delta^{(2)} : Z_\delta = Z_\alpha, \delta \in \Delta(\beta)\rangle$ so that $[C_\beta, W^\beta] = [C_\beta, V_\alpha^{(2)}] \leq Z_\alpha$. Then $[W^\beta, V_{\alpha'-2}] \leq Z_{\alpha'-3} \cap Z_\alpha$ and by Lemma 5.23, $W^\beta \trianglelefteq R_\beta Q_\alpha$. If $Z_\beta \leq Z_{\alpha'-3} = Z_{\alpha'-1}$, then $Z_{\alpha'-1} = Z_\beta \times R = Z_{\alpha+2} \leq V_\beta$, a contradiction. Thus, $W^\beta = V_\beta(W^\beta \cap Q_{\alpha'})$. If $[W^\beta \cap Q_{\alpha'}, V_{\alpha'}] \leq W^\beta$, then $V_\alpha^{(2)} \leq W^\beta \trianglelefteq L_\beta = \langle V_{\alpha'}, Q_\alpha, R_\beta\rangle$ and $V_\beta^{(3)} = W^\beta \leq Q_{\alpha'-2}$, a contradiction. Thus, $W^\beta \cap Q_{\alpha'} \not\leq Q_{\alpha'+1}$ for some $\alpha'+1 \in \Delta(\alpha')$ and since $Z_{\alpha'+1}Z_{\alpha'-1} = V_{\alpha'} \not\leq Q_\beta$, $(\alpha'+1, \beta)$ is a critical pair.

Since $V_{\alpha+3} \leq Q_{\alpha'+1}$, $[V_{\alpha'+1}^{(2)} \cap Q_{\alpha+3}, V_{\alpha+3}] \leq Z_{\alpha'+1} \cap Z_{\alpha+3} = Z_{\alpha'+1} \cap R = \{1\}$ and $V_{\alpha'+1}^{(2)} \cap Q_{\alpha+3} = Z_{\alpha'+1}(V_{\alpha'+1}^{(2)} \cap C_\beta)$. Furthermore, $[V_{\alpha'+1}^{(2)} \cap C_\beta, W^\beta \cap Q_{\alpha'}] \leq V_{\alpha'+1}^{(2)} \cap Z_\alpha$ and since $Z_\beta \not\leq V_{\alpha'}^{(3)}$, we have that $W^\beta \cap Q_{\alpha'}$ centralizes $(V_{\alpha'+1}^{(2)} \cap Q_{\alpha+3})/Z_{\alpha'+1}$. Thus, $V_{\alpha'+1}^{(2)} \not\leq Q_{\alpha+3}$ and $V_{\alpha'+1}^{(2)}/Z_{\alpha'+1}$ is an FF-module. By Lemma 5.33, $O^p(R_\alpha)$ centralizes $V_\alpha^{(2)}$ and since $V_\beta^{(3)}$ does not centralize $V_{\alpha'-2}$, it follows from Lemma 5.22 that

(5) $Z_{\alpha'-2} \neq Z_{\alpha'-4} = R$.

Suppose that $[V_\beta^{(3)}, Q_\beta]V_\beta/V_\beta$ contains a non-central chief factor for L_β. In particular, $[Q_\beta, V_\alpha^{(2)}] \not\leq V_\beta$, and since $V_\alpha^{(2)}/Z_\alpha$ is an FF-module, we deduce that $|V_\alpha^{(2)}| = p^5$. The non-central chief factor within $[V_\beta^{(3)}, Q_\beta]V_\beta/V_\beta$, U/V say, is an FF-module for $\overline{L_\beta}$ and $L_\beta/C_{L_\beta}(U/V) \cong \mathrm{SL}_2(p)$. Set $R_1 := C_{L_\beta}(U/V)$ and $R_2 := C_{L_\beta}(V_\beta^{(3)}/([V_\beta^{(3)}, Q_\beta]V_\beta))$, noticing that also $L_\beta/R_2 \cong \mathrm{SL}_2(p)$. If $R_1 \neq R_\beta$, and employing Proposition 2.47 (iii) when $p = 3$, we conclude that $L_\beta = \langle R_1, R_\beta, S\rangle$. Similarly, if $R_2 \neq R_\beta$ then $L_\beta = \langle R_2, R_\beta, S\rangle$.

Suppose that $R_1 \neq R_\beta$. Then $[V_\alpha^{(2)}, Q_\beta]V_\beta \trianglelefteq R_1$ and $[V_\alpha^{(2)}, Q_\beta, Q_\beta] \leq V_\beta$ so that $[V_\alpha^{(2)}, Q_\beta, Q_\beta] \trianglelefteq L_\beta = \langle R_1, R_\beta, S\rangle$. Since $[V_\alpha^{(2)}, Q_\beta, Q_\beta] \leq Z_\alpha$, we have that $[V_\alpha^{(2)}, Q_\beta, Q_\beta] = Z_\beta$. Setting C^α to be the preimage in $V_\alpha^{(2)}$ of $C_{V_\alpha^{(2)}/Z_\alpha}(O^p(L_\alpha))$, we have that $C^\alpha \leq V_\beta[V_\alpha^{(2)}, Q_\beta]$ and so $[C^\alpha, Q_\beta] = Z_\beta$. As in Lemma 7.26 (where C^α is defined slightly differently), we see that $|Q_\alpha/C_{Q_\alpha}(C^\alpha)| = p^2$ and $C_{Q_\alpha}(C^\alpha) \leq Q_\beta$. Now, $V_\beta \leq Q_{\alpha'-2}$ and so $[V_\beta, C^{\alpha'-1}] \leq Z_{\alpha'-2} \cap Z_{\alpha+2} = \{1\}$, for otherwise $Z_{\alpha+2} = Z_{\alpha'-1}$. But then, $V_\beta \leq C_{Q_{\alpha'-1}}(C^{\alpha'-1}) \leq Q_{\alpha'}$, a contradiction. Thus, $R_1 = R_\beta$.

Suppose that $R_2 \neq R_\beta$. Then $V_\alpha^{(2)}[V_\beta^{(3)}, Q_\beta] \trianglelefteq R_2$ and so $[V_\alpha^{(2)}, Q_\beta][V_\beta^{(3)}, Q_\beta, Q_\beta] \trianglelefteq L_\beta = \langle R_1, R_2, S\rangle$. Since $[V_\alpha^{(2)}, Q_\beta, Q_\beta] \leq Z_\alpha$, we have that $[V_\beta^{(3)}, Q_\beta, Q_\beta] \leq V_\beta$ and so $[V_\alpha^{(2)}, Q_\beta]V_\beta \trianglelefteq L_\beta$. But then $[V_\beta^{(3)}, Q_\beta]V_\beta = [V_\alpha^{(2)}, Q_\beta]V_\beta$ is centralized by Q_α, modulo V_β, and so $([V_\beta^{(3)}, Q_\beta]V_\beta)/V_\beta$ does not contain a non-central chief factor for L_β. Thus, $R_2 = R_\beta$. But now $O^p(R_\beta)$ centralizes $V_\beta^{(3)}$ and Lemma 5.22 applied to $Z_{\alpha'-1} = Z_{\alpha'-3}$ gives $V_{\alpha'} \leq V_{\alpha'-1}^{(2)} = V_{\alpha'-3}^{(2)} \leq Q_\beta$, a contradiction.

Therefore, we may assume that

(6) $([V_\beta^{(3)}, Q_\beta]V_\beta)/V_\beta$ does not contain a non-central chief factor for L_β and $[V_\alpha^{(2)}, Q_\beta]V_\beta \trianglelefteq L_\beta$.

As before, since $[V_\alpha^{(2)}, Q_\beta, Q_\beta] \leq Z_\alpha$, we have that $[V_\alpha^{(2)}, Q_\beta, Q_\beta] = Z_\beta$ and either $|V_\alpha^{(2)}| = p^4$; or $[C^\alpha, Q_\beta] = Z_\beta$ for C^α as defined above. In the latter case, we again see that $V_\beta \leq C_{Q_{\alpha'-1}}(C^{\alpha'-1}) \leq Q_{\alpha'}$, a contradiction. Thus, $|V_\alpha^{(2)}| = p^4$, $[V_\alpha^{(2)}, Q_\beta] \leq V_\beta$ and $[V_\beta^{(3)}, Q_\beta] = V_\beta$. Since $O^p(R_\beta)$ does not centralize $V_\beta^{(3)}$, by Lemma 5.35, $V_\beta^{(3)}/V_\beta$ is a quadratic 2F-module for $\overline{L_\beta}$. Moreover, since $V_\alpha^{(2)}$ generates $V_\beta^{(3)}$, is $G_{\alpha,\beta}$-invariant and has order p modulo V_β, comparing with Lemma 2.54 and using that $|S/Q_\beta| = p$, it follows that $p = 2$ and $L_\beta/C_{L_\beta}(V_\beta^{(3)}/V_\beta) \cong \mathrm{Dih}(10)$ or $(3 \times 3) : 2$.

Now, $C_{L_\beta}(V_\beta^{(3)}/V_\beta)$ normalizes $V_\alpha^{(2)}$ so that $[V_\alpha^{(2)}, C_\beta] \leq Z_\alpha$ is also normalized by $C_{L_\beta}(V_\beta^{(3)}/V_\beta)$. Since R_β normalizes Z_α, if $L_\beta = \langle S, R_\beta, C_{L_\beta}(V_\beta^{(3)}/V_\beta)\rangle$ then $[V_\alpha^{(2)}, C_\beta] = Z_\beta$ and $[V_\beta^{(3)}, C_\beta] = Z_\beta$. But then $R = [V_{\alpha'}, V_\beta] \leq [V_{\alpha'-2}^{(3)}, V_\beta] = Z_{\alpha'-2}$, a contradiction. Thus $L_\beta/C_{L_\beta}(V_\beta^{(3)}/V_\beta) \cong (3\times 3) : 2$ and $C_{L_\beta}(V_\beta^{(3)}/V_\beta) \leq R_\beta$. Then $V_{\alpha'-1}^{(2)} < \langle (V_{\alpha'-3}^{(2)})^{R_\beta S}\rangle =: W$ and $|W/V_\beta| = 4$. But now, $[W, V_\beta^{(3)}] \leq [W, Q_{\alpha'-3}] \leq Z_{\alpha'-3}$ and $[V_{\alpha'-2}^{(3)} \cap Q_\beta, V_\beta] \leq Z_\beta \cap V_{\alpha'-2} = \{1\}$ and $[V_{\alpha'-2}^{(3)} \cap Q_\beta, V_\beta^{(3)}] \leq V_\beta \cap V_{\alpha'-2}^{(3)} = Z_{\alpha+2} \leq V_{\alpha'-3}^{(2)}$. Therefore, $[V_\beta^{(3)}, V_{\alpha'-2}^{(3)}] \leq V_{\alpha'-3}^{(2)}$, a final contradiction since $V_\beta^{(3)}/V_\beta$ is not dual to an FF-module. □

Lemma 7.28 *Suppose that $C_{V_\beta}(V_{\alpha'}) = V_\beta \cap Q_{\alpha'}$ and $b > 5$. If $V_{\alpha'} \not\leq Q_\beta$ and $V_\alpha^{(2)} \leq Q_{\alpha'-2}$ then $Z_{\alpha'-1} \leq V_\beta^{(3)} \leq Q_{\alpha'-1}$; $Z_{\alpha'} \not\leq V_\alpha^{(2)}$; $V_\beta^{(3)}/V_\beta$ contains a unique non-central chief factor for $\overline{L_\beta}$ which, as an $\overline{L_\beta}$-module, is an FF-module; and $O^p(R_\beta)$ centralizes $V_\beta^{(3)}$.*

Proof By Lemma 7.27, we have that $V_\alpha^{(2)} \leq Q_{\alpha'-1}$. If $V_\beta^{(3)} \leq Q_{\alpha'-1}$, then $V_\beta^{(3)} = V_\beta(V_\beta^{(3)} \cap Q_{\alpha'})$. Since $O^p(L_\beta)$ does not centralize $V_\beta^{(3)}/V_\beta$, $Z_{\alpha'} = [V_\beta^{(3)} \cap Q_{\alpha'}, V_{\alpha'}] \leq V_\beta^{(3)}$ so that $Z_{\alpha'-1} = Z_{\alpha'}R \leq V_\beta^{(3)}$. Even still, $V_\beta^{(3)}/V_\beta$ contains a unique non-central chief factor for $\overline{L_\beta}$ which is an FF-module and by Lemma 5.35, $O^p(R_\beta)$ centralizes $V_\beta^{(3)}$. If $Z_{\alpha'} \leq V_\alpha^{(2)}$ or $[V_\alpha^{(2)} \cap Q_{\alpha'}, V_{\alpha'}] = \{1\}$, then $V_\alpha^{(2)} \trianglelefteq L_\beta = \langle V_{\alpha'}, Q_\alpha, R_\beta\rangle$, a contradiction. Hence, the conclusion of the lemma holds when $V_\beta^{(3)} \leq Q_{\alpha'-1}$.

Aiming for a contradiction, we assume for the remainder of this proof that $V_\beta^{(3)} \not\leq Q_{\alpha'-1}$. Indeed, by Lemma 7.27, we have that $V_\beta^{(3)} \cap Q_{\alpha'-2} \not\leq Q_{\alpha'-1}$. Since $R \leq Z_{\alpha'-1}$ and $R \neq Z_{\alpha'}$, it follows that $V_\beta^{(3)}$ does not centralize $Z_{\alpha'}$. Hence, as $b > 5$ and $V_\beta^{(3)}$ is abelian, we conclude that $[V_\beta^{(3)} \cap \cdots \cap Q_{\alpha'}, V_{\alpha'}] = \{1\}$. In particular,

$[V_\alpha^{(2)} \cap Q_{\alpha'}, V_{\alpha'}] = \{1\}$ and so $[V_{\alpha'}, V_\alpha^{(2)}] = R \leq V_\alpha^{(2)}$. Additionally, since $V_\beta^{(3)}$ centralizes $Z_{\alpha'-2}$, we have that $R = Z_{\alpha'-2}$ intersects Z_β trivially.

We set $W^\beta = \langle V_\delta^{(2)} : Z_\lambda = Z_\alpha, \lambda \in \Delta(\beta)\rangle$ noting that $W^\beta \trianglelefteq R_\beta Q_\alpha$ by Lemma 5.23. For such a $\lambda \in \Delta(\beta)$, (λ, α') is a critical pair. Suppose that $V_\lambda^{(2)} \not\leq Q_{\alpha'-2}$. Then $\{1\} \neq [V_{\lambda-1}, V_{\alpha'-2}] \leq Z_\lambda \cap Z_{\alpha'-3} = Z_\alpha \cap Z_{\alpha'-3}$ so that $Z_\beta \leq Z_{\alpha'-3}$ and so $Z_{\alpha'-3} = Z_{\alpha'-2} \times Z_\beta = Z_{\alpha+2}$. Now, there is $\alpha'+1 \in \Delta(\alpha')$ such that $(\alpha'+1, \beta)$ is a critical pair. As in Lemma 7.27, if $V_{\alpha'+1}^{(2)} \not\leq Q_{\alpha+3}$ then $Z_{\alpha'} = [V_{\alpha'+2}, V_{\alpha+3}] \leq V_{\alpha+3}$, a contradiction as $V_{\alpha+3}$ is centralized by $V_\beta^{(3)}$. Thus, $V_{\alpha'+1}^{(2)} \leq Q_{\alpha+2}$ and since $V_{\alpha'}^{(3)} \cap Q_{\alpha+3} \leq Q_{\alpha+2}$, applying Lemma 7.27 and the results in the first paragraph of this proof, $O^p(R_{\alpha'})$ centralizes $V_{\alpha'}^{(3)}$. But then $V_\alpha^{(2)} \trianglelefteq L_\beta = \langle V_{\alpha'}, Q_\alpha, R_\beta\rangle$, a contradiction.

Thus, $W^\beta \leq Q_{\alpha'-2}$, and so every $\lambda \in \Delta(\beta)$ with $Z_\lambda = Z_\alpha$ provides a critical pair (λ, α') satisfying the same hypothesis as (α, α'). By an observation in Lemma 7.27 this yields that $W^\beta \leq Q_{\alpha'-1}$ and $W^\beta = V_\beta(W^\beta \cap Q_{\alpha'})$. Then $V_{\alpha'}$ centralizes W^β/V_β so that $W^\beta \trianglelefteq L_\beta = \langle V_{\alpha'}, R_\beta, Q_\alpha\rangle$. Since $V_{\alpha'}$ centralizes W^β/V_β, it follows that $V_\alpha^{(2)} \trianglelefteq L_\beta$, a contradiction. This completes the proof. □

Lemma 7.29 *Suppose that $C_{V_\beta}(V_{\alpha'}) = V_\beta \cap Q_{\alpha'}$ and $b > 5$. If $V_\alpha^{(2)} \not\leq Q_{\alpha'-2}$ and $|V_\beta| \neq q^3$, then we may assume that $[V_\alpha^{(2)}, Z_{\alpha'-1}] \neq \{1\}$.*

Proof Suppose that $|V_\beta| \neq q^3$. By Lemmas 7.25 and 7.26, we may assume that for any critical pair $(\alpha^*, \alpha^{*\prime})$, $V_{\alpha^*}^{(2)} \not\leq Q_{\alpha^{*\prime}-2}$. In particular, there is an infinite path $(\alpha', \alpha'-1, \alpha'-2, \ldots, \beta, \alpha, \alpha-1, \alpha-2, \ldots)$ such that $(\alpha-2k, \alpha'-2k)$ is a critical pair for all $k \geqslant 0$. For $2k > b$, we have that $Z_{\alpha'-2k-1} \neq Z_{\alpha'-2k-3}$ and so we can arrange that for our chosen critical pair (α, α') we have that $Z_{\alpha'-1} \neq Z_{\alpha'-3}$. If $[V_\alpha^{(2)}, Z_{\alpha'-1}] = \{1\}$, then $V_\alpha^{(2)}$ centralizes $Z_{\alpha'-1}Z_{\alpha'-3}$ and by Lemma 5.32, we have a contradiction. □

Notice that by Lemmas 7.25 and 7.26, whenever $|V_\beta| \neq q^3$ we have that $V_\lambda^{(2)} \not\leq Q_{\lambda+b-2}$ for any critical pair $(\lambda, \lambda+b)$ with $\lambda \in \alpha^G$. Moreover, as demonstrated in Lemma 7.29, we may iterate backwards through critical pairs far enough that the conclusion of Lemma 7.29 holds for all critical pairs beyond a certain point. The net result of this is that whenever $|V_\beta| \neq q^3$, we may assume that we have a critical pair (α, α') with $V_\alpha^{(2)} \not\leq Q_{\alpha'-2}$ and $[V_\alpha^{(2)}, Z_{\alpha'-1}] \neq \{1\}$, and for all $k \geqslant 0$ we also have that $(\alpha-2k, \alpha'-2k)$ is a critical pair with $V_{\alpha-2k}^{(2)} \not\leq Q_{\alpha'-2-2k}$ and $[V_{\alpha-2k}^{(2)}, Z_{\alpha'-1-2k}] \neq \{1\}$. We will use this fact in the following two lemmas.

Lemma 7.30 *Suppose that $C_{V_\beta}(V_{\alpha'}) = V_\beta \cap Q_{\alpha'}$ and $b = 7$. If $V_\alpha^{(2)} \not\leq Q_{\alpha'-2}$, then $|V_\beta| = q^3$.*

Proof Suppose that $b = 7$. By Lemmas 7.25 and 7.26, we may consider a critical pair (α, α') iterated backwards so that $(\alpha+2, \alpha'+2)$ is also a critical pair. Suppose first that $V^\alpha \not\leq Q_{\alpha'-2}$. Then $[V^\alpha, V_{\alpha'-2}] \leq Z_\alpha$ and so $[V^\alpha, V_{\alpha'-2}] =$

$Z_\beta \le V_{\alpha'-2}$. Since $Z_{\alpha+2} \not\le Q_{\alpha'+2}$ and $b > 5$, we have that $Z_\beta = Z_{\alpha+3}$ intersects $Z_{\alpha'-2}$ trivially. But now, $Z_{\alpha+3}Z_{\alpha+3}^g Z_{\alpha'-2} = Z_{\alpha'-3}Z_{\alpha'-3}^g$ is normalized by $L_{\alpha'-2} = \langle V^\alpha, (V^\alpha)^g, R_{\alpha'-2}\rangle$ for some appropriately chosen $g \in L_{\alpha'-2}$, so that $V_{\alpha'-2} = Z_{\alpha'-3}Z_{\alpha'-3}^g$ is of order q^3, a contradiction. Thus, we may assume that $V^\alpha \le Q_{\alpha'-2}$.

If $V^\alpha \not\le Q_{\alpha'-1}$, then $Z_{\alpha'-2} = [V^\alpha, V_{\alpha'-2}] \le Z_\alpha$ and $Z_{\alpha'-2} = Z_\beta$. Moreover, for some $\alpha - 2 \in \Delta^{(2)}(\alpha)$ with $(\alpha-2, \alpha'-2)$ a critical pair, $V_{\alpha-2}^{(2)}$ centralizes $Z_{\alpha'-2}$ and since $V_{\alpha-2}^{(2)}$ does not centralize $Z_{\alpha'-3}$, we deduce that $Z_{\alpha'-2} = Z_{\alpha+3} = Z_\beta$. Now, $[V^{\alpha'-1}, V_\beta] \le Z_{\alpha'-1}$ and since V^α does not centralize $Z_{\alpha'-1}$, $[V^{\alpha'-1}, V_\beta] \le Z_{\alpha'-2} = Z_\beta$ and $V^{\alpha'-1} \le Q_\beta$. If $V^{\alpha'-1} \le Q_\alpha$, then $[V^{\alpha'-1}, V^\alpha] \le Z_\alpha$ so that $[V^{\alpha'-1}, V^\alpha] = Z_\beta = Z_{\alpha'-2}$ and V^α centralizes $V^{\alpha'-1}/Z_{\alpha'-1}$, a contradiction since $V^{\alpha'-1}/Z_{\alpha'-1}$ contains a non-central chief factor for $L_{\alpha'-1}$. Thus, $S = V^{\alpha'-1}Q_\alpha$ and $V_{\alpha'-1}^{(2)} \cap Q_\beta = V^{\alpha'-1}(V_{\alpha'-1}^{(2)} \cap Q_\alpha)$. Since $Z_\alpha \not\le V_{\alpha'-1}^{(2)}$, $[V_{\alpha'-1}^{(2)} \cap Q_\alpha, V^\alpha] = Z_\beta = Z_{\alpha'-2}$ and it follows that $V_{\alpha'-1}^{(2)}/V^{\alpha'-1}$ is an FF-module for $\overline{L_{\alpha'-1}}$. Similarly, $[V^{\alpha'-1} \cap Q_\alpha, V^\alpha] = Z_{\alpha'-2}$ and $V^{\alpha'-1}/Z_{\alpha'-1}$ is an FF-module for $\overline{L_{\alpha'-1}}$. Then Lemma 5.33 and Lemma 5.22 applied to $Z_\beta = Z_{\alpha+3}$ implies that $V_\beta = V_{\alpha+3} \le Q_{\alpha'}$, a contradiction.

Thus, $V^\alpha = Z_\alpha(V^\alpha \cap Q_{\alpha'})$. Suppose that $V_{\alpha'} \le Q_\beta$ and again let $(\alpha-2, \alpha'-2)$ be a critical pair. Since V^α/Z_α contains a non-central chief factor, $Z_{\alpha'} \le V^\alpha$ and $Z_{\alpha'} \not\le Z_\alpha$. Then $Z_{\alpha'} = Z_{\alpha'-2}$, otherwise $[V_\alpha^{(2)}, Z_{\alpha'-1}] = \{1\}$. But now, since $b > 5$, $V_{\alpha-2}^{(2)}$ centralizes $Z_{\alpha'-2} \le V^\alpha$ and since $[V_{\alpha-2}^{(2)}, Z_{\alpha'-3}] \ne \{1\}$, it follows that $Z_{\alpha'-2} = Z_{\alpha'-4} = Z_{\alpha+3}$. Since $R = [V_{\alpha'}, V_\beta] = Z_\beta \le V_{\alpha'}$, as $Z_{\alpha+2} \not\le Q_{\alpha'+2}$, we must have that $Z_{\alpha+3} = Z_\beta$. But then $R = Z_\beta = Z_{\alpha'}$, a contradiction.

Finally, we have that $V^\alpha \le Q_{\alpha'-1}$ and $V_{\alpha'} \not\le Q_\beta$. Set $U^\beta = \langle V^\delta : Z_\delta = Z_\alpha, \delta \in \Delta(\beta)\rangle$. Then (δ, α') is a critical pair for all such $\delta \in \Delta(\beta)$ and so $V^\delta \le Q_{\alpha'-1}$ for all such δ. By Lemma 5.23, $R_\beta Q_\alpha$ normalizes U^β. Now, $U^\beta V_\beta = V_\beta(U^\beta \cap Q_{\alpha'})$ and either $Z_{\alpha'} \le V_\beta^{(3)}$; or $V_{\alpha'}$ centralizes $U^\beta V_\beta/V_\beta$. In the former case, since $V_\beta^{(3)}$ does not centralize $Z_{\alpha'-1}$, $Z_{\alpha'} = Z_{\alpha'-2}$. Iterating backwards through critical pairs, this eventually implies that $Z_{\alpha'} = Z_\beta$ and again, $V_{\alpha'}$ centralizes $U^\beta V_\beta/V_\beta$. Thus, in all cases, $U^\beta V_\beta \trianglelefteq L_\beta = \langle V_{\alpha'}, R_\beta, Q_\alpha\rangle$ and since $V_{\alpha'}$ centralizes $U^\beta V_\beta/V_\beta$, $O^p(L_\beta)$ centralizes $U^\beta V_\beta/V_\beta$. Then $V^\alpha V_\beta \trianglelefteq L_\beta$, a final contradiction is provided by Lemma 5.32. □

Lemma 7.31 *Suppose that $C_{V_\beta}(V_{\alpha'}) = V_\beta \cap Q_{\alpha'}$ and $b > 5$. If $V_\alpha^{(2)} \not\le Q_{\alpha'-2}$, then $|V_\beta| = q^3$.*

Proof By Lemma 7.30, we may assume that $b > 7$. In the following, the aim will be to prove that $Z_{\alpha'-2} = Z_{\alpha'-4}$ for then extending far enough backwards along the critical path, by Lemma 7.29, we can manufacture a situation in which (α, α') is a critical pair, $Z_{\alpha'-1-2k} \ne Z_{\alpha'-3-2k}$ for all $k \geqslant 0$ and $Z_{\alpha'} = Z_{\alpha'-2} = \cdots = Z_{\alpha+3} = Z_\beta$. Throughout we consider a critical pair (α, α') iterated backwards far enough so that $(\alpha+2, \alpha'+2)$ is also a critical pair.

Suppose first that $V_\beta^{(3)} \cap Q_{\alpha'-2} \not\leq Q_{\alpha'-1}$. Then $Z_{\alpha'-2} = [V_\beta^{(3)} \cap Q_{\alpha'-2}, Z_{\alpha'-1}] \leq V_\beta^{(3)}$ is centralized by $V_{\alpha-2}^{(2)}$ since $b > 7$. Since $V_{\alpha-2}^{(2)}$ does not centralizes $Z_{\alpha'-3}$, we have that $Z_{\alpha'-2} = Z_{\alpha'-4}$, as desired. Thus, $V_\beta^{(3)} \cap Q_{\alpha'-2} = V_\beta(V_\beta^{(3)} \cap Q_{\alpha'})$. If $Z_{\alpha'} = [V_\beta^{(3)} \cap Q_{\alpha'}, V_{\alpha'}] \leq V_\beta^{(3)}$ then, as $V_\beta^{(3)}$ does not centralize $Z_{\alpha'-1}$, we deduce that $Z_{\alpha'} = Z_{\alpha'-2} \leq V_\beta^{(3)}$. Similarly to the above, using $b > 7$, we have that $Z_{\alpha'-2} = Z_{\alpha'-4}$, as desired. Thus

(1) if $Z_{\alpha'-2} \neq Z_{\alpha'-4}$ then $[V_\beta^{(3)} \cap Q_{\alpha'}, V_{\alpha'}] = \{1\}$.

Suppose that $V_{\alpha'} \leq Q_\beta$. Then, by the above, $V_\alpha^{(2)} \cap Q_{\alpha'-2} = Z_\alpha(V_\alpha^{(2)} \cap Q_{\alpha'})$ and $[V_\alpha^{(2)} \cap Q_{\alpha'}, V_{\alpha'}] = \{1\}$, a contradiction since both $V_\alpha^{(2)}/V^\alpha$ and V^α/Z_α contain a non-central chief factor. Thus

(2) if $Z_{\alpha'-2} \neq Z_{\alpha'-4}$ then $V_{\alpha'} \not\leq Q_\beta$ and $V_\beta^{(3)}/V_\beta$ contains a unique non-central chief factor which is an FF-module for $\overline{L_\beta}$.

By Lemma 5.35, $O^p(R_\beta)$ centralizes $V_\beta^{(3)}$. If $V^\alpha \leq Q_{\alpha'-2}$, then $V^\alpha V_\beta = V_\beta(V^\alpha V_\beta \cap Q_{\alpha'})$ and it follows that $V^\alpha V_\beta \trianglelefteq L_\beta = \langle V_{\alpha'}, Q_\alpha, R_\beta \rangle$, a contradiction by Lemma 5.32. Therefore, $V^\alpha \not\leq Q_{\alpha'-2}$ and since $V_{\alpha'-2} \leq Q_\alpha$, we have that $[V^\alpha, V_{\alpha'-2}] = Z_\beta \neq Z_{\alpha'-2}$.

Suppose that $b = 9$ and consider the critical pair $(\alpha - 2, \alpha' - 2)$. Then, as $V_{\alpha'-4} \leq Q_{\alpha-2}$, we have that $[V^{\alpha-2}, V_{\alpha'-4}] \leq Z_{\alpha-1}$. Suppose that $Z_{\alpha-1} = [V^{\alpha-2}, V_{\alpha'-4}] \leq V_{\alpha'-4}$. Since $Z_\alpha, Z_{\alpha+2} \not\leq V_{\alpha'-4}$, we must have that $Z_{\alpha-1} = Z_\beta = Z_{\alpha+3} = Z_{\alpha'-6}$. But then, $[V^{\alpha-2}, V_{\alpha'-4}] = Z_{\alpha'-6}$ and $Z_{\alpha'-5}Z_{\alpha'-5}^g \trianglelefteq L_{\alpha'-4} = \langle V^{\alpha-2}, (V^{\alpha-2})^g, R_{\alpha'-4} \rangle$ for some appropriately chosen $g \in L_{\alpha'-4}$. Then $V_{\alpha'-4} = Z_{\alpha'-5}Z_{\alpha'-5}^g$ is of order q^3, a contradiction. Thus, $[V^{\alpha-2}, V_{\alpha'-4}] = \{1\}$ so that $V^{\alpha-2}V_{\alpha-1} = V_{\alpha-1}(V^{\alpha-2}V_{\alpha-1} \cap Q_{\alpha'-2})$ and since $V^{\alpha-1}V_{\alpha-1} \not\trianglelefteq L_{\alpha-1}$, it follows that $Z_{\alpha'-2} \leq V_{\alpha-1}^{(3)}$. Then $V_{\alpha-2}^{(2)}$ centralizes $Z_{\alpha'-2}$ and so $Z_{\alpha'-2} = Z_{\alpha'-4}$, as desired. Thus

(3) if $Z_{\alpha'-2} \neq Z_{\alpha'-4}$ then $b > 9$.

Since $V_{\alpha'} \not\leq Q_\beta$, there is $\lambda \in \Delta(\alpha')$ such that (λ, β) is a critical pair with $V_\beta \not\leq Q_{\alpha'}$ and $V_\lambda^{(2)} \not\leq Q_{\alpha+3}$. In particular, since $b > 5$, $V_\lambda^{(2)}$ centralizes $Z_\beta \leq V_{\alpha'-2}$ and $Z_\beta = Z_{\alpha+3}$. Now, $(\alpha - 2, \alpha' - 2)$ is also a critical pair from which we deduce in a similar fashion that $Z_\beta = Z_{\alpha-1}$. By this logic, we could have chosen the original critical pair (α, α') such that $Z_{\alpha'} = \cdots = Z_\beta$, as desired.

In all cases we have reduced to the situation where $Z_{\alpha'-2} = Z_{\alpha'-4}$. By a previous observation we may now assume that

($\star$) (α, α') is a critical pair such that $Z_{\alpha'} = Z_{\alpha'-2} = \cdots = Z_\beta = Z_{\alpha-1} = \ldots$ and $Z_{\alpha'-1-2k} \neq Z_{\alpha'-3-2k}$ for any $k \geqslant 0$

Now, $[V_{\alpha'-2}, V^\alpha] \leq [Q_\alpha, V^\alpha] \leq Z_\alpha$ so that $[V_{\alpha'-2}, V^\alpha] = Z_\beta = Z_{\alpha'-2}$ and $V^\alpha \leq Q_{\alpha'-2}$. Moreover, $V_{\alpha'} \not\leq Q_\beta$, otherwise $R = Z_\beta = Z_{\alpha'}$ and $O^p(L_{\alpha'})$ centralizes $V_{\alpha'}$.

Suppose that $V^\alpha \not\leq Q_{\alpha'-1}$. Then $V_\beta \leq Q_{\alpha'-1}$ and so $[V_\beta, V^{\alpha'-1}] \leq Z_{\alpha'-1}$ and since $V^\alpha \not\leq Q_{\alpha'-1}$, we must have that $[V^{\alpha'-1}, V_\beta] = Z_{\alpha'-2} = Z_\beta$ and $V^{\alpha'-1} \leq Q_\beta$. Moreover, $V^{\alpha'-1} \not\leq Q_\alpha$, else $[V^\alpha, V^{\alpha'-1}] = Z_\beta = Z_{\alpha'-2} \leq Z_{\alpha'-1}$ and $V^\alpha \leq Q_{\alpha'-1}$. Thus, $[V_\alpha^{(2)} \cap Q_{\alpha'-2}, V^{\alpha'-1}] = [V^\alpha(V_\alpha^{(2)} \cap Q_{\alpha'-1}), V^{\alpha'-1}] \leq V^\alpha Z_{\alpha'-2} = V^\alpha$. It follows that both $V_\alpha^{(2)}/V^\alpha$ and V^α/Z_α are FF-modules for $\overline{L_\alpha}$ and by Lemmas 5.33 and 5.22, we have that $Z_\beta = Z_{\alpha-3}$ implies that $V_\beta = V_{\alpha+3} \leq Q_{\alpha'}$, a contradiction. Thus

(4) $V^\alpha V_\beta = V_\beta(V^\alpha V_\beta \cap Q_{\alpha'})$

As in the $b = 7$ case, again set $U^\beta = \langle V^\delta : Z_\lambda = Z_\alpha, \lambda \in \Delta(\beta)\rangle \trianglelefteq R_\beta Q_\alpha$ so that (λ, α') is a critical pair for all such λ and, by the above, $V^\lambda \leq Q_{\alpha'-1}$. Then, $U^\beta V_\beta \trianglelefteq L_\beta = \langle V_{\alpha'}, R_\beta, Q_\alpha\rangle$ and since $V_{\alpha'}$ centralizes $U^\beta V_\beta/V_\beta$, $O^p(L_\beta)$ centralizes $U^\beta V_\beta/V_\beta$ and $V_\beta V^\alpha \trianglelefteq L_\beta$. A contradiction is provided by Lemma 5.32. □

As a consequence of the work up to Lemma 7.31, we may assume that whenever $b > 5$, we have that $|V_\beta| = q^3$.

Lemma 7.32 *Suppose that $C_{V_\beta}(V_{\alpha'}) = V_\beta \cap Q_{\alpha'}$ and $b > 5$. If $V_\alpha^{(2)} \not\leq Q_{\alpha'-2}$ then either:*

(i) $R = Z_{\alpha'-2} \leq Z_{\alpha+2} \cap Z_{\alpha'-1}$; *or*
(ii) $Z_{\alpha'-1} = Z_{\alpha'-3} = R \times Z_\beta$ *and* $V_{\alpha'} \leq Q_\beta$.

Proof By Lemma 7.31, we have that $|V_\beta| = q^3$, so that $R = [V_{\alpha'}, V_\beta] \leq Z_{\alpha'-1} \cap Z_{\alpha+2}$. Moreover, R is centralized by $V_\alpha^{(2)}$ and as $V_\alpha^{(2)} \not\leq Q_{\alpha'-2}$, we conclude that $R \leq Z_{\alpha'-3}$. If $R \neq Z_{\alpha'-2}$, then it follows that $Z_{\alpha'-1} = Z_{\alpha'-3}$.

If $R \neq Z_{\alpha'-2}$ and $V_{\alpha'} \not\leq Q_\beta$, then as $R \cap Z_\beta = \{1\}$ and since $[V_{\alpha'-2}, V_\alpha^{(2)}] \leq Z_\alpha \cap Q_{\alpha'} = Z_\beta$, we must have that $Z_\beta = [V_{\alpha'-2}, V_\alpha^{(2)}] \leq Z_{\alpha'-3} = Z_{\alpha'-1}$ and $Z_{\alpha'-1} = R \times Z_\beta \leq V_\beta$. Thus, $V_\beta^{(3)} \cap Q_{\alpha'-2} \leq Q_{\alpha'-1}$, $V_\beta^{(3)} \cap Q_{\alpha'-2} = V_\beta(V_\beta^{(3)} \cap Q_{\alpha'})$ and since $Z_{\alpha'} \leq Z_{\alpha'-1} \leq V_\beta$, $V_\beta^{(3)}/V_\beta$ contains a unique non-central chief factor for L_β which is an FF-module. Then, by Lemma 5.35 and Lemma 5.22, $Z_{\alpha'-1} = Z_{\alpha'-3}$ implies that $V_{\alpha'} \leq V_{\alpha'-1}^{(2)} = V_{\alpha'-3}^{(2)} \leq Q_\alpha$, a contradiction. Hence, if $R \neq Z_{\alpha'-2}$ then $V_{\alpha'} \leq Q_\beta$. □

Lemma 7.33 *Suppose that $C_{V_\beta}(V_{\alpha'}) = V_\beta \cap Q_{\alpha'}$ and $b > 5$. Then there exists a critical pair $(\alpha^*, \alpha^{*\prime})$ such that $V_{\alpha^*}^{(2)} \leq Q_{\alpha^{*\prime}-2}$.*

Proof Aiming for a contradiction, assume otherwise. Since $V_\alpha^{(2)} \not\leq Q_{\alpha'-2}$, there is another critical pair $(\alpha - 2, \alpha' - 2)$ and we may assume recursively, that there is a path $(\alpha', \alpha'-1, \ldots, \alpha, \alpha-1, \alpha-2, \alpha-3, \ldots)$ such that $(\alpha-2k, \alpha'-2k)$ is a critical pair satisfying $V_{\alpha-2k}^{(2)} \not\leq Q_{\alpha'-2k-2}$ for all $k \geqslant 0$. Set $R_k := [V_{\alpha-2k+1}, V_{\alpha'-2k}]$ for each critical pair $(\alpha - 2k, \alpha' - 2k)$. In particular, $R = R_0$.

Choose $k \geqslant (b-1)/2$ and suppose that $Z_{\alpha'-2k-1} = Z_{\alpha'-2k-3}$. Then as $k \geqslant (b-1)/2$, we have that $2k+3 \geqslant b+2$ and so, by assumption, $(\alpha'-2k-3, \alpha'-2k-3+b)$ is a critical pair, a contradiction. Thus, for $k \geqslant (b-1)/2$, we may assume that for every critical pair $(\alpha - 2k, \alpha' - 2k)$, we have that $R_k = Z_{\alpha'-2k-2} \leq Z_{\alpha-2k+2}$. Now, if $R_k \neq Z_{\alpha-2k+3}$, then $Z_{\alpha-2k+2} \cap Q_{\alpha'-2k+2} > Z_{\alpha-2k+3}$ a contradiction as $k \geqslant 1$ and $(\alpha - 2k + 2, \alpha' - 2k + 2)$ is a critical pair. Thus, we may assume that $Z_{\alpha'-2k-2} = Z_{\alpha-2k+3}$ for sufficiently large k. Then, $R_k = R_{k+1}$ for otherwise $R_k R_{k+1} \leq Z_{\alpha-2k+2} \cap Q_{\alpha'-2k+2} > Z_{\alpha-2k+3}$ since $b > 5$. In particular, $Z_{\beta-2k} = Z_{\alpha-1-2k}$ and $(\alpha-(b-1)-2k, \beta-2k)$ is a critical pair with $R_{\frac{b-1}{2}-k} = Z_{\beta-2k-2} = Z_{\alpha-1-2k} = Z_{\beta-2k}$. But then $O^p(L_{\beta-2k})$ centralizes $V_{\beta-2k}/Z_{\beta-2k}$, a contradiction. □

We aim to show that $b \leqslant 5$, and by Lemma 7.33, we can fix some pair (α, α') with $V_\alpha^{(2)} \leq Q_{\alpha'-2}$. We start with the case where $V_{\alpha'} \leq Q_\beta$.

Lemma 7.34 *Suppose that* $C_{V_\beta}(V_{\alpha'}) = V_\beta \cap Q_{\alpha'}$ *and* $b > 5$. *Assume that* $V_{\alpha'} \leq Q_\beta$ *and* $V_\alpha^{(2)} \leq Q_{\alpha'-2}$. *Then* $V_\alpha^{(2)} \leq Q_{\alpha'-1}$.

Proof Suppose for a contradiction that $V_\alpha^{(2)} \not\leq Q_{\alpha'-1}$. Then, as $R \leq Z_{\alpha'-1}$ and R is centralized by $V_\alpha^{(2)}$, we conclude that $R = Z_{\alpha'-2}$. Let $\alpha - 1 \in \Delta(\alpha)$ such that $V_{\alpha-1} \not\leq Q_{\alpha'-1}$. If $Z_{\alpha'-1} \leq Q_{\alpha-1}$, then $Z_{\alpha'-2} = [V_{\alpha-1}, Z_{\alpha'-1}] = Z_{\alpha-1}$ from which it follows that $Z_{\alpha-1} = Z_\beta$. Then, recalling that $O^p(R_\alpha)$ centralizes $V_\alpha^{(2)}$ by Lemma 7.25, by Lemma 5.22 we have that $V_{\alpha-1} = V_\beta \leq Q_{\alpha'-1}$, a contradiction. Thus,

(1) $(\alpha' - 1, \alpha - 1)$ is a critical pair and $Z_{\alpha-1} \neq Z_\beta$.

Note that if $Z_{\alpha'-2} = Z_{\alpha'-4}$, then $Z_{\alpha'-1} \leq V_{\alpha'-2} = V_{\alpha'-4}$ is centralized by $V_\alpha^{(2)}$, a contradiction. Thus, $Z_{\alpha'-3}$ is centralized by $V_\alpha^{(4)}$ and either $Z_{\alpha'-3} = Z_{\alpha'-5}$ or $V_{\alpha'-4} = Z_{\alpha'-3}Z_{\alpha'-5}$ and $V_\alpha^{(4)} \leq Q_{\alpha'-3}$. Assume that $Z_{\alpha'-3} = Z_{\alpha'-5}$. Notice that $Z_{\alpha'-1} \leq V_{\alpha'-3}^{(2)}$ and $[V_\alpha^{(2)}, V_{\alpha'-5}^{(2)}] = \{1\}$, and so by Lemmas 5.22 and 5.35, there is not a unique non-central chief factor within $V_{\alpha-1}^{(3)}/V_{\alpha-1}$ which is an FF-module. If $V_{\alpha-1}^{(3)} \leq Q_{\alpha'-4}$ then $V_{\alpha-1}^{(3)} \leq Q_{\alpha'-3}$ and $V_{\alpha-1}^{(3)} \cap Q_{\alpha'-2} = V_{\alpha-1}(V_{\alpha-1}^{(3)} \cap Q_{\alpha'-1})$ has index q in $V_{\alpha-1}^{(3)}$ and is centralized, modulo $V_{\alpha-1}$, by $Z_{\alpha'-1}$, yielding a contradiction. Thus, there is $\alpha - 4 \in \Delta^{(3)}(\alpha - 1)$ such that $(\alpha - 4, \alpha' - 4)$ is a critical pair. Then $\{1\} \neq [V_{\alpha'-4}, V_{\alpha-3}] \leq Z_{\alpha-2} \cap Z_{\alpha'-5}$.

If $[V_{\alpha'-4}, V_{\alpha-3}] = Z_{\alpha-1}$ then $Z_\alpha = [V_{\alpha'-4}, V_{\alpha-3}] \times Z_\beta$ and since $b > 5$ so that $V_{\alpha'-4} \leq Q_{\alpha'}$, we have that $Z_\alpha \leq Q_{\alpha'}$, a contradiction. Hence, we deduce that $[V_{\alpha'-4}, V_{\alpha-3}] \neq Z_{\alpha-1}$. Since $b > 5$, $Z_{\alpha-2} = Z_{\alpha-1}[V_{\alpha'-4}, V_{\alpha-3}] \leq Q_{\alpha'-1}$ and since $(\alpha' - 1, \alpha - 1)$ is a critical pair, we conclude that $Z_\alpha = Z_{\alpha-2}$. In particular, $[V_{\alpha'-4}, V_{\alpha-3}] = Z_\beta$. Now, $V_{\alpha'-4}^{(3)} \cap Q_{\alpha-1} \leq Q_{\alpha-2}$ and $[V_{\alpha'-4}^{(3)} \cap Q_{\alpha-1}, V_{\alpha-1}] \leq Z_{\alpha-2} = Z_\alpha$. Unless $b = 7$, we deduce that $[V_{\alpha'-4}^{(3)} \cap Q_{\alpha-1}, V_{\alpha-1}] = Z_\beta \leq V_{\alpha'-4}$

and so $V_{\alpha'-4}^{(3)}/V_{\alpha'-4}$ contains a unique non-central chief factor, a contradiction. Thus,

(2) $Z_{\alpha'-2} \neq Z_{\alpha'-4}$ and if $Z_{\alpha'-3} = Z_{\alpha'-5}$ then $b = 7$.

Note that $C_{Q_{\alpha-1}}(V_{\alpha-1}^{(3)}) = V_{\alpha-1}^{(3)}(C_{Q_{\alpha-1}}(V_{\alpha-1}^{(3)}) \cap Q_{\alpha'-4})$ and since $Z_{\alpha+2} = Z_{\alpha'-3}$, we conclude that $C_{Q_{\alpha-1}}(V_{\alpha-1}^{(3)}) = V_{\alpha-1}^{(3)}(C_{Q_{\alpha-1}}(V_{\alpha-1}^{(3)}) \cap Q_{\alpha'-1})$ from which we deduce that $O^p(L_{\alpha-1})$ centralizes $C_{Q_{\alpha-1}}(V_{\alpha-1}^{(3)})/V_{\alpha-1}^{(3)}$. An application of coprime action and the three subgroups lemma yields that elements of $L_{\alpha-1}$ which have order coprime to p act faithfully on $V_{\alpha-1}^{(3)}$. Moreover, an index q^2 subgroup of $V_{\alpha-1}^{(3)}$ is centralized modulo $V_{\alpha-1}$ by $Z_{\alpha'-1}$ and as $[V_{\alpha-1}^{(3)}, Z_{\alpha'-1}, Z_{\alpha'-1}] = [V_{\alpha-1}^{(3)}, V_{\alpha'-4}^{(3)}, V_{\alpha'-4}^{(3)}] = \{1\}$, an application of Lemma 2.55 reveals that $q = p$. Indeed, Lemma 2.35 implies that $q \in \{2, 3\}$. Set $R_1 := C_{L_\beta}(V_\beta^{(3)}/V_\beta)Q_\beta$ so that by Lemma 2.41, Propositions 2.46, 2.47, and 2.51, L_β/R_1 is known.

Assume that $L_\beta = R_1 R_\beta S$. Then $V_\alpha^{(2)} \trianglelefteq R_1 S$ and $[V_\alpha^{(2)}, C_\beta] \trianglelefteq R_1 S$. But $[V_\alpha^{(2)}, C_\beta] \leq Z_\alpha$ and so is normalized by $R_\beta S$. Hence, $[V_\alpha^{(2)}, C_\beta] \trianglelefteq L_\beta$ and is contained in Z_α from which we deduce that $[V_\beta^{(3)}, C_\beta] \leq Z_\beta$. But now, $Z_{\alpha'-2} = [Z_{\alpha'-1}, V_{\alpha-1}] = [V_{\alpha'-4}^{(3)}, V_{\alpha-1}] \leq Z_{\alpha'-4}$, a contradiction.

Assume now that $V_2 := [V_\beta^{(3)}, Q_\beta]V_\beta/V_\beta$ contains a non-central chief factor for L_β and write $R_2 :== C_{L_\beta}(V_2)$. Write $R_3 := C_{L_\beta}(V_\beta^{(3)}/[V_\beta^{(3)}, Q_\beta])$ so that $L_\beta/R_i \cong \mathrm{SL}_2(p)$ for $i \in \{1, 2, 3\}$. Then $V_\alpha^{(2)}[V_\beta^{(3)}, Q_\beta] \trianglelefteq R_3 S, [V_\alpha^{(2)}, Q_\beta]V_\beta \trianglelefteq \langle R_2, R_3, S\rangle$ and we conclude by Proposition 2.46 and Proposition 2.47 that either $\langle R_2, R_3, S\rangle = L_\beta$ or $R_2 = R_3$. The former case leads to a contradiction and so $R_2 = R_3 = R_1$. But now, since $R_\beta \neq R_1$, we conclude by Propositions 2.46 and 2.47 that $L_\beta = R_1 R_\beta S$, a contradiction. Hence, $[V_\beta^{(3)}, Q_\beta]V_\beta/V_\beta$ is centralized by L_β and since $[V_\alpha, Q_\beta]V_\beta$ has index p in $V_\alpha^{(2)}$, Lemma 2.54 implies that L_β/R_1 is isomorphic to one of $(3 \times 3) : 2$ or $\mathrm{Dih}(10)$. Since $L_\beta \neq R_1 R_\beta S$ and we have that $R_1 \leq R_\beta$ so that $R_1 = Q_\beta$. Thus, we conclude that

(3) if $Z_{\alpha'-3} = Z_{\alpha'-5}$ then $\overline{L_\beta} \cong (3 \times 3) : 2$.

We remark that for $\gamma \in \Delta(\alpha - 1)$, by an earlier observation in the proof, if $V_\gamma^{(2)} \not\leq Q_{\alpha'-4}$ then $Z_\gamma = Z_\alpha$. Since $V_\alpha^{(2)} \leq Q_{\alpha'-2}$ it follows $V_\gamma^{(2)} \leq Q_{\alpha'-4}$ for seven of the nine neighbors of $\alpha - 1$. Write $C^{\alpha-1}$ for the preimage in $V_{\alpha-1}^{(3)}$ of $C_{V_{\alpha-1}^{(3)}/[V_{\alpha-1}^{(3)}, Q_{\alpha-1}]V_{\alpha-1}}(L_{\alpha-1})$ so that $|V_{\alpha-1}^{(3)}/C^{\alpha-1}| = 2^4$ and $V_\alpha^{(2)}C^{\alpha-1}/C^{\alpha-1}$. Then $V_{\alpha-1}^{(3)}/C^{\alpha-1} = V_1 \times V_2$ where $|V_i| = 2^2$ and $V_i \cap V_\gamma^{(2)}C^{\alpha-1}/C^{\alpha-1} = \{1\}$ for all $\gamma \in \Delta(\alpha - 1)$.

Now if $V_\gamma C^{\alpha-1} = V_\mu C^{\alpha-1}$ for $\gamma, \mu \in \Delta(\alpha - 1)$ then $V_\gamma C^{\alpha-1} \trianglelefteq \langle Q_{\alpha-1}, Q_\gamma, Q_\mu\rangle$. One may verify that if a subgroup of $\overline{L_\beta}$ which is isomorphic to $\mathrm{Sym}(3)$ normalizes a subgroup of $V_{\alpha-1}^{(3)}/C^{\alpha-1}$ of order 2, then such a subgroup is contained in V_1 or V_2. It follows that $\gamma = \mu$ in this case. We have that $V_{\alpha-1}^{(3)}/C^{\alpha-1}$

has 15 involutions: three in V_1, three in V_2 and nine corresponding to $V_\gamma C^{\alpha-1}/C^{\alpha-1}$ for $\gamma \in \Delta(\alpha-1)$. Form X generated by $C^{\alpha-1}$ and $V_\lambda^{(2)}$ for the seven neighbors of $\alpha-1$ with $V_\lambda^{(2)} \leq Q_{\alpha'-2}$. Since $X \leq Q_{\alpha'-4}$, we must have that $|X/C^{\alpha-1}| = 2^3$. But then $X/C^{\alpha-1} \cap V_i \neq \{1\}$ for some $i \in \{1,2\}$, a contradiction for then X/C^α contains more than seven involutions. Therefore, we have forced a contradiction. Hence,

(4) $Z_{\alpha'-2} \neq Z_{\alpha'-4}$, $Z_{\alpha'-3} \neq Z_{\alpha'-5}$ and $V_\alpha^{(4)} \leq Q_{\alpha'-3}$.

It follows that $Z_{\alpha'-2} \leq [V_\alpha^{(4)}, V_{\alpha'-2}] \leq Z_{\alpha'-3}$. If $Z_{\alpha'-2} = [V_\alpha^{(4)}, V_{\alpha'-2}]$, then $V_\alpha^{(4)} = V_\alpha^{(2)}(V_\alpha^{(4)} \cap Q_{\alpha'})$ and since $Z_{\alpha'} \not\leq V_\alpha^{(4)}$, otherwise $V_\alpha^{(2)}$ centralizes $Z_{\alpha'-1} = Z_{\alpha'} \times R$, it follows that $V_{\alpha'}$ centralizes $V_\alpha^{(4)}/V_\alpha^{(2)}$, a contradiction. Thus, $[V_\alpha^{(4)}, V_{\alpha'-2}] = Z_{\alpha'-3}$. Since $V_\alpha^{(4)} \cap Q_{\alpha'-2} = V_\alpha^{(2)}(V_\alpha^{(4)} \cap Q_{\alpha'})$, we have that $V_\alpha^{(4)}/V_\alpha^{(2)}$ contains a unique non-central chief factor and by Lemma 5.34, $O^p(R_\alpha)$ centralizes $V_\alpha^{(4)}$. Furthermore, since $V_{\alpha-1}^{(3)} \not\leq Q_{\alpha'-2}$, otherwise $Z_{\alpha'-1}$ centralizes $V_{\alpha-1}^{(3)}/V_{\alpha-1}$, we may suppose that $Z_{\alpha'-3} = [V_{\alpha-1}^{(3)}, V_{\alpha'-2}]$.

Suppose first that $b > 9$. Then, $V_\alpha^{(6)}$ centralizes $Z_{\alpha'-3} \leq V_{\alpha-1}^{(3)}$ and so centralizes $Z_{\alpha'-4}Z_{\alpha'-6}$. If $Z_{\alpha'-4} = Z_{\alpha'-6}$, then by Lemma 5.22 we have that $Z_{\alpha'-1} \leq V_{\alpha'-4}^{(3)} = V_{\alpha'-6}^{(3)}$ is centralized by $V_\alpha^{(2)}$, a contradiction. Thus, $V_\alpha^{(6)}$ centralizes $Z_{\alpha'-5}$ and so either $Z_{\alpha'-5} = Z_{\alpha'-7}$ or $V_\alpha^{(6)}$ centralizes $Z_{\alpha'-3}Z_{\alpha'-5}Z_{\alpha'-7} = V_{\alpha'-6}V_{\alpha'-4}$. In the latter case, $V_\alpha^{(6)} = V_\alpha^{(4)}(V_\alpha^{(6)} \cap Q_{\alpha'-2})$ and since $Z_{\alpha'} \not\leq V_\alpha^{(6)}$, we conclude that $O^p(L_\alpha)$ centralizes $V_\alpha^{(6)}/V_\alpha^{(4)}$, a contradiction. Thus, $Z_{\alpha'-5} = Z_{\alpha'-7}$ and as $Z_{\alpha'-1} \leq V_{\alpha'-5}^{(4)}$ and $V_\alpha^{(2)}$ centralizes $V_{\alpha'-7}^{(4)}$, by Lemmas 5.22, 5.35, and 5.36, we need only show that both $V_\beta^{(5)}/V_\beta^{(3)}$ and $V_\beta^{(3)}/V_\beta$ contain a unique non-central chief factor which is an FF-module for $\overline{L_\beta}$. We may prove it for any $\lambda \in \beta^G$ and, following the steps in an earlier part of this proof, we infer that $V_\beta^{(3)}/V_\beta$ satisfies the required condition. By the steps above, $V_{\alpha-1}^{(3)} \not\leq Q_{\alpha'-2}$. Then, as $V_{\alpha'-4} = Z_{\alpha'-3}Z_{\alpha'-7}$ is centralized by $V_{\alpha-1}^{(5)}$, $V_{\alpha-1}^{(5)} \cap Q_{\alpha'-6} = V_{\alpha-1}^{(3)}(V_{\alpha-1}^{(3)} \cap Q_{\alpha'-2})$ and since $V_{\alpha'-2} \not\leq Q_{\alpha-1}$ and $Z_{\alpha'-2} \leq V_{\alpha-1}^{(3)}$, $V_{\alpha-1}^{(5)}/V_{\alpha-1}^{(3)}$ contains a unique non-central chief factor and satisfies the required conditions. This provides the contradiction. Therefore

(5) $b \leqslant 9$.

Suppose that $b = 7$. Then $C_{Q_\alpha}(V_\alpha^{(4)}) \leq Q_{\alpha+4} = Q_{\alpha'-3}$. Thus, $V_\alpha^{(4)}C_{Q_\alpha}(V_\alpha^{(4)}) = V_\alpha^{(4)}(V_\alpha^{(4)}C_{Q_\alpha}(V_\alpha^{(4)}) \cap Q_{\alpha'})$ and since $Z_{\alpha'} \not\leq C_{Q_\alpha}(V_\alpha^{(2)}) \geq C_{Q_\alpha}(V_\alpha^{(4)})$, $O^p(L_\alpha)$ centralizes $V_\alpha^{(4)}C_{Q_\alpha}(V_\alpha^{(4)})/V_\alpha^{(4)}$. Then for $r \in O^p(R_\alpha)$ of order coprime to p, $[r, Q_\alpha, V_\alpha^{(4)}] = \{1\}$ by the three subgroup lemma and so $[Q_\alpha, r] = [Q_\alpha, r, r, r] \leq [C_{Q_\alpha}(V_\alpha^{(4)}), r, r] \leq [V_\alpha^{(4)}, r] = \{1\}$ so that $R_\alpha = Q_\alpha$ and $\overline{L_\alpha} \cong \mathrm{SL}_2(q)$. We may assume that $V_{\alpha-1}^{(3)} \leq Q_{\alpha'-4}$, $V_{\alpha-1}^{(3)} \not\leq Q_{\alpha'-2}$ and $O^p(R_{\alpha-1})$ centralizes $V_{\alpha-1}^{(3)}$. Moreover, $Z_{\alpha'-3} = [V_{\alpha'-2}, V_{\alpha-1}^{(3)}] \leq V_{\alpha-1}^{(3)}$

and so $Z_{\alpha'-3}$ is centralized by $C_{Q_{\alpha-1}}(V_{\alpha-1}^{(3)})$. Since $Z_{\alpha'-3} \neq Z_{\alpha+2}$, otherwise by Lemma 5.22, $Z_\alpha \leq V_{\alpha+2}^{(2)} = V_{\alpha'-3}^{(2)} \leq Q_{\alpha'}$, we have that $C_{Q_{\alpha-1}}(V_{\alpha-1}^{(3)})$ centralizes $V_{\alpha+3}$. It follows that $C_{Q_{\alpha-1}}(V_{\alpha-1}^{(3)}) = V_{\alpha-1}^{(3)}(C_{Q_{\alpha-1}}(V_{\alpha-1}^{(3)}) \cap Q_{\alpha'-2})$ and so $O^p(L_{\alpha-1})$ centralizes $C_{Q_{\alpha-1}}(V_{\alpha-1}^{(3)})/V_{\alpha-1}^{(3)}$. Now, letting $r \in O^p(R_{\alpha-1})$ of order coprime to p, $[r, Q_{\alpha-1}, V_{\alpha-1}^{(3)}] = \{1\}$ by the three subgroup lemma and $[Q_{\alpha-1}, r] = [Q_{\alpha-1}, r, r, r] = [C_{Q_{\alpha-1}}(V_{\alpha-1}^{(3)}), r, r] = [V_{\alpha-1}^{(3)}, r] = \{1\}$ so that $R_{\alpha-1} = Q_{\alpha-1}$ and $\overline{L_{\alpha-1}} \cong \mathrm{SL}_2(q)$. Thus, G has a weak BN-pair of rank 2 and by [27], no examples exist.

Suppose that $b = 9$. Then $C_{Q_\alpha}(V_\alpha^{(4)}) \leq Q_{\alpha+4} = Q_{\alpha'-5}$. Moreover, $Z_{\alpha'-5} \neq Z_{\alpha'-3} \leq V_\alpha^{(4)}$ so that $C_{Q_\alpha}(V_\alpha^{(4)}) \leq Q_{\alpha'-3}$ and $C_{Q_\alpha}(V_\alpha^{(4)}) = V_\alpha^{(4)}(C_{Q_\alpha}(V_\alpha^{(4)}) \cap Q_{\alpha'-2})$ and it follows that $O^p(L_\alpha)$ centralizes $C_{Q_\alpha}(V_\alpha^{(4)})/V_\alpha^{(4)}$. As in the $b = 7$ case, we get that $\overline{L_\alpha} \cong \mathrm{SL}_2(q)$. Since $Z_{\alpha'-3} \leq V_{\alpha-1}^{(3)}$, $Z_{\alpha'-4}$ is centralized by $C_{Q_{\alpha-1}}(V_{\alpha-1}^{(3)})$ and $Z_{\alpha'-6} = Z_{\alpha+3}$ is centralized by $C_{Q_{\alpha-1}}(V_{\alpha-1}^{(3)})$ from which it follows that $C_{Q_{\alpha-1}}(V_{\alpha-1}^{(3)})$ centralizes $Z_{\alpha+4} = Z_{\alpha'-5}$. Continuing as above, we see that $C_{Q_{\alpha-1}}(V_{\alpha-1}^{(3)}) = V_{\alpha-1}^{(3)}(C_{Q_{\alpha-1}}(V_{\alpha-1}^{(3)}) \cap Q_{\alpha'-2})$ and $O^p(L_{\alpha-1})$ centralizes $C_{Q_{\alpha-1}}(V_{\alpha-1}^{(3)})/V_{\alpha-1}^{(3)}$ and an application of the three subgroup lemma and coprime action yields that $\overline{L_{\alpha-1}} \cong \mathrm{SL}_2(q)$ and G has a weak BN-pair of rank 2. By Delgado and Stellmacher [27], no examples exist and the proof is complete. □

Lemma 7.35 *Suppose that $C_{V_\beta}(V_{\alpha'}) = V_\beta \cap Q_{\alpha'}$ and $b > 5$. If $V_{\alpha'} \leq Q_\beta$ and $V_\alpha^{(2)} \leq Q_{\alpha'-2}$ then $V_\alpha^{(4)} \not\leq Q_{\alpha'-4}$.*

Proof By Lemma 7.34, we may suppose that $V_\alpha^{(2)} \leq Q_{\alpha'-1}$. Note that by Lemma 7.25, $Z_{\alpha'-1} = Z_{\alpha'} \times Z_\beta \leq V_\alpha^{(2)} \leq Z(V_\alpha^{(4)})$. Aiming for a contradiction, suppose that $V_\alpha^{(4)} \leq Q_{\alpha'-4}$ throughout. If $Z_{\alpha'-1} \neq Z_{\alpha'-3}$, then $V_\alpha^{(4)} \cap Q_{\alpha'-3} = V_\alpha^{(2)}(V_\alpha^{(4)} \cap Q_{\alpha'})$ and since $Z_{\alpha'} \leq V_\alpha^{(2)}$, $V_\alpha^{(4)}$ does not centralize $Z_{\alpha'-3}$. But $Z_{\alpha'-2} \leq Z_{\alpha'-1}$ so that $Z_{\alpha'-2}Z_{\alpha'-4}$ is centralized by $V_\alpha^{(4)}$ and $Z_{\alpha'-2} = Z_{\alpha'-4}$. Now, both $V_\alpha^{(4)}/V_\alpha^{(2)}$ and $V_\alpha^{(2)}/Z_\alpha$ contain unique non-central chief factors and by Lemmas 5.33 and 5.34, we deduce that $O^p(R_\alpha)$ centralizes $V_\alpha^{(4)}$. Therefore, applying Lemma 5.22 to $Z_{\alpha'-2} = Z_{\alpha'-4}$, we conclude that $V_{\alpha'} \leq V_{\alpha'-2}^{(3)} = V_{\alpha'-4}^{(3)}$ is centralized by Z_α, a contradiction. Thus,

(1) $Z_{\alpha'-1} = Z_{\alpha'-3}$ and $V_\alpha^{(4)} \not\leq Q_{\alpha'-2}$.

In particular, it follows again by Lemma 5.34 that $O^p(R_\alpha)$ centralizes $V_\alpha^{(4)}$ and so, similarly to the above, $Z_{\alpha'-2} \neq Z_{\alpha'-4}$. Moreover, by Lemma 5.22, since $V_{\alpha'} \leq V_{\alpha'-1}^{(2)}$ and $V_{\alpha'-3}^{(2)} \leq Q_\alpha$, we have that $O^p(R_\beta)$ does not centralize $V_\beta^{(3)}$. In particular, $Z_{\alpha'-1} \cap Z_{\alpha+2} = R = Z_\beta$ for otherwise $V_{\alpha'}^{(3)} \cap Q_{\alpha+3} \leq Q_{\alpha+2}$, $[V_{\alpha'-3}^{(3)} \cap Q_{\alpha+3}, V_\beta] \leq Z_{\alpha+2} = Z_{\alpha'-1} \leq V_{\alpha'}$ and $V_{\alpha'}^{(3)}/V_{\alpha'}$ contains a unique non-central chief factor which is an FF-module, and we would have a contradiction by Lemma 5.35.

Suppose first that $b = 7$. Then $Z_\beta Z_{\alpha+3} \leq Z_{\alpha+2} \cap Z_{\alpha'-3}$ and so either $Z_\beta = Z_{\alpha+3}$ or $Z_{\alpha'-1} = Z_{\alpha'-3} = Z_{\alpha+2}$. The latter case yields an immediate contradiction, while in the former case, Lemma 5.22 implies that $V_\beta = V_{\alpha+3} \leq Q_{\alpha'}$, another contradiction. Therefore

(2) $b > 7$.

Assume that for $\alpha - 4 \in \Delta^{(4)}(\alpha)$, whenever $Z_{\alpha-4} \not\leq Q_{\alpha'-2}$ we conclude that $Z_\beta = Z_{\alpha-1}$. Choose $\delta \in \Delta(\alpha)$ such that $Z_\delta \neq Z_\beta$ so that $V_\delta^{(3)} \leq Q_{\alpha'-2}$. Moreover, $V_\delta^{(3)}$ centralizes $Z_{\alpha'-1} \leq V_\alpha^{(2)}$ and $[V_\delta^{(3)}, V_{\alpha'}] = [V_\alpha^{(2)}, V_{\alpha'}][V_\delta^{(3)} \cap Q_{\alpha'}, V_{\alpha'}] \leq V_\alpha^{(2)}$. Thus, $V_\delta^{(3)} \trianglelefteq L_\alpha = \langle V_{\alpha'}, R_\alpha, Q_\delta \rangle$, a contradiction. Thus, we may assume that

$(\star)$ there exists $\alpha - 4 \in \Delta^{(4)}(\alpha)$ with $Z_{\alpha-4} \not\leq Q_{\alpha'-2}$ and $Z_\beta \neq Z_{\alpha-1}$.

Suppose that $V_{\alpha'-2} \not\leq Q_{\alpha-1}$. Since $V_\alpha^{(2)} \leq Q_{\alpha'-2}$, it follows that $Z_{\alpha'-2} = [V_\alpha^{(2)}, V_{\alpha'-2}] = Z_\beta$. Moreover, there is $\lambda \in \Delta(\alpha' - 2)$ such that $(\lambda, \alpha - 1)$ is a critical pair with $V_{\alpha-1} \leq Q_{\alpha'-2}$. If $V_\lambda^{(2)} \leq Q_\beta$, then by Lemma 7.34 $V_\lambda^{(2)} \leq Q_\alpha$ and $Z_\alpha \leq V_\lambda^{(2)}$, a contradiction since $b > 5$. Thus, $V_\lambda^{(2)} \not\leq Q_\beta$ and $(\lambda + 2, \beta)$ is also a critical pair. Moreover, $\{1\} \neq [V_\beta, V_{\lambda+1}] \leq Z_{\alpha+2} \cap Z_\lambda$. Since $Z_\lambda \not\leq Q_{\alpha-1}$ and $Z_{\alpha'-2} \leq V_\alpha^{(2)}$, it follows that $[V_\beta, V_{\lambda+1}] = Z_{\alpha'-2} = Z_\beta$. But then $V_{\lambda+1} \leq Q_\beta$, a contradiction. Thus, $V_{\alpha'-2} \leq Q_{\alpha-1}$ and $[V_{\alpha'-2}, V_{\alpha-1}] = \{1\}$, otherwise $Z_{\alpha-1} = [V_{\alpha'-2}, V_{\alpha-1}] = Z_{\alpha'-2}$ and since $Z_\alpha \not\leq V_{\alpha'-2}$, $Z_{\alpha-1} = Z_\beta$, a contradiction. Therefore

(3) $V_{\alpha'-2} \leq Q_{\alpha-2}$.

Suppose that $[V_{\alpha'-2}, V_{\alpha-3}] = Z_\beta$ so that $Z_{\alpha'-2} \cap Z_\beta = \{1\}$. As $Z_\beta \leq Z_{\alpha-2}$ and $Z_\beta \cap Z_{\alpha-1} = \{1\}$, $Z_\alpha = Z_{\alpha-2}$. Immediately, we have that $[V_\alpha^{(2)}, V_{\alpha'-2}] \leq Z_{\alpha'-2} \cap Z_\alpha = \{1\}$ so that $V_\alpha^{(2)} \leq C_{\alpha'-2}$.

Choose $\lambda \in \Delta(\alpha' - 2)$ such that $Z_\lambda \neq Z_{\alpha'-1}$ and set $W^{\alpha'-2} := \langle V_\delta^{(2)} : Z_\delta = Z_\lambda, \delta \in \Delta(\alpha' - 2)\rangle$. Then, for $\delta \in \Delta(\alpha' - 2)$ with $Z_\delta = Z_\lambda$, we have that $[V_\beta, V_\delta^{(2)}] \leq Z_\delta \cap Z_{\alpha+2}$. Since $Z_{\alpha+2} \leq Z(V_\alpha^{(4)})$, we have that $Z_\delta \cap Z_{\alpha+2} \leq Z_{\alpha'-2}$, otherwise $V_\alpha^{(4)}$ centralizes $V_{\alpha'-2} = Z_\delta Z_{\alpha'-1}$. Now, if $V_\delta^{(2)} \not\leq Q_\beta$ then there is $\delta + 1 \in \Delta(\delta)$ with $[V_{\delta+1}, V_\beta] = Z_{\alpha'-2} \neq Z_{\delta+1}$. It follows that $(\alpha, \delta + 1)$ is a critical pair with $V_\alpha^{(2)} \leq Q_{\alpha'-2}$ and $V_{\delta+1} \not\leq Q_\beta$. Hence, by Lemma 7.28, we have that $Z_\delta \leq V_\beta^{(3)} \leq V_\alpha^{(4)}$. Since $b > 7$, $V_\alpha^{(4)}$ is abelian so that $V_\alpha^{(4)}$ centralizes $V_{\alpha'-2} = Z_\delta Z_{\alpha'-3}$, a contradiction as $V_\alpha^{(4)} \not\leq Q_{\alpha'-2}$.

Hence, $V_\delta^{(2)} \leq Q_\beta$ and so $[V_\beta, V_\delta^{(2)}]$ is either trivial or equal to Z_β. Since $Z_{\alpha'-2} \neq Z_\beta$, we have that $[V_\beta, V_\delta^{(2)}] = \{1\}$. Then $[V_\alpha^{(2)}, V_\lambda^{(2)}] \leq Z_\lambda \cap Z_\alpha$ and for a similar reason as before, $[V_\alpha^{(2)}, V_\lambda^{(2)}] = \{1\}$. It follows that $W^{\alpha'-2} \leq Q_{\alpha-2}$ and $Z_\beta \leq [W^{\alpha'-2}, V_{\alpha-3}] \leq Z_{\alpha-2} = Z_\alpha$. Since $Z_\alpha \not\leq V_{\alpha'-2}^{(3)}$, we have that $[W^{\alpha'-2}, V_{\alpha-3}] = Z_\beta \leq V_{\alpha'-2}$ and $V_{\alpha-3}$ centralizes $W^{\alpha'-2}/V_{\alpha'-2}$. But now, by Lemma 5.23, $W^{\alpha'-2} \trianglelefteq L_{\alpha'-2} = \langle V_{\alpha-3}, R_{\alpha'-2}, Q_\lambda\rangle$. Since $V_{\alpha-3}$ centralizes $W^{\alpha'-2}/V_{\alpha'-2}$, it follows that $V_\lambda^{(2)} \trianglelefteq L_{\alpha'-2}$, a contradiction. Thus

(4) $Z_\beta \neq [V_{\alpha'-2}, V_{\alpha-3}] \le Z_{\alpha-2} \cap Z_{\alpha'-3}$.

Then $Z_\alpha \neq Z_{\alpha-2}$, else $Z_\beta < Z_\beta[V_{\alpha'-2}, V_{\alpha-3}] \le Z_\alpha \cap Z_{\alpha'-3}$, an obvious contradiction. Still, $Z_\alpha < Z_\alpha[V_{\alpha'-2}, V_{\alpha-3}] \le V_{\alpha-1} \cap Z_\alpha Z_{\alpha'-1}$. As $V_\beta \le C_{\alpha'-2}$, it follows that $Z_\beta \le [V_\beta, V_{\alpha'-2}^{(3)}] \le Z_{\alpha+2} \cap V_{\alpha'-2}$. Since $Z_{\alpha+2} \cap Z_{\alpha'-1} = Z_\beta$, we have that $Z_{\alpha+2} \cap V_{\alpha'-2} = Z_\beta$, otherwise $Z_{\alpha'-1}(Z_{\alpha+2} \cap Z_{\alpha+2} \le V_\alpha^{(2)}$ and $C_{V_{\alpha'-2}}(V_\alpha^{(4)}) > Z_{\alpha'-3}$, a contradiction. Thus, $[V_\beta, V_{\alpha'-2}^{(3)}] = Z_\beta$ and $V_{\alpha'-2}^{(3)} \le Q_\beta$.

Then $V_{\alpha'-2}^{(3)} \cap Q_\alpha$ centralizes $Z_\alpha[V_{\alpha'-2}, V_{\alpha-3}] \le V_{\alpha-1}$ and so $V_{\alpha'-2}^{(3)} \cap Q_\alpha \le Q_{\alpha-2}$. Then $[V_{\alpha'-2}, V_{\alpha-3}] \le [V_{\alpha'-2}^{(3)} \cap Q_\alpha, V_{\alpha-3}] \le Z_{\alpha-2}$. If $[V_{\alpha'-2}^{(3)} \cap Q_\alpha, V_{\alpha-3}] = [V_{\alpha'-2}, V_{\alpha-3}]$, then $V_{\alpha'-2}^{(3)}/V_{\alpha'-2}$ contains a unique non-central chief factor which is an FF-module. By Lemma 5.35, $O^p(R_{\alpha'-2})$ centralizes $V_{\alpha'-2}^{(3)}$ and Lemma 5.22 applied to $Z_{\alpha'-1} = Z_{\alpha'-3}$ implies that $V_{\alpha'} \le V_{\alpha'-1}^{(2)} = V_{\alpha'-3}^{(2)} \le Q_\alpha$, a contradiction. Thus, $Z_{\alpha-1} \le Z_{\alpha-2} \le V_{\alpha'-2}^{(3)}$ and since $b > 5$, we have that $Z_\beta = Z_{\alpha-1}$, a final contradiction by the choice of $\alpha - 4$. □

By Lemma 7.35, whenever $b > 5$ and $V_{\alpha'} \le Q_\beta$, we may assume that there is a critical pair $(\alpha - 4, \alpha' - 4)$. In the following lemma, we let $(\alpha - 4, \alpha' - 4)$ be such a pair and investigate the action of $V_{\alpha'-4}$ on $V_{\alpha-3}$ and vice versa.

Lemma 7.36 *Suppose that $C_{V_\beta}(V_{\alpha'}) = V_\beta \cap Q_{\alpha'}$ and $b > 5$. If $V_{\alpha'} \le Q_\beta$ and $V_\alpha^{(2)} \le Q_{\alpha'-2}$ then $b > 7$, $Z_\alpha \neq Z_{\alpha-2}$, $O^p(R_\beta)$ centralizes $V_\beta^{(3)}$ and setting $R^\dagger := [V_{\alpha'-4}, V_{\alpha-3}]$, either:*

(i) *$R^\dagger = Z_{\alpha-1} = Z_\beta$; or*
(ii) *$R^\dagger \neq Z_{\alpha-1}$.*

Proof By Lemmas 7.34 and 7.35, $V_\alpha^{(2)} \le Q_{\alpha'-1}$, $Z_{\alpha'-1} = Z_{\alpha'} \times Z_\beta \le V_\alpha^{(2)} \le Z(V_\alpha^{(4)})$, $V_\alpha^{(4)} \not\le Q_{\alpha'-4}$ and there is a critical pair $(\alpha - 4, \alpha' - 4)$. Set $R^\dagger := [V_{\alpha'-4}, V_{\alpha-3}] \le Z_{\alpha'-5} \cap Z_{\alpha-2}$. By assumption, $R^\dagger \not\le Z_{\alpha'-4}$.

Suppose first that $R^\dagger = Z_{\alpha-1} \le Z_{\alpha'-5}$. Then, as $b > 5$, $Z_{\alpha-1} = Z_\beta$ so that by Lemma 5.22, $V_{\alpha-1} = V_\beta$. Then $[V_{\alpha'-4}^{(3)}, V_{\alpha-1}] = [V_{\alpha'-4}^{(3)}, V_\beta] = \{1\}$ and so $V_{\alpha'-4}^{(3)} \le Q_{\alpha-2}$. Moreover, $V_{\alpha'-4} \not\le Q_{\alpha-3}$, else $Z_{\alpha-3} = R^\dagger = Z_{\alpha-1}$ and by Lemma 5.22, $V_{\alpha-3} = V_{\alpha-1} \le Q_{\alpha'-4}$, a contradiction as $(\alpha - 4, \alpha' - 4)$ is a critical pair. Then $V_{\alpha'-4}(V_{\alpha'-4}^{(3)} \cap Q_{\alpha-3} \cap Q_{\alpha-4})$ is an index q subgroup of $V_{\alpha'-4}^{(3)}$ which is centralized, modulo $V_{\alpha'-4}$, by $Z_{\alpha-4}$ and so, $V_{\alpha'-4}^{(3)}/V_{\alpha'-4}$ contains a unique non-central chief factor and by Lemma 5.35, and conjugacy, $O^p(R_\beta)$ centralizes $V_\beta^{(3)}$. In particular, applying Lemma 5.22, if $Z_\alpha = Z_{\alpha-2}$ then $Z_{\alpha-4} \le V_{\alpha-2}^{(2)} = V_\alpha^{(2)} \le Q_{\alpha'-4}$, a contradiction. Hence

(1) if $b > 7$ and $R^\dagger = Z_{\alpha-1}$ then (i) holds.

Assume now that $R^\dagger \neq Z_{\alpha-1}$ so that $Z_{\alpha-1}R^\dagger \leq Z_{\alpha-2}$ properly contains $Z_{\alpha-1}$ and is centralized by $V_{\alpha'-4}^{(3)}$. If $Z_\alpha = Z_{\alpha-2}$ then $Z_\beta R^\dagger \leq Z_\alpha \cap Q_{\alpha'}$ and we deduce that $R^\dagger = Z_\beta \leq V_{\alpha'-4}$. Now, $V_{\alpha'-4}^{(3)}$ centralizes $Z_\alpha = Z_{\alpha-2}$ so that $V_{\alpha'-4}^{(3)} \cap Q_{\alpha-1} \leq Q_{\alpha-2}$ and $[V_{\alpha'-4}^{(3)} \cap Q_{\alpha-1}, V_{\alpha-3}] \leq Z_{\alpha-2} \cap V_{\alpha'-4}^{(3)}$. Indeed, unless $b = 7$, $Z_{\alpha-2} \cap V_{\alpha'-4}^{(3)} \leq Z_\alpha \cap Q_{\alpha'} = Z_\beta \leq V_{\alpha'-4}$ and $V_{\alpha'-4}^{(3)}/V_{\alpha'-4}$ contains a unique non-central chief factor. By Lemma 5.35, $O^p(R_\beta)$ centralizes $V_\beta^{(3)}$ and Lemma 5.22 applied to $Z_\alpha = Z_{\alpha-2}$ gives a contradiction as above. Therefore

(2) if $b > 7$ and $R^\dagger \neq Z_{\alpha-1}$ then $Z_\alpha \neq Z_{\alpha-2}$.

If $R^\dagger \neq Z_{\alpha-1}$ and $b > 7$ then it follows that $V_{\alpha'-4}^{(3)}$ centralizes $V_{\alpha-1}$ and $V_{\alpha'-4}^{(3)} \cap Q_{\alpha-3} \cap Q_{\alpha-4}$ is an index q^2 subgroup of $V_{\alpha'-4}^{(3)}$ centralized by $Z_{\alpha-4}$. Hence, $V_{\alpha'-4}^{(3)}$ contains only two non-central chief factors for $L_{\alpha'-4}$, one in $V_{\alpha'-4}$ and one in $V_{\alpha'-4}^{(3)}/V_{\alpha'-4}$. Moreover, both non-central chief factors are FF-modules for $\overline{L_{\alpha'-4}}$ and by Lemma 5.35, and conjugacy, we have that $O^p(R_\beta)$ centralizes $V_\beta^{(3)}$. Thus

(3) if $b > 7$ and $R^\dagger \neq Z_{\alpha-1}$ then (ii) holds.

To complete the proof it remains to prove that $b > 7$, so assume that $b = 7$ for the remainder of the proof.

Suppose first that $R = Z_\beta = Z_{\alpha'-2}$. Since $Z_\beta \neq Z_{\alpha+3} = Z_{\alpha'-4}$, for otherwise by Lemma 5.22, $V_\beta = V_{\alpha+3} \leq Q_{\alpha'}$, we may assume that $Z_{\alpha+2} = Z_\beta \times Z_{\alpha+3} = Z_{\alpha'-2} \times Z_{\alpha'-4} = Z_{\alpha'-3}$. If $O^p(R_\beta)$ centralizes $V_\beta^{(3)}$ then Lemma 5.22 applied to $Z_{\alpha+2} = Z_{\alpha'-3}$ implies that $Z_\alpha \leq V_{\alpha+2}^{(2)} = V_{\alpha'-3}^{(2)} \leq Q_{\alpha'}$, a contradiction. Hence

(4) if $R = Z_\beta = Z_{\alpha'-2}$ then $Z_{\alpha+2} = Z_{\alpha'-3}$ and $O^p(R_\beta)$ acts non-trivially on $V_\beta^{(3)}$.

But now, $V_{\alpha'}^{(3)} \cap Q_{\alpha+3}$ centralizes $Z_{\alpha+2} = Z_{\alpha'-3}$ and $[V_{\alpha'}^{(3)} \cap Q_{\alpha+3} \cap Q_\beta, V_\beta] \leq Z_\beta = Z_{\alpha'-2} \leq V_{\alpha'}$. Moreover, $[V_{\alpha'}^{(3)}, V_\beta, V_\beta] \leq [V_{\alpha'}^{(3)}, V_{\alpha+3}^{(3)}, V_{\alpha+3}^{(3)}] = \{1\}$ and applying Lemmas 2.35 and 2.55, we deduce that $O^p(R_\beta)$ centralizes $V_\beta^{(3)}$ unless perhaps, $q \in \{2, 3\}$.

Now, $[V_{\alpha'}^{(3)} \cap Q_{\alpha+3}, V_\beta] \leq Z_{\alpha+2} \cap V_{\alpha'}^{(3)}$. In particular, we deduce that $Z_{\alpha'-3} \neq Z_{\alpha'-1}$ for otherwise $V_{\alpha'}^{(3)}/V_{\alpha'}$ contains a unique non-central chief factor for $L_{\alpha'}$ and by Lemma 5.35, $O^p(R_{\alpha'})$ centralizes $V_{\alpha'}^{(3)}$. But then, recalling from Lemma 7.25 that $Z_{\alpha'-1} \leq V_\alpha^{(2)}$, we have that $V_{\alpha'-2} = Z_{\alpha'-1}Z_{\alpha'-3} = Z_{\alpha'-1}Z_{\alpha+2} \leq V_\alpha^{(2)}$. Since $V_{\alpha'-2} \leq Q_{\alpha'}$, $Z_\alpha \not\leq V_{\alpha'-2}$ and so $Z_\alpha V_{\alpha'-2}$ is a subgroup of $V_\alpha^{(2)}$ of order p^4. Now, $V_\alpha^{(2)}/Z_\alpha$ is a FF-module for $\overline{L_\alpha}$ and V_β/Z_α has order p and generates $V_\alpha^{(2)}/Z_\alpha$, we infer that $p^4 \leqslant |V_\alpha^{(2)}| \leqslant p^5$. If $|V_\alpha^{(2)}| = p^4$, then $[V_\alpha^{(2)}, V_{\alpha'}] = [V_{\alpha'-2}Z_\alpha, V_{\alpha'}] = Z_\beta$, a contradiction by Lemma 7.25. Thus, $|V_\alpha^{(2)}| = p^5$ and the preimage of $C_{V_\alpha^{(2)}/Z_\alpha}(O^p(L_\alpha))$ in $V_\alpha^{(2)}$, which we write as C^α, has order p^3. By the action of Q_β on $V_\alpha^{(2)}$, we must have that $C^\alpha V_\beta \le [V_\alpha^{(2)}, Q_\beta]V_\beta$. Moreover, since $Z_\alpha = Z(Q_\alpha)$, we must have that $[Q_\alpha, C^\alpha] = Z_\alpha$

If $[V_\beta^{(3)}, Q_\beta]V_\beta/V_\beta$ is centralized by $O^p(L_\beta)$ then we have that $C^\alpha V_\beta \trianglelefteq L_\beta$. But then $Z_\beta \le [C^\alpha V_\beta, Q_\beta] \le Z_\alpha$ so that $[C^\alpha V_\beta, Q_\beta] = Z_\beta$. Then, we deduce that $C_{Q_\alpha}(C^\alpha) \le Q_\beta$ for otherwise $Z_\alpha = [Q_\alpha, C^\alpha] = [Q_\alpha \cap Q_\beta, C^\alpha] \le Z_\beta$, a contradiction. But now, as $C^{\alpha'-1}V_{\alpha'-2} \trianglelefteq L_{\alpha'-2}$, V_β centralizes $C^{\alpha'-1} \le C^{\alpha'-3}V_{\alpha'-2}$ so that $V_\beta \le C_{Q_{\alpha'-1}}(C^{\alpha'-1}) \le Q_{\alpha'}$, a contradiction. Thus

(5) if $R = Z_\beta = Z_{\alpha'-2}$ then $q \in \{2, 3\}$ and $[V_\beta^{(3)}, Q_\beta]V_\beta/V_\beta$ contains a non-central chief factor for L_β.

Moreover, since $V_{\alpha'}^{(3)} \cap Q_{\alpha+3} \le Q_{\alpha+2}$, an index p^2 subgroup of $V_{\alpha'}^{(3)}/V_{\alpha'}$ is centralized by Z_α and we conclude that $V_\beta^{(3)}/V_\beta$ contains two non-central chief factors for L_β, one in $V_\beta^{(3)}/[V_\beta^{(3)}, Q_\beta]V_\beta$ by Lemma 5.8 and one in $[V_\beta^{(3)}, Q_\beta]V_\beta/V_\beta$. Both are FF-modules for $\overline{L_\beta}$. Notice that $[V_\alpha^{(2)}, Q_\beta, Q_\beta] \le Z_\alpha$ so that $[V_\beta^{(3)}, Q_\beta, Q_\beta] \le V_\beta$. Write $R_1 := C_{L_\beta}([V_\beta^{(3)}, Q_\beta]V_\beta/V_\beta)$ and $R_2 := C_{L_\beta}(V_\beta^{(3)}/[V_\beta^{(3)}, Q_\beta]V_\beta)$ so that $L_\beta/R_1 \cong L_\beta/R_2 \cong L_\beta/R_\beta \cong \mathrm{SL}_2(p)$. Indeed, either $p \in \{2, 3\}$ and $L_\beta = \langle R_1, R_2, S\rangle$ by Proposition 2.47 (ii) or $R_1 = R_2$. In the former case, we have that $V_\alpha^{(2)}[V_\beta^{(3)}, Q_\beta] \trianglelefteq R_2S$ so that $[V_\alpha^{(2)}[V_\beta^{(3)}, Q_\beta], Q_\beta]V_\beta = [V_\alpha^{(2)}, Q_\beta]V_\beta \trianglelefteq R_2S$. But $[V_\alpha^{(2)}, Q_\beta]V_\beta \trianglelefteq R_1S$ so that $[V_\beta^{(3)}, Q_\beta]V_\beta = [V_\alpha^{(2)}, Q_\beta]V_\beta \trianglelefteq L_\beta$, impossible as then $[V_\beta^{(3)}, Q_\beta]V_\beta/V_\beta$ is centralized by Q_α, and so centralized by $O^p(L_\beta)$. Thus, $R_1 = R_2$ and as $O^p(R_\beta)$ does not centralize $V_\beta^{(3)}$ and R_β normalizes $Q_\alpha \cap Q_\beta$, we satisfy the hypothesis of Proposition 5.30 with $\lambda = \beta$. Since $b = 7$, outcome of Proposition 5.30 holds and we have that $V_\alpha^{(4)} \le \langle Z_\beta^X\rangle \le Z(O_p(X))$. In particular, $V_\alpha^{(4)}$ is abelian, and by conjugacy $Z_\alpha \le V_{\alpha'-3}^{(4)} \ge V_{\alpha'}$, impossible since $[Z_\alpha, V_{\alpha'}] \ne \{1\}$. Thus, we have that

(6) $Z_{\alpha'-2} \ne Z_\beta$ and $Z_{\alpha'-2} < Z_{\alpha'-2}Z_\beta \le Z_{\alpha'-1}$.

If $Z_{\alpha'-2} \not\le Z_{\alpha+2}$, then $Z_{\alpha+2} < Z_{\alpha+2}Z_{\alpha'-2} \le V_{\alpha+3} = V_{\alpha'-4}$ and $Z_{\alpha+2}Z_{\alpha'-2} \le V_\alpha^{(2)}$ is centralized by $V_\alpha^{(4)}$, a contradiction by Lemma 7.35. But then, since $Z_{\alpha'-2}Z_{\alpha'-4} \le Z_{\alpha+2} \cap Z_{\alpha'-3}$, we either get that $Z_{\alpha+2} = Z_{\alpha'-3}$ or $Z_{\alpha'-2} = Z_{\alpha'-4}$. In the former case, since $Z_\beta Z_{\alpha'-2} \le Z_{\alpha'-3} \cap Z_{\alpha'-1}$, we conclude that $Z_{\alpha'-1} = Z_{\alpha'-3} = Z_{\alpha+2}$. Then, $[V_{\alpha'}^{(3)} \cap Q_{\alpha+3}, V_\beta] \le Z_{\alpha+2} \le V_{\alpha'}$ and by Lemma 5.35, $O^p(R_{\alpha'})$ centralizes $V_{\alpha'}^{(3)}$. Applying Lemma 5.22 gives $V_\beta \le V_{\alpha+2}^{(2)} \le V_{\alpha'-1}^{(2)} \le Q_{\alpha'}$, a contradiction. Hence, $Z_{\alpha+2} \cap Z_{\alpha'-3} = Z_{\alpha'-2}$ and $Z_{\alpha'-2} = Z_{\alpha'-4}$. In particular, $Z_{\alpha+2} = Z_\beta \times Z_{\alpha+3} = Z_\beta \times Z_{\alpha'-2} = Z_{\alpha'-1}$ and $[V_{\alpha'}^{(3)} \cap Q_{\alpha+3}, V_\beta] \le Z_{\alpha+2} \le V_{\alpha'}$ and by Lemma 5.35, $O^p(R_\beta)$ centralizes $V_\beta^{(3)}$. Indeed, $Z_\alpha \ne Z_{\alpha-2}$ and $Z_{\alpha'-1} \ne Z_{\alpha'-3}$, else by Lemma 5.22, $Z_{\alpha-4} \le V_{\alpha-2}^{(2)} = V_\alpha^{(2)} \le Q_{\alpha'-4}$ and $V_{\alpha'} \le V_{\alpha'-1}^{(2)} = V_{\alpha'-3}^{(2)} \le Q_\alpha$ respectively.

We will show that whenever $(\alpha-4, \alpha'-4)$ is a critical pair, we have that $Z_\beta = Z_{\alpha-1}$. Choose $\alpha-4$ such that $Z_{\alpha-4} \not\le Q_{\alpha'-4}$. By the above, since $Z_\alpha \ne Z_{\alpha-2}$, assuming $Z_\beta \ne Z_{\alpha-1}$, we deduce that (ii) holds and $R^\dagger := [V_{\alpha-3}, V_{\alpha'-4}] \not\le Z_\alpha$. But $R^\dagger \le Z_{\alpha+2} \le V_\beta$ so that $Z_\alpha R^\dagger \le V_\beta \cap V_{\alpha-1}$ and we deduce that $V_\beta = Z_\alpha Z_{\alpha-2} = V_{\alpha-1}$. Then, if $Z_\beta \ne Z_{\alpha-1}$, $V_\beta \trianglelefteq L_\alpha = \langle Q_\beta, Q_{\alpha-1}, R_\alpha \rangle$, a contradiction. Therefore, we have shown that

(7) whenever $Z_{\alpha-4} \not\le Q_{\alpha'-4}$, $Z_\beta = Z_{\alpha-1}$.

Choose $\delta \in \Delta(\alpha)$ such that $Z_\delta \ne Z_\beta$ so that $V_\delta^{(3)} \le Q_{\alpha'-4}$. Suppose that $V_\delta^{(3)} \not\le Q_{\alpha'-3}$. There is $\delta-2 \in \Delta^{(2)}(\delta)$ such that $Z_{\alpha'-4} = [V_{\delta-2}, Z_{\alpha'-3}] \le Z_{\delta-1}$ and since $Z_{\alpha'-2} = Z_{\alpha'-4} = Z_{\alpha+3}$, $Z_{\alpha'-2} \le V_\beta \cap V_\delta$. If $Z_{\alpha'-2} \le Z_\alpha$, then $Z_\alpha = Z_\beta \times Z_{\alpha'-2} = Z_{\alpha'-1}$, a clear contradiction. Thus, $V_\beta = Z_{\alpha'-2}Z_\alpha = V_\delta$. But $Z_\beta \ne Z_\delta$ so that $V_\beta \trianglelefteq L_\alpha = \langle Q_\beta, Q_\delta, R_\alpha \rangle$, a contradiction.

Hence, $V_\delta^{(3)} \le Q_{\alpha'-3}$ and since $Z_{\alpha'-3} \ne Z_{\alpha'-1} = Z_{\alpha+2}$, $V_\delta^{(3)}$ centralizes $V_{\alpha'-2}$ and $V_\delta^{(3)} \le Q_{\alpha'-1}$. Setting $W^\alpha := \langle V_\lambda^{(3)} : Z_\lambda = Z_\delta, \lambda \in \Delta(\alpha) \rangle$, we have that $W^\alpha = V_\alpha^{(2)}(W^\alpha \cap Q_{\alpha'})$ and as $Z_{\alpha'} \le V_\alpha^{(2)}$, $V_{\alpha'}$ centralizes $W^\alpha/V_\alpha^{(2)}$. Moreover, since $R_\alpha Q_\delta$ normalizes W^α by Lemma 5.23, $W^\alpha \trianglelefteq L_\alpha = \langle V_{\alpha'}, Q_\delta, R_\alpha \rangle$. Since $V_{\alpha'}$ centralizes $W^\alpha/V_\alpha^{(2)}$, $O^p(L_\alpha)$ centralizes $W^\alpha/V_\alpha^{(2)}$ and $V_\delta^{(3)} \trianglelefteq L_\alpha$, a final contradiction. Hence, $b > 7$, completing the proof. □

Lemma 7.37 *Suppose that $C_{V_\beta}(V_{\alpha'}) = V_\beta \cap Q_{\alpha'}$ and $b > 5$. If $V_\alpha^{(2)} \le Q_{\alpha'-2}$, then $V_{\alpha'} \not\le Q_\beta$.*

Proof Aiming for a contradiction, we assume throughout that $V_{\alpha'} \le Q_\beta$. Since $V_{\alpha'} \le Q_\beta$, by Lemma 7.36 we may assume that $b > 7$ throughout. Recall from Lemma 7.25 that $Z_{\alpha'-1} \le V_\alpha^{(2)} \le Z(V_\alpha^{(4)})$. Notice that by Lemma 7.36, we have that $O^p(R_\beta)$ centralizes $V_\beta^{(3)}$ and by Lemma 5.22, if $Z_{\alpha'-1} = Z_{\alpha'-3}$ then $V_{\alpha'} \le V_{\alpha'-1}^{(2)} = V_{\alpha'-3}^{(2)} \le Q_\alpha$, a contradiction. Hence, we may assume that $Z_{\alpha'-1} \ne Z_{\alpha'-3}$ throughout the remainder of the proof. We fix $\alpha-4 \in \Delta^{(4)}(\alpha)$ with $(\alpha-4, \alpha'-4)$ a critical pair.

Suppose first that $Z_{\alpha'-2} \ne Z_{\alpha'-4}$ so that $Z_{\alpha'-3} = Z_{\alpha'-2} \times Z_{\alpha'-4}$ is centralized by $V_\alpha^{(4)}$. Then, $V_{\alpha'-2} = Z_{\alpha'-1}Z_{\alpha'-3}$ is centralized by $V_\alpha^{(4)}$ so $V_\alpha^{(4)} \cap Q_{\alpha'-4} = V_\alpha^{(2)}(V_\alpha^{(4)} \cap Q_{\alpha'})$ and since $Z_{\alpha'} \le V_\alpha^{(2)}$, it follows from Lemma 5.34 that $O^p(R_\alpha)$ centralizes $V_\alpha^{(4)}$. In particular, we deduce that $Z_\beta \ne Z_{\alpha-1}$, otherwise by Lemma 5.22 we have that $V_{\alpha-3} \le V_{\alpha-1}^{(3)} = V_\beta^{(3)} \le Q_{\alpha'-4}$, a contradiction. Furthermore, as $V_\alpha^{(4)} \not\le Q_{\alpha'-4}$ we have that $Z_{\alpha'-3} = Z_{\alpha'-5}$.

By Lemma 7.36, $Z_\alpha \ne Z_{\alpha-2}$, $O^p(R_\beta)$ centralizes $V_\beta^{(3)}$ and as $Z_{\alpha-1} \ne Z_\beta$, and again setting $R^\dagger := [V_{\alpha'-4}, V_{\alpha-3}]$, we have that $Z_{\alpha-1} < R^\dagger Z_{\alpha-1} \le Z_{\alpha-2}$ and $R^\dagger Z_{\alpha-1}$ is centralized by $V_{\alpha'-4}^{(3)}$. Thus, $V_{\alpha'-4}^{(3)} \le Q_{\alpha-2}$. Notice that, as $b > 7$, if $Z_{\alpha-2} \le V_{\alpha'-4}^{(3)}$ then $Z_{\alpha-1} \le V_{\alpha'-4}^{(3)} \le Q_{\alpha'}$ and we conclude that $Z_{\alpha-1} = Z_\beta$, a contradiction. Thus, $Z_{\alpha-2} \not\le V_{\alpha'-4}^{(3)}$.

If $V_{\alpha'-4} \not\leq Q_{\alpha-3}$ then $R^\dagger \neq Z_{\alpha-3}$ and $V_{\alpha'-4}^{(3)} = V_{\alpha'-4}(V_{\alpha'-4}^{(3)} \cap Q_{\alpha-3})$. Then $Z_{\alpha-3} = [V_{\alpha-3}, (V_{\alpha'-4}^{(3)} \cap Q_{\alpha-3})]$ for otherwise, $O^p(L_{\alpha'-4})$ centralizes $V_{\alpha'-4}^{(3)}/V_{\alpha'-4}$. But then $Z_{\alpha-2} = R^\dagger \times Z_{\alpha-3} \leq V_{\alpha'-4}^{(3)}$, a contradiction. Thus, $V_{\alpha'-4} \leq Q_{\alpha-3}$, $R^\dagger = Z_{\alpha-3}$ and $Z_{\alpha-3} \leq [V_{\alpha'-4}^{(3)}, V_{\alpha-3}] \leq Z_{\alpha-2} \cap V_{\alpha'-4}^{(3)} = Z_{\alpha-3}$ so that $[V_{\alpha'-4}^{(3)}, V_{\alpha-3}] = Z_{\alpha-3}$ and $V_{\alpha'-4}^{(3)} = V_{\alpha'-4}(V_{\alpha'-4}^{(3)} \cap Q_{\alpha-4})$. But then $O^p(L_{\alpha'-4})$ centralizes $V_{\alpha'-4}^{(3)}/V_{\alpha'-4}$, another contradiction.

Therefore, $Z_{\alpha'-2} = Z_{\alpha'-4}$ and by Lemma 5.22, $V_{\alpha'-2} = V_{\alpha'-4}$ so that $V_\alpha^{(4)} \cap Q_{\alpha'-4} \cap Q_{\alpha'-3} \leq Q_{\alpha'-2}$. Since $Z_{\alpha'-1}$ is centralized by $V_\alpha^{(4)}$, $V_\alpha^{(4)} \cap Q_{\alpha'-4} \cap Q_{\alpha'-3} = V_\alpha^{(2)}(V_\alpha^{(4)} \cap Q_{\alpha'})$. If $V_\alpha^{(4)}/V_\alpha^{(2)}$ contains a unique non-central chief factor which is an FF-module for $\overline{L_\alpha}$, then by Lemma 5.22, $V_{\alpha'} \leq V_{\alpha'-2}^{(3)} = V_{\alpha'-4}^{(3)} \leq Q_\alpha$, a contradiction. Thus, $V_\alpha^{(4)} \not\leq Q_{\alpha'-4}$ and $V_\alpha^{(4)} \cap Q_{\alpha'-4} \not\leq Q_{\alpha'-3}$.

Since $b > 7$, $Z_{\alpha'-4} = Z_{\alpha'-2} \leq Z_{\alpha'-1} \leq V_\alpha^{(2)} \leq Z(V_\alpha^{(6)})$. If $Z_{\alpha'-4} = Z_{\alpha'-6}$, then by Lemma 5.22, $V_{\alpha'-4} = V_{\alpha'-6}$ is centralized by $V_\alpha^{(4)}$, a contradiction. Thus, $Z_{\alpha'-5}Z_{\alpha'-7}$ is centralized by $V_\alpha^{(6)}$. If $Z_{\alpha'-5} \neq Z_{\alpha'-7}$ then $V_\alpha^{(6)} \leq Q_{\alpha'-5}$ and $V_\alpha^{(6)} = V_\alpha^{(4)}(V_\alpha^{(6)} \cap Q_{\alpha'})$. But then $O^p(L_\alpha)$ centralizes $V_\alpha^{(6)}/V_\alpha^{(4)}$, and we have a contradiction. Thus, $Z_{\alpha'-5} = Z_{\alpha'-7}$. But now, as $O^p(R_\beta)$ centralizes $V_\beta^{(3)}$ by Lemma 7.36, by Lemma 5.22 we have that $V_{\alpha'-4} \leq V_{\alpha'-5}^{(2)} = V_{\alpha'-7}^{(2)}$ is centralized by $V_\alpha^{(4)}$, a final contradiction. □

Proposition 7.38 *Suppose that $C_{V_\beta}(V_{\alpha'}) = V_\beta \cap Q_{\alpha'}$. Then $b \leqslant 7$.*

Proof By Lemma 7.37, we have that $V_{\alpha'} \not\leq Q_\beta$. Applying Lemma 7.28, we have that $Z_{\alpha'-1} \leq V_\beta^{(3)} \leq Q_{\alpha'-1}$ and $O^p(R_\beta)$ centralizes $V_\beta^{(3)}$. In particular, if $Z_{\alpha'-1} = Z_{\alpha'-3}$, then $V_{\alpha'} \leq V_{\alpha'-1}^{(2)} = V_{\alpha'-3}^{(2)}$ is centralized by Z_α, a contradiction. Hence, $V_{\alpha'-2} = Z_{\alpha'-1}Z_{\alpha'-3}$. Aiming for a contradiction, assume throughout that $b > 7$.

Suppose first that $V_\beta^{(5)} \leq Q_{\alpha'-4}$. Then, since $Z_{\alpha'-1} \leq V_\beta^{(3)} \leq Z(V_\beta^{(5)})$, $V_\beta^{(5)} \cap Q_{\alpha'-3}$ centralizes $V_{\alpha'-2}$ and so $V_\beta^{(5)} \cap Q_{\alpha'-3} = V_\beta^{(3)}(V_\beta^{(5)} \cap Q_{\alpha'})$. Since $Z_{\alpha'} \leq V_\beta^{(3)}$, $V_\beta^{(5)} \not\leq Q_{\alpha'-3}$. Moreover by Lemma 5.36, we have that $O^p(R_\beta)$ centralizes $V_\beta^{(5)}$ and so $V_\alpha^{(4)} \not\leq Q_{\alpha'-3}$, else $V_\alpha^{(4)} \trianglelefteq L_\beta = \langle V_{\alpha'}, Q_\alpha, R_\beta \rangle$. Thus, there is $\alpha - 4 \in \Delta^{(4)}(\alpha)$ such that $Z_{\alpha'-4} = [Z_{\alpha-4}, Z_{\alpha'-3}]$ and since $Z_{\alpha'-2} \leq Z_{\alpha'-1} \leq V_\beta^{(3)}$, we deduce that

(1) if $V_\beta^{(5)} \leq Q_{\alpha'-4}$ then $Z_{\alpha'-2} = Z_{\alpha'-4}$.

Suppose that $Z_{\alpha'-3} \not\leq Q_{\alpha-3}$. Then $(\alpha' - 3, \alpha - 3)$ is a critical pair with $V_{\alpha-3} \leq Q_{\alpha'-4}$. By Lemma 7.37, $V_{\alpha'-3}^{(2)} \not\leq Q_{\alpha-1}$ and either $Z_\alpha = Z_{\alpha-2}$ or $Z_{\alpha-1} = [V_{\alpha'-4}, V_{\alpha-3}] = Z_{\alpha'-4}$. In the former case it follows from Lemma 5.22 that $Z_{\alpha-4} \leq V_{\alpha-2}^{(2)} = V_\alpha^{(2)} \leq Q_{\alpha'-3}$, a contradiction. In the latter case, we have that $Z_\beta = Z_{\alpha-1} = Z_{\alpha'-4} = Z_{\alpha'-2}$. Then $R \cap Z_{\alpha'-2} = \{1\}$, so that $Z_{\alpha'} \leq Z_{\alpha'-1} = R \times Z_{\alpha'-2} \leq V_\beta$ and $V_{\alpha'}$ centralizes $V_\beta^{(3)}/V_\beta$, a contradiction.

Thus, $Z_{\alpha'-3} \le Q_{\alpha-3}$ and $Z_{\alpha'-4} = Z_{\alpha-3}$. If $Z_{\alpha-3} \le Z_\alpha$, then $Z_{\alpha-3} = Z_\beta = Z_{\alpha'-4} = Z_{\alpha'-2}$. But then $R \cap Z_{\alpha'-2} = \{1\}$ and $Z_{\alpha'-1} = R \times Z_\beta$ so that $Z_{\alpha'-1} \le V_\beta$ and $V_{\alpha'}$ centralizes $V_\beta^{(3)}/V_\beta$, a contradiction. Thus, $V_{\alpha-1} = Z_\alpha Z_{\alpha-3}$ is centralized by $V_{\alpha'-3}^{(2)}$ so that $V_{\alpha'-3}^{(2)} \le Q_{\alpha-2}$. Then, $Z_{\alpha-3} \le [V_{\alpha'-3}^{(2)}, V_{\alpha-3}] \le Z_{\alpha-2}$ and since $V_{\alpha-3}$ does not centralize $V_{\alpha'-3}^{(2)}/Z_{\alpha'-3}$, we may assume that $Z_{\alpha-2} \le V_{\alpha'-3}^{(2)}$. Still, $[V_{\alpha'-3}^{(2)} \cap Q_{\alpha-3}, V_{\alpha-3}] \le Z_{\alpha'-3}$ and it follows from Lemma 5.33 then $O^p(R_\alpha)$ centralizes $V_\alpha^{(2)}$. Since $Z_{\alpha'-2} = Z_{\alpha'-4}$, Lemma 5.22 implies that $V_{\alpha'-2} = V_{\alpha'-4}$. Moreover, since $V_{\alpha'-4}$ is not centralized by $V_\beta^{(5)}$ but $Z_{\alpha'-1}Z_{\alpha'-5} \le V_{\alpha'-4}$ and $Z_{\alpha'-1}Z_{\alpha'-5}$ is centralized by $V_\beta^{(5)}$, it follows that

(2) if $V_\beta^{(5)} \le Q_{\alpha'-4}$ then $Z_{\alpha'-1} = Z_{\alpha'-5}$.

Now, if $Z_{\alpha'-4} = Z_{\alpha'-6}$ then Lemma 5.22 implies that $Z_{\alpha'-3} \le V_{\alpha'-4} = V_{\alpha'-6}$ is centralized by $V_\alpha^{(4)}$, a contradiction. Thus $Z_{\alpha'-5} = Z_{\alpha'-4} \times Z_{\alpha'-6}$ is centralized by $V_{\alpha-4}^{(2)}$ since $Z_{\alpha'-4} = Z_{\alpha-3}$. Moreover, $Z_{\alpha'-5} \ne Z_{\alpha'-7}$, otherwise Lemma 5.22 implies that $Z_{\alpha'-3} \le V_{\alpha'-5}^{(2)} = V_{\alpha'-7}^{(2)}$ is centralized by $V_\alpha^{(4)}$. Hence, $V_{\alpha-4}^{(2)}$ centralizes $V_{\alpha'-6}$ and $V_{\alpha-4}^{(2)} \le Q_{\alpha'-5}$. If $V_{\alpha-4}^{(2)} \le Q_{\alpha'-4}$, then $V_{\alpha-4}^{(2)} = Z_{\alpha-4}(V_{\alpha-4}^{(2)} \cap Q_{\alpha'-3})$ is centralized, modulo $Z_{\alpha-4}$, by $Z_{\alpha'-3}$ so that $O^p(L_{\alpha-4})$ centralizes $V_{\alpha-4}^{(2)}/Z_{\alpha-4}$, a contradiction. Then $V_{\alpha-4}^{(2)} \not\le Q_{\alpha'-4}$ and $[V_{\alpha-4}^{(2)}, V_{\alpha'-4}] \not\le Z_{\alpha'-4}$. Since $Z_{\alpha'-4} = Z_{\alpha-3} \le V_{\alpha-4}^{(2)}$, we assume that $Z_{\alpha'-5} \le V_{\alpha-4}^{(2)}$.

Now, $V_{\alpha-4}^{(4)}$ centralizes $Z_{\alpha'-6} \le Z_{\alpha'-5}$ and either $Z_{\alpha'-6} = Z_{\alpha'-8}$; or $V_{\alpha-4}^{(4)}$ centralizes $Z_{\alpha'-5}Z_{\alpha'-7}$. In the latter case, we may assume that $Z_{\alpha'-5} \ne Z_{\alpha'-7}$ for the same reason as above, and so either $V_{\alpha-4}^{(4)} \le Q_{\alpha'-5}$ and $O^p(L_{\alpha-4})$ centralizes $V_{\alpha-4}^{(4)}/V_{\alpha-4}^{(2)}$, a contradiction; or $Z_{\alpha'-7} = Z_{\alpha'-9}$, $O^p(R_\beta)$ centralizes $V_\beta^{(5)}$ and $Z_{\alpha'-3} \le V_{\alpha'-7}^{(4)} = V_{\alpha'-9}^{(4)}$ is centralized by $V_\alpha^{(4)}$, another contradiction. Thus

(3) if $V_\beta^{(5)} \le Q_{\alpha'-4}$ then $Z_{\alpha'-6} = Z_{\alpha'-8}$.

Additionally, we have that $V_{\alpha'-6} = V_{\alpha'-8}$. Suppose that $V_{\alpha-4}^{(4)} \le Q_{\alpha'-8}$. Then $[V_{\alpha-4}^{(4)} \cap Q_{\alpha'-7}, V_{\alpha'-6}] = [V_{\alpha-4}^{(4)} \cap Q_{\alpha'-7}, V_{\alpha'-8}] = Z_{\alpha'-8} = Z_{\alpha'-6}$ and $V_{\alpha-4}^{(4)} \cap Q_{\alpha'-7} \le Q_{\alpha'-6}$. But $V_{\alpha-4}^{(4)} \cap Q_{\alpha'-7}$ centralizes $Z_{\alpha'-5}$ so that $V_{\alpha-4}^{(4)} \cap Q_{\alpha'-7} = V_{\alpha-4}^{(2)}(V_{\alpha-4}^{(4)} \cap Q_{\alpha'-4})$ and by Lemma 5.34, $O^p(R_{\alpha-4})$ centralizes $V_{\alpha-4}^{(4)}$. But now, Lemma 5.22 applied to $Z_{\alpha'-2} = Z_{\alpha'-4}$ implies that $V_{\alpha'} \le V_{\alpha'-2}^{(3)} = V_{\alpha'-4}^{(3)} \le Q_\alpha$, a contradiction.

Thus, we have shown that there is a critical pair $(\alpha - 8, \alpha' - 8)$. Since $Z_{\alpha'-5}Z_{\alpha'-9} \le V_{\alpha'-8}$ is centralized by $V_{\alpha-4}^{(4)}$, we get that $Z_{\alpha'-1} = Z_{\alpha'-5} = Z_{\alpha'-9}$. We claim that the pair $(\alpha - 8, \alpha' - 8)$ satisfies $V_{\alpha'-8} \not\le Q_{\alpha-7}$ and $V_{\alpha-8}^{(2)} \le Q_{\alpha'-10}$. By Lemma 7.37, if $V_{\alpha'-8} \le Q_{\alpha-7}$ then we have that $V_{\alpha-8}^{(2)} \not\le Q_{\alpha'-10}$. More generally, if $V_{\alpha-8}^{(2)} \not\le Q_{\alpha'-10}$ then as $Z_{\alpha'-9} = Z_{\alpha'-5} \le V_{\alpha-4}^{(2)}$ is centralized by

$V_{\alpha-8}^{(2)}$ since $b > 7$, we deduce that $Z_{\alpha'-9} = Z_{\alpha'-11}$. Then applying Lemma 5.22 gives $V_{\alpha'-8} \le V_{\alpha'-9}^{(2)} = V_{\alpha'-11}^{(2)}$ is centralized by $V_{\alpha-4}^{(4)}$, a contradiction. All in, we have that shown that

(4) if $V_\beta^{(5)} \le Q_{\alpha'-4}$ then $Z_{\alpha'-1} = Z_{\alpha'-5} = Z_{\alpha'-9}$,$O^p(R_\alpha)$ centralizes $V_\alpha^{(2)}$ and there is a critical pair $(\alpha - 8, \alpha' - 8)$ with $V_{\alpha'-8} \not\le Q_{\alpha-7}$, $V_{\alpha-8}^{(2)} \le Q_{\alpha'-10}$, $Z_{\alpha'-2} = Z_{\alpha'-4}$ and $Z_{\alpha'-6} = Z_{\alpha'-8}$.

Suppose now that $V_\beta^{(5)} \not\le Q_{\alpha'-4}$. Since $Z_{\alpha'-2} \le Z_{\alpha'-1}$ is centralized by $V_\beta^{(5)}$, it follows that either $Z_{\alpha'-2} = Z_{\alpha'-4}$; or $Z_{\alpha'-3} = Z_{\alpha'-5}$. In the latter case, we have that $V_\beta^{(5)} \cap Q_{\alpha'-4}$ centralizes $V_{\alpha'-2}$ so that $V_\beta^{(5)} \cap Q_{\alpha'-4} = V_\beta^{(3)}(V_\beta^{(5)} \cap \cdots \cap Q_{\alpha'})$ and Lemma 5.36 implies that $O^p(R_\beta)$ centralizes $V_\beta^{(5)}$. But then Lemma 5.22 applied to $Z_{\alpha'-3} = Z_{\alpha'-5}$ gives $V_{\alpha'} \le V_{\alpha'-3}^{(4)} = V_{\alpha'-5}^{(4)} \le Q_\alpha$, a contradiction. Thus,

(5) if $V_\beta^{(5)} \not\le Q_{\alpha'-4}$ then $Z_{\alpha'-2} = Z_{\alpha'-4}$.

Suppose that $V_\beta^{(5)} \not\le Q_{\alpha'-4}$ and observe that if $V_{\alpha'-4} \le Q_{\beta-4}$ then $V_{\beta-5}^{(2)} \not\le Q_{\alpha'-6}$ by Lemma 7.37. Assume that $V_{\beta-5}^{(2)} \not\le Q_{\alpha'-6}$. If $Z_{\alpha'-5} = Z_{\alpha'-7}$ then by Lemma 5.22 we deduce that $V_{\alpha'-4} \le V_{\alpha'-5}^{(2)} = V_{\alpha'-7}^{(2)}$. Since $b \geqslant 9$ we have that $V_\beta^{(5)}$ centralizes $V_{\alpha'-7}^{(2)}$ and as $V_\beta^{(5)} \not\le Q_{\alpha'-4}$ we have a contradiction. Hence, $Z_{\alpha'-5} \ne Z_{\alpha'-7}$ and $V_{\beta-5}^{(2)}$ does not centralize $Z_{\alpha'-5}$. Note that for $\lambda \in \Delta^{(3)}(\beta)$, $V_{\alpha'-4} \le Q_\lambda$ so that $[V_{\alpha'-4}, V_\lambda^{(2)}] \le Z_\lambda \le V_\beta^{(3)}$. Hence, we deduce that $[V_\beta^{(5)}, V_{\alpha'-4}] \le Z_{\alpha'-5} \cap V_\beta^{(3)}$. Since $Z_{\alpha'-4} = Z_{\alpha'-2} \le V_\beta^{(3)}$, we conclude that $Z_{\alpha'-5} = Z_{\alpha'-4}[V_\beta^{(5)}, V_{\alpha'-4}] \le V_\beta^{(3)}$. Since $V_{\beta-5}^{(2)}$ does not centralize $Z_{\alpha'-5}$, we deduce that $b = 9$.

We now prove, generally, that $b > 9$. We have that $Z_{\alpha'-2} = Z_{\alpha'-4}$. Since $V_{\alpha'} \not\le Q_\beta$, there is a critical pair $(\alpha' + 1, \beta)$ with $V_\beta \not\le Q_{\alpha'}$. Then $[V_{\alpha+3}, V_{\alpha'+1}^{(2)}] \le Z_{\alpha'+1} \cap Z_{\alpha+4} \le Z_{\alpha'}$. In particular, if $V_{\alpha'+1}^{(2)} \not\le Q_{\alpha+3}$, then $Z_{\alpha'} \le Z_{\alpha+4} = Z_{\alpha'-5}$. Furthermore, by Lemma 7.32, $R = Z_{\alpha+3} \le Z_{\alpha'-1} \cap Z_{\alpha+4}$. Since $R \ne Z_{\alpha'}$, we deduce that $Z_{\alpha'-1} = Z_{\alpha'-5}$. Since $Z_{\alpha'-1}$ is centralized by $V_\beta^{(3)}$ we conclude that $V_\beta^{(5)} \le Q_{\alpha'-4}$ and so by earlier work, $R = Z_{\alpha+3} = Z_\beta$, a contradiction. Hence, $V_{\alpha'+1}^{(2)} \le Q_{\alpha+3}$ and so $(\alpha' + 1, \beta)$ satisfies the same hypothesis as (α, α'). Applying the arguments in this proof, we deduce that $Z_{\alpha+3} = Z_{\alpha+5}$. Hence, $Z_{\alpha'-2} = Z_{\alpha'-4} = Z_{\alpha+5} = Z_{\alpha+3}$.

If $V_\beta^{(5)} \le Q_{\alpha'-4}$ then $Z_{\alpha+3} = Z_{\alpha'-6} = Z_{\alpha'-8} = Z_\beta$. But then $R \ne Z_{\alpha'-2}$, $Z_{\alpha'-1} = Z_{\alpha'-2} \times R = Z_{\alpha+2}$ and $[V_{\alpha'}, V_\beta^{(3)}] = Z_{\alpha'-1} \le V_\beta$, a contradiction for then $O^p(L_\beta)$ centralizes $V_\beta^{(3)}/V_\beta$. Hence, $V_\beta^{(5)} \not\le Q_{\alpha'-4}$ and we must have that $V_{\beta-5}^{(2)} \not\le Q_{\alpha'-6}$. Since $[V_{\beta-4}, V_{\alpha'-4}] \ne Z_{\alpha'-4} = Z_{\alpha'-6}$, Lemma 7.32 reveals that

$Z_{\alpha'-5} = Z_{\alpha+2}$, a contradiction since then $V_{\beta-5}^{(2)}$ centralizes $Z_{\alpha'-5}$. Thus, we have shown that

(6) $b \geqslant 11$ and if $V_\beta^{(5)} \not\leq Q_{\alpha'-4}$ then $V_{\alpha'-4} \not\leq V_{\beta-4}$ and $V_{\beta-5}^{(2)} \leq Q_{\alpha'-6}$.

We have demonstrated, regardless of the hypothesis on $V_\beta^{(5)}$, that

(7) $Z_{\alpha'-2-4k} = Z_{\alpha'-4-4k}$ for $k \geqslant 0$, and there are suitable critical pairs to iterate upon (either $(\beta - 4k - 1, \alpha' - 4k)$ or $(\alpha - 8k, \alpha' - 8k)$).

We observe that if $O^p(R_\alpha)$ centralizes $V_\alpha^{(2)}$, then using Lemma 5.22 and $Z_{\alpha'-2} = Z_{\alpha'-4}$, we have that $Z_{\alpha'-1}Z_{\alpha'-5} \leq V_{\alpha'-4}$ is centralized by $V_\beta^{(5)}$ and we conclude that $Z_{\alpha'-1} = Z_{\alpha'-5}$. We also note that if any of the critical pairs we construct satisfy $V_\beta^{(5)} \leq Q_{\alpha'-4}$ then $O^p(R_\alpha)$ centralizes $V_\alpha^{(2)}$.

Now, if $O^p(R_\alpha)$ centralizes $V_\alpha^{(2)}$ then it follows that $Z_{\alpha'-1} = Z_{\alpha'-1-4k}$ for all $k \geqslant 0$. Since $Z_\alpha \neq Z_{\alpha'-1}$ we deduce that $Z_{\alpha+2} = Z_{\alpha'-1}$. But then $Z_{\alpha'-1} \leq V_\beta \leq V_\alpha^{(2)}$, against Lemma 7.28. Hence, we have shown that

(8) $O^p(R_\alpha)$ does not centralize $V_\alpha^{(2)}$ and for every critical pair $(\alpha^*, \alpha^{*\prime})$ we construct iteratively, we have that $V_{\alpha^{*\prime}} \not\leq Q_{\beta^*}$, $V_{\beta^*}^{(3)} \leq Q_{\alpha^{*\prime}-1}$, $V_{\beta^*}^{(5)} \not\leq Q_{\alpha^{*\prime}-4}$ and $Z_{\alpha^{*\prime}-2} = Z_{\alpha^{*\prime}-4}$.

Since $b \geqslant 11$, $V_\beta^{(7)}$ centralizes $Z_{\alpha'-4} \leq Z_{\alpha'-1} \leq V_\beta^{(3)}$. Unless $Z_{\alpha'-4} = Z_{\alpha'-6}$, $[V_\beta^{(7)}, Z_{\alpha'-5}] = \{1\}$. Notice that if $Z_{\alpha'-5} = Z_{\alpha'-7}$, then Lemma 5.22 implies that $V_{\alpha'-4} \leq V_{\alpha'-5}^{(2)} = V_{\alpha'-7}^{(2)}$ is centralized by $V_\beta^{(5)}$, a contradiction. Thus, $V_\beta^{(7)}$ centralizes $V_{\alpha'-6} = Z_{\alpha'-5}Z_{\alpha'-7}$. But then $V_\beta^{(7)} = V_\beta^{(5)}(V_\beta^{(7)} \cap Q_{\alpha'-4})$ and $V_\beta^{(5)} \cap Q_{\alpha'-4} \leq Q_{\alpha'-3}$, otherwise $V_\beta^{(7)} = V_\beta^{(5)}(V_\beta^{(7)} \cap Q_{\alpha'})$ so that $O^p(L_\beta)$ centralizes $V_\beta^{(7)}/V_\beta^{(5)}$, another contradiction. Then, $V_\beta^{(5)} \cap Q_{\alpha'-4} = V_\beta^{(3)}(V_\beta^{(5)} \cap \dots \cap Q_{\alpha'})$ and Lemma 5.36 implies that $O^p(R_\beta)$ centralizes $V_\beta^{(5)}$. In particular, $V_\alpha^{(4)} \not\leq Q_{\alpha'-4}$ for otherwise $V_\alpha^{(4)} \trianglelefteq L_\beta = \langle V_{\alpha'}, Q_\alpha, R_\beta \rangle$, a contradiction. We have shown that

(9) whenever $Z_{\alpha^{*\prime}-4} \neq Z_{\alpha^{*\prime}-6}$, we have that $O^p(R_\beta)$ centralizes $V_\beta^{(5)}$, $V_{\alpha^*}^{(4)} \not\leq Q_{\alpha^{*\prime}-4}$ and $V_{\alpha^*-4}^{(2)} \leq Q_{\alpha^{*\prime}-6}$.

Suppose that $Z_{\alpha'-4} \neq Z_{\alpha'-6}$ so that there is a critical pair $(\alpha - 4, \alpha' - 4)$ and $O^p(R_\beta)$ centralizes $V_\beta^{(5)}$. Since $V_{\alpha'} \not\leq Q_\beta$, there is $\alpha' + 1 \in \Delta(\alpha')$ such that $(\alpha' + 1, \beta)$ is a critical pair. Suppose, in addition, that $V_{\alpha'+1}^{(2)}$ centralizes Z_β. Since $Z_{\alpha+2} = Z_\beta \times R \neq Z_{\alpha+4}$, we have that $V_{\alpha'+1}^{(2)}$ centralizes $V_{\alpha+3}$ and $V_{\alpha'+1}^{(2)} = Z_{\alpha'+1}(V_{\alpha'+1}^{(2)} \cap Q_\beta)$. In particular, $V_{\alpha'+1}^{(2)} \cap Q_\beta \not\leq Q_\alpha$, otherwise $V_{\alpha'+1}^{(2)}$ is normalized by $L_{\alpha'} = \langle V_\beta, Q_{\alpha'+1}, R_{\alpha'} \rangle$. But now, $[V_\alpha^{(2)} \cap Q_{\alpha'} \cap Q_{\alpha'+1}, V_{\alpha'+1}^{(2)} \cap Q_\beta] \leq Z_{\alpha'+1} \cap V_\alpha^{(2)}$ and since $Z_{\alpha'} \not\leq V_\alpha^{(2)}$ by Lemma 7.28, we infer that

$[V_\alpha^{(2)} \cap Q_{\alpha'} \cap Q_{\alpha'+1}, V_{\alpha'+1}^{(2)} \cap Q_\beta] = \{1\}$ and $V_\alpha^{(2)}/Z_\alpha$ is an FF-module for $\overline{L_\alpha}$, a contradiction by Lemma 5.33.

Thus, to show that $Z_{\alpha'-4} = Z_{\alpha'-6}$, it suffices to prove that Z_β is centralized by $V_{\alpha'}^{(3)}$. Since $V_\alpha^{(4)} \not\leq Q_{\alpha'-4}$, $\{1\} \neq [V_{\alpha-3}, V_{\alpha'-4}] \leq Z_{\alpha-2} \cap V_{\alpha'-4}$. If $[V_{\alpha-3}, V_{\alpha'-4}] = Z_{\alpha-1}$, then $Z_{\alpha-1} = Z_\beta \leq V_{\alpha'-4}$, for otherwise $Z_\alpha \leq Q_{\alpha'}$. Since $b > 7$, this leads to a contradiction. Thus, $Z_{\alpha-1} < [V_{\alpha-3}, V_{\alpha'-4}]Z_{\alpha-1} \leq Z_{\alpha-2} \neq Z_\alpha$ and $V_{\alpha'-4}^{(3)}$ centralizes $V_{\alpha-1} = Z_{\alpha-2}Z_\alpha$. Thus, since $V_{\alpha'-4}^{(3)}/V_{\alpha'-4}$ contains a non-central chief factor, $[V_{\alpha-3}, V_{\alpha'-4}] < [V_{\alpha-3}, V_{\alpha'-4}^{(3)}] \leq Z_{\alpha-2}$ so that $Z_{\alpha-2} \leq V_{\alpha'-4}^{(3)}$. In particular, $Z_{\alpha-1} \leq V_{\alpha'-4}^{(3)}$ and since $b > 7$, we have that $Z_{\alpha-1} = Z_\beta \leq V_{\alpha'-4}^{(3)}$. Since $b > 9$, $V_{\alpha'}^{(3)}$ centralizes $V_{\alpha'-4}^{(3)}$ so that $V_{\alpha'+1}^{(2)}$ centralizes Z_β, as required. Thus, we have shown that

(10) $Z_{\alpha'-2} = Z_{\alpha'-4} = Z_{\alpha'-6}$ and there is a critical pair $(\beta - 5, \alpha' - 4)$

Then, as $[V_{\beta-4}, V_{\alpha'-4}] \neq Z_{\alpha'-6}$ and $Z_{\alpha'-5} \neq Z_{\alpha'-7}$, $V_{\beta-5}^{(2)} \leq Q_{\alpha'-6}$ and by Lemma 7.37, we have that $V_{\alpha'-4} \not\leq Q_{\beta-4}$. In particular, $(\beta - 5, \alpha' - 4)$ satisfies the same hypothesis as (α, α') and applying the same methodology as above, we infer that $Z_{\alpha'-6} = Z_{\alpha'-8} = Z_{\alpha'-10}$. Applying this iteratively, we deduce that $Z_{\alpha'-2} = \dots = Z_\beta$. In particular, $Z_{\alpha'-2} = Z_\beta \neq R \leq Z_{\alpha'-1} \cap Z_{\alpha+2}$ so that $Z_{\alpha'-1} = Z_{\alpha+2}$. But then $[V_\beta^{(3)}, V_{\alpha'}] = Z_{\alpha'-1} = Z_{\alpha+2} \leq V_\beta$ and $O^p(L_\beta)$ centralizes $V_\beta^{(3)}/V_\beta$, a final contradiction. □

Proposition 7.39 *Suppose that $C_{V_\beta}(V_{\alpha'}) = V_\beta \cap Q_{\alpha'}$. Then $b \leqslant 5$.*

Proof By Lemma 7.37 and Proposition 7.38, we have that $V_{\alpha'} \not\leq Q_\beta$, $O^p(R_\beta)$ centralizes $V_\beta^{(3)}$ and $b = 7$. Since $V_{\alpha'-2} = Z_{\alpha'-1}Z_{\alpha'-3} \leq V_\beta^{(3)}$ and $V_\beta^{(3)}$ is abelian, we have that $C_{Q_\beta}(V_\beta^{(3)}) = V_\beta^{(3)}(C_{Q_\beta}(V_\beta^{(3)}) \cap Q_{\alpha'})$ and since $Z_{\alpha'} \leq V_\beta^{(3)}$, $O^p(L_\beta)$ centralizes $C_{Q_\beta}(V_\beta^{(3)})/V_\beta^{(3)}$. In particular, $O^p(R_\beta)$ centralizes $C_{Q_\beta}(V_\beta^{(3)})$. But now, by the three subgroup lemma, for $r \in O^p(R_\beta)$ of order coprime to p, $[r, Q_\beta, V_\beta^{(3)}] = \{1\}$ and r centralizes Q_β. Thus, $R_\beta = Q_\beta$ and $\overline{L_\beta} \cong \mathrm{SL}_2(q)$. In particular, for $\mu, \delta \in \Delta(\lambda)$ where $\lambda \in \beta^G$, whenever $Z_\mu \cap Z_\delta > Z_\lambda$ we have that $\mu = \delta$.

Let $\alpha' + 1 \in \Delta(\alpha')$ such that $Z_{\alpha'+1} \not\leq Q_\beta$. Then, $V_\alpha^{(2)} \cap Q_{\alpha'} \not\leq Q_{\alpha'+1}$, for otherwise $V_{\alpha'}$ normalizes $V_\alpha^{(2)}$, a contradiction for then $L_\beta = \langle V_{\alpha'}, Q_\alpha, Q_\beta \rangle$ normalizes $V_\alpha^{(2)}$. Notice that $[V_{\alpha'+1}^{(2)}, V_{\alpha+3}] \leq Z_{\alpha+4} \cap Z_{\alpha'+1}$. Since $(\alpha' + 1, \beta)$ is a critical pair, we have that $Z_{\alpha+4} \cap Z_{\alpha'+1} = Z_{\alpha'-3} \cap Z_{\alpha'+1} \leq Z_{\alpha'}$. But if $Z_{\alpha'} \leq Z_{\alpha'-3}$, since $Z_{\alpha'-1} \neq Z_{\alpha'-3}$, we deduce that $Z_{\alpha'} = Z_{\alpha'-2} \neq R$. Then $R \neq Z_{\alpha+3}$ for otherwise $Z_{\alpha'-1} = Z_{\alpha'-2}R = Z_{\alpha'-2}Z_{\alpha'-4} = Z_{\alpha'-3}$. Thus, $V_{\alpha'+1}^{(2)}$ centralizes $Z_{\alpha+3}R$, $Z_{\alpha+3} < Z_{\alpha+3}R$ and since $Z_{\alpha+2} \neq Z_{\alpha+4}$, we deduce that $[V_{\alpha'+1}^{(2)}, V_{\alpha+3}] = \{1\}$. Thus, whether $Z_{\alpha'} \leq Z_{\alpha+4}$ or not, $V_{\alpha'+1}^{(2)} \leq Q_{\alpha+2}$ and $V_{\alpha'+1}^{(2)} = Z_{\alpha'+1}(V_{\alpha'+1}^{(2)} \cap Q_\beta)$. Since $V_{\alpha'+1}^{(2)} \ntrianglelefteq L_{\alpha'} = \langle V_\beta, Q_{\alpha'+1}, Q_{\alpha'} \rangle$, we may assume that $V_{\alpha'+1}^{(2)} \cap Q_\beta \not\leq Q_\alpha$ and $Z_\beta \not\leq V_{\alpha'+1}^{(2)}$. But now, $[V_{\alpha'+1}^{(2)} \cap Q_\beta, V_\alpha^{(2)} \cap Q_{\alpha'} \cap$

$Q_{\alpha'+1}] \leq Z_{\alpha'+1} \cap V_\alpha^{(2)}$. If $Z_{\alpha'} \cap V_\alpha^{(2)} \neq \{1\}$, then since $Z_{\alpha'} V_\alpha^{(2)} \trianglelefteq L_\beta = \langle V_{\alpha'}, Q_\alpha \rangle$, we deduce that $V_\alpha^{(2)}$ has index strictly less than q in $V_\beta^{(3)}$ and is centralized, modulo V_β, by Q_α, a contradiction by Lemma 2.41 since $V_\beta^{(3)}/V_\beta$ contains a non-central chief factor for L_β. Hence, $[V_{\alpha'+1}^{(2)} \cap Q_\beta, V_\alpha^{(2)} \cap Q_{\alpha'} \cap Q_{\alpha'+1}] = \{1\}$ and $V_\alpha^{(2)}/Z_\alpha$ is an FF-module for $\overline{L_\alpha}$. Then by Lemma 5.33, $O^p(R_\alpha)$ centralizes $V_\alpha^{(2)}$.

It follows from the arguments above, that if $Z_{\alpha+3} = R \neq Z_{\alpha'-2}$, then $Z_{\alpha'-1} = Z_{\alpha'-3}$ and we have a contradiction. Similarly, $Z_{\alpha'-2} = R \neq Z_{\alpha+3}$ yields $Z_{\alpha+2} = Z_{\alpha+4}$, another contradiction. Suppose that $Z_{\alpha+3} \neq R \neq Z_{\alpha'-2}$. In particular, $R \not\leq Z_{\alpha'-3}$. But now, $Z_{\alpha'-3} < RZ_{\alpha'-3} \leq V_{\alpha'-2} \cap V_{\alpha'-4}$ and $V_{\alpha'-2} = V_{\alpha'-4}$. If $Z_{\alpha'-2} \neq Z_{\alpha'-4}$ then $L_{\alpha'-3} = \langle R_{\alpha'-3}, Q_{\alpha'-2}, Q_{\alpha'-4} \rangle$ normalizes $V_{\alpha'-2}$, a contradiction. Thus, $Z_{\alpha'-2} = Z_{\alpha'-4} = Z_{\alpha+3}$ so that $Z_{\alpha'-1} = RZ_{\alpha'-2} = RZ_{\alpha+3} = Z_{\alpha+2} \leq V_\beta$ from which it follows that $V_{\alpha'}$ centralizes $V_\beta^{(3)}/V_\beta$, a contradiction. Thus

(1) $R = Z_{\alpha'-2} = Z_{\alpha'-4} = Z_{\alpha+3}$.

By Lemma 5.22, we conclude that $V_{\alpha'-2} = V_{\alpha'-4}$. We may assume that $V_\alpha^{(4)}$ does not centralize $Z_{\alpha'-3}$, for otherwise $V_\alpha^{(4)}$ centralizes $V_{\alpha'-2} = V_{\alpha'-4} = Z_{\alpha'-3}Z_{\alpha+2}$, $V_\alpha^{(4)} = V_\beta^{(3)}(V_\alpha^{(4)} \cap Q_{\alpha'})$ and $V_\alpha^{(4)} \trianglelefteq L_\beta = \langle V_{\alpha'}, Q_\alpha \rangle$. Choose $\alpha-4 \in \Delta^{(4)}(\alpha)$ such that $[Z_{\alpha-4}, Z_{\alpha'-3}] \neq \{1\}$. If $Z_{\alpha'-3} \leq Q_{\alpha-3}$, then $Z_{\alpha-3} = [Z_{\alpha-4}, Z_{\alpha'-3}] = Z_{\alpha+3}$ so that $Z_\beta \cap Z_{\alpha-3} = \{1\}$. Now, $Z_{\alpha-3} \leq V_{\alpha-1} \cap V_\beta$ and since $Z_{\alpha-3} \cap Z_\alpha \leq Q_{\alpha'} \cap Z_\alpha = Z_\beta$, we deduce that $V_{\alpha-1} = Z_\alpha Z_{\alpha-3} = V_\beta$. Since $O^p(R_\alpha)$ centralizes $V_\alpha^{(2)}$, $Z_{\alpha-1} = Z_\beta$, for otherwise $V_\beta \trianglelefteq L_\alpha = \langle R_\alpha, Q_{\alpha-1}, Q_\beta \rangle$.

If $Z_{\alpha'-3} \not\leq Q_{\alpha-3}$ then $(\alpha'-3, \alpha-3)$ is a critical pair. By Lemma 7.37, we may assume that either $(\alpha'-3, \alpha-3)$ satisfies the same hypothesis as (α, α'), in which case $Z_{\alpha-1} = Z_\beta$; or $V_{\alpha'-3}^{(2)} \not\leq Q_{\alpha-1}$ and by Lemma 7.32, either $[V_{\alpha'-4}, V_{\alpha-3}] = Z_{\alpha-1} \leq Z_{\alpha+2}$, and again $Z_{\alpha-1} = Z_\beta$, or $Z_{\alpha-2} = Z_\alpha$, and by Lemma 5.22, we have a contradiction. Thus, we have shown that

(2) whenever there is $Z_{\alpha-4}$ such that $Z_{\alpha-4}$ does not centralizes $Z_{\alpha'-3}$, we have $Z_{\alpha-1} = Z_\beta$.

Choose $\lambda \in \Delta(\alpha)$ such that $Z_\lambda \neq Z_\beta$ so that $V_\lambda^{(3)}$ centralizes $Z_{\alpha'-3}$. Then $V_\lambda^{(3)}$ centralizes $V_{\alpha'-4} = V_{\alpha'-2}$ so that $V_\lambda^{(3)} = V_\beta(V_\lambda^{(3)} \cap Q_{\alpha'})$. Then, $V_\lambda^{(3)} V_\beta^{(3)} \trianglelefteq L_\beta = \langle Q_\alpha, V_{\alpha'} \rangle$. In particular, $[C_\beta, V_\lambda^{(3)} V_\beta^{(3)}]$ is a normal subgroup L_β contained in $[C_\beta, V_\beta^{(3)}][Q_\alpha, V_\lambda^{(3)}]$. Noticing that $[V_{\alpha'+1}^{(2)} \cap Q_\beta, V_\alpha^{(2)}] = [V_{\alpha'+1} \cap Q_\beta, V_\beta(V_\alpha^{(2)} \cap Q_{\alpha'})] = Z_\beta R = Z_{\alpha+2}$, we have that $[S, V_\alpha^{(2)}] \leq V_\beta$ and $|V_\alpha^{(2)}| = q^4$. But then $[Q_\beta, V_\beta^{(3)}] = V_\beta$ and since $[V_{\alpha'}, V_\beta^{(3)}] = Z_{\alpha'-1} \leq V_{\alpha+3} \leq V_{\alpha+2}^{(2)}$, we must have that $|V_\beta^{(3)}| = q^5$ and $[Q_\alpha, V_\beta^{(3)}] = V_\alpha^{(2)}$. Thus, $V_\beta < [C_\beta, V_\lambda^{(3)} V_\beta^{(3)}] \leq V_\alpha^{(2)}$ and it follows that $V_\alpha^{(2)} = V_\beta[C_\beta, V_\lambda^{(3)} V_\beta^{(3)}] \trianglelefteq L_\beta$, a contradiction. □

In conjunction with the results in earlier sections, we have now proved that Hypothesis 5.1 implies that $b \leqslant 5$. In the next lemmas and proposition, we show this bound is tight by witnessing an example with $b = 5$. In [27] and [26], this configuration is shown to be parabolic isomorphic to F_3. In our case, it is demonstrated in [79] that this leads to an exotic fusion system. The presence of this fusion system may go some way to explaining why it is so difficult to uniquely determine F_3 from a purely 3-local perspective.

Proposition 7.40 *Suppose that $C_{V_\beta}(V_{\alpha'}) = V_\beta \cap Q_{\alpha'}$ and $b = 5$. Then $V_{\alpha'} \not\leq Q_\beta$.*

Proof Assume that $V_{\alpha'} \leq Q_\beta$. If $V_\alpha^{(2)} \leq Q_{\alpha'-2}$, then it follows from Lemma 7.25 that $|V_\beta| = q^3$ and $O^p(R_\alpha)$ centralizes $V_\alpha^{(2)}$. Then, $Z_\beta = R \leq Z_{\alpha'-1}$ and since $V_\beta \neq V_{\alpha'-2}$, by Lemma 5.22, we may assume that $Z_\beta \neq Z_{\alpha'-2}$ so that $Z_{\alpha'-1} = Z_{\alpha+2}$. But now, $V_{\alpha'}^{(3)} \cap Q_{\alpha'-2} \leq Q_{\alpha+2}$ so that $[V_{\alpha'}^{(3)} \cap Q_{\alpha'-2}, V_\beta] \leq Z_{\alpha+2} = Z_{\alpha'-1} \leq V_{\alpha'}$. By Lemma 5.35, $O^p(R_{\alpha'})$ centralizes $V_{\alpha'}^{(3)}$ and Lemma 5.22 applied to $Z_{\alpha'-1} = Z_{\alpha+2}$ implies that $V_\beta \leq V_{\alpha+2}^{(2)} = V_{\alpha'-1}^{(2)} \leq Q_{\alpha'}$, a contradiction.

Suppose now that $V_{\alpha'} \leq Q_\beta, |V_\beta| = q^3$ and $V_\alpha^{(2)} \not\leq Q_{\alpha'-2}$. If $Z_\beta = R \neq Z_{\alpha'-2}$ then, as above, $Z_{\alpha'-1} = Z_{\alpha+2}$ and $[V_{\alpha'}^{(3)} \cap Q_{\alpha'-2}, V_\beta] \leq Z_{\alpha+2} = Z_{\alpha'-1} \leq V_{\alpha'}$. Then $O^p(R_{\alpha'})$ centralizes $V_{\alpha'}^{(3)}$ and Lemma 5.22 provides a contradiction. Thus, $Z_\beta = Z_{\alpha'-2} \neq Z_{\alpha'}$. But now, $[V_{\alpha'-2}, V_\alpha^{(2)}] \leq Z_\alpha \cap Z_{\alpha+2} = Z_\beta = Z_{\alpha'-2}$ and $V_\alpha^{(2)} \leq Q_{\alpha'-2}$, a contradiction.

Thus, if $V_{\alpha'} \leq Q_\beta$ then $|V_\beta| \neq q^3$. Notice that if $Z_{\alpha'-2} = Z_\beta$, then $Z_\beta Z_\beta^g Z_{\alpha'} = Z_{\alpha'-1} Z_{\alpha'-1}^g$ is of order q^3 and normalized by $L_{\alpha'} = \langle V_\beta, V_\beta^g, R_{\alpha'} \rangle$, for some appropriately chosen $g \in L_{\alpha'}$, a contradiction. Now, if $Z_{\alpha'-2} \leq V^\alpha$, then $V_\beta = Z_{\alpha+2} Z_\alpha C_{V_\beta}(O^p(L_\beta)) \leq V^\alpha$. But then $V^\alpha = V_\alpha^{(2)}$ and we have a contradiction. Since $[Q_\alpha, V_\alpha^{(2)}] \leq V^\alpha$ and $V_{\alpha'-2} \leq Q_\alpha$, it follows that $V_\alpha^{(2)} \cap Q_{\alpha'-2}$ centralizes $V_{\alpha'-2}$ and $V_\alpha^{(2)} \cap Q_{\alpha'-2} \leq Q_{\alpha'-1}$. Since both $V_\alpha^{(2)}/V^\alpha$ and V^α/Z_α have non-central chief factors, $[V_\alpha^{(2)} \cap Q_{\alpha'-2} \cap Q_{\alpha'}, V_{\alpha'}] = Z_{\alpha'} \leq V_\alpha^{(2)}$ and both $V_\alpha^{(2)}/V^\alpha$ and V^α/Z_α are FF-modules for $\overline{L_\alpha}$. Then by Lemma 5.33, we have that $O^p(R_\alpha)$ centralizes $V_\alpha^{(2)}$ and by Lemma 5.22, $Z_{\alpha'} \neq Z_{\alpha'-2}$ and $Z_{\alpha'-1} \leq V_\alpha^{(2)}$. Since $V_\alpha^{(2)} \not\leq Q_{\alpha'-2}$, and $V_\alpha^{(2)}$ centralizes $Z_{\alpha'-1} Z_{\alpha+2}$. By Lemma 5.32, we deduce that $Z_{\alpha'-1} = Z_{\alpha+2}$. But now $[V_\beta, V_{\alpha'}] = Z_\beta \leq Z_{\alpha'-1}$ and $Z_{\alpha'-1} Z_{\alpha'-1}^g$ is of order q^3 and normalized by $L_{\alpha'} = \langle V_\beta, V_\beta^g, R_{\alpha'} \rangle$, a final contradiction. □

Proposition 7.41 *Suppose that $C_{V_\beta}(V_{\alpha'}) = V_\beta \cap Q_{\alpha'}$ and $b = 5$. Then $|V_\beta| = q^3$, $R = Z_{\alpha'-2} \neq Z_\beta \neq Z_{\alpha'} \neq R$ and $Z_{\alpha'-1} \neq Z_{\alpha+2}$.*

Proof By Proposition 7.40, we have that $V_{\alpha'} \not\leq Q_\beta$ for all critical pairs (α, α'), so that $Z_{\alpha'} \neq R \neq Z_\beta$. Fix $\alpha'+1 \in \Delta(\alpha')$ such that $Z_{\alpha'+1} \not\leq Q_\beta$. In particular, $(\alpha'+1, \beta)$ is a critical pair satisfying the same hypothesis as (α, α').

Assume first that $|V_\beta| = q^3$. Then $R \leq Z_{\alpha'-1} \cap Z_{\alpha+2}$. If $Z_{\alpha'-1} = Z_{\alpha+2}$ then $V_\beta^{(3)} \cap Q_{\alpha'-2} \leq Q_{\alpha'-1}$, $[V_\beta^{(3)} \cap Q_{\alpha'-2}, V_{\alpha'}] \leq Z_{\alpha'-1} \leq V_\beta$ and it follows that $V_\beta^{(3)}/V_\beta$ contains a unique non-central chief factor for L_β which, as a $\overline{L_\beta}$-module, is an FF-module. Then, Lemmas 5.35 and 5.22 applied to $Z_{\alpha'-1} = Z_{\alpha+2}$ gives

$V_{\alpha'} \leq V^{(2)}_{\alpha'-1} = V^{(2)}_{\alpha+2} \leq Q_\beta$, a contradiction. Now, $R \leq Z_{\alpha'-1} \cap Z_{\alpha+2}$ and so $R = Z_{\alpha'-2}$, otherwise $Z_{\alpha'-1} = Z_{\alpha+2}$. Hence, aiming for a contradiction, we may assume that $|V_\beta| \neq q^3$ for the remainder of the proof.

Suppose first that $Z_{\alpha'} = Z_{\alpha'-2}$. Then $[V^\alpha, V_{\alpha'-2}] \leq Z_\alpha \cap V_{\alpha'-2} \leq Z_\beta \leq Z_{\alpha+2}$. In particular, if $V^\alpha \not\leq Q_{\alpha'-2}$ then $Z_\beta \neq Z_{\alpha'-2}$ and $Z_\beta Z_\beta^g Z_{\alpha'-2} = Z_{\alpha+2} Z_{\alpha+2}^g$ is of order q^3 and normalized by $L_{\alpha'-2} = \langle V^\alpha, (V^\alpha)^g, R_{\alpha'-2}\rangle$ for some appropriately chosen $g \in L_{\alpha'-2}$, a contradiction. Thus, $V^\alpha \leq Q_{\alpha'-2}$. Then, either $V^\alpha = Z_\alpha(V^\alpha \cap Q_{\alpha'})$, or $V^\alpha \not\leq Q_{\alpha'-1}$ and $Z_\beta = [V^\alpha, Z_{\alpha'-1} = Z_{\alpha'-2}$. Assume that $V^\lambda \leq Q_{\alpha'-1}$ for all $\lambda \in \Delta(\beta)$ with $Z_\lambda = Z_\alpha$. Forming $U^\beta := \langle V^\lambda : \lambda \in \Delta(\beta), Z_\lambda = Z_\alpha\rangle$ so that $R_\beta Q_\alpha$ normalizes U^β by Lemma 5.23, we have that $U^\beta = Z_\alpha(U^\beta \cap Q_{\alpha'})$ and as $Z_{\alpha'} = Z_{\alpha'-2} \leq V_\beta$, $[V_{\alpha'}, U^\beta V_\beta] \leq V_\beta$ so that $U^\beta V_\beta \trianglelefteq L_\beta = \langle V_{\alpha'}, Q_\alpha, R_\beta\rangle$ and $O^p(L_\beta)$ centralizes $U^\beta V_\beta / V_\beta$. But then, $V^\alpha V_\beta \trianglelefteq L_\beta$, a contradiction by Lemma 5.32. Thus, we deduce that $Z_{\alpha'} = Z_{\alpha'-2} = Z_\beta$ and $V^\lambda \not\leq Q_{\alpha'-1}$ for some $\lambda \in \Delta(\beta)$ with $Z_\lambda = Z_\alpha$. Then, using that $[V^{\alpha'-1}, V_\beta] \leq Z_{\alpha'-1}$ and V_β is centralized by V^λ, it follows that $[V^{\alpha'-1}, V_\beta] \leq Z_{\alpha'-2} = Z_\beta$ so that $V^{\alpha'-1} \leq Q_\beta$. Then $[V^{\alpha'-1} \cap Q_\lambda, V^\lambda] \leq Z_\lambda = Z_\alpha$ and since $Z_\alpha \not\leq V^{(2)}_{\alpha'-1}$, we have that $[V^{\alpha'-1} \cap Q_\lambda, V^\lambda] \leq Z_\beta \leq Z_{\alpha'-1}$. Since $V^{\alpha'-1}/Z_{\alpha'-1}$ contains a non-central chief factor, $S = V^{\alpha'-1} Q_\alpha$ and $V^{\alpha'-1}/Z_{\alpha'-1}$ is an FF-module for $\overline{L_{\alpha'-1}}$. Then $V^{(2)}_{\alpha'-1} \cap Q_\beta = V^{\alpha'-1}(V^{(2)}_{\alpha'-1} \cap Q_\alpha)$ and since $Z_\alpha \not\leq V^{(2)}_{\alpha'-1}$, $V^{(2)}_{\alpha'-1}/V^{\alpha'-1}$ is also an FF-module for $\overline{L_{\alpha'-1}}$. Then Lemmas 5.33 and 5.22 applied to $Z_\beta = Z_{\alpha'-2} = Z_{\alpha'}$ gives $V_{\alpha'} = V_\beta$, a contradiction. Hence, we have shown that

(1) $Z_{\alpha'} \neq Z_{\alpha'-2}$.

Since $(\alpha'+1, \beta)$ is also critical pair, we deduce by a similar argument that $Z_\beta \neq Z_{\alpha'-2} \neq Z_{\alpha'}$. As in Proposition 7.40, this implies that $Z_{\alpha'-2} \not\leq V^\alpha$ so that $V^{(2)}_\alpha \cap Q_{\alpha'-2} \leq Q_{\alpha'-1}$. Moreover, $[V^\alpha, V_{\alpha'-2}] \leq Z_\alpha \cap V_{\alpha'-2} = Z_\beta$ and if $V^\alpha \not\leq Q_{\alpha'-2}$, then $Z_\beta Z_\beta^g Z_{\alpha'-2} = Z_{\alpha+2} Z_{\alpha+2}^g$ is of order q^3 and normalized by $L_{\alpha'-2} = \langle V^\alpha, (V^\alpha)^g, R_{\alpha'-2}\rangle$ for some appropriately chosen $g \in L_{\alpha'-2}$, a contradiction. Thus, $[V^\alpha, V_{\alpha'-2} = \{1\}$ and $V^\alpha = Z_\alpha(V^\alpha \cap Q_{\alpha'})$. Set $U_\beta := \langle (V^\alpha)^{G_\beta}\rangle$. Then $[U_\beta, V_{\alpha'-2}] \leq [U_\beta, C_\beta] \cap V_{\alpha'-2} \leq V_{\alpha'-2} \cap V_\beta$.

Assume that $U_\beta \not\leq Q_{\alpha'-2}$ and so there is some $\beta - 3 \in \Delta^{(3)}(\beta)$ with $(\beta-3, \alpha'-2)$ a critical pair. Indeed, we must have that $V_\beta \cap V_{\alpha'-2} \leq Z_{\alpha+2} C_{V_\beta}(O^p(L_\beta)) \cap Z_{\alpha+2} C_{V_{\alpha'-2}}(O^p(L_{\alpha'-2}))$. But then $Z_\beta \geq [Q_{\alpha+2}, V_{\alpha'-2} \cap V_\beta] \leq Z_{\alpha'-2}$ and we deduce that $V_{\alpha'-2} \cap V_\beta \leq \Omega(Z(Q_{\alpha+2})) \cap V_\beta = Z_{\alpha+2}$. But then, $Z_{\alpha+2} Z_{\alpha+2}^g$ is of order q^3 and normalized by $L_{\alpha'-2} = \langle U_\beta, U_\beta^g, R_{\alpha'-2}\rangle$, for some appropriate $g \in L_{\alpha'-2}$, another contradiction. Thus

(2) $U_\beta \leq Q_{\alpha'-2}$ and $[U_\beta, V_{\alpha'-2}] \leq Z_{\alpha'-2}$.

Suppose that $U_\beta \not\leq Q_{\alpha'-1}$ so that $V^\mu \not\leq Q_{\alpha'-1}$ for some $\mu \in \Delta(\beta)$. By a similar argument to the above, we also have that $[V^{\alpha'-1}, V_\beta] = \{1\}$ and $V^{\alpha'-1} \leq Q_\mu$. Then $\{1\} \neq [V^\mu, V^{\alpha'-1}] \leq Z_\mu \cap V^{\alpha'-1}$. Notice that $Z_\beta \not\leq V^{\alpha'-1}$ for otherwise $V_{\alpha'-2} = Z_{\alpha'-1} Z_{\alpha+2} C_{V_{\alpha'}}(O^p(L_{\alpha'-2})) \leq V^{\alpha'-1}$. Thus, $Z_\mu = [V^{\alpha'-1}, V^\mu] \times Z_\beta$

centralizes $V_{\alpha'}$ and since $R \neq \{1\}$, it follows that $Z_\mu = Z_{\alpha+2}$. Since $Z_{\alpha'-2} \leq V^{\alpha'-1}$ and $Z_\beta \not\leq V^{\alpha'-1}$, we have that $[V^\mu, V^{\alpha'-1}] = Z_{\alpha'-2} \leq Z_{\alpha'-1}$, a contradiction since $V^\mu \not\leq Q_{\alpha'-1}$. Thus

(3) $U_\beta \leq Q_{\alpha'-1}$.

Since $V^\alpha V_\beta \ntrianglelefteq L_\beta$, U_β/V_β contains a non-central chief factor for L_β and we conclude that $[U_\beta \cap Q_{\alpha'}, V_{\alpha'}] = Z_{\alpha'} \leq U_\beta$ and $Z_{\alpha'} \cap V_\beta = \{1\}$. But now, $Z_{\alpha'-2} \neq Z_{\alpha'}$ and $V_{\alpha'-2} = Z_{\alpha'-1}Z_{\alpha+2}C_{V_{\alpha'-2}}(O^p(L_{\alpha'-2})) \leq U_\beta$.

Suppose that $[V_\alpha^{(2)}, Z_{\alpha'-1}] \neq \{1\}$. Then there is $\alpha - 1 \in \Delta(\alpha)$ such that $[V_{\alpha-1}, Z_{\alpha'-1}] \neq \{1\}$. If $Z_{\alpha'-1} \leq Q_{\alpha-1}$, then $Z_{\alpha-1} = [Z_{\alpha'-1}, V_{\alpha-1}] \leq [V_{\alpha'-2}, V_\alpha^{(2)}]$. Since $Z_\alpha \not\leq V_{\alpha'-2}$, it follows that $Z_{\alpha-1} = Z_\beta$. Then $V_{\alpha-1} \not\leq Q_{\alpha'-2}$ since $Z_{\alpha'-2} \neq Z_\beta$. But then $[V_{\alpha'-2}, V_\alpha^{(2)}] \leq Z_\beta Z_{\alpha'-2}$ and if $V_\alpha^{(2)} \not\leq Q_{\alpha'-2}$, then $Z_{\alpha+2}Z_{\alpha+2}^g$ is of order q^3 and normalized by $L_{\alpha'-2} = \langle V_\alpha^{(2)}, (V_\alpha^{(2)})^g, R_{\alpha'-2}\rangle$ for some appropriately chosen $g \in L_{\alpha'-2}$, a contradiction. Thus, $Z_{\alpha'-1} \not\leq Q_{\alpha-1}$ and $(\alpha' - 1, \alpha - 1)$ is a critical pair.

Since $Z_{\alpha'-2} \neq Z_\beta$, $(\alpha' - 1, \alpha - 1)$ satisfies the same hypothesis as (α, α') and so we see that $V_\beta \leq U_{\alpha'-2}$. But then $R = [V_\beta, V_{\alpha'}] \leq [U_{\alpha'-2}, C_{\alpha'-2}] \leq V_{\alpha'-2}$ and $R \leq V_\beta \cap V_{\alpha'-2} \leq Z_{\alpha+2}C_{V_\beta}(O^p(L_\beta)) \cap Z_{\alpha+2}C_{V_{\alpha'-2}}(O^p(L_{\alpha'-2}))$. Similarly to before, this implies that $|V_\beta| = q^3$, and we have a contradiction. Thus

(4) $[V_\alpha^{(2)}, Z_{\alpha'-1}] = \{1\}$.

Since $Z_{\alpha'-1} \neq Z_{\alpha'-3}$, else $Z_{\alpha'} \cap V_\beta \neq \{1\}$, it follows that $V_\alpha^{(2)}$ centralizes $V_{\alpha'-2}$ and $V_\alpha^{(2)} \leq Q_{\alpha'-1}$. In particular, this holds for any $\lambda \in \Delta(\beta)$ with $Z_\lambda = Z_\alpha$. Forming $W^\beta := \langle V_\lambda^{(2)} : Z_\lambda = Z_\alpha, \lambda \in \Delta(\beta)\rangle$, we have that $W^\beta U_\beta/U_\beta$ is centralized by $V_{\alpha'}$, and by Lemma 5.23, is normalized by $R_\beta Q_\alpha$. But then $W^\beta U_\beta \trianglelefteq L_\beta = \langle V_{\alpha'}, R_\beta, Q_\alpha\rangle$ and since $V_{\alpha'}$ centralizes $W^\beta U_\beta/U_\beta$ we deduce that $V_\beta^{(3)} = V_\alpha^{(2)}U_\beta \trianglelefteq L_\beta$. Now, $R = [V_\beta, V_{\alpha'}] \leq [V_\beta, V_{\alpha'-1}^{(2)}U_{\alpha'-2}] = [V_\beta, V_{\alpha+2}^{(2)}U_{\alpha'-2}] = [V_\beta, U_{\alpha'-2}]$ and since $V_\beta \leq C_{\alpha'-2}$, it follows that $R \leq V_\beta \cap V_{\alpha'-2}$, which again implies that $|V_\beta| = q^3$, a final contradiction. □

For the remainder of the analysis when $b = 5$, we set $W^\beta := \langle V_\lambda^{(2)} : \lambda \in \Delta(\beta), Z_\lambda = Z_\alpha\rangle \trianglelefteq R_\beta Q_\alpha$.

Proposition 7.42 *Suppose that $C_{V_\beta}(V_{\alpha'}) = V_\beta \cap Q_{\alpha'}$ and $b = 5$. Then $O^p(L_\beta)$ centralizes $[V_\beta^{(3)}, Q_\beta]V_\beta/V_\beta$.*

Proof Aiming for a contradiction, suppose throughout that $[V_\beta^{(3)}, Q_\beta]V_\beta/V_\beta$ contains a non-central chief factor for L_β.

In addition, first assume that $Z_{\alpha'} \cap [V_\beta^{(3)}, Q_\beta]V_\beta = \{1\}$. Notice that $[W^\beta, Q_\beta] = [W^\beta, (Q_\alpha \cap Q_\beta)][W^\beta, (Q_\beta \cap Q_{\alpha+2})] \leq Z_\alpha[Q_{\alpha+2}, Q_{\alpha+2}] \leq Q_{\alpha'-2}$. Now, $[W^\beta \cap Q_{\alpha'-2}, Z_{\alpha'-1}] \leq Z_{\alpha'-2} \cap [W^\beta, Z_{\alpha'-1}] \leq Z_{\alpha'-2} \cap Z_\alpha = \{1\}$ and so $[W^\beta, Q_\beta, V_{\alpha'}] \leq Z_{\alpha'-1} \cap [V_\beta^{(3)}, Q_\beta] \leq Z_{\alpha'-2} \leq V_\beta$. In particular, it follows that $[W^\beta, Q_\beta]V_\beta \trianglelefteq L_\beta = \langle V_{\alpha'}, Q_\alpha, R_\beta \rangle$ and $[W^\beta, Q_\beta]V_\beta = [V_\beta^{(3)}, Q_\beta]V_\beta$. But then, $[[V_\beta^{(3)}, Q_\beta]V_\beta, V_{\alpha'}] \leq V_\beta$, a contradiction since $[V_\beta^{(3)}, Q_\beta]V_\beta/V_\beta$ contains a non-central chief factor.

Thus, $Z_{\alpha'} \cap [V_\beta^{(3)}, Q_\beta]V_\beta \neq \{1\}$. But then $V_{\alpha'-2} \cap [V_\beta^{(3)}, Q_\beta]V_\beta > Z_{\alpha+2}$. If $W^\beta \leq Q_{\alpha'-2}$ then $[W^\beta, Z_{\alpha'-1}] \leq Z_{\alpha'-2} \cap Z_\alpha = \{1\}$ so that $W^\beta[V_\beta^{(3)}, Q_\beta]$ is normalized by $L_\beta = \langle V_{\alpha'}, Q_\alpha, R_\beta \rangle$ so that $V_\beta^{(3)} = W^\beta[V_\beta^{(3)}, Q_\beta]$. But then $V_{\alpha'}$ centralizes $V_\beta^{(3)}/[V_\beta^{(3)}, Q_\beta]$, a contradiction by Lemma 5.8. Hence, there is $\lambda \in \Delta(\beta)$ with (λ, α') critical and $V_\lambda^{(2)} \not\leq Q_{\alpha'-2}$. Thus, $(\lambda - 2, \alpha' - 2)$ is a critical pair so that by Proposition 7.41, $V_{\alpha'-2} \not\leq Q_{\lambda-1}$. Now, for all $\mu \in \Delta(\beta)$ with $Z_\mu \neq Z_\lambda$, $[V_\mu^{(2)}, Q_\beta]V_\beta = [V_\mu^{(2)}, Q_\mu \cap Q_\beta][V_\mu^{(2)}, Q_\lambda \cap Q_\beta] \leq Z_\mu[Q_\lambda, Q_\lambda] \leq Q_{\lambda-1}$. On the other hand, if $Z_\mu = Z_\lambda$ for $\mu \in \Delta(\beta)$ then $V_\mu^{(2)} \leq Q_{\lambda-1}$ for otherwise there is $\mu - 2 \in \Delta^{(2)}(\mu)$ with $(\mu - 2, \lambda - 1)$ a critical pair, and as $Z_\mu = Z_\lambda$ this contradicts Proposition 7.41. Hence, $[V_\beta^{(3)}, Q_\beta]V_\beta \leq Q_{\lambda-1}$. But then, $V_{\alpha'-2} \leq Q_{\lambda-1}$, against Proposition 7.40. □

Lemma 7.43 *Suppose that $C_{V_\beta}(V_{\alpha'}) = V_\beta \cap Q_{\alpha'}$ and $b = 5$. Then $[V_\beta^{(3)}, Q_\beta]V_\beta \leq Z(V_\beta^{(3)})$,. $V_{\alpha'}$ acts quadratically on $V_\beta^{(3)}/V_\beta$, $p \in \{2, 3\}$ and for $V := V_\alpha^{(2)}/Z_\alpha$ either:*

(i) *S acts quadratically on V;*
(ii) *$V = [V, R_\alpha]$; or*
(iii) *$V = C_V(R_\alpha)$.*

Moreover, $\overline{L_\beta} \cong \mathrm{SL}_2(q)$.

Proof Since $O^p(L_\beta)$ centralizes $[V_\beta^{(3)}, Q_\beta]V_\beta/V_\beta$, $[V_\lambda^{(2)}, Q_\beta]V_\beta \trianglelefteq L_\beta$ for any $\lambda \in \Delta(\beta)$. It follows that $[V_\beta^{(3)}, Q_\beta]V_\beta = [V_\lambda^{(2)}, Q_\beta]V_\beta$ for any $\lambda \in \Delta(\beta)$. But $V_\lambda^{(2)}$ is elementary abelian so that $[V_\beta^{(3)}, Q_\beta]V_\beta \leq Z(V_\beta^{(3)})$. Moreover, $[V_\beta^{(3)}, Q_\beta, V_\beta^{(3)}] = \{1\}$ and it follows from the three subgroup lemma that $[V_\beta^{(3)}, V_\beta^{(3)}, Q_\beta] = \{1\}$ so that $[V_\beta^{(3)}, V_\beta^{(3)}] \leq Z(Q_\beta) \cap V_\beta = Z_\beta$. Since $V_\beta V_{\alpha'} \leq V_{\alpha'-2}^{(3)}$, it follows by conjugacy that $V_\beta^{(3)}$ is non-abelian and so $[V_\beta^{(3)}, V_\beta^{(3)}] = Z_\beta$. Then $[V_\beta^{(3)}, V_{\alpha'}, V_{\alpha'}] \leq [V_\beta^{(3)}, V_{\alpha'-2}^{(3)}, V_{\alpha'-2}^{(3)}] \leq Z_{\alpha'-2} \leq V_\beta$, and $V_\beta^{(3)}/V_\beta$ admits quadratic action.

Now, $C_{Q_\beta}(V_\beta^{(3)})$ centralizes $V_{\alpha'-2} \leq V_\beta^{(3)}$ and since $Z_{\alpha'} \not\leq Z(V_\beta^{(3)})$ (else $O^p(L_\beta)$ centralizes $V_\beta^{(3)}/Z(V_\beta^{(3)})$ and $V_\beta^{(3)}$ is abelian), we have that $V_{\alpha'}$ centralizes $C_{Q_\beta}(V_\beta^{(3)})/V_\beta$ and so $O^p(L_\beta)$ centralizes $C_{Q_\beta}(V_\beta^{(3)})V_\beta^{(3)}/V_\beta^{(3)}$. An application of the three subgroups lemma and coprime action yields that $O^p(C_{L_\beta}(V_\beta^{(3)}))$ centralizes Q_β so that $C_{L_\beta}(V_\beta^{(3)})$ is a p-group. Moreover, $C_\beta = V_\beta^{(3)}(C_\beta \cap Q_{\alpha'-2})$ and $[C_\beta \cap Q_{\alpha'-2}, V_{\alpha'-2}^{(3)}] \leq [V_{\alpha'-2}^{(3)}, Q_{\alpha'-2}] \leq V_{\alpha+2}^{(2)} \leq V_\beta^{(3)}$ so that $O^p(L_\beta)$

centralizes $C_\beta/V_\beta^{(3)}$. But then $O^p(R_\beta)$ centralizes $Q_\beta/V_\beta^{(3)}$. Indeed, $V_\beta^{(3)}/Z_\beta = [V_\beta^{(3)}/Z_\beta, O^p(R_\beta)] \times C_{V_\beta^{(3)}/Z_\beta}(O^p(R_\beta))$. Therefore, $[O^p(R_\beta), V_\beta^{(3)}, Q_\beta] \leq Z_\beta$ by the three subgroup lemma, and $[V_\beta^{(3)}, O^p(R_\beta), C_{V_\beta^{(3)}}(O^p(R_\beta))] = \{1\}$.

If $[V_\beta^{(3)}, O^p(R_\beta)] \not\leq Q_{\alpha'-2}$, then $[V_{\alpha'-2}^{(3)} \cap Q_\beta, [V_\beta^{(3)}, O^p(R_\beta)]] \leq Z_\beta \leq V_{\alpha'-2}$ and we deduce that $V_{\alpha'-2}^{(3)}/V_{\alpha'-2}$ contains a unique non-central chief factor which is an FF-module for $\overline{L_{\alpha'-2}}$. Then Lemma 5.35 implies that $O^p(R_\beta)$ centralizes $V_\beta^{(3)}$ and an application of the three subgroup lemma yields that $O^p(R_\beta)$ centralizes Q_β and $\overline{L_\beta} \cong \mathrm{SL}_2(q)$. If $[V_\beta, O^p(R_\beta)] \leq Q_{\alpha'-2}$ then $Z_{\alpha'-2} \leq V_\beta \leq C_{V_\beta^{(3)}}(O^p(R_\beta))$ so that $[V_\beta, O^p(R_\beta)] \leq C_{\alpha'-2}$. On the other hand, if $Z_{\alpha'} \leq [V_\beta, O^p(R_\beta)]$, then $C_{V_\beta^{(3)}}(O^p(R_\beta))$ centralizes $V_{\alpha'-2} = Z_{\alpha+2}Z_{\alpha'}$ and $V_\beta^{(3)} \leq C_{\alpha'-2}$, a contradiction. Thus, $[V_\beta, O^p(R_\beta), V_{\alpha'-2}^{(3)}] \leq V_{\alpha'-2} \cap [V_\beta, O^p(R_\beta)] = Z_\beta$ so that $O^p(L_\beta)$ centralizes $[V_\beta, O^p(R_\beta)]$. Hence, $O^p(R_\beta)$ centralizes $V_\beta^{(3)}$ and the three subgroup lemma yields that $R_\beta = Q_\beta$ and $\overline{L_\beta} \cong \mathrm{SL}_2(q)$.

Writing $Q := Q_\beta \cap O^p(L_\beta)$, we have that $[V_\alpha^{(2)}, Q, Q] \leq [V_\beta^{(3)}, Q, Q] \leq V_\beta$. By coprime action, and setting $V := V_\alpha^{(2)}/Z_\alpha$, we have that $V = [V, R_\alpha] \times C_V(R_\alpha)$ and either $V_\beta/Z_\alpha \leq [V, R_\alpha]$ or Q acts quadratically on $[V, R_\alpha]$. Similarly, either $V_\beta/Z_\alpha \leq C_V(R_\alpha)$ or Q acts quadratically on $C_V(R_\alpha)$. Since both $[V, R_\alpha]$ and $C_V(R_\alpha)$ are normalized by G_α, and V_β/Z_α generates V, we have shown that either Q acts quadratically on V, $V = [V, R_\alpha]$ or $V = C_V(R_\alpha)$.

In all cases, Q acts cubically on V and so if $p \geqslant 5$ the Hall-Higman theorem yields that $O^p(R_\alpha)$ centralizes $V_\alpha^{(2)}$. Since Q centralizes $C_\beta/V_\beta^{(3)}$, $[C_{Q_\alpha}(V_\alpha^{(2)}), Q, Q] \leq [C_\beta, Q, Q] \leq [V_\beta^{(3)}, Q] \leq V_\alpha^{(2)}$ and a standard argument implies that $O^p(R_\alpha)$ centralizes $C_{Q_\alpha}(V_\alpha^{(2)})$. A final application of the three subgroup lemma yields that $O^p(R_\alpha)$ centralizes Q_α, G has a weak BN-pair of rank 2 and [27] provides a contradiction. Hence, $p \in \{2, 3\}$. □

Proposition 7.44 *Suppose that* $C_{V_\beta}(V_{\alpha'}) = V_\beta \cap Q_{\alpha'}$ *and* $b = 5$. *Then* $p = 3$ *and* G *is parabolic isomorphic to* F_3.

Proof Aiming for a contradiction, we assume throughout that $R_\alpha > Q_\alpha$, for otherwise G has a weak BN-pair of rank 2 and the proposition holds by Delgado and Stellmacher [27]. Note that $[V_\beta^{(3)} \cap Q_{\alpha'-2}, V_{\alpha'-2}] \leq Z_{\alpha'-2} \cap [V_\beta^{(3)}, V_\beta^{(3)}] = Z_{\alpha'-2} \cap Z_\beta = \{1\}$ so that $V_\beta^{(3)} \cap Q_{\alpha'-2} \leq C_{\alpha'-2}$. Then $V_\beta^{(3)} \cap Q_{\alpha'-2} = V_\beta(V_\beta^{(3)} \cap Q_{\alpha'})$ so that a subgroup of index at most q^2 in $V_\beta^{(3)}$ is centralized, modulo V_β, by $V_{\alpha'}$. In particular, $V_\beta^{(3)}/[V_\beta^{(3)}, Q_\beta]V_\beta$ is a quadratic 2F-module for $\overline{L_\beta} \cong \mathrm{SL}_2(q)$. Write A_β to be the preimage in $V_\beta^{(3)}$ of $C_{V_\beta^{(3)}/[V_\beta^{(3)}, Q_\beta]V_\beta}(O^p(L_\beta))$ so that $V_\beta^{(3)}/A_\beta$ is a direct sum of natural $\mathrm{SL}_2(q)$-modules by Lemma 2.42.

By the 2-transitivity of $\overline{L_\beta} \cong \mathrm{SL}_2(q)$, we have that $V_\beta^{(3)} = V_\alpha^{(2)} V_{\alpha+2}^{(2)}$. Since $V_\alpha^{(2)} \cap A_\beta = V_{\alpha+2}^{(2)} \cap A_\beta$, and an easy counting argument, we conclude that $A_\beta = V_\alpha^{(2)} \cap V_{\alpha+2}^{(2)}$. Let U be the preimage in $V_\beta^{(3)}$ of a natural $\mathrm{SL}_2(q)$-module in $V_\beta^{(3)}/A_\beta$. If $C_U(V_\alpha^{(2)}) > U \cap V_\alpha^{(2)}$ then since $V_\alpha^{(2)}$ is abelian and $C_U(V_\alpha^{(2)})$ is normalized by $L_\beta \cap G_{\alpha\beta}$, we deduce that $[U, V_\alpha^{(2)}] = \{1\}$. Since $U \trianglelefteq L_\beta$ and $V_\beta^{(3)}$ is non-abelian, we deduce that $U < V_\beta^{(3)}$ and $U \leq Z(V_\beta^{(3)})$. In particular, $[U, V_{\alpha'-2}] = \{1\}$. If $Z_{\alpha'} \leq U$, then $Z_{\alpha'-1} \leq U$ and so $V_\beta^{(3)}$ centralizes $V_{\alpha'-2} = Z_{\alpha+2} Z_{\alpha'-1}$. But then $V_\beta^{(3)}$ contains a unique non-central chief factor, a contradiction. Hence, $V_{\alpha'}$ centralizes U/V_β, another contradiction. Thus, we have show that $C_U(V_\alpha^{(2)}) = U \cap V_\alpha^{(2)}$ from which it follows that $C_{V_\beta^{(3)}}(V_\alpha^{(2)}) = V_\alpha^{(2)}$. Then, $[C_{Q_\alpha}(V_\alpha^{(2)}), Q] \leq [C_\beta, Q] \cap C_{Q_\alpha}(V_\alpha^{(2)}) \leq C_{V_\beta^{(3)}}(V_\alpha^{(2)}) \leq V_\alpha^{(2)}$. By the three subgroups lemma, p'-order elements of L_α acts faithfully on $V_\alpha^{(2)}$, and by assumption we have that $V \neq C_V(R_\alpha)$. By Lemma 7.43,

(1) either V is a quadratic module, or $V = [V, R_\alpha]$.

Assume that $V_\beta^{(3)}/A_\beta$ is a direct sum of two natural $\mathrm{SL}_2(q)$-modules. Write $V_1 = \langle V_{\alpha'-2}^{L_\beta} \rangle A_\beta$ so that by Lemma 2.22, $|V_1/A_\beta| = q^2$. Set $V_2 \trianglelefteq L_\beta$ such that $A_\beta < V_2 < V_\beta^{(3)}$ and $V_2 \cap V_1 = A_\beta$. If $V_1 \not\leq Q_{\alpha'-2}$, then $V_{\alpha'-2}^{(3)}$ centralizes a subgroup of $V_\beta^{(3)}/V_1$ of index strictly less than q, a contradiction. Hence, $V_1 \leq Q_{\alpha'-2}$ so that $V_1 \leq C_{\alpha'-2}$ and as $A_\beta \leq Z(V_\beta^{(3)})$, V_1 is elementary abelian. Note that there is $V_\lambda \not\leq V_1$ for some $\lambda \in \Delta(\alpha+2)$ and we may as well assume that $V_2 = \langle V_\lambda^{L_\lambda} \rangle A_\beta$ and V_2 is also elementary abelian. Furthermore, $[V_2 \cap Q_{\alpha'-2}, V_{\alpha'-2}^{(3)}] \leq V_\beta$ and $V_\beta^{(3)} = V_2(V_\beta^{(3)} \cap Q_{\alpha'-2})$.

Let $D := [V_{\alpha'-1}^{(2)}, V_2]V_{\alpha'-2}$ so that $[V_2, D] \leq V_{\alpha'-2}$ and $[Q_{\alpha'-1}, D] \leq V_{\alpha'-2}$. Therefore, $D \trianglelefteq \langle V_2, Q_{\alpha'-1} \rangle = L_{\alpha'-2}$. Then, as $D \leq V_{\alpha'-1}^{(2)}$, $D \leq V_{\alpha+2}^{(2)} \cap V_{\alpha'-1}^{(2)} = A_{\alpha'-2}$. Now, $[V_{\alpha+2}^{(2)}, V_2] \leq Z_{\alpha+2}^{(2)} \leq A_{\alpha'-2}$ and so we deduce that $[V_2, V_{\alpha'-2}^{(3)}] = [V_2, V_{\alpha+2}^{(2)} V_{\alpha'-1}^{(2)}] \leq A_{\alpha'-2}$, a contradiction since $V_2 \not\leq Q_{\alpha'-2}$. Therefore

(2) $V_\beta^{(3)}/A_\beta$ is a natural $\mathrm{SL}_2(q)$-modules for $\overline{L_\beta}$.

Let $x \in Q := Q_\beta \cap O^p(L_\beta)$ with $x \notin Q_\alpha$. Since $[x, A_\beta] \leq V_\beta$ and A_β has index q in $V_\alpha^{(2)}$, we observe that, modulo Z_α, x centralizes a subgroup of $V_\alpha^{(2)}$ of index at most q^2. Moreover this subgroup contains V_β and is contained in A_β and so is centralized by $O^p(L_\beta) \cap G_{\alpha,\beta}$. But $O^p(L_\beta) \cap G_{\alpha,\beta}$ acts irreducibly on $Q_\beta/Q_\beta \cap Q_\alpha \cong Q/Q \cap Q_\alpha$ from which it follows that S centralizes a subgroup of $V := V_\alpha^{(2)}/Z_\alpha$ of index at most q^2.

Suppose that S acts quadratically on V. Then by Proposition 2.51 and Lemma 2.55, using that S/Q_α is elementary abelian, we deduce that $L_\alpha/C_{L_\alpha}(V)$

is isomorphic to one of $\mathrm{SL}_2(q)$, $(Q_8 \times Q_8) : 3$, $\mathrm{SU}_3(2)'$, $\mathrm{Dih}(10)$ or $(3 \times 3) : 2$ where $p \in \{2, 3\}$ as appropriate. Note that $V_\alpha^{(2)} \cap A_\beta$ is normalized by $G_{\alpha,\beta}$ and of index p and we deduce that $L_\alpha/C_{L_\alpha}(V) \not\cong \mathrm{SU}_3(2)'$. If $q > p$ then a consideration of the Schur multiplier of $\mathrm{PSL}_2(q)$ reveals that $\overline{L_\alpha} \cong \mathrm{SL}_2(q)$, a contradiction since $R_\alpha \neq Q_\alpha$. Hence, $q = p$. Assume $q = 3$. If $L_\alpha/C_{L_\alpha}(V) \cong (Q_8 \times Q_8) : 3$, then using that $R_\alpha \cap C_{L_\alpha}(V) = Q_\alpha$, we deduce that $R_\alpha/Q_\alpha = Q_8 \times Q_8$. Meanwhile, if $L_\alpha/C_{L_\alpha}(V) \cong \mathrm{SL}_2(3)$ then Proposition 2.47 yields that $\overline{L_\alpha} \cong (Q_8 \times Q_8) : 3$. In either case, $Z(\overline{L_\alpha})$ contains an elementary abelian subgroup of order 4 and there is $t \in L_\alpha \cap G_{\alpha,\beta}$ an involution with $[t, L_\alpha] \le Q_\alpha$. Since $\overline{L_\beta} \cong \mathrm{SL}_2(3)$, we can choose t such that $[t, L_\beta] \le Q_\beta$, a contradiction by Proposition 5.5 (v). Hence, $q = 2$.

If $L_\alpha \neq R_\alpha C_{L_\alpha}(V)S$ then $C_{L_\alpha}(V) \le R_\alpha$ and $\overline{L_\alpha} \cong (3 \times 3) : 2$. Since $\overline{L_\beta} \cong \mathrm{Sym}(3)$ in this case, we may apply [19, Theorem B] and since $(3 \times 3) : 2$ is not a homomorphic image of $\mathrm{Sym}(4)$, we force a contradiction.

Form $X := \langle G_\beta, R_\alpha S\rangle$ and let K be the largest subgroup of S which is normalized by X. Then $Z_\beta = \langle Z_\beta^X\rangle \le Z(K)$ and by construction, $Z_\alpha \not\le K$, otherwise $V_\beta^{(3)} \le \langle Z_\alpha^X\rangle \le Z(K)$ is abelian, a contradiction. Even still, $[K, Z_\alpha] = \{1\}$ and taking normal closures under X, we deduce that $K \le C_{Q_\beta}(V_\beta^{(3)})$. But $O^p(L_\beta)$ centralizes $C_{Q_\beta}(V_\beta^{(3)})/V_\beta$ and so L_β/K is of characteristic p. Assume that $R_\alpha S/K$ is not of characteristic 2 so that $O^2(R_\alpha S)$ acts non-trivially on K. Since $Z_\alpha \not\le K$, K is not self-centralizing and we may assume that $C_S(K) \le Q_\alpha$ and $C_S(K) \not\le Q_\beta$. If $C_S(K)^x \cap Q_\beta \not\le Q_\alpha$ for some $x \in L_\beta$, then $[C_S(K)^x \cap Q_\beta, K] = \{1\}$ so that $[O^2(R_\alpha S), K] \le [\langle (C_S(K)^x \cap Q_\beta)^{R_\alpha S}\rangle, K] = \{1\}$, a contradiction. Thus, $\langle (C_S(K) \cap Q_\beta)^{L_\beta}\rangle \le Q_\alpha$ and so $[O^2(L_\beta), Q_\beta] \le [\langle C_S(K)^{L_\beta}\rangle, Q_\beta] \le \langle (C_S(K) \cap Q_\beta)^{L_\beta}\rangle \le Q_\alpha$ and $Q_\alpha \cap Q_\beta \trianglelefteq L_\beta$, a contradiction by Proposition 5.27. Thus, the triple $(G_\beta/K, R_\alpha G_{\alpha,\beta}/K, G_{\alpha,\beta}/K)$ satisfies Hypothesis 5.1 and assuming that G is a minimal counterexample to Theorem C, and using that Q_β/K contains three non-central chief factors for $\overline{L_\beta}$ and $q = 2$, we obtain a contradiction. Hence

(3) S does not act quadratically on V, $V_\beta^{(3)}/A_\beta$ is a natural $\mathrm{SL}_2(q)$-module for $\overline{L_\beta}$ and $q > 2$.

Let $Z_\alpha < U < V_\alpha^{(2)}$ with $U \trianglelefteq L_\alpha$. Since $U < V_\alpha^{(2)}$, $U \cap V_\beta = Z_\alpha$. Thus, if U/Z_α contains a non-central chief factor for L_α, by Lemma 2.41 we deduce that $V_\alpha^{(2)} = U(V_\alpha^{(2)} \cap A_\beta)$. But then $[V_\alpha^{(2)}, Q, Q] = [U, Q, Q] \le V_\beta \cap U = Z_\alpha$, a contradiction since V is not quadratic. Hence, $U/Z_\alpha \le C_V(O^p(L_\alpha)) \le C_V(R_\alpha) = \{1\}$ and V is an irreducible faithful module for $L_\alpha/C_{L_\alpha}(V)$.

For $v \in V_\alpha^{(2)} \setminus A_\beta$, we have that $[v, Q_\beta]V_\beta \trianglelefteq O^p(L_\beta)$. But for R a Hall p'-subgroup of $L_\beta \cap G_{\alpha,\beta}$, we have that $V_\alpha^{(2)} = \langle v(V_\alpha^{(2)} \cap A_\beta)^R\rangle$ so that $[V_\alpha^{(2)}, Q_\beta] = [v(V_\alpha^{(2)} \cap A_\beta), Q_\beta] = [v, Q_\beta]V_\beta$ and V is a nearly quadratic module for $\overline{L_\alpha}$. An appeal to Theorem 2.58 yields that $L_\alpha/C_{L_\alpha}(V) \cong (\mathrm{P})\mathrm{SL}_2(q)$; $L_\alpha/C_{L_\alpha}(V) \cong (2 \times \mathrm{Sym}(4))'$; or $L_\alpha/C_{L_\alpha}(V)$ has a unique component, and it has order coprime to 3.

If $L_\alpha/C_{L_\alpha}(V) \cong (2 \times \mathrm{Sym}(4))'$ then $|Z(L_\alpha/C_{L_\alpha}(V))| = 2$ using that $R_\alpha \cap C_{L_\alpha}(V) = Q_\alpha$, we ascertain that $Z(\overline{L_\alpha})$ is elementary abelian of order 4. Thus, there is $t \in L_\alpha \cap G_{\alpha,\beta}$ an involution with $[t, L_\alpha] \le Q_\alpha$. Since $\overline{L_\beta} \cong \mathrm{SL}_2(3)$, we can choose t such that $[t, L_\beta] \le Q_\beta$, a contradiction by Proposition 5.5 (v).

If $L_\alpha/C_{L_\alpha}(V)$ has a component K of $3'$-order then $K \cong \mathrm{Sz}(r)$ for some $r \geqslant 8$. Since the outerautomorphism group of K is cyclic, we deduce that $q = 3$ and that $L_\alpha/C_{L_\alpha}(V) \cong Sz(r) : 3$. Then, one can calculate (e.g. using MAGMA [14]) that $\mathrm{Sz}(8) : 3$ is generated by two conjugate 3-elements and so a subgroup of this shape centralizes a subgroup of index 3^4 in V, a contradiction since 7 divides the order of $\mathrm{Sz}(8)$ but not the order of $\mathrm{GL}(4, 3)$.

Thus, $L_\alpha/C_{L_\alpha}(V) \cong \mathrm{(P)SL}_2(q)$. Since $L_\alpha/R_\alpha \cong \mathrm{SL}_2(q)$, applying Lemma 5.33 and Proposition 2.47 (iii) when $q = 3$ and an argument utilizing the Schur multiplier of $\mathrm{PSL}_2(q)$ when $q > 3$, we deduce that

(4) $q = 3$, $\overline{L_\alpha} \cong (2^2 \times Q_8) : 3$ where $\overline{C_{L_\alpha}(V)} \cong Q_8$ and $\overline{R_\alpha} \cong 2^2$, and V is an irreducible 2F-module for $L_\alpha/C_{L_\alpha}(V) \cong \mathrm{PSL}_2(3)$.

Let K be the largest subgroup of S normalized by both L_β and $R_\alpha S$. Then $Z_\beta \le K \ne \{1\}$. Moreover, $Z_\alpha \not\le K$, for otherwise, $Z_\alpha \le Z(K)$ and taking respective normal closures yields $V_\beta^{(3)} \le Z(K)$ is abelian, a contradiction. Since Z_α centralizes K, we infer that $[K, V_\beta^{(3)}] = \{1\}$ and $K \le C_{Q_\beta}(V_\beta^{(3)})$. Since coprime elements of $O^3(L_\beta)$ act faithfully Q_β/C_β, we conclude that L_β/K is of characteristic 3.

Assume that $R_\alpha S/K$ is not of characteristic 3 so that $O^3(L)$ acts non-trivially on K. Since $Z_\alpha \not\le K$, K is not self-centralizing and we may assume that $C_S(K) \le Q_\alpha$ and $C_S(K) \not\le Q_\beta$. If $C_S(K)^x \cap Q_\beta \not\le Q_\alpha$ for some $x \in L_\beta$, then $[C_S(K)^x \cap Q_\beta, K] = \{1\}$ so that $[O^3(L), K] \le [\langle (C_S(K)^x \cap Q_\beta)^{R_\alpha S}\rangle, K] = \{1\}$, a contradiction. Thus, $\langle (C_S(K) \cap Q_\beta)^{L_\beta}\rangle \le Q_\alpha$ and so $[O^3(L_\beta), Q_\beta] \le [\langle C_S(K)^{L_\beta}\rangle, Q_\beta] \le \langle (C_S(K) \cap Q_\beta)^{L_\beta}\rangle \le Q_\alpha$ and $Q_\alpha \cap Q_\beta \trianglelefteq L_\beta$, a contradiction by Proposition 5.27. Thus, the triple $(L_\beta/K, R_\alpha G_{\alpha\beta}/K, G_{\alpha,\beta}/K)$ satisfies Hypothesis 5.1 and assuming that G is a minimal counterexample to Theorem C, and using that Q_β/K contains three non-central chief factors for $\overline{L_\beta}$ and $O^{3'}(R_\alpha S/Q_\alpha) \cong \mathrm{PSL}_2(3)$, we have a final contradiction. □

Finally, to end this section we may assume that $b = 3$. Unfortunately, most of the techniques introduced earlier in this section are not applicable in this setting and so the methodology for this case is different from the rest of this section. The aim throughout will be to show that $R_\beta = Q_\beta$ and $R_\alpha = Q_\alpha$ for then an appeal to [27] yields $p = 2$ and G is parabolic isomorphic to M_{12} or $\mathrm{Aut}(\mathrm{M}_{12})$.

Proposition 7.45 *Suppose that $C_{V_\beta}(V_{\alpha'}) = V_\beta \cap Q_{\alpha'}$ and $b = 3$. Then $R_\beta = Q_\beta$, $\overline{L_\beta} \cong \mathrm{SL}_2(q)$ and $O^p(L_\beta)$ centralizes C_β/V_β.*

Proof Suppose first that $V_{\alpha'} \le Q_\beta$. Then $R := [V_\beta, V_{\alpha'}] = Z_\beta \le Z_{\alpha+2} = Z_{\alpha'-1}$ so that $L_\beta = \langle V_\beta, V_\beta^g, R_{\alpha'}\rangle$ normalizes $Z_{\alpha+2}Z_{\alpha+2}^g$ for some suitably chosen $g \in L_{\alpha'}$. Thus, $|V_{\alpha'}| = q^3 = |V_\beta|$. Moreover, since $V_\beta \not\le Q_{\alpha'}$ and $[C_{\alpha'}, V_\beta] \le Z_{\alpha+2} \le V_{\alpha'}$, we deduce that $O^p(L_{\alpha'})$ centralizes $C_{\alpha'}/V_{\alpha'}$. Then for any $r \in R_{\alpha'}$ of order coprime to p, we have that $[r, Q_{\alpha'}] \le C_{\alpha'}$ by the three subgroup lemma so that

$[r, Q_{\alpha'}] = \{1\}$ by coprime action. Hence, $R_{\alpha'} = Q_{\alpha'}$, $\overline{L_{\alpha'}} \cong \mathrm{SL}_2(q)$ and in this case the result holds by conjugacy.

If $V_{\alpha'} \not\leq Q_\beta$, then $C_{\alpha'} = V_{\alpha'}(C_{\alpha'} \cap Q_\beta)$ so that $[V_\beta, C_{\alpha'}] \leq RZ_\beta \leq V_{\alpha'}$ and again, $O^p(L_{\alpha'})$ centralizes $C_{\alpha'}/V_{\alpha'}$. As above, this implies that $R_{\alpha'} = Q_{\alpha'}$, $\overline{L_{\alpha'}} \cong \mathrm{SL}_2(q)$ and the result holds by conjugacy. □

Proposition 7.46 *Suppose that $C_{V_\beta}(V_{\alpha'}) = V_\beta \cap Q_{\alpha'}$ and $b = 3$. Then $|V_\beta| = q^3$, Q_β/C_β is a natural module for $\overline{L_\beta} \cong \mathrm{SL}_2(q)$, and there exists a critical pair (α, α') with $Z_{\alpha'} \cap Z_\beta = \{1\}$.*

Proof As described in Proposition 7.45, the proof when $V_{\alpha'} \leq Q_\beta$ is straightforward. Suppose that $V_{\alpha'} \not\leq Q_\beta$ and $Z_{\alpha'} \neq Z_\beta$. Since $Z_{\alpha'-1}$ is a natural module for $L_{\alpha'-1}/R_{\alpha'-1} \cong \mathrm{SL}_2(q)$, we have that $Z_\beta \cap Z_{\alpha'} = \{1\}$. Moreover, since $V_{\alpha'} \not\leq Q_\beta$, we deduce that $Q_{\alpha'-1} = V_{\alpha'}V_\beta(Q_{\alpha'-1} \cap Q_{\alpha'} \cap Q_\beta)$. Then $[V_{\alpha'} \cap V_\beta, Q_{\alpha'-1}] \leq Z_{\alpha'} \cap Z_\beta = \{1\}$ so that $V_{\alpha'} \cap V_\beta \leq \Omega(Z(Q_{\alpha'-1}))$. Note that by Proposition 5.27, $Q_\beta \cap O^p(L_\beta) \not\leq Q_{\alpha'-1}$ so that $(Q_\beta \cap O^p(L_\beta))Q_{\alpha'-1} \in \mathrm{Syl}_p(G_{\beta,\alpha'-1})$ and $\Omega(Z(Q_{\alpha'-1})) \cap C_{V_\beta}(O^p(L_\beta)) = Z_\beta$. Hence, $[V_{\alpha'}, V_\beta] \leq V_{\alpha'} \cap V_\beta = Z_{\alpha'-1}$ and we conclude that $|V_\beta| = q^3$.

Thus, we may assume that for every critical pair (α, α'), we have that $V_{\alpha'} \not\leq Q_\beta$ and $Z_{\alpha'} = Z_\beta$. Since $Z_\beta \ntrianglelefteq L_{\alpha'-1}$, there is $\lambda \in \Delta(\alpha' - 1)$ such that $Z_\lambda \cap Z_\beta = \{1\}$. Moreover, by assumption, $V_\lambda \leq Q_\beta$ and $V_\beta \leq Q_\lambda$ so that $[V_\lambda, V_\beta] \leq Z_\beta \cap Z_\lambda = \{1\}$. Then, $[C_\beta \cap Q_\lambda, V_\lambda] \leq [C_\beta, C_\beta] \cap Z_\lambda$. If $Z_\lambda \cap \Phi(C_\beta) \neq \{1\}$, then since $Z_{\alpha+2} = Z_\lambda \times Z_\beta$, $\Phi(C_\beta) \cap Z_{\alpha+2} > Z_\beta$ and $V_\beta \leq \Phi(C_\beta)$. But $O^p(L_\beta)$ centralizes C_β/V_β, a contradiction by coprime action. Therefore, $C_\beta \cap Q_\lambda = C_\beta \cap C_\lambda$ is of index at most q in C_β and C_λ. By the same reasoning, $C_{\alpha'} \cap Q_\lambda = C_{\alpha'} \cap C_\lambda$ and since $V_{\alpha'} \leq C_\lambda$ and $V_{\alpha'} \not\leq C_\beta$, we deduce that $C_\beta \not\leq Q_\lambda$, $C_\beta \cap C_\lambda$ is proper in C_β, and $C_\lambda = V_{\alpha'}V_\beta(C_\lambda \cap C_{\alpha'} \cap C_\beta)$.

Since $V_\beta V_\lambda \leq C_\beta \cap C_\lambda$, we have that $C_\beta \cap C_\lambda \trianglelefteq \langle Q_{\alpha+2}, O^p(L_\lambda), O^p(L_\beta)\rangle = \langle L_\beta, L_\lambda\rangle$. If $C_\lambda \cap C_\beta$ is not elementary abelian then $Z_{\alpha+2} = Z_\beta \times Z_\lambda \leq \Phi(C_\beta \cap C_\lambda) \leq \Phi(C_\beta)$ and $V_\beta \leq \Phi(C_\beta)$, a contradiction for then $O^p(L_\beta)$ centralizes C_β/V_β. Hence, $C_\beta \cap C_\lambda$ is elementary abelian. Then $\Omega(Z(C_\lambda)) = C_\lambda \cap C_\beta \cap C_{\alpha'}$ and $C_\lambda = V_\beta V_{\alpha'}\Omega(Z(C_\lambda))$. But then $Z_\lambda \leq [C_\lambda, C_\lambda] = [V_\beta, V_{\alpha'}] = R$. By the structure of $Z_{\alpha'-1}$, there is $\mu \in \Delta(\alpha' - 1)$ such that $Z_\mu \cap Z_\beta = \{1\}$ and $Z_{\alpha'-1} = Z_\mu \times Z_\lambda$. Repeating the above argument, we deduce that $Z_{\alpha'-1} \leq R$. But since $R = [C_\lambda, C_\lambda] \trianglelefteq L_\lambda$, we deduce that $V_\lambda \leq R$, a clear contradiction. Thus, there is a pair (α, α') satisfying the required properties and we deduce that $|V_\beta| = q^3$ in all cases. In particular, $C_\beta = Q_\alpha \cap Q_{\alpha+2}$ has index q^2 in Q_β and Q_β/C_β is a natural $\mathrm{SL}_2(q)$-module for $\overline{L_\beta}$. □

For the remainder of this section, we arrange that (α, α') is a critical pair with $Z_{\alpha'} \neq Z_\beta$.

Proposition 7.47 *Suppose that $C_{V_\beta}(V_{\alpha'}) = V_\beta \cap Q_{\alpha'}$ and $b = 3$. Then $V_\alpha^{(2)} \not\leq Q_\beta$, $V_\alpha^{(2)} \cap Q_\beta \not\leq C_\beta$, $\Phi(V_\alpha^{(2)}) = Z_\alpha$ and $C_{L_\alpha}(V_\alpha^{(2)}/Z_\alpha) = Q_\alpha$*

Proof Note that if $V_\alpha^{(2)} \leq Q_\beta$, then $V_\lambda^{(2)} \leq Q_\beta$ for all $\lambda \in \Delta(\beta)$. Since α' is conjugate to β, we have that $V_\mu^{(2)} \leq Q_{\alpha'}$ for all $\mu \in \Delta(\alpha')$. But then, $V_\beta \leq V_{\alpha'-1}^{(2)} \leq Q_{\alpha'}$, a contradiction. Now, as $Q_\alpha \cap Q_\beta \ntrianglelefteq L_\beta$ by Proposition 5.27, we have that $O^p(L_\beta)$ does not centralize Q_β/C_β. In particular, $V_\alpha^{(2)} \cap Q_\beta \geq [V_\alpha^{(2)}, Q_\beta] \nleq C_\beta$.

Since $V_\alpha^{(2)}$ is non-abelian, otherwise by conjugacy $V_{\alpha'} \leq V_{\alpha+2}^{(2)}$ centralizes V_β, we have that $[V_\alpha^{(2)}, V_\alpha^{(2)}] = Z_\alpha$. But now, $V_\alpha^{(2)}$ is generated by V_δ for $\delta \in \Delta(\alpha)$ so that $V_\alpha^{(2)}/Z_\alpha$ is elementary abelian and $\Phi(V_\alpha^{(2)}) = Z_\alpha$.

Let $r \in C_{L_\alpha}(V_\alpha^{(2)}/Z_\alpha)$ have order coprime to p. Then by coprime action, r centralizes $V_\alpha^{(2)}$. Applying the three subgroup lemma, we have that r centralizes $Q_\alpha/C_{Q_\alpha}(V_\alpha^{(2)})$. But $C_{Q_\alpha}(V_\alpha^{(2)}) \leq C_\beta$ so that for $Q := Q_\beta \cap O^p(L_\beta)$, $Q \nleq Q_\alpha$ and $[Q, C_{Q_\alpha}(V_\alpha^{(2)})] \leq C_{Q_\alpha}(V_\alpha^{(2)}) \cap V_\beta = Z_\alpha$ and we deduce that r centralizes Q_α/Z_α and $r = 1$. Hence, $C_{L_\alpha}(V_\alpha^{(2)}/Z_\alpha) = Q_\alpha$, and the proof is complete. □

Proposition 7.48 *Suppose that* $C_{V_\beta}(V_{\alpha'}) = V_\beta \cap Q_{\alpha'}$ *and* $b = 3$. *Then either* $Q_\alpha = V_\alpha^{(2)} = [V_\alpha^{(2)}, O^p(L_\alpha)]$, $Z(Q_\alpha) = Z_\alpha$, $C_{V_\alpha^{(2)}/Z_\alpha}(O^p(L_\alpha)) = \{1\}$, $L_\beta = \langle V_\alpha^{(2)}, V_{\alpha+2}^{(2)}\rangle$, $C_\beta = V_\alpha^{(2)} \cap V_{\alpha+2}^{(2)}$ *and* $\Phi(C_\beta) \leq Z_\beta$; *or G is parabolic isomorphic to* Aut(M_{12}).

Proof Note that the configurations in Theorem C with $b = 3$ (that is, the amalgams which are parabolic isomorphic to M_{12} and Aut(M_{12})) satisfy the conclusion of the lemma. Hence, assuming that G does not satisfy the conclusion of the lemma, we may also suppose that G is a minimally chosen counterexample to Theorem C.

Since $Z(Q_\alpha)$ centralizes V_β, $Z(Q_\alpha) \leq C_\beta$ and $Q_\beta \cap O^p(L_\beta)$ centralizes $Z(Q_\alpha)/Z_\alpha$ since $Z(Q_\alpha) \cap V_\beta = Z_\alpha$. Hence, L_α centralizes $Z(Q_\alpha)/Z_\alpha$. Indeed, $Z(Q_\alpha)$ is elementary abelian, for otherwise L_α would centralize $Z(Q_\alpha)/\Phi(Z(Q_\alpha))$. Now, $Z(Q_\alpha) = Z_\alpha(Z(Q_\alpha) \cap Q_{\alpha'})$ and so we need only demonstrate that $Z(Q_\alpha) \cap Q_{\alpha'} = Z_\beta$ to show that $Z(Q_\alpha) = Z_\alpha$

For this, we observe that $Z(Q_\alpha) \cap Q_{\alpha'}$ has exponent p and if $Z(Q_\alpha) \cap Q_{\alpha'} \leq C_{\alpha'}$, then $Z(Q_\alpha) \cap Q_{\alpha'}$ is centralized by $\langle V_{\alpha'}, Q_\alpha\rangle$. If $V_{\alpha'} \leq Q_\beta$, then $S = V_{\alpha'}Q_\alpha$ centralizes $Z(Q_\alpha) \cap Q_{\alpha'} = \Omega(Z(S)) = Z_\beta$ whereas if $V_{\alpha'} \nleq Q_\beta$, then $L_\beta = \langle V_{\alpha'}, Q_\alpha\rangle$ centralizes $Z(Q_\alpha) \cap Q_{\alpha'} = \Omega(Z(L_\beta)) = Z_\beta$. Thus, we may assume that $Z(Q_\alpha) \cap Q_{\alpha'} \nleq C_{\alpha'}$. But $Z(Q_\alpha)V_\beta \trianglelefteq L_\beta$ so that $Z(Q_\alpha)V_\beta = Z(Q_{\alpha+2})V_\beta$ and $Z(Q_\alpha) \cap Q_{\alpha'} \leq Z(Q_{\alpha+2})V_\beta \cap Q_{\alpha'} = Z(Q_{\alpha+2})(V_\beta \cap Q_{\alpha'}) \leq C_{\alpha'}$, a contradiction. Therefore, $Z(Q_\alpha) = Z_\alpha$.

Since $Q_\alpha = V_\alpha^{(2)}C_\beta$, $Q_\beta \cap O^p(L_\beta)$ centralizes $Q_\alpha/V_\alpha^{(2)}$ and $[Q_\alpha, O^p(L_\alpha)] \leq [V_\alpha^{(2)}, O^p(L_\alpha)]$. Furthermore, if $V_\beta \cap [V_\alpha^{(2)}, O^p(L_\alpha)] > Z_\alpha$, then by the $G_{\alpha,\beta}$-irreducibility of V_β/Z_α, $V_\beta \leq [V_\alpha^{(2)}, O^p(L_\alpha)] = V_\alpha^{(2)}$. Otherwise, by Lemma 2.30 we have that $C_{V_\alpha^{(2)}/Z_\alpha}(O^p(L_\alpha))$ is non-trivial. Write C^α for the preimage in $V_\alpha^{(2)}$ of this group. Note that, by definition, $V_\alpha^{(2)} = [V_\alpha^{(2)}, O^p(L_\alpha)]V_\beta$ so that $[V_\alpha^{(2)}, Q_\beta] = [V_\alpha^{(2)}, O^p(L_\alpha), Q_\beta]$. In particular, since C^α is $G_{\alpha,\beta}$-invariant, if $C^\alpha \nleq C_\beta$ then $V_\alpha^{(2)} \cap Q_\beta = (C^\alpha \cap Q_\beta)(V_\alpha^{(2)} \cap C_\beta$ and $[V_\alpha^{(2)} \cap Q_\beta, Q_\beta] \leq C^\alpha([V_\alpha^{(2)}, O^p(L_\alpha)] \cap V_\beta) \leq C^\alpha$ and by Lemma 2.41, $R_\alpha = Q_\alpha$. Then [27] yields a contradiction.

Hence, $C^\alpha \leq C_\beta$ so that $\{1\} = [V_\beta, C^\alpha]^{G_\alpha} = [V_\alpha^{(2)}, C^\alpha]$. Then, by the three subgroups lemma and using that $[Q_\alpha, O^p(L_\alpha)] \leq V_\alpha^{(2)}$, $[Q_\alpha, C^\alpha, O^p(L_\alpha)] = \{1\}$ and since $[Z_\alpha, O^p(L_\alpha)] \neq \{1\}$, $C^\alpha \leq Z(Q_\alpha) = Z_\alpha$, a contradiction. Thus, $V_\alpha^{(2)} = [V_\alpha^{(2)}, O^p(L_\alpha)]$.

Set $L_\beta^* := \langle V_\alpha^{(2)}, V_{\alpha+2}^{(2)} \rangle$ so that $L_\beta = L_\beta^* C_\beta$, $L_\beta^* \trianglelefteq L_\beta$ and $O^p(L_\beta) \leq L_\beta^*$. Since $V_\alpha^{(2)} \cap Q_\beta \cap Q_{\alpha+2} = V_\alpha^{(2)} \cap C_\beta \trianglelefteq G_\beta$, we have $V_\alpha^{(2)} \cap Q_\beta \cap Q_{\alpha+2} \leq V_{\alpha+2}^{(2)}$ so that $V_\alpha^{(2)} \cap Q_\beta \cap Q_{\alpha+2} = V_\alpha^{(2)} \cap V_{\alpha+2}^{(2)} = V_{\alpha+2}^{(2)} \cap Q_\beta \cap Q_\alpha$. Indeed, it follows that $C_\beta \cap L_\beta^* = V_\alpha^{(2)} \cap V_{\alpha+2}^{(2)}$, $\Phi(C_\beta \cap L_\beta^*) \leq Z_\alpha \cap Z_{\alpha+2} = Z_\beta$ and $V_\alpha^{(2)}$ has index q in a Sylow p-subgroup of L_β^*. We note that $L_\beta^* < L_\beta$ for otherwise the conclusion of the lemma holds, and we have a contradiction since G was an assumed minimal counterexample.

Since $Q_\alpha = V_\alpha^{(2)}(Q_\alpha \cap C_\beta)$ and $[Q_\beta \cap O^p(L_\beta), C_\beta] \leq V_\beta \leq V_\alpha^{(2)}$ we deduce that $O^p(L_\alpha)$ centralizes $Q_\alpha/V_\alpha^{(2)}$. Moreover, $S/V_\alpha^{(2)} = (Q_\alpha/V_\alpha^{(2)}) : (V_{\alpha+2}^{(2)} \cap Q_\beta)V_\alpha^{(2)}/V_\alpha^{(2)}$ and by Gaschütz' theorem [49, (3.3.2)], $L_\alpha/V_\alpha^{(2)}$ splits over $Q_\alpha/V_\alpha^{(2)}$. Thus, writing $S^* = S \cap L_\beta^* = V_\alpha^{(2)}(V_{\alpha+2}^{(2)} \cap Q_\beta)$, we have that $S^* = V_\alpha^{(2)}(Q_\beta \cap O^p(L_\beta)) \trianglelefteq G_{\alpha,\beta}$ and $S^* \in \mathrm{Syl}_p(\langle (S^*)^{G_\alpha} \rangle)$.

Set $L_\alpha^* := \langle (S^*)^{G_\alpha} \rangle$ so that $V_\alpha^{(2)} = O_p(L_\alpha^*)$. Since $S^* \trianglelefteq G_{\alpha,\beta}$ and $L_\lambda^* = O^p(L_\lambda)S^*$, L_λ^* is normalized by $G_{\alpha,\beta}$ and $N_{L_\mu^*}(S^*)$ for $\{\lambda, \mu\} = \{\alpha, \beta\}$. Indeed, since no non-trivial subgroup of $G_{\alpha,\beta}$ is normal in G, no non-trivial subgroup $G_{\alpha,\beta}^* := N_{L_\alpha^*}(S^*)N_{L_\beta^*}(S^*)$ is normal in $\langle L_\alpha^*, L_\beta^* \rangle$. Writing $G_\lambda^* = L_\lambda^* G_{\alpha,\beta}^*$, and using that $L_\lambda/Q_\lambda \cong L_\lambda^*/O_p(L_\lambda^*)$, for $\lambda \in \{\alpha, \beta\}$ the tuple $(G_\alpha^*, G_\beta^*, G_{\alpha,\beta}^*)$ satisfies Hypothesis 5.1. Since $L_\beta^* < L_\beta$ and G was an assumed minimal counterexample, we conclude that $(G_\alpha^*, G_\beta^*, G_{\alpha,\beta}^*)$ is one of the configurations listed in Theorem C. In particular, $p = 2$, $G_\lambda/Q_\lambda \cong G_\lambda^*/O_2(G_\lambda^*) \cong \mathrm{Sym}(3)$ for $\lambda \in \{\alpha, \beta\}$ and $(G_\alpha^*, G_\beta^*, G_{\alpha,\beta}^*)$ is parabolic isomorphic to one of M_{12} or $\mathrm{Aut}(\mathrm{M}_{12})$. But then, G has a weak BN-pair of rank 2 so itself is captured in Theorem C and satisfies the conclusion of the lemma. This contradiction completes the proof. □

Whenever G is assumed to not be parabolic isomorphic to $\mathrm{Aut}(\mathrm{M}_{12})$, we will utilize the Proposition 7.48 without reference.

Lemma 7.49 *Suppose that $C_{V_\beta}(V_{\alpha'}) = V_\beta \cap Q_{\alpha'}$, $b = 3$ and G is not parabolic isomorphic to* M_{12} *or* $\mathrm{Aut}(\mathrm{M}_{12})$*. Then $[Q_\beta, C_\beta] = V_\beta$.*

Proof Note that $Q_\beta = C_\beta[Q_\beta, O^p(L_\beta)]$ so that $[Q_\beta, C_\beta] \leq [C_\beta, C_\beta][O^p(L_\beta), C_\beta] = V_\beta$. Aiming for a contradiction, assume that $C_{Q_\beta}(C_\beta/Z_\alpha) > V_\alpha^{(2)} \cap Q_\beta$ and set $V := V_\alpha^{(2)}/Z_\alpha$. Then as $C_{Q_\beta}(C_\beta/Z_\alpha)$ is invariant under $G_{\alpha,\beta}$, we have that $[Q_\beta, C_\beta] \leq Z_\alpha$ and $[Q_\beta, C_\beta] = Z_\beta$. Then for $v \in V_\alpha^{(2)} \setminus Q_\beta$, we have that $[v, Q_\beta]C_\beta = [V_\alpha^{(2)}, Q_\beta]C_\beta$ has index q in $V_\alpha^{(2)}$. In particular, if Q_β acts quadratically on V, then an index q subgroup of V is centralized by Q_β and a combination of Lemma 2.41 and [27] yields a contradiction.

If V is not irreducible for $\overline{L_\alpha}$, then for some $Z_\alpha < U < V_\alpha^{(2)}$ with $U \trianglelefteq L_\alpha$, since $V_\alpha^{(2)} = [V_\alpha^{(2)}, O^p(L_\alpha)]$ and $C_V(O^p(L_\alpha)) = \{1\}$, U is non-trivial. Note that if $U \not\leq Q_\beta$, then $V_\alpha^{(2)} \cap Q_\beta = [U, Q_\beta]C_\beta$ and so a UC_β has index strictly less than q in $V_\alpha^{(2)}$, a contradiction by Lemma 2.41. But then $U \leq Q_\beta$ and by Lemma 2.41, $|UC_\beta/C_\beta| = q$ and $V_\alpha^{(2)} \cap Q_\beta = [U, Q_\beta]C_\beta$. Then by Lemma 2.41, both $V_\alpha^{(2)}$ and U/Z_α are natural modules for $\overline{L_\alpha}$. Since Q_β is not quadratic on V, $q > 2$ and using a standard argument involving the Schur multiplier of $\mathrm{PSL}_2(q)$, we conclude by Proposition 2.47 that $\overline{L_\alpha} \cong (Q_8 \times Q_8) : 3$. But then, there is $t \in L_\alpha \cap G_{\alpha,\beta}$ an involution with $[t, L_\alpha] \leq Q_\alpha$ and, since $\overline{L_\beta} \cong \mathrm{SL}_2(3)$, we can choose t such that $[t, L_\beta] \leq Q_\beta$, a contradiction by Proposition 5.5 (v).

Hence, if Q_β does not act quadratically on V, then V is a faithful irreducible nearly quadratic module for $\overline{L_\alpha}$. Using that $L_\alpha/R_\alpha \cong \mathrm{SL}_2(q)$ and $\overline{L_\alpha}$ acts faithfully on $V_\alpha^{(2)}/Z_\alpha$, and comparing with the list in Theorem 2.58 we conclude that either $R_\alpha = Q_\alpha$, $q = 3$ and $\overline{L_\alpha} \cong (2 \wr \mathrm{Sym}(4))'$ or $q = 3$ and $\overline{L_\alpha}$ contains a component of order coprime to 3. In the second case, we note that $\overline{L_\alpha}$ has no quotient isomorphic to $\mathrm{SL}_2(3)$, while in the third case we observe, arguing in a similar fashion to the proof of Proposition 7.44, that $\overline{L_\alpha} \cong \mathrm{Sz}(r) : 3$ which also has no quotient isomorphic to $\mathrm{SL}_2(3)$. Hence. $R_\alpha = Q_\alpha$ and [27] yields a contradiction. Hence, $[Q_\beta, C_\beta] = V_\beta$. □

Lemma 7.50 *Suppose that* $C_{V_\beta}(V_{\alpha'}) = V_\beta \cap Q_{\alpha'}$, $b = 3$ *and* G *is not parabolic isomorphic to* M_{12} *or* $\mathrm{Aut}(\mathrm{M}_{12})$*. Then* $q = p$.

Proof Let $x \in Q_\beta \setminus (Q_\beta \cap V_\alpha^{(2)})$. Write C for the preimage in C_β of $C_{C_\beta/Z_\alpha}(x)$ so that, as $V_\beta \leq C$, we have that $C \trianglelefteq L_\beta$. Then, C is normalized by $L_\beta \cap G_{\alpha,\beta}$ so that $[C, Q_\beta] = [C, V_\alpha^{(2)} \cap Q_\beta][C, \langle x^{L_\beta \cap G_{\alpha,\beta}} \rangle] \leq Z_\alpha$ from which we infer that $[C, Q_\beta] = Z_\beta$. In particular, $C_{C_\beta/Z_\alpha}(x) < C_\beta/Z_\alpha$ and $C/Z_\beta = Z(Q_\beta/Z_\beta)$. Set $V := V_\alpha^{(2)}/Z_\alpha$ and let C_x be the preimage in $V_\alpha^{(2)}$ of $C_V(x)$ so that $C_x \cap C_\beta = C$. By the action of S on Q_β/C_β, we have that $C_x \leq Q_\beta$ so that $|C_x/C| \leqslant q$. Note that if $q = p$, then $C_V(x) = C_V(Q_\beta)$.

Aiming for a contradiction, suppose that $q > p$ and there is $A \leq S/Q_\alpha$ with $|A| \geqslant p^2$ and $C_V(A) = C_V(B)$ for all $B \leq A$ with $[A : B] = p$. Using Proposition 2.8, we have that $O_{p'}(\overline{L_\alpha}) = \langle C_{O_{p'}(\overline{L_\alpha})}(B) : [A : B] = p \rangle$ so that $C_V(A)$ is normalized by $AO_{p'}(\overline{L_\alpha})$. Set $H_\alpha = \langle A^{AO_{p'}(\overline{L_\alpha})} \rangle$ so that $[H_\alpha, C_V(A)] = \{1\}$. Then, by coprime action, we get that $V = [V, O_{p'}(H_\alpha)] \times C_V(O_{p'}(H_\alpha))$ is an A-invariant decomposition and as $C_V(A) \leq C_V(O_{p'}(H_\alpha))$, it follows that $O_{p'}(H_\alpha)$ centralizes V so that $O_{p'}(H_\alpha) = \{1\}$ and $H_\alpha = A \trianglelefteq AO_{p'}(\overline{L_\alpha})$. Then, $[\langle A^{\overline{L_\alpha}} \rangle, O_{p'}(\overline{L_\alpha})] = [A, O_{p'}(\overline{L_\alpha})] = \{1\}$ and we have that $O_{p'}(\overline{L_\alpha}) \leq Z(\overline{L_\alpha})$. Then [27] gives a contradiction since $q > p$. Hence

(1) if $q > p$, then there is a maximal subgroup A of Q_β with $C_V(A) > C_V(Q_\beta) \geq C/Z_\alpha$.

Then, for C_A the preimage in $V_\alpha^{(2)}$ of $C_V(A)$, we have that $[C_A, V_\alpha^{(2)}, A] \le [Z_\alpha, A] = Z_\beta$, $[A, C_A, V_\alpha^{(2)}] = \{1\}$ and by the three subgroups lemma, $[V_\alpha^{(2)}, A, C_A] \le Z_\beta$. Since $([V_\alpha^{(2)}, A] \cap C_\beta)V_\beta \trianglelefteq L_\beta$, we have that $Z_\beta = [([V_\alpha^{(2)}, A] \cap C_\beta)V_\beta, C_A C_\beta]^{L_\beta} = [([V_\alpha^{(2)}, A] \cap C_\beta)V_\beta, \langle C_A^{L_\beta}\rangle C_\beta] = [([V_\alpha^{(2)}, A] \cap C_\beta)V_\beta, Q_\beta]$. In particular, $([V_\alpha^{(2)}, A] \cap C_\beta)V_\beta \le C$.

It follows that $[V_\alpha^{(2)}, Q_\beta] \cap C_\beta \le C$ so that $[V, Q_\beta, Q_\beta, Q_\beta] = \{1\}$ and by Corollary 2.60, if $p \geqslant 5$ then $R_\alpha = Q_\alpha$. In this scenario, [27] provides a contradiction. Hence, $p \leqslant 3$. Moreover, $[V_\alpha^{(2)}, Q_\beta]C = [V_\alpha^{(2)}, y]C$ for all $y \in Q_\beta \setminus (V_\alpha^{(2)} \cap Q_\beta)$. Indeed, $[V, y, y] = [V, Q_\beta, y]$. Hence if y is quadratic on V, then $[V, Q_\beta, Q_\beta] = \{1\}$ and $C_V(Q_\beta) = C_V(A)$ for all maximal subgroups of A of Q_β which contain $V_\alpha^{(2)} \cap Q_\beta$, a contradiction. Since involutions in S/Q_α always act quadratically on V when $p = 2$, we conclude that

(2) $p = 3$.

Now, $C_V(x) > C_V(B) > \cdots > C_V(A) > C_V(Q_\beta) \geqslant C/Z_\alpha$ for a maximal chain of subgroups $\langle x\rangle < B < \cdots < A < Q_\beta$, where we identify A with some subgroup of Q_β not contained in $Q_\alpha \cap Q_\beta$. We deduce that $C_x C$ has index at most 3 in $[V_\alpha^{(2)}, Q_\beta]C$. Then, the commutation map $\theta : [V_\alpha^{(2)}, Q_\beta]C/Z_\alpha \to [V_\alpha^{(2)}, Q_\beta]C/Z_\alpha$ such that $[v, q]cZ_\alpha\theta = [[v, q]cZ_\alpha, x]$ has image $[[V_\alpha^{(2)}, Q_\beta], x]Z_\alpha/Z_\alpha$ and kernel C_x/Z_α. Hence, $|[V, Q_\beta, x]| = 3$. Note that $[V, Q_\beta, x]$ is normalized by $S = V_\alpha^{(2)}Q_\beta$ and the three subgroups lemma yields that $[V, Q_\beta, x] = [V, x, Q_\beta]$. But using that $[V_\alpha^{(2)}, Q_\beta]C = [V_\alpha^{(2)}, x]C$, we have that $[V, x, Q_\beta] = [V, Q_\beta, Q_\beta]$ has order 3 and $[V, Q_\beta, Q_\beta] = [V, y, y]$ for all $y \in Q_\beta \setminus (V_\alpha^{(2)} \cap Q_\beta)$. By coprime action, $O_{3'}(\overline{L_\alpha}) = \langle C_{O_{3'}(\overline{L_\alpha})}(yQ_\alpha) : y \in Q_\beta \setminus (V_\alpha^{(2)} \cap Q_\beta)\rangle$ so that $O_{3'}(\overline{L_\alpha})\overline{G_{\alpha,\beta}}$ normalizes $[V, Q_\beta, Q_\beta]$.

Form $H = \langle \overline{S}^{O_{3'}(\overline{L_\alpha})\overline{G_{\alpha,\beta}}}\rangle$ so that $O_{p'}(H) \le \overline{R_\alpha}$ since $q > 3$. By coprime action, $V = [V, O_{3'}(H)] \times C_V(O_{3'}(H))$ and we write V_1, V_2 for the preimage in $V_\alpha^{(2)}$ of $[V, O_{3'}(H)], C_V(O_{3'}(H))$ respectively. Note that $[V, Q_\beta, Q_\beta] \le C_V(O_{3'}(H))$ so that $[V_1, Q_\beta, Q_\beta] \le Z_\alpha$.

Assume first that $V_1 \not\le Q_\beta$. Since V_1 is $G_{\alpha,\beta}$-invariant, $V_\alpha^{(2)} = V_1 C_\beta$ so that $[V_\alpha^{(2)}, Q_\beta, Q_\beta] = [V_1, Q_\beta, Q_\beta]Z_\beta \le Z_\alpha$, a contradiction.

Suppose now that $V_1 \le Q_\beta$ but $V_1 \not\le C_\beta$. Since $V_\alpha^{(2)} \not\le Q_\beta$, $V_2 \not\le Q_\beta$ and $V_\alpha^{(2)} = V_2 C_\beta$. Then $V_1 \le [V_2, Q_\beta]C_\beta$ so that $[V_1, Q_\beta] \le V_1 \cap [V_2, Q_\beta, Q_\beta]V_\beta \le V_\beta$. But then $[V_\alpha^{(2)} \cap Q_\beta, Q_\beta] = [V_1, Q_\beta][C_\beta, Q_\beta] = V_\beta$ so that $\Phi(Q_\beta) = V_\beta$. Then $\{1\} \ne [V, Q_\beta, Q_\beta] \le V_\beta/Z_\alpha$ and since V_β/Z_α is irreducible under the action of $G_{\alpha,\beta}$, we conclude that $[V, Q_\beta, Q_\beta] = V_\beta/Z_\alpha$ has order 3, a contradiction since $q > 3$. Therefore

(3) $V_1 \le C_\beta$.

Indeed, $[V_1, V_\beta] = \{1\}$. But then $[V_\alpha^{(2)}, O_{3'}(H), V_\beta] \leq [V_1, V_\beta] = \{1\}$, $[V_\alpha^{(2)}, V_\beta, O_{3'}(H)] = [Z_\alpha, O_{3'}(H)] = \{1\}$ and the three subgroup lemma yields that $[V_\beta, O_{3'}(H)] \leq Z(V_\alpha^{(2)} = Z_\alpha$. But then $V_\alpha^{(2)} = V_2 C_\beta$ so that $[V_\alpha^{(2)}, Q_\beta] \leq V_2$ and $O_{3'}(H)$ centralizes V. Hence, $O_{3'}(H) = \{1\}$, $\overline{S} \trianglelefteq H$, $[\overline{S}, O_{3'}(H)] = \{1\}$ and $\overline{L_\alpha}$ is central extension of $\mathrm{PSL}_2(q)$ by a $3'$-group. Since $q > 3$, we have that $R_\alpha = Q_\alpha$ and [27] provides the desired contradiction. □

Proposition 7.51 *Suppose that $C_{V_\beta}(V_{\alpha'}) = V_\beta \cap Q_{\alpha'}$ and $b = 3$. Then $p = 2$ and G is parabolic isomorphic to* M_{12} *or* $\mathrm{Aut}(\mathrm{M}_{12})$.

Proof Aiming for a contradiction, we assume throughout that G is not parabolic isomorphic to M_{12} or $\mathrm{Aut}(\mathrm{M}_{12})$. By Lemma 7.50, we have that $q = p$. For $Q := [Q_\beta, O^p(L_\beta)]$, we have that $|Q/[Q, Q]V_\beta| = p^2$ by coprime action using the central involution of $\overline{L_\beta}$ when $p \geqslant 5$, and the p-solvability of L_β otherwise. Then, since $Q_\beta = QC_\beta$ and $[Q, C_\beta] \leq V_\beta$, we have that $\Phi(Q_\beta) = \Phi(Q)V_\beta$ and $\Phi(Q)V_\beta = C_\beta \cap Q$. Note that by [61, Lemma 2.73], we have that $[Q, Q]V_\beta/V_\beta$ has order at most p and Q/V_β is either elementary abelian or an extraspecial group.

Assume that Q/V_β is non-abelian. Then $(V_\alpha^{(2)} \cap Q)/V_\beta$ is elementary abelian of index p in Q/V_β and it follows that Q/V_β is dihedral when $p = 2$; and is extraspecial of exponent p when p is odd. Since $\overline{L_\beta}$ acts faithfully on Q/V_β, we conclude that p is odd. Furthermore, since Q/Z_β has class at most 3, applying the Hall-Higman theorem to the action of Q of V, we deduce that either $p \in \{3, 5\}$, or $R_\alpha = Q_\alpha$. In the latter case, [27] yields a contradiction. Therefore

(1) if Q/V_β is non-abelian then $p \in \{3, 5\}$.

As in the proof of Lemma 7.50, we set C_x to be the preimage in $V_\alpha^{(2)}$ of $C_{V_\alpha^{(2)}/Z_\alpha}(Q_\beta)$ and we deduce that $C_x \leq V_\alpha^{(2)} \cap Q_\beta$. Furthermore, $[C_x C_\beta, Q_\beta] = V_\beta$ and as Q_β/V_β is non-abelian, $[V_\alpha^{(2)} \cap Q_\beta, Q_\beta] > V_\beta$ and $C_x \leq C_\beta$. Then by the proof of Lemma 7.50, $C_x = C < C_\beta$ where $C/Z_\beta = Z(Q_\beta/Z_\beta)$. Indeed, C_x has index at least p^3 in $V_\alpha^{(2)}$ Moreover, the commutation homomorphism $\theta : (V_\alpha^{(2)} \cap Q_\beta)/Z_\alpha \to (V_\alpha^{(2)} \cap Q_\beta)/Z_\alpha$ such that $vZ_\alpha\theta = [vZ_\alpha, x]$ has image contained in $\Phi(Q_\beta)V_\beta/Z_\alpha$ which has order p^2 and kernel C_x from which we deduce that

(2) if Q/V_β is non-abelian then $C_x = C$ has index p in C_β.

Since p is odd, and $V_\alpha^{(2)}$ has class 2 and is generated by elements of order p, [33, Lemma 10.14] yields that $V_\alpha^{(2)} = \Omega(V_\alpha^{(2)})$ is of exponent p. In particular, $(V_\lambda^{(2)} \cap Q_\beta)/Z_\beta$ is of exponent p for all $\lambda \in \Delta(\beta)$ and we deduce that $\Omega(Q_\beta/Z_\beta) = Q_\beta/Z_\beta$. Since $|Q/Z_\alpha| \in \{3^4, 5^4\}$ and $|Q/Z_\beta| \in \{3^5, 5^5\}$, we will make liberal use of the Small Groups library in MAGMA [14] to derive a contradiction when $\Phi(Q_\beta) > V_\beta$.

Assume first that $Z(Q/Z_\alpha) = V_\beta/Z_\alpha$. Then $\Phi(Q)/Z_\beta \not\leq Z(Q_\beta/Z_\beta)$ so that $C_\beta = \Phi(Q)C$. Write $D = C_C(O^p(L_\beta))$. Then $C = V_\beta D$ and $[Q_\beta, D] =$

$[QC, D] = D'$. Note that $D \leq Q_{\alpha'}$ and $[D, V_{\alpha'}] \leq Z_{\alpha'} \cap D = \{1\}$ so that $D \leq C_\beta \cap C_{\alpha'}$. In particular, $\Phi(D) \leq \Phi(C_\beta) \cap \Phi(C_{\alpha'}) = Z_\beta \cap Z_{\alpha'} = \{1\}$ and D is centralized by $V_{\alpha'}Q_\beta$. Since D is of exponent p, $D = Z_\beta$, $C = V_\beta$, $|C_\beta/V_\beta| = p$ and $|Q| = |Q_\beta| = p^6$. If $p = 5$, then $\overline{L_\alpha}$ is a subgroup of $\mathrm{SL}_4(5)$ with a strongly 5-embedded subgroup and some quotient by a $5'$-group isomorphic to $\mathrm{SL}_2(5)$. Computing in MAGMA [14], we have that $R_\alpha = Q_\alpha$, and a contradiction follows from [27]. Therefore

(3) if Q/V_β is non-abelian and $Z(Q/Z_\alpha) = V_\beta/Z_\alpha$ then $p = 3$.

Now, $|\Phi(Q_\beta)Z_\alpha/Z_\alpha| = 9$ and we deduce that $V_\beta/Z_\alpha < \Phi(Q_\beta)Z_\alpha/Z_\alpha = \Phi(Q_\beta/Z_\alpha)$ and Q_β/Z_α has maximal class. Moreover, $(V_\alpha^{(2)} \cap Q_\beta)/Z_\alpha$ is elementary abelian of index p in Q_β/Z_α so that Q_β/Z_α is isomorphic to $3 \wr 3 = \mathit{SmallGroup}(3^4, 7)$. Furthermore, we infer that $V_\beta/Z_\beta = Z(Q_\beta/Z_\beta)$ has order 9, $\Phi(Q_\beta/Z_\beta)$ has order 27, $\Omega(Q_\beta/Z_\beta) = Q_\beta/Z_\beta$ and $Q_\beta/Z_\lambda \cong Q_\beta/Z_\alpha$ for any $\lambda \in \Delta(\beta)$. Calculating in MAGMA [14], the only possibility is $Q_\beta/Z_\beta \cong \mathit{SmallGroup}(3^5, 3)$. Note that $L_\beta/C_{Q_\beta}(Q/Z_\beta)$ embeds a subgroup of $\mathrm{Aut}(Q/Z_\beta)$. Even better, $L_\beta/C_{Q_\beta}(Q/Z_\beta)$ embeds as a subgroup of the normal closure of a Sylow 3-subgroup in $\mathrm{Aut}(Q/Z_\beta)$. But in this case, $\mathrm{Aut}(Q/Z_\beta)$ has normal Sylow 3-subgroup, a contradiction. Therefore

(4) if Q/V_β is non-abelian then $Z(Q/Z_\alpha) = V_\beta\Phi(Q)/Z_\alpha$ has index p^2 in Q/Z_α.

In particular, Q has class 2 and so acts cubically on $V_\alpha^{(2)}/Z_\alpha$. By Corollary 2.60, either $p = 3$ or $R_\alpha = Q_\alpha$. In the latter case, [27] gives a contradiction. Note that $Z_\alpha \geq [\Phi(Q)V_\beta, Q] \trianglelefteq L_\beta$ so that $\Phi(Q)V_\beta/Z_\beta = Z(Q/Z_\beta)$. If $V_\beta \leq \Phi(Q)$ then $\Phi(Q)/Z_\beta = \Phi(Q/Z_\beta) = Z(Q/Z_\beta)$ has order 27, and $\Omega(Q/Z_\beta) = Q/Z_\beta$. No such group exists.

Hence, $\Phi(Q) \cap V_\beta = Z_\beta$ so that $\Phi(Q/Z_\beta) = 3$, $Z(Q/Z_\beta) = 27$ and $Q/Z_\beta \cong \mathit{SmallGroup}(3^5, 62)$. Then $Q/Z_\alpha \cong \mathit{SmallGroup}(81, 12)$ has four distinct elementary abelian subgroups of order 3^3. Since one of these groups is $(V_\alpha^{(2)} \cap Q)/Z_\alpha$ which is normalized by $L_\beta \cap G_{\alpha,\beta}$, at least one other is also normalized by $L_\beta \cap G_{\alpha,\beta}$. But $Q/\Phi(Q)V_\beta \cong Q_\beta/C_\beta$ is a natural $\mathrm{SL}_2(3)$-module for $\overline{L_\beta}$ and, as such, has a unique subspace of order 3 normalized by $\overline{L_\beta \cap G_{\alpha,\beta}}$, a contradiction. Thus

(5) $\Phi(Q_\beta) = V_\beta$.

Setting $V := V_\alpha^{(2)}/Z_\alpha$, Q_β acts cubically on V and Corollary 2.60 yields that $p \in \{2, 3\}$ or $R_\alpha = Q_\alpha$. In the latter case, [27] provides a contradiction. We again recall the notations C and C_x from Lemma 7.50. We may assume that $R_\alpha \neq Q_\alpha$ otherwise the result holds by [27]. In particular, by Lemma 2.41 we deduce that

(6) V is not an FF-module.

Then, forming the commutation homomorphism $\theta : (V_\alpha^{(2)} \cap Q_\beta)/Z_\alpha \to (V_\alpha^{(2)} \cap Q_\beta)/Z_\alpha$ such that $vZ_\alpha\theta = [vZ_\alpha, x]$ has image V_β/Z_α and kernel C_x/Z_α. Thus, C_x has index p in $V_\alpha^{(2)} \cap Q_\beta$ and index p^2 in $V_\alpha^{(2)}$. Note that $C_x \neq C_\beta$ for otherwise

$[Q_\beta, C_\beta] = [\langle x\rangle(V_\alpha^{(2)} \cap Q_\beta), C_\beta] \leq Z_\alpha$, a contradiction by Lemma 7.49. Indeed, $V_\alpha^{(2)} \cap Q_\beta = C_x C_\beta$.

Since $V = [V, O^p(L_\alpha)]$, if V is not irreducible for $\overline{L_\alpha}$ then by Lemma 2.41, V contains two non-central chief factors for $\overline{L_\alpha}$ and both are natural $\mathrm{SL}_2(p)$-modules. By Propositions 2.46 and 2.47, we conclude that $\overline{L_\alpha} \cong (3 \times 3) : 2$ or $(Q_8 \times Q_8) : 3$ and $|V| = p^4$. Instead (or perhaps simultaneously), if x acts quadratically on V then both V and $\overline{L_\alpha}$ are described by Proposition 2.51. Since $R_\alpha \neq Q_\alpha$ we either have that $\overline{L_\alpha} \cong (3 \times 3) : 2$ or $\mathrm{SU}_3(2)'$ and $p = 2$, or $p = 3$ and $\overline{L_\alpha} \cong (Q_8 \times Q_8) : 3$.

If $\overline{L_\alpha} \cong \mathrm{SU}_3(2)'$ or $(Q_8 \times Q_8) : 3$ then there is $t \in L_\alpha \cap G_{\alpha,\beta}$ non-trivial with $[t, L_\alpha] \leq Q_\alpha$ and, since $\overline{L_\beta} \cong \mathrm{SL}_2(p)$, we can choose t such that $[t, L_\beta] \leq Q_\beta$, a contradiction by Proposition 5.5 (v). Hence, $p = 2$, $\overline{L_\alpha} \cong (3 \times 3) : 2$ and as $V = [V, O^2(L_\alpha)]$, $|V| = 2^4$ and $|S| = 2^7$. By Proposition 2.47, there is $P_\alpha \leq L_\alpha$ such that $P_\alpha/Q_\alpha \cong \mathrm{Sym}(3)$, $L_\alpha = P_\alpha R_\alpha$ and we may choose P_α such that neither V_β nor C_β are normal in P_α. It follows that no subgroup of S is normal in both P_α and L_β so that (P_α, L_β, S) satisfies Hypothesis 5.1. Since we could have chosen G minimally as a counterexample, and as $|S| = 2^7$, we deduce that (P_α, L_β, S) is parabolic isomorphic to $\mathrm{Aut}(\mathrm{M}_{12})$. But then one can calculate, e.g. using MAGMA [14], that $|\mathrm{Aut}(Q_\alpha)|_3 = 3$, a contradiction. Hence, we have that

(7) V is irreducible, x is not quadratic on V and $p = 3$.

But then $[V_\alpha^{(2)}, x] \not\leq C_x$ and we deduce that $V_\alpha^{(2)} \cap Q_\beta = [V_\alpha^{(2)}, x]C_x$. Then, V is a nearly quadratic module for $\overline{L_\alpha}$ and an appeal to Theorem 2.58, along with a justification as in the proof of Lemma 7.49, yields a contradiction. This completes the proof. □

7.3 $b = 1$

From this point on, restating parts of Lemma 7.1, we may assume the following:

- $b = 1$ so that $Z_\alpha \not\leq Q_\beta$;
- $\Omega(Z(S)) = Z_\beta = \Omega(Z(L_\beta))$; and
- $Z(L_\alpha) = \{1\}$.

Proposition 7.52 *Suppose that p is odd. Then $\overline{L_\beta} \cong \mathrm{SL}_2(p^n)$ or $(\mathrm{P})\mathrm{SU}_3(p^n)$, or $|S/Q_\beta| = 3$.*

Proof Since $Q_\beta/\Phi(Q_\beta)$ is a faithful $\overline{L_\beta}$ module and $[Q_\beta, Z_\alpha, Z_\alpha] = \{1\}$, Lemma 2.35 implies that either the result holds; or that $L_\beta/C_{L_\beta}(U/V) \cong 4 \circ 2_-^{1+4}.\mathrm{Alt}(6)$ for some L_β-chief factor U/V with $U < V \leq Q_\beta$. In the latter case, we have that $|Z_\alpha Q_\beta/Q_\beta| = 3$, impossible since $G_{\alpha,\beta}$ normalizes Z_α and acts irreducibly on S/Q_β. Hence, the result. □

Proposition 7.53 *Suppose that $p \geqslant 5$. Then G has a weak BN-pair of rank 2 and is locally isomorphic to H where $F^*(H) = \mathrm{PSp}_4(p^n)$, $\mathrm{PSU}_4(p^n)$ or $\mathrm{PSU}_5(p^n)$.*

Proof Let K_β be a critical subgroup of Q_β. By Theorem 2.59, $O^p(L_\beta)$ acts faithfully on $K_\beta/\Phi(K_\beta)$. Assume that $K_\beta \leq Q_\alpha$. Since $\overline{L_\beta} \cong \mathrm{SL}_2(p^n)$ or $(\mathrm{P})\mathrm{SU}_3(p^n)$, we have that $[K_\beta, O^p(L_\beta)] \leq [K_\beta, \langle Z_\alpha^{L_\beta}\rangle] = \{1\}$, a contradiction. Hence, $K_\beta \not\leq Q_\alpha$, $[Q_\alpha, K_\beta, K_\beta, K_\beta] = \{1\}$ and K_β acts cubically on Q_α.

Since $Q_\alpha/\Phi(Q_\alpha)$ is a faithful $\overline{L_\alpha}$-module which admits cubic action, we may apply Corollary 2.60 so that $\overline{L_\alpha} \cong (\mathrm{P})\mathrm{SL}_2(p^r)$ or $(\mathrm{P})\mathrm{SU}_3(p^r)$; or $p = 5$, $\overline{L_\alpha} \cong 3\cdot\mathrm{Alt}(6)$ or $3\cdot\mathrm{Alt}(7)$, and for W some composition factor of $Q_\alpha/\Phi(Q_\alpha)$, $|W| \geqslant 5^6$. If $\overline{L_\alpha} \cong (\mathrm{P})\mathrm{SL}_2(p^r)$ or $(\mathrm{P})\mathrm{SU}_3(p^r)$ then G has a weak BN-pair of rank 2 and by Delgado and Stellmacher [27], G is locally isomorphic to H where $F^*(H) = \mathrm{PSp}_4(p^{n+1})$, $\mathrm{PSU}_4(p^n)$ or $\mathrm{PSU}_5(p^n)$ for $n \geqslant 1$. Thus it remains to check that $\overline{L_\alpha} \not\cong 3\cdot\mathrm{Alt}(6)$ or $3\cdot\mathrm{Alt}(7)$ and so assume that $p = 5$ and $|S/Q_\alpha| = 5$. Since Q_β is not centralized by Z_α, $\overline{L_\beta} \cong \mathrm{SL}_2(5)$ and Q_β contains exactly one non-central chief factor for L_β, which is isomorphic to a natural $\mathrm{SL}_2(5)$-module. Since $Z(L_\alpha) = \{1\}$, Z_α contains a non-central chief factor for L_α. Moreover, Z_α also admits cubic action, and so L_α/R_α is also determined by Corollary 2.60. Hence, $R_\alpha = Q_\alpha$, Z_α is a faithful $\overline{L_\alpha}$-module and $|Z_\alpha| \geqslant 5^6$.

Suppose that $Z_\alpha \cap Q_\beta \leq Q_\lambda$ for all $\lambda \in \Delta(\beta)$. Since $L_\beta = \langle Z_\lambda, Q_\beta : \lambda \in \Delta(\beta)\rangle$, it follows that $Z_\alpha \cap Q_\beta$ is centralized by $O^p(L_\beta)$. Since $Q_\alpha \cap Q_\beta \ntrianglelefteq L_\beta$, $O^p(L_\beta) \cap Q_\beta \not\leq Q_\alpha$ and so $[Z_\alpha, Q_\beta, Q_\beta \cap O^p(L_\beta)] = \{1\}$ and Z_α is a quadratic module, a contradiction to Lemma 2.35. Thus, $Z_\alpha \cap Q_\beta \not\leq Q_{\alpha+2}$ for some $\alpha+2 \in \Delta(\beta)$ and $Z_\alpha \cap Q_\beta \cap Q_{\alpha+2}$ has index at most 25 in Z_α. If $Z_{\alpha+2} \cap Q_\beta \leq Q_\alpha$ then $[Z_{\alpha+2}, Z_\alpha, Z_\alpha] = \{1\}$ and so, $Z_\alpha \cap Q_\beta$ acts quadratically on $Z_{\alpha+2}$ and since $\alpha+2$ is conjugate to α, we have a contradiction. Thus, $Z_{\alpha+2} \cap Q_\beta \not\leq Q_\alpha$. But now, $\overline{L_\alpha}$ is generated by two conjugates of $(Z_{\alpha+2} \cap Q_\beta)Q_\alpha/Q_\alpha$, and as an index 25 subgroup of Z_α is centralized by $Z_{\alpha+2} \cap Q_\beta$ and $Z(L_\alpha) = \{1\}$, we have that $|Z_\alpha| \leqslant 5^4$, a contradiction. □

Given the above proposition, we suppose that $p \in \{2, 3\}$ for the remainder of this section.

Notation 7.54 We introduce some notation specific to the case where $b = 1$.

- F_β is a normal subgroup of G_β which satisfies $[F_\beta, O^p(L_\beta)] \neq \{1\}$ and is minimal by inclusion with respect to adhering to these conditions.
- $W_\beta := \langle (Z_\alpha \cap Q_\beta)^{G_\beta} \rangle$.
- $D_\beta := C_{Q_\beta}(O^p(L_\beta))$.

Lemma 7.55 *The following hold:*

(i) $F_\beta \not\leq Q_\alpha$;
(ii) $F_\beta = [F_\beta, O^p(L_\beta)] \leq O^p(L_\beta)$; *and*
(iii) *for any p-subgroup $U \trianglelefteq L_\alpha$ with $U \not\leq Q_\beta$, we have $[F_\beta, Q_\beta] \leq U$.*

Proof We have that $[F_\beta, O^p(L_\beta)] \leq O^p(L_\beta)$ and by coprime action $[F_\beta, O^p(L_\beta), O^p(L_\beta)] = [F_\beta, O^p(L_\beta)]$. By minimality of F_β, $F_\beta = [F_\beta, O^p(L_\beta)]$ and (ii) holds. If $F_\beta \leq Q_\alpha$, then $[F_\beta, S]$ is strictly contained in

F_β and normalized by $G_\beta = \langle Z_\alpha^{L_\beta}\rangle G_{\alpha,\beta}$ and, by minimality, $[F_\beta, S] \leq D_\beta$. But then $[F_\beta, L_\beta] \leq D_\beta$, a contradiction. Hence, $F_\beta \not\leq Q_\alpha$ and (i) holds.

Let $H_\beta := \langle (U \cap F_\beta)^{G_\beta}\rangle \trianglelefteq G_\beta$ for $U \trianglelefteq L_\alpha$ with $U \not\leq Q_\beta$. By minimality of F_β, either $H_\beta = F_\beta$ or $H_\beta \leq D_\beta$. Suppose the latter. Then $[F_\beta, U] \leq F_\beta \cap U \leq H_\beta \leq D_\beta$ so that $[F_\beta, \langle U^{G_\beta}\rangle] \leq D_\beta$. Now, $F_\beta = [F_\beta, O^p(L_\beta)] \leq [F_\beta, \langle U^{G_\beta}\rangle G_{\alpha,\beta}] \leq D_\beta[F_\beta, G_{\alpha,\beta}]$. Then, by minimality of F_β, $F_\beta/F_\beta \cap D_\beta$ is an irreducible $G_{\alpha,\beta}$-module so that $[S, F_\beta] \leq D_\beta$. As above, this implies that $[F_\beta, L_\beta] \leq D_\beta$, a contradiction. Thus, $H_\beta = F_\beta$. Now, $[U \cap F_\beta, Q_\beta] \leq [F_\beta, Q_\beta] \leq D_\beta$ and so $[U \cap F_\beta, Q_\beta] \trianglelefteq G_\beta$. But then $[U \cap F_\beta, Q_\beta] = [U \cap F_\beta, Q_\beta]^{G_\beta} = [\langle (U \cap F_\beta)^{G_\beta}\rangle, Q_\beta] = [F_\beta, Q_\beta]$ and $U \geq [U \cap F_\beta, Q_\beta] = [F_\beta, Q_\beta]$. Then (iii) holds, completing the proof. □

Lemma 7.56 *Suppose that $m_p(S/Q_\alpha) = 1$ and $p \in \{2, 3\}$. Then $p = 3$, $\overline{L_\beta} \cong \mathrm{SL}_2(3)$, Z_α is an irreducible 2F-module for $\overline{L_\alpha}$ and Q_α is elementary abelian.*

Proof Assume that $m_p(S/Q_\alpha) = 1$. Since W_β is generated by elements of order p and $m_p(S/Q_\alpha) = 1$, $|W_\beta Q_\alpha/Q_\alpha| = p$ and Z_α centralizes an index p subgroup of W_β. Since $[Z_\alpha, Q_\beta] \leq W_\beta$, W_β contains all non-central chief factors for L_β in Q_β and so, $W_\beta/C_{W_\beta}(O^p(L_\beta))$ is the unique non-central chief factor for L_β inside Q_β. Moreover, $W_\beta/C_{W_\beta}(O^p(L_\beta))$ is a natural $\mathrm{SL}_2(p)$-module for $\overline{L_\beta} \cong \mathrm{SL}_2(p)$ and $L_\beta = \langle Q_\alpha, Q_\beta, Z_{\alpha+2}\rangle$ for some $\alpha + 2 \in \Delta(\beta)$. Then $Z_\alpha \cap Q_\beta \leq (Z_\alpha \cap W_\beta)(Z_{\alpha+2} \cap W_\beta) \trianglelefteq L_\beta$ and so $W_\beta = (Z_\alpha \cap W_\beta)(Z_{\alpha+2} \cap W_\beta)$.

Suppose first that W_β is abelian. Then, as $Z_\alpha \cap Q_\beta \leq W_\beta$, an index p subgroup of Z_α is centralized by W_β and as $Z(L_\alpha)$, applying Lemma 2.41 we have deduce Z_α is a natural $\mathrm{SL}_2(p)$-module. But then $Z_\alpha \cap Q_\beta = Z_\beta$ and $W_\beta = Z_\beta$, a contradiction. Thus, W_β is non-abelian.

Since W_β is non-abelian and $W_\beta \cap Q_\alpha \cap Q_{\alpha+2}$ has index p^2 in W_β, $W_\beta \cap Q_\alpha \cap Q_{\alpha+2} = Z(W_\beta) = W_\beta \cap D_\beta$. Note that $Z_\alpha \cap Z(W_\beta)$ has index p in $Z_\alpha \cap W_\beta$ and since $O^p(L_\beta)$ acts trivially on $Z(W_\beta)$, we see that $Z_\alpha \cap Z(W_\beta) = Z_{\alpha+2} \cap Z(W_\beta)$. Since $W_\beta = (Z_\alpha \cap W_\beta)(Z_{\alpha+2} \cap W_\beta)$, it follows that $Z_\alpha \cap Z(W_\beta)$ has index p^2 in W_β and so $Z(W_\beta) \leq Z_\alpha$ and $Z(W_\beta)$ has exponent p. Notice that every element of W_β lies in $(Z_\lambda \cap W_\beta)\Omega(Z(W_\beta))$ for some $\lambda \in \Delta(\beta)$, and that $(Z_\lambda \cap W_\beta)\Omega(Z(W_\beta))$ is of exponent p, from which it follows that W_β is of exponent p. In particular, since W_β is not elementary abelian, $p \neq 2$. Therefore, $\Omega(Z(W_\beta))$ has index 9 in Z_α, Z_α is 2F-module and since $[Z_\alpha, W_\beta] \not\leq \Omega(Z(W_\beta))$ and S/Q_α has a unique element of order 3, Z_α does not admit quadratic action by any element $x \in S \setminus Q_\alpha$.

Now, by minimality of F_β, $\Phi(F_\beta) \leq Q_\alpha$ so that $F_\beta(Q_\alpha \cap Q_\beta) = W_\beta(Q_\alpha \cap Q_\beta)$ since S/Q_α has a unique subgroup of order p. Then $[F_\beta, Z_\alpha] = [W_\beta, Z_\alpha]$. Moreover, $F_\beta = [F_\beta, O^p(L_\beta)] \leq [F_\beta, Z_\alpha]^{L_\beta} \leq W_\beta$ and since F_β contains a non-central chief factor for L_β, $W_\beta = F_\beta Z(W_\beta)$. Then, since $[F_\beta, Q_\alpha] = [F_\beta, Z_\alpha(Q_\alpha \cap Q_\beta)] \leq Z_\alpha$ by Lemma 7.55, it follows that $O^3(L_\alpha)$ centralizes Q_α/Z_α. In particular, every $3'$-element of L_α acts non-trivially on Z_α.

Let $U < Z_\alpha$ be a non-trivial subgroup of Z_α which is normal in L_α. If $C_S(U) \not\leq Q_\alpha$, then $O^3(L_\alpha)$ centralizes U and as $U \trianglelefteq S$, $U \cap Z_\beta \neq \{1\}$ and $Z(L_\alpha) \neq \{1\}$, a contradiction. If $U \not\leq Q_\beta$, then $Z_\alpha = U(Z_\alpha \cap Q_\beta)$ and by Lemma 7.55, it follows that $[F_\beta, Z_\alpha] \le U$ so that $[O^3(L_\alpha), Z_\alpha] \le U$ and $C_{Z_\alpha}(O^3(L_\alpha)) \neq \{1\}$ by Lemma 2.30. But then $Z(L_\alpha) \ge Z_\beta \cap C_{Z_\alpha}(O^3(L_\alpha)) \neq \{1\}$, a contradiction. Thus, $U \le Q_\beta$ and as Z_α is 2F, we may assume that both Z_α/U and U are FF-modules for $\overline{L_\alpha}$ and by Proposition 2.47 (ii), either $\overline{L_\alpha} \cong \mathrm{SL}_2(3)$ or $(Q_8 \times Q_8) : 3$. If $\overline{L_\alpha} \cong \mathrm{SL}_2(3)$, then G has a weak BN-pair of rank 2 and by [27], we have a contradiction. If $\overline{L_\alpha} \cong (Q_8 \times Q_8) : 3$, since $|\mathrm{Out}(\overline{L_\beta})| = 2$ and a Hall $3'$-subgroup of $L_\alpha \cap G_{\alpha,\beta}$ is isomorphic to an elementary abelian group of order 4, it follows that there is an involution $t \in G_{\alpha,\beta}$ such that $[L_\alpha, t] \le Q_\alpha$ and $[L_\beta, t] \le Q_\beta$, a contradiction by Proposition 5.5 (v).

Thus, we have that Z_α is an irreducible 2F-module. Since Z_α is irreducible and $Z_\alpha \not\le \Phi(Q_\alpha)$, $Z_\alpha \cap \Phi(Q_\alpha) = Z_\beta \cap \Phi(Q_\alpha) = \{1\}$ so that $\Phi(Q_\alpha) = \{1\}$ and Q_α is elementary abelian. □

Proposition 7.57 *Suppose that $m_p(S/Q_\alpha) = 1$ and $p \in \{2, 3\}$. Then $p = 3$, $Z_\alpha = Q_\alpha$ is an irreducible 2F-module for $\overline{L_\alpha}$ and one of the following holds:*

(i) *G has a weak BN-pair of rank 2 and G is locally isomorphic to H where $F^*(H) \cong \mathrm{PSp}_4(3)$;*
(ii) *$|S| = 3^5$, $\overline{L_\alpha} \cong \mathrm{Alt}(5)$, Z_α is the non-trivial composition factor of the* GF(3)-*permutation module, $\overline{L_\beta} \cong \mathrm{SL}_2(3)$ and $Q_\beta \cong 3 \times 3^{1+2}_+$;*
(iii) *$|S| = 3^5$, $\overline{L_\alpha} \cong O^{3'}(2 \wr \mathrm{Sym}(4))$, Z_α is a reflection module, $\overline{L_\beta} \cong \mathrm{SL}_2(3)$ and $Q_\beta \cong 3 \times 3^{1+2}_+$; or*
(iv) *$|S| = 3^6$, $\overline{L_\alpha} \cong O^{3'}(2 \wr \mathrm{Sym}(5))$, Z_α is a reflection module, $\overline{L_\beta} \cong \mathrm{SL}_2(3)$ and $Q_\beta \cong 3 \times 3 \times 3^{1+2}_+$.*

Proof By Lemma 7.56, Z_α is the unique non-central chief factor for L_α in Q_α and Q_α is elementary abelian. Moreover, $W_\beta/\Omega(Z(W_\beta))$ is the unique non-central chief factor for L_β inside Q_β, and is a natural $\mathrm{SL}_2(3)$-module for $\overline{L_\beta} \cong \mathrm{SL}_2(3)$.

Suppose first that $|Z_\alpha| = 3^3$. Then $\overline{L_\alpha}$ is isomorphic to a subgroup X of $\mathrm{GL}_3(3)$ which has a strongly 3-embedded subgroup. One can check that the only groups which satisfy $X = O^{3'}(X)$ are $\mathrm{PSL}_2(3)$, $\mathrm{SL}_2(3)$ and $13 : 3$. In the first two cases, G has a weak BN-pair of rank 2 and comparing with [27], we have that $\overline{L_\alpha} \cong \mathrm{PSL}_2(3)$ and G is locally isomorphic to H, where $F^*(H) \cong \mathrm{PSp}_4(3)$, and (i) holds. Suppose that $\overline{L_\alpha} \cong 13 : 3$ and let $t_\beta \in L_\beta \cap G_{\alpha,\beta}$ be an involution. Then $t_\beta \in G_\alpha$ and writing $\overline{t_\beta} := t_\beta Q_\alpha/Q_\alpha$, $\overline{t_\beta}$ acts on $\overline{L_\alpha}$ and inverts $\overline{S} = Q_\beta Q_\alpha/Q_\alpha$, a contradiction since any involutary automorphism of $13 : 3$ centralizes a Sylow 3-subgroup. Thus, for the remainder of the proof, we may assume that

(1) $|Z_\alpha| > 3^3$.

Let $t_\beta \le G_{\alpha,\beta} \cap L_\beta$ be an involution. Then, using coprime action, $[t_\beta, Q_\alpha] \le W_\beta$ and $[t_\beta, C_{W_\beta}(O^3(L_\beta))] = \{1\}$. In particular, it follows that t_β centralizes an index 3 subgroup of Q_α. Let $L^* := \langle t_\beta^{G_\alpha} \rangle$ and $\overline{L^*} = L^* Q_\alpha/Q_\alpha \le \overline{G_\alpha}$. Since $\overline{L^*} \trianglelefteq \overline{G_\alpha}$, we have that $[\overline{L^*}, \overline{L_\alpha}] \le \overline{L^*}$. Note that $\overline{t_\beta}$ inverts $W_\beta Q_\alpha/Q_\alpha \cong W_\beta/W_\beta \cap Q_\alpha$ and so $W_\beta Q_\alpha/Q_\alpha = [W_\beta Q_\alpha/Q_\alpha, \overline{t_\beta}] \le [\overline{L_\alpha}, \overline{L^*}] \le \overline{L^*}$.

If $\overline{G_\alpha}$ is not 3-solvable, then $\overline{L_\alpha}/O_{3'}(\overline{L_\alpha})$ is a non-abelian finite simple group and since $\overline{L^*} \trianglelefteq \overline{G_\alpha}$, we have that $\overline{L_\alpha} \leq \overline{L^*}$. If $\overline{G_\alpha}$ is 3-solvable, let O_α be the preimage of $O_{3'}(\overline{L_\alpha})$ in L_α. By coprime action, we have that $Q_\alpha = [Q_\alpha, O_\alpha] \times C_{Q_\alpha}(O_\alpha)$ is an S-invariant decomposition. Since Z_α is irreducible, we infer that $[Q_\alpha, O_\alpha] = [Z_\alpha, O_\alpha] = Z_\alpha$ and as $Z_\beta \leq Z_\alpha$, it follows that $C_{Q_\alpha}(O_\alpha) = \{1\}$ and $Q_\alpha = Z_\alpha$. If $|S/Q_\alpha| > 3$, then $W_\beta \leq \Phi(Q_\beta)(Z_\alpha \cap Q_\beta)$ and it follows from the Dedekind modular law that $W_\beta = \Phi(Q_\beta)(Z_\alpha \cap Q_\beta) \cap W_\beta = (Z_\alpha \cap Q_\beta)(\Phi(Q_\beta) \cap W_\beta)$. Since W_β contains all non-central chief factors for L_β inside Q_β, $\Phi(Q_\beta) \cap W_\beta \leq Z(W_\beta)$ so that $W_\beta = (Z_\alpha \cap Q_\beta)Z(W_\beta)$, a contradiction. Thus, $|S/Q_\alpha| = 3$ and, again, $\overline{L_\alpha} \leq \overline{L^*}$.

Since $\Omega(S/Q_\alpha)$ does not act quadratically on Z_α, $\overline{L^*}$ is not generated by transvections and as $|Z_\alpha| \geqslant 3^4$, we may apply the main result of [78]. Using that S/Q_α is cyclic, we have that

(2) $\overline{L^*}$ is isomorphic to the reduction modulo 3 of a finite irreducible reflection group of degree n in characteristic 0, and $3^4 \leqslant |Z_\alpha| \leqslant 3^5$.

Suppose that there is $t_\alpha \in L^* \cap G_{\alpha,\beta}$ an element of order 4 with $t_\alpha^2 Q_\alpha \in Z(\overline{L^*})$. Then $t_\alpha \in G_\beta$ and t_α acts on $\overline{L_\beta}$. We may assume that t_α^2 acts non-trivially on $\overline{L_\beta}$ for otherwise $t_\alpha^2 Q_\alpha$ is centralized by $\overline{L_\alpha}$ and $t_\alpha^2 Q_\beta$ is centralized by $\overline{L_\beta}$, a contradiction by Proposition 5.5 (v). But t_α normalizes S/Q_β and so either t_α inverts S/Q_β or centralizes S/Q_β. In either case, t_α^2 centralizes S/Q_β and by Proposition 2.16 (viii), t_α^2 acts trivially on $\overline{L_\beta}$, a contradiction.

Upon comparing the groups listed in [78] and the orders of $\mathrm{GL}_4(3)$ and $\mathrm{GL}_5(3)$ we have that $|S/Q_\alpha| = 3$ and

(3) $\overline{L^*} \cong G(1,1,5)$, $G(2,1,4)$, $G(2,2,4)$, $G(2,1,5)$ or $G(2,2,5)$ (in the Todd-Shepherd enumeration convention).

If $\overline{L^*} \cong G(1,1,5) \cong \mathrm{Sym}(5)$, then $\overline{L_\alpha} \cong \mathrm{Alt}(5)$. Then G is determined in [44] and outcome (ii) follows in this case. Thus, $O_2(\overline{L^*}) \neq \{1\}$ and writing O_α for the preimage of $O_2(\overline{L^*})$ in G_α, we have by coprime action that $Q_\alpha = [Q_\alpha, O_\alpha] \times C_{Q_\alpha}(O_\alpha)$. Since Z_α is irreducible and is the unique non-central chief factor within Q_α, $Q_\alpha = [Q_\alpha, O_\alpha] = Z_\alpha$. In particular, $W_\beta = Q_\beta$, $|Q_\beta| \leqslant 3^5$ and Q_β/Z_β is a natural $\mathrm{SL}_2(3)$-module for $\overline{L_\beta}$.

Now, $G(2,1,4) \cong 2 \wr \mathrm{Sym}(4)$ and $G(2,2,4)$ is isomorphic to an index 2 subgroup of $G(2,1,4)$. Therefore, if $|Z_\alpha| = 3^4$, $\overline{L_\alpha} \cong O^{3'}(2 \wr \mathrm{Sym}(4))$ and the action of $\overline{L_\alpha}$ is determined uniquely up to conjugacy in $\mathrm{GL}_4(3)$. We observe that W_β is exponent 3, and so we can construct a complement to Z_α is S, so the S splits over Z_α. By Gaschütz' theorem [49, (3.3.2)], L_α splits over Z_α. Indeed, it follows in this case that S is isomorphic to a Sylow 3-subgroup of $\mathrm{Alt}(12)$. Furthermore, Q_β has exponent 3 and is of order 3^4, and $Z(Q_\beta) = Z_\beta$ is elementary abelian of order 9. Thus, $Q_\beta \cong 3_+^{1+2} \times 3$.

Finally, $G(2,1,5) \cong 2 \wr \mathrm{Sym}(5)$ and $G(2,2,5)$ is isomorphic to an index 2 subgroup of $G(2,1,5)$. Therefore, if $|Z_\alpha| = 3^5$, then $\overline{L_\alpha} \cong O^{3'}(2 \wr \mathrm{Sym}(5))$ and the action of $\overline{L_\alpha}$ is determined uniquely up to conjugacy in $\mathrm{GL}_5(3)$. As in the previous case, we may construct a complment to Z_α in S and then Gaschütz' theorem yields that L_α splits over Z_α. Indeed, it follows in this case that S is isomorphic to a Sylow 3-subgroup of Alt(15). Furthermore, Q_β has exponent 3 and is of order 3^5, and $Z(Q_\beta) = Z_\beta$ is elementary abelian of order 27. Indeed, $Q_\beta \cong 3_+^{1+2} \times 3 \times 3$. □

Remark Note that the modules occurring in the above proposition are some of the nearly quadratic modules described in Theorem 2.58. Indeed, one could prove the above proposition using this characterization along with some knowledge of the quasisimple groups which have a Sylow 3-subgroup of order 3.

Lemma 7.58 *Assume that* $m_p(S/Q_\alpha) \geqslant 2$. *Then there is* $\alpha + 2 \in \Delta(\beta)$ *such that* $Z_\alpha \cap Q_\beta \not\leq Q_{\alpha+2}$ *and* $Z_{\alpha+2} \cap Q_\beta \not\leq Q_\alpha$.

Proof Suppose that $Z_\alpha \cap Q_\beta \leq Q_\lambda$ for all $\lambda \in \Delta(\beta)$. Then $Z_\alpha \cap Q_\beta$ is centralized by $\langle Z_\alpha^{G_\beta}\rangle$ and so normalized by $G_\beta = \langle Z_\alpha^{G_\beta}\rangle G_{\alpha,\beta}$. But then, Z_α centralizes $Q_\beta/Z_\alpha \cap Q_\beta$ and $Z_\alpha \cap Q_\beta$, and we have a contradiction by coprime action arguments. Thus, there is some $\lambda \in \Delta(\beta)$ with $Z_\alpha \cap Q_\beta \not\leq Q_\lambda$.

Assume first that $q_\beta = p$ and choose $\alpha + 2 \in \Delta(\beta)$ such that $Z_\alpha \cap Q_\beta \not\leq Q_{\alpha+2}$. Aiming for a contradiction, we suppose that $Z_{\alpha+2} \cap Q_\alpha \leq Q_\alpha$. Then $Z_{\alpha+2} \cap Q_\beta$ has index p in $Z_{\alpha+2}$ is centralized by $Z_\alpha \cap Q_\beta$. Thus, Lemma 2.41 implies that $L_{\alpha+2}/R_{\alpha+2} \cong \mathrm{SL}_2(p)$. By conjugacy, and since $Z(L_\alpha) = \{1\}$, we have that Z_α is a natural $\mathrm{SL}_2(p)$-module for L_α/R_α so that $Z_\alpha \cap Q_\beta = Z_\beta \leq Q_{\alpha+2}$, a contradiction.

Hence, we may assume that $q_\beta \geqslant p^2$ and by Proposition 7.52 when $p = 3$ and Proposition 2.15 when $p = 2$, we have that $\overline{L_\beta}/O_{p'}(\overline{L_\beta})$ is isomorphic to a rank 1 group of Lie type. If $p = 3$, then $\overline{L_\beta}$ is isomorphic to a rank 1 group of Lie type, and so $\overline{L_\beta}$ is 2-transitive on neighbors of β. Hence, since $Z_\alpha \cap Q_\alpha \not\leq Q_\lambda$ for some $\lambda \in \Delta(\beta)$, we must also have that $Z_\lambda \cap Q_\beta \not\leq Q_\alpha$, and we label $\lambda = \alpha + 2$ to complete the proof.

If $p = 2$ then, since $\overline{L_\beta}$ is generated by involutions, there is $T \in \mathrm{Syl}_2(L_\beta)$ and $t \in T$ such that

$$(Z_\alpha \cap Q_\beta)^t = Z_\alpha^t \cap Q_\beta = Z_{\alpha\cdot t} \cap Q_\beta \not\leq Q_\alpha$$

and $t^2 \in Q_\beta \leq G_{\alpha,\beta}$. Since $t^2 \in Q_\beta$, we have that $(Z_\alpha \cap Q_\beta)^{t^2} = Z_\alpha \cap Q_\beta \not\leq Q_\alpha^t = Q_{\alpha\cdot t}$. Setting $\alpha + 2 = \alpha \cdot t$ we have $Z_{\alpha+2} \cap Q_\beta \not\leq Q_\alpha$ and $Z_\alpha \cap Q_\beta = (Z_{\alpha+2} \cap Q_\beta)^t \not\leq Q_\alpha^t = Q_{\alpha+2}$, hence the result. □

We write $r_\alpha := |(Z_\alpha \cap Q_\beta)Q_{\alpha+2}/Q_{\alpha+2}|$ and $r_{\alpha+2} := |(Z_{\alpha+2} \cap Q_\beta)Q_\alpha/Q_\alpha|$. Since both (α, β) and $(\alpha + 2, \beta)$ are critical pairs, without loss of generality, we assume throughout the remainder of this section that $r_{\alpha+2} \geqslant r_\alpha$.

Lemma 7.59 *Assume that* $m_p(S/Q_\alpha) \geqslant 2$ *and write* $\widetilde{L_\alpha} := L_\alpha/R_\alpha$. *Then either*

(i) $r_{\alpha+2} = r_\alpha = p$;
(ii) $O_{p'}(\widetilde{L_\alpha})$ *is central in* $\widetilde{L_\alpha}$; *or*
(iii) *there is a subgroup* C *of* $\widetilde{L_\alpha}$ *of order* p^2 *such that* $C_{Z_\alpha}(C)$ *has index at most* $q_\beta p^2$ *in* Z_α.

Proof Assume throughout that $r_{\alpha+2} > p$ with the aim of demonstrating that we are in case (ii) or (iii). Let $A = (Z_{\alpha+2} \cap Q_\beta)Q_\alpha/Q_\alpha$ so that A is elementary abelian of order $r_{\alpha+2} \geqslant r_\alpha$ and $C_{Z_\alpha}(A)$ has at most index $q_\beta r_\alpha$ in Z_α. Since $r_{\alpha+2} > p$, using Proposition 2.8, we have that $O_{p'}(\widetilde{L_\alpha}) = \langle C_{O_{p'}(\widetilde{L_\alpha})}(B) : [A : B] = p\rangle$. Assume that $C_{Z_\alpha}(A) = C_{Z_\alpha}(B)$ for all $B \leq A$ with $|A/B| = p$. Then $C_{Z_\alpha}(A)$ is normalized by $\widetilde{A}O_{p'}(\widetilde{L_\alpha})$. Set $H_\alpha = \langle \widetilde{A}^{O_{p'}(\widetilde{L_\alpha})}\rangle$ so that $[H_\alpha, C_{Z_\alpha}(\widetilde{A})] = \{1\}$. Then, by coprime action, we get that $Z_\alpha = [Z_\alpha, O_{p'}(H_\alpha)] \times C_{Z_\alpha}(O_{p'}(H_\alpha))$ is an A-invariant decomposition and as $C_{Z_\alpha}(A) \leq C_{Z_\alpha}(O_{p'}(H_\alpha))$, it follows that $O_{p'}(H_\alpha)$ centralizes Z_α so that $O_{p'}(H_\alpha) = \{1\}$ and $H_\alpha = \widetilde{A} \trianglelefteq \widetilde{A}O_{p'}(\widetilde{L_\alpha})$. Then, $[\langle A^{\widetilde{L_\alpha}}\rangle, O_{p'}(\widetilde{L_\alpha})] = [A, O_{p'}(\widetilde{L_\alpha})] = \{1\}$ and we have that $O_{p'}(\widetilde{L_\alpha}) \leq Z(\widetilde{L_\alpha})$.

So assume now that $C_{Z_\alpha}(A) < C_{Z_\alpha}(B)$ for some $B \leq A$ with $|A/B| = p$. Applying the same line of reasoning as above to B in place of A, and continuing recursively, we either satisfy case (ii) or ultimately descend to a group $C \leq A$ of order p^2 such that $C_{Z_\alpha}(C)$ has index at most $q_\beta r_\alpha p^2/r_{\alpha+2} \leqslant q_\beta p^2$ in Z_α, which is (iii). □

Proposition 7.60 *Suppose that $m_p(S/Q_\alpha) \geqslant 2$. Then S/Q_α is elementary abelian and $S = Q_\alpha Q_\beta$.*

Proof Aiming for a contradiction, we assume that S/Q_α is not elementary abelian. Since $m_p(S/Q_\alpha) \geqslant 2$, by Proposition 2.15 and using that S/Q_α is not elementary abelian, we have that $\overline{L_\alpha}/O_{p'}(\overline{L_\alpha})$ is isomorphic to a simple group of Lie type. We analyze the cases below.

Case (1): $\overline{L_\alpha}/O_{2'}(\overline{L_\alpha}) \cong \mathrm{Sz}(2^n)$.

Assume that $q_\beta > q_\alpha$ so that $F_\beta \cap Q_\alpha$ has index strictly less than q_β in F_β and is centralized by Z_α. By Lemma 2.41, we have a contradiction. Thus, if $\overline{L_\alpha}/O_{2'}(\overline{L_\alpha}) \cong \mathrm{Sz}(2^n)$ then we have that $q_\beta \leqslant q_\alpha$.

Applying Proposition 2.31, we see that in $q_\alpha^{\frac{4}{3}} \leqslant 2q_\beta \leqslant 2q_\alpha$ in outcome (i) of Lemma 7.59 and $q_\alpha^2 \leqslant 2^2 q_\beta \leqslant 2^2 q_\alpha$ in outcome (iii) of Lemma 7.59. Since $q_\alpha \geqslant 8$, outcome (iii) of Lemma 7.59 provides a contradiction. Suppose that outcome (i) holds so that $r_{\alpha+2} = r_\alpha = 2$. Then $q_\alpha = q_\beta = 8$ so that, by Lemma 2.41, $Q_\beta/\Phi(Q_\beta)$ contains a unique non-central chief factor, which is a natural module for $\overline{L_\beta} \cong \mathrm{SL}_2(8)$. Hence, $Q_\beta = (Q_\beta \cap Q_\alpha)(Q_\beta \cap Q_{\alpha+2})\Phi(Q_\beta)$ so that $Q_\beta = (Q_\beta \cap Q_\alpha)(Q_\beta \cap Q_{\alpha+2})$. In particular, $(Q_\alpha \cap Q_\beta)Q_{\alpha+2} \in \mathrm{Syl}_2(L_{\alpha+2})$. Then $\Phi(Q_\alpha)Z_\alpha(Q_\alpha \cap Q_\beta \cap Q_{\alpha+2})$ has index q_α in Q_α and $\Phi(Q_\alpha)(\Phi(Q_\alpha)Z_\alpha \cap Q_\beta \cap Q_{\alpha+2})$ has index q_α in $\Phi(Q_\alpha)Z_\alpha$. Applying Proposition 2.31, we have that $O^p(L_\alpha)$ centralizes $Q_\alpha/\Phi(Q_\alpha)Z_\alpha$ and $\Phi(Q_\alpha)Z_\alpha/\Phi(Q_\alpha)$ so that $O^p(L_\alpha)$ centralizes $Q_\alpha/\Phi(Q_\alpha)$, a contradiction.

Thus, if $\overline{L_\alpha}/O_{2'}(\overline{L_\alpha}) \cong \mathrm{Sz}(2^n)$ then $q_\beta \leqslant q_\alpha$, outcome (ii) of Lemma 7.59 holds, and $L_\alpha/R_\alpha \cong \mathrm{Sz}(q_\alpha)$. Since $Z_\alpha \cap Q_\beta$ is $G_{\alpha,\beta}$-invariant of order q_β, we have that $q_\alpha = q_\beta$ by [34, Theorem 2.8.11]. Then applying Lemma 2.41 we conclude that $Q_\beta/\Phi(Q_\beta)$ contains a unique non-central chief factor, which is a natural module for $\overline{L_\beta} \cong \mathrm{SL}_2(q_\beta)$. Now, $Q_\alpha = Z_\alpha(Q_\alpha \cap Q_\beta)$ and by Lemma 7.55, we deduce that

$O^2(L_\alpha)$ centralizes Q_α/Z_α so that $R_\alpha = Q_\alpha$. But then $(G_\alpha, G_\beta, G_{\alpha,\beta})$ is a weak BN-pair of rank 2, a contradiction by [27].

Case (2): $\overline{L_\alpha}/O_{3'}(\overline{L_\alpha}) \cong \mathrm{Ree}(q_\alpha)$.

Assume first that $S \neq Q_\alpha Q_\beta$ so that $\overline{L_\beta} \cong \mathrm{SU}_3(q_\beta)$ and $Q_\alpha Q_\beta$ has index q_β^2 in S. Thus, either $q_\alpha = q_\beta$ or $q_\alpha = q_\beta^2$. In the former case, we have that $Q_\beta \cap Q_\alpha$ has index q_β in Q_β and is centralized by Z_α. Applying Lemma 2.41 gives a contradiction. In the latter case, we get that q_α is a square, another contradiction since q_α is an odd power of 3. Hence, $S = Q_\alpha Q_\beta$.

Now, by Lemma 7.55 we have that $[Q_\beta, F_\beta] \leq Z_\alpha$ and by the structure of S/Q_α, we deduce that $F_\beta Q_\alpha/Q_\alpha = Z(S/Q_\alpha)$ has order q_α. Then $F_\beta \cap Q_\alpha$ is an index q_α subgroup of F_β which is centralized by Z_α and we deduce that $q_\beta \leqslant q_\alpha$. If outcome (ii) of Lemma 7.59 does not hold then applying Proposition 2.32, we see that $q_\alpha^2 \leqslant 3^2 q_\beta \leqslant 3^2 q_\alpha$. Since q_α is an odd power of 3, this implies that $q_\alpha = q_\beta = 3$. However, we now form $H = \langle Z_{\alpha+2} \cap Q_\beta, (Z_{\alpha+2} \cap Q_\beta), Q_\alpha \rangle$ with $\overline{H} O_{3'}(\overline{L_\alpha})/O_{3'}(\overline{L_\alpha}) \cong \mathrm{PSL}_2(8)$ so that $|Z_\alpha/C_{Z_\alpha}(H)| \leqslant 3^6$, a contradiction by Proposition 2.32. Thus, outcome (ii) of Lemma 7.59 holds and $L_\alpha/R_\alpha \cong \mathrm{Ree}(q_\alpha)$.

Notice that if $\overline{L_\beta} \cong \mathrm{SL}_2(q_\beta)$ then applying Lemma 7.55 we have that $O^3(L_\alpha)$ centralizes Q_α/Z_α and $R_\alpha = Q_\alpha$. Then $(G_\alpha, G_\beta, G_{\alpha,\beta})$ is a weak BN-pair of rank 2 and [27] provides a contradiction. Thus, $\overline{L_\beta} \cong \mathrm{SU}_3(q_\beta)$. Then $\Phi(Q_\beta)(Q_\beta \cap Q_\alpha)$ has index q_α in Q_β and is centralized, modulo $\Phi(Q_\beta)$, by Z_α. Applying Proposition 2.17 and using that the minimum GF(3)-representation of $\mathrm{SU}_3(3^n)$ is $6n$, we deduce that $q_\alpha \geqslant q_\beta^2$. But then $\Phi(Q_\alpha)(Q_\beta \cap Q_\alpha)$ is an index q_β^2 subgroup of Q_α which is centralized, modulo $\Phi(Q_\alpha)$, by F_β by Lemma 7.55. Lemma 2.41 yields a contradiction in this case.

Case (3): $\overline{L_\alpha}/O_{p'}(\overline{L_\alpha}) \cong \mathrm{PSU}_3(q_\alpha)$.

We have that $F_\beta \cap Q_\alpha$ has index q_α in F_β and is centralized by Z_α. By Lemma 2.41, we deduce that $q_\beta \leqslant q_\alpha$. Suppose that outcome (ii) of Lemma 7.59 does not hold. Then applying Propositions 2.31 and 2.32, we see that $q_\alpha^{\frac{3}{2}} \leqslant p q_\beta \leqslant p q_\alpha$ in outcome (i) of Lemma 7.59 and $q_\alpha^2 \leqslant p^2 q_\beta \leqslant p^2 q_\alpha$ in outcome (iii) of Lemma 7.59. In either case, we deduce that $2 < q_\alpha = q_\beta \leqslant p^2$. Reapplying Lemma 2.41 to the action of Z_α on F_β we deduce that $L_\beta/C_{L_\beta}(F_\beta/F_\beta \cap D_\beta) \cong \mathrm{SL}_2(q_\beta)$, $F_\beta/F_\beta \cap D_\beta$ is natural $\mathrm{SL}_2(q_\beta)$-module, and $S = Q_\alpha Q_\beta$. Then, applying Lemma 7.55 we have that $O^p(L_\alpha)$ centralizes Q_α/Z_α and $R_\alpha = Q_\alpha$. Since $O_{p'}(\overline{L_\alpha}) \neq \{1\}$, using coprime action and that $Z_\beta \leq Z_\alpha$, we deduce that $Q_\alpha = Z_\alpha$. Moreover, $F_\beta \cap D_\beta \leq Z_\alpha$ and since $F_\beta/F_\beta \cap D_\beta$ is natural $\mathrm{SL}_2(q_\beta)$-module, $F_\beta = (F_\beta \cap Z_\alpha)(F_\beta \cap Z_{\alpha+2})$. Now, $W_\beta = (Z_\alpha \cap Q_\beta)(Z_{\alpha_2} \cap Q_\beta) \trianglelefteq L_\beta$ and since S/Q_α is non-abelian, it follows that $r_\alpha = r_{\alpha+2} = q_\alpha = q_\beta$. But now, since $S = Q_\alpha Q_\beta$, we have that $W_\beta = \Phi(Q_\beta)(W_\beta \Phi(Q_\beta) \cap Q_\alpha)$ and so Z_α centralizes Q_β/W_β and $W_\beta/\Phi(Q_\beta) \cap W_\beta$. By coprime action, we conclude that $O^p(L_\beta)$ centralizes $Q_\beta/\Phi(Q_\beta)$, an obvious contradiction.

Hence, we are in outcome (ii) of Lemma 7.59 and L_α/R_α is isomorphic to $\mathrm{SU}_3(q_\alpha)$ or $\mathrm{PSU}_3(q_\alpha)$. Our first aim will be to show that $q_\alpha = q_\beta$ so suppose for a contradiction that $q_\beta < q_\alpha$. In particular, we see that $q_\alpha \geqslant p^2$. Note that $Z_\alpha \cap Q_\beta \leq F_\beta$ so that $r_\alpha \leq q_\alpha$. Since $Z_\alpha \cap Q_{\alpha+2}$ has index $r_\alpha q_\beta$ in Z_α and $r_\alpha q_\beta \leqslant q_\alpha q_\beta \leqslant q_\alpha^2$. If $r_{\alpha+2} = p$ then $r_\alpha = p$ and by Propositions 2.31 and 2.32 we see that $pq_\beta \geqslant q_\alpha^{\frac{3}{2}} \geqslant pq_\alpha > pq_\beta$, a contradiction. Thus, $r_{\alpha+2} \geqslant p^2$ and by Propositions 2.31 and 2.32, we see that $q_\alpha q_\beta \geqslant r_\alpha q_\beta \geqslant q_\alpha^2 > q_\alpha q_\beta$, another contradiction. Hence, we must have that $q_\alpha = q_\beta$.

Now, applying the above reasoning, we see that either $q_\alpha = q_\beta = p^2$ and $r_{\alpha+2} = r_\alpha = p$, or $q_\alpha = q_\beta = r_\alpha = r_{\alpha+2} > 2$. In the latter case, using that $\overline{L_\alpha}$ is generated by three conjugates of $\overline{Z_{\alpha+2} \cap Q_\beta}$ and that $Z(L_\alpha) = \{1\}$, we have that $|Z_\alpha| = q^6$ and Z_α is a natural module for $L_\alpha/R_\alpha \cong \mathrm{SU}_3(q_\alpha)$, impossible since $Z_\alpha \cap Q_\beta$ is a $G_{\alpha,\beta}$-invariant subgroup of index q_α.

In the former case, we apply Lemma 2.41 to the action of Z_α on F_β to deduce that $L_\beta/C_{L_\beta}(F_\beta/F_\beta \cap D_\beta) \cong \mathrm{SL}_2(q_\beta)$ and $F_\beta/F_\beta \cap D_\beta$ is natural $\mathrm{SL}_2(q_\beta)$-module. Then $Q_\beta/\Phi(Q_\beta)$ is a faithful, quadratic 2F-module for $\overline{L_\beta}$ so that $\overline{L_\beta} \cong \mathrm{SL}_2(q_\beta)$ by Lemma 2.55. But then, applying Lemma 7.55, we have that $O^p(L_\alpha)$ centralizes Q_α/Z_α and $R_\alpha = Q_\alpha$. Hence, $(G_\alpha, G_\beta, G_{\alpha,\beta})$ is a weak BN-pair of rank 2 and a comparison with [27] provides a final contradiction. This completes the proof. □

Proposition 7.61 *Suppose that $m_p(S/Q_\alpha) \geqslant 2$, $m_p(S/Q_\beta) \geqslant 2$ and $p \in \{2, 3\}$. Then one of the following holds:*

(i) *G has a weak BN-pair of rank 2 and G is locally isomorphic to H where $F^*(H) \cong \mathrm{PSU}_4(p^{n+1})$, $\mathrm{PSU}_5(2^{n+1})$, $\mathrm{PSU}_5(3^n)$ or $\mathrm{PSp}_4(3^{n+1})$ for $n \geqslant 1$; or*
(ii) *$p = 3$, $|S| = 3^7$, $\overline{L_\alpha} \cong \mathrm{M}_{11}$, $Z_\alpha = Q_\alpha$ is the "code" module for $\overline{L_\alpha}$, $\overline{L_\beta} \cong \mathrm{SL}_2(9)$ and $Q_\beta \cong 3_+^{1+4}$.*

Proof We have that S/Q_α is elementary abelian by Proposition 7.60. Assume first that $q_\alpha > q_\beta$. Since $m_p(S/Q_\beta) > 1$, by the groups listed in Proposition 2.15 we deduce that either $q_\alpha > q_\beta \geqslant p^2$, or $\overline{L_\beta} \cong \mathrm{SU}_3(3)$. In the latter case, we have that $\Phi(Q_\alpha)(Q_\alpha \cap Q_\beta)$ is a subgroup of Q_α of index 9 which, by Lemma 7.55, is centralized, modulo $\Phi(Q_\alpha)$, by F_β. Since $m_p(S/Q_\alpha) \geqslant p^2$, by Lemma 2.41 we have that $\overline{L_\alpha} \cong \mathrm{SL}_2(9)$ and $(G_\alpha, G_\beta, G_{\alpha,\beta})$ is a weak BN-pair of rank 2. Applying [27], G is locally isomorphic to H where $F^*(H) \cong \mathrm{PSU}_5(3)$, and we are in (i).

Suppose now that $q_\alpha > q_\beta \geqslant p^2$. Assume first that $\overline{L_\beta}/O_{2'}(\overline{L_\beta}) \cong \mathrm{Sz}(q_\beta)$. Then $\Phi(Q_\alpha)(Q_\alpha \cap Q_\beta)$ has index q_β in Q_α and is centralized, modulo $\Phi(Q_\alpha)$, by F_β. Then Lemma 2.41 yields a contradiction. Assume now that $\overline{L_\beta}/O_{p'}(\overline{L_\beta}) \cong \mathrm{PSU}_3(q_\beta)$. Then applying Lemma 2.41 to $Q_\alpha/\Phi(Q_\alpha)$ in a similar manner as before, we conclude that $q_\beta^2 \geqslant q_\alpha$. Since $Q_\alpha \cap Q_\beta$ has index q_α in Q_β and is centralized by Z_α, applying Propositions 2.31 and 2.32 yields $q_\alpha = q_\beta^2$. Then Lemma 2.41 gives that $Q_\alpha/\Phi(Q_\alpha)$ contains a unique non-central chief factor for L_α which, as an $\overline{L_\alpha}$-module, is isomorphic to a natural module for $\overline{L_\alpha} \cong \mathrm{SL}_2(q_\alpha)$. Moreover, $\Phi(Q_\beta)(Q_\beta \cap Q_\alpha)$ has index q_β^2 in Q_β and is centralized, modulo $\Phi(Q_\beta)$, by Z_α and so applying Lemma 2.55, we have that $\overline{L_\beta} \cong \mathrm{SU}_3(q_\beta)$. Therefore, $(G_\alpha, G_\beta, G_{\alpha,\beta})$ is a weak BN-pair of rank 2 and comparing with [27], G is locally isomorphic to H where $F^*(H) \cong \mathrm{PSU}_5(p^n)$ where $p \in \{2, 3\}$ and $n > 1$.

Note that if $q_\alpha \leqslant q_\beta$ then since Z_α centralizes $Q_\alpha \cap Q_\beta$, Lemma 2.41 implies that $\overline{L_\beta} \cong \mathrm{SL}_2(q_\beta)$ and $q_\alpha = q_\beta$. Thus, in total, we may assume that

(1) $q_\beta \leqslant q_\alpha$ and $\overline{L_\beta}/O_{p'}(\overline{L_\beta}) \cong \mathrm{PSL}_2(q_\beta)$.

In particular, since $Q_\alpha = Z_\alpha(Q_\alpha \cap Q_\beta)$, we have by Lemma 7.55 that $O^p(L_\alpha)$ centralizes Q_α/Z_α and $R_\alpha = Q_\alpha$.

Now, if $\overline{L_\beta} \not\cong \mathrm{SL}_2(q_\beta)$ then comparing with Proposition 7.52 and Lemma 2.55, we infer that $p = 2$ and $q_\beta^2 < q_\alpha$. Since $q_\beta > p$, we have that $q_\alpha > p^4$ and Proposition 2.15, using that S/Q_α is elementary abelian, implies that $\overline{L_\alpha}/O_{p'}(\overline{L_\alpha}) \cong \mathrm{PSL}_2(q_\alpha)$. Applying Proposition 2.31, we see that $q_\alpha^{\frac{2}{3}} \leqslant pq_\beta$ in outcome (i) of Lemma 7.59 and $q_\alpha \leqslant p^2 q_\beta$ in outcome (iii) of Lemma 7.59. Both of these outcomes yield contradictions.

In outcome (ii) of Lemma 7.59 we have that $\overline{L_\alpha}$ is isomorphic to a central extension of $\mathrm{PSL}_2(q_\alpha)$ and we deduce that $|Z_\alpha| \leqslant (q_\beta r_\alpha)^2 \leqslant (q_\beta q_\alpha)^2 < q_\alpha^3$. Applying Lemma 2.43, and using that $Z_\alpha \cap Q_\beta$ is a $G_{\alpha,\beta}$-invariant subgroup of Z_α of index $q_\beta < q_\alpha^{\frac{1}{2}}$, we deduce that $q_\alpha = q_\beta^3$ and Z_α is a triality module for $\overline{L_\alpha}$. However, since $F_\beta \leq O^2(L_\beta)$ and $F_\beta' \leq D_\beta$, we have that $[Z_\alpha, F_\beta, F_\beta, F_\beta] \leq [F_\beta', F_\beta] = \{1\}$, a contradiction to the module structure of Z_α. Thus

(2) $\overline{L_\beta} \cong \mathrm{SL}_2(q_\beta)$.

Note that $Q_\alpha \cap Q_\beta \cap Q_{\alpha+2}$ is normalized by $L_\beta = \langle Z_\alpha Z_{\alpha+2}, Q_\beta \rangle$ and centralized by Z_α and so we must have that $D_\beta = Q_\alpha \cap Q_\beta \cap Q_{\alpha+2}$ has index q_α^2 in Q_β. In particular, $Z_\alpha \cap D_\beta$ has index at most $q_\beta q_\alpha$ in Z_α and is centralized by $S = Q_\alpha(Q_\beta \cap O^p(L_\beta))$. We have by Lemma 2.41 that either Z_α contains a unique non-central chief factor for L_α, or Z_α contains two non-central chief factors and $q_\alpha = q_\beta$. In the latter case, by Lemma 2.41, both chief factors are natural $\mathrm{SL}_2(q_\alpha)$ modules. Writing R_1 and R_2 for their centralizers in L_α, we have that $L_\alpha/R_1 \cong L_\alpha/R_2 \cong \mathrm{SL}_2(q_\alpha)$ and $R_1 \cap R_2 = Q_\alpha$. Since $q_\alpha > 3$, it follows that $\overline{L_\alpha} \cong \mathrm{SL}_2(q_\alpha)$ in this case. By [27], no configurations exist with this structure. Hence, Z_α contains a unique non-central chief factor for L_α. By Lemma 7.55, this chief factor is also the unique non-central chief factor for L_α within Q_α and $R_\alpha = Q_\alpha$.

Since $Z(L_\alpha) = \{1\}$, Lemma 2.30 yields that Z_α is irreducible and since $O^p(L_\alpha)$ centralizes Q_α/Z_α, $\Phi(Q_\alpha) = \{1\}$ and Q_α is elementary abelian. Since $D_\beta \leq Q_\alpha$, D_β is centralized by $S = F_\beta Q_\alpha$ so that $D_\beta = Z_\beta$. Furthermore, $Q_\beta = F_\beta(Q_\alpha \cap Q_\beta \cap Q_{\alpha+2}) = F_\beta Z_\beta$ and since $F_\beta = [F_\beta, O^p(L_\beta)]$ we deduce by Lemma 2.42 that Q_β/Z_β is a direct product of n natural $\mathrm{SL}_2(q_\beta)$-modules and $q_\alpha = q_\beta^n$.

We aim to show that $\overline{L_\alpha}$ is quasisimple. Assume first that $p = 2$ and $q_\alpha > q_\beta$. If outcome (iii) of Lemma 7.59 occurs, then applying Proposition 2.31 we deduce that $4q_\beta \geqslant q_\alpha$ as since $q_\beta \geqslant 4$ and $q_\beta^2 \leqslant q_\alpha$, we must have that $q_\alpha = 2^4$ and $q_\beta = 4$. Then, for C as described in Lemma 7.59, set P_α generated by Q_α and three conjugates of C so that P_α centralizes a subgroup of Z_α of index at most $(4q_\beta)^3$, $\overline{P_\alpha}/O_{2'}(\overline{P_\alpha}) \cong \mathrm{PSL}_2(2^4)$ and $O_{2'}(\overline{P_\alpha}) \neq \{1\}$. But $(4q_\beta)^3 = q_\alpha^3$ and Lemma 2.44 provides a contradiction. If outcome (i) holds, then applying Proposition 2.31 we deduce that $2q_\beta \geqslant q_\alpha^{\frac{2}{3}}$ and since $q_\beta \geqslant 4$ and $q_\beta^2 \leqslant q_\alpha$, we must have that $q_\alpha = q_\beta^2$ and $q_\beta = \{4, 8\}$. Set P_α such that P_α/Q_α is generated by 4 conjugate involutions with the property that P_α centralizes a subgroup of Z_α of index at most $2^4 q_\beta^4$, $\overline{P_\alpha}/O_{2'}(\overline{P_\alpha}) \cong \mathrm{PSL}_2(q_\alpha)$ and $O_{2'}(\overline{P_\alpha}) \neq \{1\}$. But $(4q_\beta)^2 \leqslant q_\alpha^{3}$ and Lemma 2.44 provides a contradiction. Hence, if $p = 2$ and $q_\alpha > q_\beta$ then outcome (ii) of Lemma 7.59 is satisfied and $\overline{L_\alpha}$ is quasisimple.

Assume now that $p = 2$ and $q_\alpha = q_\beta$. If Q_β is elementary abelian, then $Z_\alpha \cap Q_\beta$ is centralized by $S = Q_\alpha Q_\beta$ so that $Z_\alpha \cap Q_\beta = Z_\beta$, a contradiction. Now, if Q_β/M is elementary abelian for all M which are maximal in Z_β, then $\Phi(Q_\beta) \leq \Phi(Z_\beta) = \{1\}$, a contradiction. Hence, there is M maximal in Z_β such that Q_β/M is non-abelian. Since $q_\alpha > 2$, this is against [27, (5.13)] and we have a final contradiction. Hence, if $p = 2$ then $\overline{L_\alpha}$ is quasisimple.

Assume that $p = 3$ and $\overline{L_\alpha}$ is not quasisimple. If Z_α admits quadratic action, then by Lemma 2.35 we have that $q_\alpha = q_\beta = 9$ and $\overline{L_\alpha} \cong 4 \circ 2_-^{1+4}.\mathrm{Alt}(6)$. Since Z_α centralizes $Q_\alpha \cap Q_\beta$, and $D_\beta = Z_\beta$, we have that Q_β/Z_β is a natural module for $\overline{L_\beta} \cong \mathrm{SL}_2(9)$. Hence, Z_β has index 3^6 in S, and so has index 3^4 in Q_α. Since Q_α is elementary abelian, $O^3(L_\alpha)$ centralizes Q_α/Z_α and $O_{3'}(\overline{L_\alpha})$ is non-trivial and acts faithfully on Q_α, a coprime action argument, using that $Z_\beta \leq Z_\alpha$ gives that $Q_\alpha = Z_\alpha$. Now, $\overline{L_\alpha}$ may be generated by two conjugate Sylow 3-subgroups and as $Z(L_\alpha) = \{1\}$, we ascertain that $|Z_\alpha| \leqslant 3^8$. We employ MAGMA [14] to verify that $\overline{L_\alpha}$ has a unique faithful GF(3)-module of dimension at most 8 (indeed, this module is the restriction to $\overline{L_\alpha}$ of the natural module for $\mathrm{SU}_4(3)$). Since the maximal codimension of a subspace of this module that is fixed by a Sylow 3-subgroup, is 6 and $|Z_\alpha/Z_\beta| = 3^4$, we have a contradiction.

Hence, no element of $\overline{S}$ acts quadratically on Z_α. Now, since Q_β is a direct sum of natural modules, for any $z \in Z_\alpha \setminus [Z_\alpha, Q_\beta]Z_\beta$, we have that $[z, Q_\beta]Z_\beta = [Z_\alpha, Q_\beta]Z_\beta$. Then, Z_α is a faithful simple nearly quadratic module for $\overline{L_\alpha}$. Applying Theorem 2.58, we have that $F^*(\overline{L_\alpha}) = Z(\overline{L_\alpha})K$ where K is a component of $\overline{L_\alpha}$. Since $m_3(S/Q_\alpha) > 1$, applying Proposition 2.15, if $\overline{L_\alpha}$ is not quasisimple then the only possibility is that $K \cong \mathrm{Sz}(2^r)$ for some odd $r > 1$. Since $\mathrm{Out}(K)$ is solvable and $\overline{L_\alpha}/O_{3'}(\overline{L_\alpha})$ is non-abelian simple, this gives a contradiction.

Thus $\overline{L_\alpha}$ is quasisimple and by Proposition 2.15, we have shown that

(3) $\overline{L_\alpha} \cong \mathrm{SL}_2(p^n), \mathrm{PSL}_2(p^n)$ for $n \geqslant 2$ and $p \in \{2, 3\}$; or $p = 3$ and $\overline{L_\alpha}$ is either isomorphic to M_{11} or a coprime central extension of $\mathrm{PSL}_3(4)$.

If $\overline{L_\alpha}/Z(\overline{L_\alpha}) \cong \mathrm{PSL}_2(q_\alpha)$ then G has a weak BN-pair of rank 2 and G is determined up to local isomorphism in [27]. Comparing with the amalgams determined there, we have that G is locally isomorphic to H where $F^*(H) \cong \mathrm{PSU}_4(p^n)$ or $\mathrm{PSp}_4(3^n)$ for $n \geqslant 2$, and we are case (i). Hence, we may assume that $p = 3$ and $q_\beta = q_\alpha = 9$.

Suppose that $\overline{L_\alpha} \cong \mathrm{M}_{11}$. Then the amalgam is described in [59] and we have (ii) as a conclusion in this case.

We may assume that $\overline{L_\alpha}$ is isomorphic to a central extension of $\mathrm{PSL}_3(4)$. Since $\overline{L_\alpha}$ is generated by two conjugate Sylow 3-subgroups and Z_β has index 3^4 in Z_α, we deduce that $|Z_\alpha| \leqslant 3^8$. In fact, the possibilities for irreducible modules for central extensions of $\mathrm{PSL}_3(4)$ imply that $|Z_\alpha| \in \{3^6, 3^8\}$. If $|Z_\alpha| = 3^8$, then $\overline{L_\alpha} \cong 4 \cdot \mathrm{PSL}_3(4)$ and we infer that $|Z_\beta| = 3^4$. Comparing with the structure of the 8-dimensional module for $\overline{L_\alpha}$, this yields a contradiction. Thus, $|Z_\alpha| = 3^6$. One can check that for each of the modules in question that $|Z(\overline{L_\alpha})| = 2$ and so by coprime action, $Q_\alpha = [Q_\alpha, Z(\overline{L_\alpha})] \times C_{Q_\alpha}(Z(\overline{L_\alpha}))$ and as $Z_\alpha = [Q_\alpha, Z(\overline{L_\alpha})]$, we deduce that $Q_\alpha = Z_\alpha$. For t an involution in the preimage in L_α of $Z(\overline{L_\alpha})$, we have that $S = Z_\alpha : C_S(t)$. Moreover, S is of order 3^8 and is isomorphic to a Sylow 3-subgroup of Suz or $\mathrm{PSp}_4(9)$. In the former case, Z_β is of order 3 a contradiction since Z_β has index 3^4 in Z_α.

Thus S is isomorphic to a Sylow 3-subgroup of $\mathrm{PSp}_4(9)$, and we calculate in $\mathrm{GL}_6(3)$ using MAGMA [14] that $\overline{G_\alpha}$ embeds as a subgroup of $2 \cdot \mathrm{PSL}_3(4).2^2$ and for any element $x \in \overline{G_\alpha}$ of order 8, $[x^4, Z_\alpha] \not\leq [S, Z_\alpha] = Z_\alpha \cap Q_\beta$. Let $t_\beta \in L_\beta \cap G_{\alpha,\beta}$ be an element of order 8, so that $t_\beta^4 Q_\beta \leq Z(\overline{L_\beta})$. But then $[t_\beta^4, Z_\alpha] \leq Z_\alpha \cap Q_\beta$ and since $t_\beta \leq G_\alpha$, we have a contradiction. □

Proposition 7.62 *Suppose that $m_p(S/Q_\alpha) \geqslant 2$, $m_p(S/Q_\beta) = 1$ and $p \in \{2, 3\}$. Then one of the following holds:*

(i) *G has a weak BN-pair of rank 2 and G is locally isomorphic to H where $F^*(H) \cong \mathrm{PSU}_4(p)$ or $\mathrm{PSU}_5(2)$;*
(ii) *$p = 3$, $|S| = 3^6$, $\overline{L_\alpha} \cong \mathrm{PSL}_2(9)$, $Z_\alpha = Q_\alpha$ is a natural $\Omega_4^-(3)$-module, $\overline{L_\beta} \cong (Q_8 \times Q_8) : 3$ and $Q_\beta \cong 3_+^{1+4}$;*
(iii) *$p = 3$, $|S| = 3^6$, $\overline{L_\alpha} \cong \mathrm{PSL}_2(9)$, $Z_\alpha = Q_\alpha$ is a natural $\Omega_4^-(3)$-module, $\overline{L_\beta} \cong 2 \cdot \mathrm{Alt}(5)$ and $Q_\beta \cong 3_+^{1+4}$;*
(iv) *$p = 3$, $|S| = 3^6$, $\overline{L_\alpha} \cong \mathrm{PSL}_2(9)$, $Z_\alpha = Q_\alpha$ is a natural $\Omega_4^-(3)$-module, $\overline{L_\beta} \cong 2_-^{1+4}.\mathrm{Alt}(5)$ and $Q_\beta \cong 3_+^{1+4}$;*
(v) *$p = 3$, $|S| = 3^7$, $\overline{L_\alpha} \cong \mathrm{M}_{11}$ and $Z_\alpha = Q_\alpha$ is the "cocode" module for $\overline{L_\alpha}$, $\overline{L_\beta} \cong \mathrm{SL}_2(3)$ and Q_β is a special group of order 3^6 isomorphic to a Sylow 3-subgroup of $\mathrm{SL}_3(9)$; or*
(vi) *$p = 3$, $|S| = 3^7$, $\overline{L_\alpha} \cong \mathrm{M}_{11}$ and $Z_\alpha = Q_\alpha$ is the "cocode" module for $\overline{L_\alpha}$, $\overline{L_\beta} \cong \mathrm{SL}_2(5)$ and Q_β is a special group of order 3^6 isomorphic to a Sylow 3-subgroup of $\mathrm{SL}_3(9)$.*

Proof Suppose that $m_p(S/Q_\alpha) \geqslant 2$ and $m_p(S/Q_\beta) = 1$. Then $\Phi(Q_\alpha)(Q_\alpha \cap Q_\beta)$ is centralized, modulo $\Phi(Q_\alpha)$, by F_β by Lemma 7.55. Observe first that if $|S/Q_\beta| \neq p$, then using that $m_p(S/Q_\alpha) > 1$ and applying Lemma 2.41, we must have that S/Q_β is generalized quaternion and by Lemma 7.55 that $Q_\alpha/\Phi(Q_\alpha)$ has an index 4 subgroup centralized by F_β. Then Lemma 2.41 yields that $\overline{L_\alpha} \cong \mathrm{PSL}_2(4)$. But then $Q_\beta/\Phi(Q_\beta)$ is a quadratic 2F-module and by Proposition 2.51, we deduce that $\overline{L_\beta} \cong \mathrm{SU}_3(2)$. Thus, $(G_\alpha, G_\beta, G_{\alpha,\beta})$ is a weak BN-pair and is locally isomorphic to H where $F^*(H) \cong \mathrm{PSU}_5(2)$, and we are in one of the possibilities in case (i). Thus, we continue with the added assumption that

(1) $|S/Q_\beta| = p$

By Lemma 7.55 we have that $O^p(L_\alpha)$ centralizes Q_α/Z_α so that $R_\alpha = Q_\alpha$. Furthermore, there is a unique non-central chief factor within Z_α for L_α and as $Z(L_\alpha) = \{1\}$, applying Lemma 2.30, we conclude that Z_α is an irreducible faithful module for $\overline{L_\alpha}$. Since $O^p(L_\alpha)$ centralizes Q_α/Z_α, we conclude that $\Phi(Q_\alpha) \cap Z_\alpha = \Phi(Q_\alpha) = \{1\}$ and Q_α is elementary abelian. Note that if $O_{p'}(\overline{L_\alpha}) \neq \{1\}$, then $Z_\beta \le Z_\alpha = [Q_\alpha, O_{p'}(\overline{L_\alpha})]$ by the irreducibility of Z_α and by coprime action, we infer that $Q_\alpha = Z_\alpha$ is elementary abelian.

We aim to show that $O_{p'}(\overline{L_\alpha}) \le Z(\overline{L_\alpha})$. Aiming for a contradiction, suppose otherwise. Assume that there does not exist $x \in S/Q_\alpha$ of order p such that $|Z_\alpha/C_{Z_\alpha}(x)| = p^2$. Applying Lemma 7.59, we conclude that there is $C \le S/Q_\alpha$ of order p^2 such that $|Z_\alpha/C_{Z_\alpha}(C)| = |Z_\alpha/C_{Z_\alpha}(x)| = p^3$ for all $x \in C^\#$. Following the methodology of Lemma 7.59 using Proposition 2.8 we deduce that $O_{p'}(\overline{L_\alpha}) \le Z(\overline{L_\alpha})$, a contradiction.

Hence, we may assume that there is $x \in S/Q_\alpha$ of order p such that $|Z_\alpha/C_{Z_\alpha}(x)| = p^2$. If x can be chosen to act quadratically on Z_α then $\overline{L_\alpha}$ is determined by Proposition 2.51. Since $m_p(S/Q_\alpha) > 1$, we have a contradiction to the assumption that $\overline{L_\alpha}$ is not quasisimple. Therefore, we have that $p = 3$ and $[Z_\alpha, x]C_{Z_\alpha}(x)$ has index 3 in Z_α. By Theorem 2.58 we have that $F^*(\overline{L_\alpha}) = Z(\overline{L_\alpha})K$ where K is a component of $\overline{L_\alpha}$. Since $m_p(S/Q_\alpha) \geqslant 2$, comparing with Proposition 2.15 we must have that K is a $3'$-group. Hence, $K \cong \mathrm{Sz}(2^r)$ for some odd $r > 1$. Since $\mathrm{Out}(K)$ is solvable and $\overline{L_\alpha}/O_{3'}(\overline{L_\alpha})$ is non-abelian simple, this gives a contradiction. Hence

(2) $O_{p'}(\overline{L_\alpha}) \le Z(\overline{L_\alpha})$, $\overline{L_\alpha}$ is quasisimple and Z_α is a faithful, irreducible module for $\overline{L_\alpha}$.

Suppose that $\overline{L_\alpha} \cong \mathrm{SL}_2(p^n)$ or $\mathrm{PSL}_2(p^n)$ for any $n > 1$. Unless $r_\alpha = p$ we have that $|Z_\alpha| \leqslant r_\alpha^2 p^2 \leqslant p^{2n+2}$ and if $r_\alpha = p$, we have that $|Z_\alpha| \leqslant p^6$. Since the minimal degree of a $\mathrm{GF}(p)$-representation of $\overline{L_\alpha}$ is $2n$ and $n \geqslant 2$, we deduce that $r_\alpha \geqslant p^{n-1}$. Setting K to be Hall p'-subgroup of $L_\alpha \cap G_{\alpha,\beta}$, it follows from Smith's theorem ([34, Theorem 2.8.11]) that $Z_\beta = C_{Z_\alpha}(S)$ and $Z_\alpha/[Z_\alpha, S]$ are irreducible and 1-dimensional as $\overline{F}K$-modules, where $\overline{F}$ is an appropriately chosen splitting field of characteristic p. But $[Z_\alpha, S] = [Z_\alpha, Q_\beta] \le Z_\alpha \cap Q_\beta$ and since $Z_\alpha \cap Q_\beta$ has index p in Z_α, $[Z_\alpha, S] = Z_\alpha \cap Q_\beta$ and $|Z_\beta| = |Z_\alpha/[Z_\alpha, S]| = p$. If $n > 2$, then $|Z_\alpha| \leqslant p^{2n+2} < p^{3n}$ and [23, Lemma 2.6] implies that Z_α is a triality module for $\overline{L_\alpha} \cong \mathrm{SL}_2(p^3)$ and $|Z_\alpha| = p^8$. Since $|Z_\alpha| \leqslant r_\alpha^2 p^2$, we have that $r_\alpha = p^3$ and $S = (Z_{\alpha+2} \cap Q_\beta)Q_\alpha$ centralizes $Z_\alpha \cap Q_\beta \cap Q_{\alpha+2}$. But then $Z_\beta = Z_\alpha \cap Q_\beta \cap Q_{\alpha+2}$ is index p^4 in Z_α. Since $|Z_\beta| = p$, we conclude that $p^5 = |Z_\alpha| = p^8$, a contradiction. Since $\overline{L_\alpha} \cong \mathrm{(P)SL}_2(q_\alpha)$ whenever $m_p(S/Q_\alpha) > 2$, we reduce to the case where

(3) $|S/Q_\alpha| = p^2$.

We have that $\overline{L_\alpha}$ is isomorphic to $\mathrm{PSL}_2(p^2)$, $\mathrm{SL}_2(p^2)$, M_{11} or a central extension of $\mathrm{PSL}_3(4)$. Since both M_{11} and central extensions of $\mathrm{PSL}_3(4)$ are generated by two conjugate Sylow 3-subgroups (or by three 3-elements), combing with the earlier analysis, we see that $|Z_\alpha| \leqslant p^6$ in all cases.

Now, $F_\beta/F_\beta \cap D_\beta$ is a quadratic 2F-module and so, by Proposition 2.51, both $\overline{L_\beta}$ and $F_\beta/D_\beta \cap F_\beta$ are determined. Checking against the list of groups provided in Proposition 2.51, either $\overline{L_\beta}$ is p-solvable or has a non-trivial center. For T_β the preimage in L_β of $O_{p'}(\overline{L_\beta})$, we have that $T_\beta > Q_\beta$ and by coprime action $Q_\beta/\Phi(Q_\beta) = [Q_\beta/\Phi(Q_\beta), T_\beta] \times C_{Q_\beta/\Phi(Q_\beta)}(T_\beta)$ where $[Q_\beta/\Phi(Q_\beta), T_\beta]$ contains all non-central chief factors in $Q_\beta/\Phi(Q_\beta)$ and $C_{Q_\beta/\Phi(Q_\beta)}(T_\beta) = C_{Q_\beta/\Phi(Q_\beta)}(O^p(L_\beta))$. In particular, $F_\beta\Phi(Q_\beta)/\Phi(Q_\beta) = [Q_\beta/\Phi(Q_\beta), T_\beta]$. Since $\Phi(Q_\beta) \leq Q_\alpha$, $[\Phi(Q_\beta), Z_\alpha] = \{1\}$ and it follows that $\Phi(Q_\beta) \leq D_\beta$ so that $Q_\beta = F_\beta D_\beta$. Since $D_\beta \leq Q_\alpha$ is elementary abelian and $F_\beta \leq O^p(L_\beta)$, $S = F_\beta Q_\alpha$ centralizes D_β so that $D_\beta = Z_\beta$. We now analyze the possible cases individually.

Case (1): $\overline{L_\alpha}$ is isomorphic to a central extension of $\mathrm{PSL}_3(4)$.

Since $|Z_\alpha| \leqslant 3^6$, we have that $|Z_\alpha| = 3^6$, $|Z(\overline{L_\alpha})| = 2$ and since $O^3(L_\alpha)$ centralizes Q_α/Z_α, an easy coprime action argument yields that $Q_\alpha = Z_\alpha$ is elementary abelian. Hence, $|S| = 3^8$ and comparing with the modules in Proposition 2.51, $|Q_\beta/Z_\beta| = 3^4$ so that $|Z_\beta| = 3^3$. Checking the appropriate irreducible GF(3)-modules associated to $\overline{L_\alpha}$, we have that $|Z_\beta| \leqslant 3^2$, a contradiction.

Case (2): $\overline{L_\alpha} \cong \mathrm{M}_{11}$.

Then $p = 3$, $|Z_\alpha| = 3^5$ and $\overline{L_\beta} \cong 2 \cdot \mathrm{Alt}(5)$, $2_-^{1+4}.\mathrm{Alt}(5)$, $\mathrm{SL}_2(3)$ or $(Q_8 \times Q_8) : 3$ by Proposition 2.51. In the first three cases, the structure of L_α and L_β is determined in [59] and outcomes (v) and (vi) follow in these cases. Suppose that $\overline{L_\beta} \cong (Q_8 \times Q_8) : 3$ with $|Q_\beta/Z_\beta| = 3^4$ and let K_β be a Hall $2'$-subgroup of $G_{\alpha,\beta} \cap L_\beta$. Then $K_\beta \leq G_\alpha$ and so K_β acts on L_α/Q_α. Since M_{11} has no outer automorphisms, if $K_\beta \not\leq L_\alpha$, then there is an involution $t \in K_\beta$ such that $[t, L_\alpha] \leq Q_\alpha$ and $[t, L_\beta] \leq Q_\beta$, a contradiction by Proposition 5.5. Thus, $K_\beta \leq L_\alpha$ and we may assume that $L_\alpha = G_\alpha$. Since $[K_\beta, Z_\alpha] \leq Z_\alpha \cap Q_\beta$ and K_β centralizes Z_β it follows that $|C_{Z_\alpha}(K_\beta)| = 3^3$. One can check that in the unique 5-dimensional GF(3)-module for M_{11} which has a 2-dimensional space fixed by a Sylow 3-subgroup, this provides a contradiction.

Case (3): $\overline{L_\alpha} \cong \mathrm{PSL}_2(p^2)$ or $\mathrm{SL}_2(p^2)$.

Again by Smith's theorem [34, Theorem 2.8.11], we have that $|Z_\beta| = p$ so that $F_\beta = Q_\beta$. By the minimality of F_β, it follows that $Z(Q_\beta) = \Phi(Q_\beta) = Z_\beta$ is of order p and Q_β is extraspecial. Since $Q_\beta \cap Q_\alpha$ is an elementary abelian subgroup of index p^2 in Q_β, we have that $|Q_\beta| = p^5$. In particular, $|S| = p^6$ and $Z_\alpha = Q_\alpha$ is of order p^4.

If $p = 2$, then $\overline{L_\beta} \cong \mathrm{Dih}(10)$, $\mathrm{Sym}(3)$ or $(3 \times 3) : 2$. In the first two cases, G has a weak BN-pair and so comparing with [27], we have that $L_\beta \cong \mathrm{Sym}(3)$ and G is locally isomorphic to H where $F^*(H) \cong \mathrm{PSU}_4(2)$., and we are in case (i) Since Q_β is extraspecial, comparing with [76], $\overline{L_\beta}$ is isomorphic to a subgroup of $O_4^+(2)$ if $Q_\beta \cong 2_+^{1+4}$; or $O_4^-(2)$ if $Q_\beta \cong 2_-^{1+4}$. Note that 9 does not divide $|O_4^-(2)|$ and so we may assume that $Q_\beta \cong 2_+^{1+4}$. Let K be a Sylow 3-subgroup of $L_\alpha \cap G_{\alpha,\beta}$. Then K acts non-trivially on Q_β and so K also embeds into $O_4^+(2)$ while normalizing

$\overline{L_\beta} \cong (3 \times 3) : 2$. But for $H \le O_4^+(2)$ with $H \cong (3 \times 3) : 2$ we have that $|N_{O_4^+(2)}(H)/H| = 2$, a contradiction.

Thus, we may assume that $p = 3$ and $\overline{L_\beta} \cong \mathrm{SL}_2(3), (Q_8 \times Q_8) : 3, 2 \cdot \mathrm{Alt}(5)$ or $2_-^{1+4}.\mathrm{Alt}(5)$. Since $|Z_\alpha| = 3^4$ and is not quadratic, we have that $\overline{L_\alpha} \cong \mathrm{PSL}_2(9)$ and Z_α is a natural $\Omega_4^-(3)$-module. If $\overline{L_\beta} \cong \mathrm{SL}_2(3)$ then G has a weak BN-pair and comparing with [27], we have that G is locally isomorphic to H where $F^*(H) \cong \mathrm{PSU}_4(3)$, and we are in case (i). If $\overline{L_\beta} \cong 2 \cdot \mathrm{Alt}(5)$ or $2_-^{1+4}.\mathrm{Alt}(5)$ then the structure of L_α and L_β is determined in [59] and we obtain conclusions (iii) and (iv). Finally, suppose that $\overline{L_\beta} \cong (Q_8 \times Q_8) : 3$. Since Q_β is extraspecial of order 3^5 and $\overline{L_\beta}$ embeds in the automorphism group of Q_β, it follows from [76] that $Q_\beta \cong 3_+^{1+4}$ and we have (ii) as a conclusion. □

We conclude this final chapter by summarizing what has been shown:

Theorem 7.63 *Suppose that $\mathcal{A} = \mathcal{A}(G_\alpha, G_\beta, G_{\alpha,\beta})$ is an amalgam satisfying Hypothesis 5.1. If $Z_{\alpha'} \le Q_\alpha$, then one of the following holds:*

(i) *G has a weak BN-pair of rank 2; or*
(ii) *$p = 3$, $b = 1$, $|S| \leqslant 3^7$ and the shapes of L_α and L_β are known.*

Appendix A

We use this appendix as an opportunity to summarize the main amalgam theoretic results in this text in the form of a table.

We mostly provide descriptions only of the "simple" amalgams i.e. those generated by $O^p(L_\alpha)$ and $O^p(L_\beta)$. However, we also include the 2-local amalgams associated to $G_2(2)$ and ${}^2F_4(2)$ because their structures are quite different from the generic structures of $G_2(2^n)$ and ${}^2F_4(2^n)$, and we would like to prevent any confusion.

In the table, we provide an almost simple group realizing each amalgam and containing S as a Sylow p-subgroup, the prime p associated to the amalgam plus any conditions related to the completion we have chosen, the critical distance b of the amalgam, the "shapes" of L_α and L_β, and where the amalgams are classified in the text.

Finally, some words on what "shape" means in the context of L_α and L_β. Our notation is reminiscent of ATLAS notation for maximal p-local subgroups and so, for $\lambda \in \{\alpha, \beta\}$, L_λ has the form $L_\lambda := q^{n_1+n_2+\cdots+n_r} : L$ where q is some prime power. Here, $q^{n_1+n_2+\cdots+n_r}$ is the largest normal p-subgroup of L_λ, so represents Q_λ, and the integers n_i indicate the chief factors for L_λ contained in Q_λ. The case where $n_i = 1$ indicates that there is a factor of size q centralized by L_λ. The case $n_i > 1$ indicates there is a non-central chief factor of order q^{n_i}. Finally the notation $(n_i + \cdots + n_j)$ with brackets means that this section of Q_λ is elementary abelian and contains multiple chief factors (of sizes $q^{n_i}, \ldots, q^{n_j}$ respectively). Of course, we do not provide all information for when sections are elementary.

As an example, for the amalgam with completion $\mathrm{PSU}_5(q)$, Q_α contains two chief factors for L_α, both of order q^4. On the other hand, in $\mathrm{PSp}_4(2^n)$, Q_α is elementary abelian. Moreover, Q_α contains a subgroup of order q centralized by L_α and the quotient of Q_α by this subgroup is a non-central chief factor for L_α of order q^2 (Table A.1).

© The Author(s), under exclusive license to Springer Nature Switzerland AG 2024

M. van Beek, *Rank 2 Amalgams and Fusion Systems*, Lecture Notes
in Mathematics 2343, https://doi.org/10.1007/978-3-031-54461-3

Table A.1 The Amalgams

Example completion	p	b	L_α	L_β	Reference
$PSL_3(q)$	any	1	$q^2 : SL_2(q)$	$q^2 : SL_2(q)$	6.7
$PSp_4(q)$	2	1	$q^{(1+2)} : SL_2(q)$	$q^{(1+2)} : SL_2(q)$	6.7
$PSp_4(q)$	odd	1	$q^3 : PSL_2(q)$	$q_+^{1+2} : SL_2(q)$	7.53, 7.57, 7.61
$PSU_4(q)$	any	1	$q^4 : PSL_2(q^2)$	$q^{1+(2+2)} : SL_2(q)$	7.53, 7.61, 7.62
$PSU_5(q)$	any	1	$q^{4+4} : SL_2(q^2)$	$q_+^{1+6} : SU_3(q)$	7.53, 7.61, 7.62
Co_2	3	1	$3^4 : PSL_2(9)$	$3_+^{1+4} : 2_-^{1+4}.Alt(5)$	7.62
Co_3	3	1	$3^5 : M_{11}$	$3_+^{1+4} : SL_2(9)$	7.61
McL	3	1	$3^4 : PSL_2(9)$	$3_+^{1+4} : SL_2(5)$	7.62
Suz	3	1	$3^5 : M_{11}$	$3^{1+1+(2+2)} : SL_2(3)$	7.62
Ly	3	1	$3^5 : M_{11}$	$3^{1+1+4} : SL_2(5)$	7.62
$PSU_6(2)$	3	1	$3^4 : PSL_2(9)$	$3^{1+(2+2)} : (Q_8 \times Q_8) : 3$	7.62
$PSU_5(2)$	3	1	$3^4 : Alt(5)$	$3^{1+1+2} : SL_2(3)$	7.57
$\Omega_8^+(2)$	3	1	$3^4 : (2 \wr Sym(4))'$	$3^{1+1+2} : SL_2(3)$	7.57
$\Omega_{10}^-(2)$	3	1	$3^5 : (2 \wr Sym(5))'$	$3^{1+1+1+2} : SL_2(3)$	7.57
$G_2(2)'$	2	2	$2^{2+2} : SL_2(2)$	$2^{1+1+2} : Sym(3)$	6.22
$G_2(2)$	2	2	$2^{2+1+2} : SL_2(q)$	$2^{1+1+1+2} : SL_2(2)$	6.22
$G_2(3)$	2	2	$2^{2+1+2} : SL_2(2)$	$2^{1+(2+2)} : (3 \times 3) : 2$	6.22
$G_2(4)$	2	2	$2^{4+2+4} : SL_2(q)$	$2^{2+4+4} : SL_2(4)$	6.23
$G_2(q)$	3	2	$q^{1+(2+2)} : SL_2(q)$	$q^{1+(2+2)} : SL_2(q)$	6.6
$G_2(q)$	$(p, 3) = 1, q \geqslant 5$	2	$q^{2+1+2} : SL_2(q)$	$q_+^{1+4} : SL_2(q)$	6.18, 6.23
$^3D_4(q)$	any	2	$q^{(2+1+1+1)+(2+2+2)} : SL_2(q)$	$q_+^{1+8} : SL_2(q^3)$	6.18, 6.23
J_2	2	2	$2^{2+(2+2)} : SL_2(2)$	$2_-^{1+4} : SL_2(4)$	6.23
$PSp_6(3)$	2	2	$2^{2+6} : SU_3(2)'$	$2^{2+(2+4)} : SL_2(4)$	6.23
Ly	5	2	$5^{2+1+2} : SL_2(5)$	$5_+^{1+4} : SL_2(9)$	6.18
HN	5	2	$5^{2+1+2} : SL_2(5)$	$5_+^{1+4} : 2_-^{1+4} : 5$	6.18
B	5	2	$5^{2+1+2} : SL_2(5)$	$5_+^{1+4} : 2_-^{1+4}.Alt(5)$	6.18
M	7	2	$7^{2+1+2} : SL_2(7)$	$7_+^{1+4} : 2.Alt(7)$	6.18
$^2F_4(2)'$	2	3	$2^{2+1+2+1+2+2} : SL_2(2)$	$2^{1+4+4} : Sz(2)$	7.24
$^2F_4(2)$	2	3	$2^{2+1+2+1+1+2+2} : SL_2(2)$	$2^{1+4+1+4} : Sz(2)$	7.24
$^2F_4(q)$	2	3	$q^{2+1+2+4+2} : SL_2(q)$	$q^{1+4+1+4} : Sz(q)$	7.23, 7.24
M_{12}	2	3	$2^{2+1+2} : SL_2(2)$	$2^{1+(2+2)} : SL_2(2)$	7.51
F_3	3	5	$3^{2+3+(2+2)} : SL_2(3)$	$3^{1+2+1+2+1+2} : SL_2(3)$	7.44

References

1. An, J., Heiko, D.: The AWC-goodness and essential rank of sporadic simple groups. J. Algebra **356**(1), 325–354 (2012)
2. Andersen, K.S., Oliver, B., Ventura, J.: Fusion systems and amalgams.Math. Z. **274**(3–4), 1119–1154 (2013)
3. Andersen, K.S., Oliver, B., Ventura, J.: Reduced fusion systems over 2- groups of small order. J. Algebra **489**, 310–372 (2017)
4. Aschbacher, M.: Finite Group Theory, vol. 10. Cambridge University Press, Cambridge (2000)
5. Aschbacher, M.: Generation of fusion systems of characteristic 2-type. Invent. Math. **180**(2), 225–299 (2010)
6. Aschbacher, M.: $S3$-free 2-fusion systems. Proc. Edinb. Math. Soc. **56**(1), 27–48 (2013)
7. Aschbacher, M.: N-groups and fusion systems. J. Algebra **449**, 264–320 (2016)
8. Aschbacher, M.: On Fusion Systems of Component Type, vol. 257. American Mathematical Society, Providence (2019). https://doi.org/10.1090/memo/1236
9. Aschbacher, M.: Quaternion Fusion Packets, vol. 765. American Mathematical Society, Providence (2021)
10. Aschbacher, M.: Walter's theorem for fusion systems. Proc. Lond. Math. Soc. **122**(4), 569–615 (2021)
11. Aschbacher, M., Kessar, R., Oliver, B.: Fusion Systems in Algebra and Topology, vol. 391. Cambridge University Press, Cambridge (2011)
12. Beisiegel, B.: Semi-extraspezielle p-groups. Math. J. **156**(3), 247–254 (1977)
13. Bender, H.: Transitive Gruppen gerader Ordnung, in denen jede Involution genau einen Punkt festläßt. J. Algebra **17**(4), 527–554 (1971)
14. Bosma, W., Cannon, J., Playoust, C.: The Magma algebra system I: the user language. J. Symb. Comput. **24**(3–4), 235–265 (1997)
15. Bray, J.N., Holt, D.F., Roney-Dougal, C.M.: The Maximal Subgroups of the Low-Dimensional Finite Classical Groups, vol. 407. Cambridge University Press, Cambridge (2013)
16. Broto, C., Møller, J., Oliver, B.: Automorphisms of Fusion Systems of Finite Simple Groups of Lie Type/Automorphisms of Fusion Systems of Sporadic Simple Groups, vol. 262. American Mathematical Society, Providence (2019). https://doi.org/10.1090/memo/1267
17. Broto, C., Møller, J., Oliver, B.: Equivalences between fusion systems of finite groups of Lie type. J. Am. Math. Soc. **25**(1), 1–20 (2012)
18. Bundy, D., Hebbinghaus, N., Stellmacher, B.: The local $C(G, T)$ theorem. J. Algebra **300**(2), 741–789 (2006)
19. Chermak, A.: Large triangular amalgams whose rank-1 kernels are not all dis- tinct. Commun. Algebra **14**(4), 667–706 (1986)

© The Author(s), under exclusive license to Springer Nature Switzerland AG 2024

M. van Beek, *Rank 2 Amalgams and Fusion Systems*, Lecture Notes
in Mathematics 2343, https://doi.org/10.1007/978-3-031-54461-3

20. Chermak, A.: Quadratic pairs without components. J. Algebra **258**(2), 442–476 (2002)
21. Chermak, A.: Quadratic pairs. J. Algebra **277**(1), 36–72 (2004)
22. Chermak, A.: Small modules for small groups. Preprint (2001)
23. Chermak, A., Delgado, A.L.: J-modules for local BN-pairs. Proc. Lond. Math. Soc. **3**(1), 69–112 (1991)
24. Conway, J.H., et al.: Atlas of Finite Groups: Maximal Subgroups and Ordinary Characters for Simple Groups. Oxford University Press, Oxford (1985)
25. Craven, D.A.: The Theory of Fusion Systems: An Algebraic Approach, vol. 131. Cambridge University Press, Cambridge (2011)
26. Delgado, A.L.: Amalgams of type F3. J. Algebra **117**(1), 149–161 (1988)
27. Delgado, A., Stellmacher, B.: "Weak (B,N)-pairs of rank 2, Groups and graphs: new results and methods. In: Delgado, A., Goldschmidt, D., Stellmacher, B. (eds.) DMV Seminar, vol. 6. Birkhaüser, Basel (1985)
28. Fan, P.S.: Amalgams of prime index. J. Algebra **98**(2), 375–421 (1986)
29. Glauberman, G.: Isomorphic subgroups of finite p-groups. II. Can. J. Math. **23**(6), 1023–1039 (1971)
30. Glauberman, G.: Prime-power factor groups of finite groups. II. Math. Z. **117**(1–4), 46–56 (1970)
31. Goldschmidt, D.M.: Automorphisms of trivalent graphs. Ann. Math. **111**(2), 377–406 (1980)
32. Gorenstein, D.: Finite Groups, vol. 301. American Mathematical Society, Providence (2007)
33. Gorenstein, D., Lyons, R., Solomon, R.: The Classification of the Finite Simple Groups, Number 2. Mathematical Surveys and Monographs, vol. 40. American Mathematical Society, Providence (1996)
34. Gorenstein, D., Lyons, R., Solomon, R.: The Classification of the Finite Simple Groups, Number 3. Mathematical Surveys and Monographs, vol. 40 American Mathematical Society, Providence (1998)
35. Grazian, V.: Fusion systems containing pearls. J. Algebra **510**, 98–140 (2018)
36. Grazian, V., Parker, C.: Saturated fusion systems on p-groups of maximal class. Memoirs Am. Math. Soc. (2022)
37. Guralnick, R.M., Lawther, R., Malle, G.: The 2F-modules for nearly simple groups. J. Algebra **307**(2), 643–676 (2007)
38. Guralnick, R.M., Malle, G.: Classification of 2F-modules, I. J. Algebra **257**(2), 348–372 (2002)
39. Guralnick, R.M., Malle, G.: Classification of 2F-modules, II. Finite Groups 2003, Proceedings of the Gainesville Conference on Finite Groups, pp. 117–184 (2004)
40. Hall, P., Higman, G.: On the p-length of p-soluble groups and reduction theorems for Burnside's problem. Proc. Lond. Math. Soc. **3**(1), 1–42 (1956)
41. Hayashi, M.: Amalgams of solvable groups. J. Fac. Sci. Univ. Tokyo Sect. IA Math. **39** (1992)
42. Henke, E.: Recognizing SL2(q) in fusion systems. J. Group Theory **13**(5), 679–702 (2010)
43. Huppert, B.: Endliche Gruppen I, vol. 134. Springer-Verlag, Berlin (2013)
44. Jianhua, H., Jixin, M., Stellmacher B.: Amalgams of rank 2 in characteristic 3 involving L2(5). Acta Math. Sinica **5**(3), 263–270 (1989)
45. Kaspczyk, J.: A characterization of the groups and by their fusion systems, $q$odd. Forum Mathematics, Sigma, vol. 10, e60 Cambridge University Press. Cambridge (2022)
46. Kleidman, P.B.: The maximal subgroups of the Chevalley groups G2(q) with q odd, the Ree groups ^{2}G2(q), and their automorphism groups. J. Algebra **117**(1), 30–71 (1988)
47. Kleidman, P.B., Wilson, R.A.: The maximal subgroups of J4. Proc. Lond. Math. Soc. **3**(3), 484–510 (1988)
48. Kleshchev, A., Zalesski, A.: Minimal polynomials of elements of order $p$in p-modular projective representations of alternating groups. Proc. Am. Math. Soc. **132**(6), 1605–1612 (2004)
49. Kurzweil, H., Stellmacher, B.: The Theory of Finite Groups: An Introduction. Springer Science & Business Media, Berlin (2006)
50. Martineau, R.P.: On 2-modular representations of the Suzuki groups. Am. J. Math. **94**(1), 55–72 (1972)

51. Meierfrankenfeld, U., Stellmacher, B.: Nearly quadratic modules. J. Algebra **319**(11), 4798–4843 (2008)
52. Meierfrankenfeld, U., Stellmacher, B., Stroth, G.: Finite Groups of Local Characteristic p, p. 155. Groups, Combinatorics & Geometry, Durham (2001/2003)
53. Meixner, T.: Groups acting transitively on locally finite classical Tits chamber systems. Finite Geometries, Buildings and Related Topics, pp. 45–65. Oxford University Press, Oxford (1990)
54. Mitchell, H.H.: Determination of the ordinary and modular ternary linear groups. Trans. Am. Math. Soc. **12**(2), 207–242 (1911)
55. Niles, R.: Pushing-up in finite groups. J. Algebra **57**(1), 26–63 (1979)
56. Oliver, B.: Reduced fusion systems over 2-groups of sectional rank at most 4, vol. 239. American Mathematical Society, Providence (2016)
57. Oliver, B., Ventura, J.: Saturated fusion systems over 2-groups. Trans. Am. Math. Soc. **361**(12), 6661–6728 (2009)
58. Onofrei, S.: Saturated fusion systems with parabolic families. J. Algebra **348**(1), 61–84 (2011)
59. Papadopoulos, P.: Some amalgams in characteristic 3 related to Co1. J. Algebra **195**(1), 30–73 (1997)
60. Parker, C., Rowley, P.: Local characteristic $p$completions of weak BN -pairs. Proc. Lond. Math. Soc. **93**(2), 325–394 (2006)
61. Parker, C., Rowley, P.: Symplectic Amalgams. Springer Science & Business Media, Berlin (2012)
62. Parker, C., Semeraro, J.: Algorithms for fusion systems with applications to p-groups of small order. Math. Comput. **90**, 2415–2461 (2021)
63. Parker, C., Semeraro, J.: Fusion systems over a Sylow p-subgroup of G2(p). Math. Z. **289**(1–2), 629–662 (2018)
64. Parker, C., Stroth, G.: Strongly p-embedded Subgroups. Pure Appl. Math. Q. **7**(3), 797–858 (2009)
65. Puig, L.: Frobenius categories. J. Algebra **303**(1), 309–357 (2006)
66. Puig, L.: Structure Locale dans les Groupes Finis. Société Mathématique de France, Marseille (1976)
67. Robinson, G.R.: Amalgams, blocks, weights, fusion systems and finite simple groups. J. Algebra **314**(2), 912–923 (2007)
68. Ruiz, A., Viruel, A.: The classification of p-local finite groups over the extraspecial group of order p^3 and exponent p. Math. Z. **248**(1), 45–65 (2004)
69. Stellmacher, B.: On the 2-local structure of finite groups. Groups Combin. Geom. **165**, 159 (1992)
70. Stellmacher, B.: Pushing up. Archiv der Math. **46**(1), 8–17 (1986)
71. Stellmacher, B., Timmesfeld, F.G.: Rank 3 Amalgams, vol. 649. American Mathematical Society, Providence (1998)
72. Stroth, G.: On groups of local characteristic p. J. Algebra **300**(2), 790–805 (2006)
73. Suzuki, M.: On a class of doubly transitive groups. Ann. Math. 105–145 (1962)
74. Timmesfeld, F.G.: On amalgamation of rank 1 parabolic groups. Geometriae Dedicata **25**(1–3), 5–70 (1988)
75. Ward, H.N.: On Ree's series of simple groups. Trans. Am. Math. Soc. **121**(1), 62–89 (1966)
76. Winter, D.L.: The automorphism group of an extraspecial p-group. Rocky Mt. J. Math. **2**(2), 159–168 (1972)
77. Zalesskii, A.E.: Minimal polynomials and eigenvalues of p-elements in representations of quasi-simple groups with a cyclic Sylow p-subgroup. J. Lond. Math. Soc. **59**(3), 845–866 (1999)
78. Zalesskii, A.E., Serezkin, V.N.: Finite linear groups generated by reflections. Math. USSR-Izv **17**, 477–503 (1981)
79. van Beek, M.:Exotic fusion systems related to sporadic simple groups. Preprint (2022). arXiv:2201.01790
80. van Beek, M.: Saturated fusion systems on Sylow p-subgroups of rank 2 simple groups of Lie type. Preprint (2023). arXiv:2302.02222

LECTURE NOTES IN MATHEMATICS

Editors in Chief: J.-M. Morel, B. Teissier;

Editorial Policy

1. Lecture Notes aim to report new developments in all areas of mathematics and their applications – quickly, informally and at a high level. Mathematical texts analysing new developments in modelling and numerical simulation are welcome.

   Manuscripts should be reasonably self-contained and rounded off. Thus they may, and often will, present not only results of the author but also related work by other people. They may be based on specialised lecture courses. Furthermore, the manuscripts should provide sufficient motivation, examples and applications. This clearly distinguishes Lecture Notes from journal articles or technical reports which normally are very concise. Articles intended for a journal but too long to be accepted by most journals, usually do not have this "lecture notes" character. For similar reasons it is unusual for doctoral theses to be accepted for the Lecture Notes series, though habilitation theses may be appropriate.

2. Besides monographs, multi-author manuscripts resulting from SUMMER SCHOOLS or similar INTENSIVE COURSES are welcome, provided their objective was held to present an active mathematical topic to an audience at the beginning or intermediate graduate level (a list of participants should be provided).

   The resulting manuscript should not be just a collection of course notes, but should require advance planning and coordination among the main lecturers. The subject matter should dictate the structure of the book. This structure should be motivated and explained in a scientific introduction, and the notation, references, index and formulation of results should be, if possible, unified by the editors. Each contribution should have an abstract and an introduction referring to the other contributions. In other words, more preparatory work must go into a multi-authored volume than simply assembling a disparate collection of papers, communicated at the event.

3. Manuscripts should be submitted either online at www.editorialmanager.com/lnm to Springer's mathematics editorial in Heidelberg, or electronically to one of the series editors. Authors should be aware that incomplete or insufficiently close-to-final manuscripts almost always result in longer refereeing times and nevertheless unclear referees' recommendations, making further refereeing of a final draft necessary. The strict minimum amount of material that will be considered should include a detailed outline describing the planned contents of each chapter, a bibliography and several sample chapters. Parallel submission of a manuscript to another publisher while under consideration for LNM is not acceptable and can lead to rejection.

4. In general, **monographs** will be sent out to at least 2 external referees for evaluation.

   A final decision to publish can be made only on the basis of the complete manuscript, however a refereeing process leading to a preliminary decision can be based on a pre-final or incomplete manuscript.

   Volume Editors of **multi-author works** are expected to arrange for the refereeing, to the usual scientific standards, of the individual contributions. If the resulting reports can be forwarded to the LNM Editorial Board, this is very helpful. If no reports are forwarded or if other questions remain unclear in respect of homogeneity etc, the series editors may wish to consult external referees for an overall evaluation of the volume.

5. Manuscripts should in general be submitted in English. Final manuscripts should contain at least 100 pages of mathematical text and should always include

   - a table of contents;
   - an informative introduction, with adequate motivation and perhaps some historical remarks: it should be accessible to a reader not intimately familiar with the topic treated;
   - a subject index: as a rule this is genuinely helpful for the reader.
   - For evaluation purposes, manuscripts should be submitted as pdf files.

6. Careful preparation of the manuscripts will help keep production time short besides ensuring satisfactory appearance of the finished book in print and online. After acceptance of the manuscript authors will be asked to prepare the final LaTeX source files (see LaTeX templates online: https://www.springer.com/gb/authors-editors/book-authors-editors/manuscriptpreparation/5636) plus the corresponding pdf- or zipped ps-file. The LaTeX source files are essential for producing the full-text online version of the book, see http://link.springer.com/bookseries/304 for the existing online volumes of LNM). The technical production of a Lecture Notes volume takes approximately 12 weeks. Additional instructions, if necessary, are available on request from lnm@springer.com.

7. Authors receive a total of 30 free copies of their volume and free access to their book on SpringerLink, but no royalties. They are entitled to a discount of 33.3 % on the price of Springer books purchased for their personal use, if ordering directly from Springer.

8. Commitment to publish is made by a *Publishing Agreement*; contributing authors of multiauthor books are requested to sign a *Consent to Publish form*. Springer-Verlag registers the copyright for each volume. Authors are free to reuse material contained in their LNM volumes in later publications: a brief written (or e-mail) request for formal permission is sufficient.

Addresses:
Professor Jean-Michel Morel, CMLA, École Normale Supérieure de Cachan, France
E-mail: moreljeanmichel@gmail.com

Professor Bernard Teissier, Equipe Géométrie et Dynamique,
Institut de Mathématiques de Jussieu – Paris Rive Gauche, Paris, France
E-mail: bernard.teissier@imj-prg.fr

Springer: Ute McCrory, Mathematics, Heidelberg, Germany,
E-mail: lnm@springer.com

Printed in the United States
by Baker & Taylor Publisher Services

Printed in the United States
by Baker & Taylor Publisher Services